The numerical control of machine tools

To my wife and colleagues

THE NUMERICAL CONTROL OF MACHINE TOOLS

Basic Principles, Systems
Analysis and Industrial Applications

Professor Dr.-Ing Wilhelm Simon

Translated by PERA

Edward Arnold London
Crane, Russak New York

Contents

I.   *Basic Theory*

1.   Statement of the problem and possible solutions —
      Outline of the theoretical treatment   1
1.1   General statement of the problem
      The Principle — Advantages of the concept —
      Criteria the solutions are required to meet   1
1.2   Basic possible solutions
      Open-loop system — Closed-loop system —
      On-off type circuit and continuous-control circuit —
      Slide displacement measurement control and workpiece
      measurement control — Inspection machines   6
1.3   Summary of Sections 1.1 and 1.2   12
1.4   Analysis of the information flow in production technology
      Working information — Dimensional information —
      Process information   12
1.5   Rationalisation of the flow of information as a new
      engineering problem
      Information flow and coding theory — Redundancy and
      the choice of a code — Pure binary code — Binary
      codes with error detection — Hamming code — BCD code —
      3-excess code — Comparison of five binary codes —
      Code conversion   15
1.6   Summary of Sections 1.4 and 1.5   29

2.   The machine tool as a link in the chain of data-
      processing systems — the fundamentals of slide
      displacement measurement   30
2.1   The machine tool as a data-processing system
      Internal and external data processing — Switching
      information — Displacement information — Classification
      of machine tools by type of control — Positioning
      control and straight-line control — Continuous-path
      control   30

2.2   The incorporation of the machine tool into a larger
      information structure
      Discussion of the system and development of model
      systems — Punched-tape control — Magnetic-tape control   40
2.3   Summary of Sections 2.1 and 2.2   43
2.4   The principles of displacement measurement
      Analogue and digital methods of operation —
      Incremental and absolute methods of measurement —
      Direct and indirect measurement of displacement   44
2.5   Summary of the principles underlying the measurement
      of displacement   51

II.   *Internal data processing*

3.   Digital displacement measuring systems and comparators
      Tabular summary of various designs   52
3.1   The digital-incremental process and its associated equipment
      Precautions against spurious pulses — Zero-point
      displacement and location   54
3.1.1 Rotary displacement measuring systems for the incremental
      process   61
3.1.2 Linear displacement measuring system for the
      incremental process   67
3.1.3 Subdivision of scales by means of electronic circuits   71
3.1.4 Methods for simplifying linear scales
      (Multiprismat method, Polygon mirror process)   72
3.1.5 Pre-selector counters as comparators for the
      incremental process — Principles of semi-conductor
      elements and integrated circuits — Numerical displays   77
3.1.6 Summary of conclusions — incremental process   94
3.2   The digital-absolute process and its associated equipment
      V-scanning — Gray code   95
3.2.1 Rotary systems for the numerical measurement of
      position   103
3.2.2 Linear systems for the numerical measurement of position   105
3.2.3 Possible designs of numerical comparator for the
      absolute process   105
3.2.4 Zero point shift   111
3.2.5 Summary of discussion of absolute process   111

4.   Analogue displacement measuring systems and
      comparators   113
4.1   Principles of measuring analogue displacements
      or angles   113

4.2   The synchro as an inductive angle measuring device 116
4.3   The linear Inductosyn 126
4.4   The Accupin 133
4.5   Visual indication of analogue displacement measurements 136
4.6   Digital/analogue converters 138
4.7   Analogue type comparators 144
4.8   Summarising remarks on the analogue process 147

5.   Special problems of numerical measurement technology
      as applied to machines
      The measuring systems compared 149
5.1   Two-part measuring systems 149
5.2   Electro-optical system for synchronous operation 152
5.3   Digital follow-on control circuit with a single
      synchro-Bendix system 153
5.4   Numerical workpiece measurement control systems 156
5.5   Numerical inspection machines 159
5.6   Comparison of numerical measuring systems used in
      conjunction with machine tools 161

6.   Drive systems 170
6.1   Stepped electro-mechanical drives
      Remote-controlled clutches 171
6.2   Infinitely-variable electric drives 174
6.2.1   The d.c. motor 175
6.2.2   Control of d.c. motors
      Ward-Leonard set — Amplifier machine — Silicon
      controlled rectifiers (thyristors) 180
6.3   Stepped and infinitely-variable electro-hydraulic drives 186
6.3.1   Pumps and auxiliary equipment (pressure and flow controllers) . 186
6.3.2   Electro-hydraulic controllers (energy switching devices)
      Non-continuous controllers — Continuous controllers 190
6.3.3   Hydraulic drive units (actuators)
      Feed drive unit using hydraulic cylinder and piston —
      Feed drive unit using hydraulic motor 195
6.3.4   Hydraulic fluids 202
6.4   Comparison of infinitely-variable electric and
      hydraulic drive systems
      Thyristors, special d.c. motors, Minertia motors,
      servo valves, axial-piston motors, Roll-Vane motors 203
6.5   Stepping motors
      Electric stepping motors — Electro-hydraulic
      stepping motors 205
6.6   Pneumo-hydraulic positioning systems with open loops 210

6.6.1　Moog system 211
6.6.2　Plessey system 216
6.7　　Summary of drive system 220

7.　　　The Machine Tool and Numerical Control 223
7.1　　Special design features required to match machine
　　　　tools to numerical control systems
　　　　Use of special machine elements 224
7.2　　Tools as information stores 230
7.2.1　Tools as form stores (rigid information stores)
　　　　Form tools and plain tools 230
7.2.2　Tools in turrets 232
7.2.3　Tools in magazines 242
7.2.4　Mechanical and manual tool changing systems
　　　　Pick-up system − Manual tool changing − Coding
　　　　the shanks of the tools − double-gripping system 245
7.2.5　Tool presetting and tool adjustment 258
7.3　　Extension of the kinematic possibilities 265
7.3.1　ISO Axis definitions − System of symbols for
　　　　machines and control systems 265
7.3.2　Multi-axis machines 268
7.3.3　NC Boring heads 271
7.3.4　NC Machining centres and machining lines 274
7.4　　Combination of NC techniques with other methods 278
7.4.1　Combination with cam controls and copying systems 278
7.4.2　2½-D Continuous-path control systems 280
7.5　　NC techniques for non-cutting machinery 280
7.5.1　NC Machines for metal forming 280
7.5.2　NC Flame-cutting machines 283
7.5.3　Other applications of NC
　　　　Drawing machines − wiring machines − spark
　　　　erosion machines 284
7.6　　Summary 288

8.　　　Computers in machine-tool control systems 292
8.1　　Calculation of corrections by means of digital
　　　　computers
　　　　Zero-point correction − Tool dimension corrections 293
8.2　　Internal interpolators 297
8.2.1　The DDA process
　　　　Linear interpolation − Circular interpolation 299
8.2.2　Direct calculation of function 311
8.2.3　Comparison of interpolation processes 317
8.3　　Thread cutting 318

Contents

8.4   Optimisation of technological working conditions
        within machines (Adaptive control, AC)   322
8.4.1   Automatic depth-of-cut adjustment for individual
        cuts on a lathe (Example)   323
8.4.2   AC and DC techniques on a milling machine (Example)   328
8.5   Design of computing circuits — Miniature computers
        and magnetic stores for internal data processing   331
8.6   Summary   336

9.   Input units and control consoles   339
9.1   Basic methods available for feeding-in work information   339
9.2   The manual input of work information   341
9.3   Punched tape as a data carrier (information carrier)
        The basic principles   346
9.3.1   Punched tape scanners —
        Series and parallel scanning — Types of scanner for
        Series-scanning — Tape transport — Rapid, error-free
        reading — ability to deal with long lengths of tape   348
9.3.2   Decoders and checking of characters   357
9.3.3   Buffer stores
        The address method — The stepping-switch method   358
9.3.4   Pneumatic scanning of punched tape and pneumatic logic
        elements   361
9.4   Magnetic tapes as information carriers on numerically controlled
        machine tools
        Storage of digital and analogue guide values   366
9.5   Direct numerical control (DNC, on-line control)
        Junction problems   371
9.6   Conclusions   376

III.   *External Data Processing*

10.   Manual and semi-automatic parts programming —
        The use of small computers for machine-orientated
        programming   379
10.1   The basic possibilities of programming   379
10.2   An adequate supply of characters   380
10.3   Special punched-tape codes which cannot be
        derived from Table 15   384
10.4   The meaning of characters and the construction
        of programmes   389
10.5   Coding of axis directions   399

10.6   Example of programming 403
10.7   Arrangement of programming area and the use of
        miniature computers 406
10.7.1 Use of typewriters 406
10.7.2 The use of auxiliary calculators 409
10.7.3 The VDF Autoprogrammer and Simulator as an example 417
10.8   Summary 420

11.   The use of computers and machine parts programming —
        Problem-orientated languages for production processes 423
11.1   Fundamentals of electronic computers 423
11.2   General discussion of problem-orientated programme
        languages 428
11.3   Programme languages for production engineering 430
11.3.1 The APT programming system 432
11.3.2 The EXAPT programming system and MINIAPT as
        examples of APT languages 438
11.3.3 SYMAP and AUTOPROG as examples of languages that
        are not related to APT 459
11.3.4 Symbol language for small computers 466
11.4   External interpolators and magnetic-tape control
        systems 471
11.5   Summary 475

IV.   *Summary*

12.   The integration of NC machines and computers in the
        production process 477
12.1   Arrangement of basic systems according to degree
        of automation 479
12.2   Detailed break-down of total time for passage
        through the system t_d 487
12.3   Introduction into the method of simultaneous
        examination 491
12.4   Higher order systems 498

V.   *Appendix*

        Examples of numerically-controlled machine tools 504
        Bibliography 527
        Index 541

Foreword to the Second Edition

The first edition of this book appeared seven years ago in the autumn of 1963. What was described in the foreword to that edition as the 'bold venture' of writing 'a book on a very new technical development which is still in a state of flux' still aptly describes the present position. The hope that seven years would see some slackening in the development of this field has proved to be false. On the contrary, in fact: the rapid progress that has been made in the related field of computer technology has resulted in a constant stream of new developments and proposals, and even an expert in this field finds difficulty in keeping up-to-date. The statement made in the foreword to the first edition that 'On the other hand, the technical, economic, and human problems involved in the most modern form of automation are so pressing that there is a growing need for a comprehensive treatment of the subject. Uncertainty and the lack of an overall grasp of the factors involved can easily lead to a dangerous under- or over-estimate of a new technique' remains as true now as it was then. As with the first edition, I hesitated a long time before attempting to 'weigh up the odds' and to write a new book. It is quite natural that in view of considerable progress that has been made in the field of numerical control since the first edition, this new edition has become an almost entirely new text book in the form of a systems analysis.

The success of the first edition, which has in the meanwhile been translated into three languages, has, however, encouraged me to write this second edition more in the nature of a basic text book for students of all ages and with varying degrees of prior knowledge of the subject, so that it is suitable for study by readers having different amounts of experience and various levels of understanding. To avoid any misunderstanding it seems, therefore, desirable to repeat some of the sentences from the first foreword which are still valid today: 'This book does not set out to explain the construction of machine tools to the machine tool designer, nor mathematical logic to the expert in data processing, but rather to open up a general view to the narrow specialist. For it is inherent in this new development that the operating process must be regarded as a self-contained whole from the systems engineering viewpoint. No responsible engineer nor plant manager can ignore this fact any longer.' The number of those who are 'responsible' and who are required to play a part in planning and decision-making processes has increased considerably since then for NC technology and the computer belong together and are influencing every aspect of work in industrial plants to an ever-increasing extent. The metal-working industry, which is affected most by these developments, employs about 40% of the gainfully-employed populations in both East and West, and hence forms the backbone of every developed industrial nation. For this reason the final chapter also contains some suggestions of ways in which it is possible to clarify the rapid and wide-spread developments in this field and of taking them into account for long-term planning.

This second edition largely reflects the state-of-the-art in 1969/70, but it also makes some mention of developments which will not find their way into industrial practice for some years to come; this is all the more reason to study these developments — which tend to be particularly cost-intensive and far-reaching in their consequences — with particular care before they are adopted for practical use.

Finally a few words about the book itself. Purely in order to ensure a uniform and easily understood treatment I again decided to write the entire book myself; although it is normal practice for books which touch on so many disciplines to be written by a team of authors. The only way in which I have found it possible to prevent the book expanding to an unwieldy size, and so ensure that it retains the characteristics of a text book, however, is to make frequent reference to various publications, manufacturers' literature etc. Without the assistance of my colleagues and specialist lecturers it would, however, have been quite impossible for me to cover the very wide range of material dealt with in this book without making major errors. I am indebted to all of them for their help. I can name only a few here otherwise the list would be too long; they must be taken as representative of the many others who have helped in this work. The following representatives of manufacturers, members of associations and standards committees, and staff members of universities and technical institutes were particularly helpful with lectures, frequent discussions and suggestions, etc.:

Dipl.-Ing. P. Boese, Dipl.-Ing. P. Brödner, Dr.-Ing. H. Fetzer, Dipl.-Ing. A.-W. Kamp, Dipl.-Ing. G. Koschnik, Dipl.-Ing. R. Langebartels, Dipl.-Ing. G. Sautter, Dr.-Ing. W. Schaller, Dipl.-Ing. H. Scheidemann.

I should also like to thank the Hanser Publishing Company and staff for their efficiency in dealing with the somewhat complicated layout of this book, especially since, because of the rapid rate of progress in this field, it is never possible to keep the interval between the final production of the manuscript and the publication of the finished book short enough.

Berlin, July 1970   W. Simon

List of firms*

Allgemeine Electricitäts Gesellschaft (AEG-Telefunken), Berlin-Frankfurt a.M.
AGIE, AG für industrielle Elektronik, CH-6616 Losone-Locarno
C. Behrens AG, D-3220 Alfeld (Leine)
Gebr. Boehringer GmbH (VDF-Drehbankfabriken e.V.), D-7320 Göppingen
Reinhard Bohle KG, Werkzeugmaschinenfabrik, D-4801 Jöllenbeck/über
    Bielefeld
Brown, Boveri & Cie (BBC), Baden (Switzerland)-Mannheim
Burgmaster-Europe, INC, D-7320 Göppingen
Burkhardt Maschinenbau GmbH, D-7417 Pfullingen
Burkhardt & Weber KG, Werkzeugmaschinenfabrik, D-7410 Reutlingen
Cincinnati Milacron Company, Cincinnati, Ohio 45209/USA
CNMP/Berthiez, F-75 Paris 16
Collet and Engelhardt, Maschinenfabrik GmbH, D-6050 Offenbach/Main
DIXI S.A./Usine 2, CH-2400 Le Locle/Switzerland
O. Dörries GmbH, Maschinenfabriken, D-5160 Düren
Droop & Rein, Werkzeugmaschinenfabrik, D-4800 Bielefeld
Ferranti Ltd, Dalkeith, Scotl./UK
Festo-Pneumatic, D-7301 Berkheim-Eßlingen
Ratier-Forest, Division Machines-Outils, F-12 Capdenac/France
Fortuna-Werke, Maschinenfabrik AG, D-7000 Stuttgart-Bad Cannstatt
Friden Inc., San Leandro, Cal./USA
Maschinenfabrik Froriep GmbH, D-4070 Rheydt
General Electric Company, Speciality Control Department, Waynesboro/Va./
    USA
Ghielmetti AG, Fabrik elektrischer Schaltapparate, CH-4500 Solothurn/
    Switzerland
Werkzeugmaschinenfabrik Gildemeister & Comp. AG, D-4800 Bielefeld
Ateliers GSP, Guillemins, Sergot et Pégard, F 92 Courbevoie/Seine
Güttinger AG für elektronische Geräte, CH-9052 Niederteufen/Switzerland
Dr. Johannes Heidenhain, D-8225 Traunreut/Obb.
Gebr. Heinemann AG, Werkzeugmaschinenfabrik, D-7742 St. Georgen/
    Schwarzwald

*Acknowledgements are made to the above firms for the illustrations

Gebr. Heller, Maschinenfabrik GmbH, D-7440 Nürtingen
Herbert-De Vlieg, Alfred Herbert Ltd., Edgwick Works, Coventry/UK
Heyligenstaedt & Comp., Werkzeugmaschinenfabrik GmbH, D-6300 Gießen
Hughes-Aircraft, Ind Syst. Div., Los Angeles, Cal./USA
Karl Hüller GmbH, Werkzeugmaschinenfabrik, D-7140 Ludwigsburg
Index-Werke KG, D-7300 Eßlingen
Kearney & Trecker Corp., Milwaukee, Wisc./USA
Kelch & Co., Werkzeugmaschinenfabrik, D-7060 Schorndorf (Württ.)
Kjellberg-Eberle GmbH, D-6000 Frankfurt/M.
Herm. Kolb, Maschinenfabrik, D-5000 Köln 30
AS-Kongsberg Vapenfabrikk, Kongsberg/Norway
Ludwigsburger Maschinenbau GmbH (Burr), D-7140 Ludwigsburg
Marwin Machine Tools Ltd., Anstey/Leicester/UK
Messer Griesheim GmbH, Schweißtechnik, D-6000 Frankfurt a.M.
Molins Machine Co. Ltd., Machine Tool Div., Saunderton, High Wycombe/
        Bucks/UK
A. Monforts Maschinenfabrik, D-4050 Mönchengladbach
Moog Ltd., Hydra-Point Division, Cheltenham/UK
Max Müller, Brinker Maschinenfabrik, D-3000 Hannover
Nixdorf Computer AG, D-4790 Paderborn
Ing. C. Olivetti & C., S.p.A., I-10090 San Bernardo d'Ivrea/Italy
Philips Gloeilampenfabriken N.V., Eindhoven/Netherlands
Pittler Maschinenfabrik AG, D-6070 Langen/Hessen
Plessey Electronics, The Plessey Comp. Ltd., Autom. Div., Poole/Dorset/UK
Pratt & Whitney Co., Inc., West Hartford, Conn./USA
Remex Electronics/Division of Ex-Ell-O-Comp., Hawthorne, Calif./USA
Scharmann & Co, D-4070 Rheydt
Schiess HG., D-4000 Düsseldorf-Oberkassel
Schweizerische Industrie Gesellschaft (SIG), CH-8212 Neuhausen am
        Rheinfall/Switzerland
Siemens AG, Berlin-München-Erlangen
Sperry Gyroscope Co. of Canada, Montreal/Canada
Standard Elektrik Lorenz (SEL), D-7000 Stuttgart-Zuffenhausen
Sundstrand Machine Tool, Belvidere, Ill./USA
The Superior Electric Company, Bristol, Conn. 06010/USA
Texas Instruments, Dallas Texas 75222/USA
Trumpf & Co., D-7000 Stuttgart-Weilimdorf
TWK-Elektronik, Kessler & Co., D-4000 Düsseldorf
Valvo-GmbH., D-2000 Hamburg 1
VEB Großmaschinenbau „8.Mai", DDR-90 Karl-Marx-Stadt
VEB Werkzeugmaschinenkombinat „Fritz Heckert", Betrieb Mikromat,
        DDR-8036 Dresden
Wadkin Ltd, Green Lane Works, Leicester/UK
Wanderer-Werke AG, D-8013 Haar bei München
Fritz Werner Werkzeugmaschinen GmbH, D-1000 Berlin 48

I   Basic Theory

1   Statement of the problem and possible solutions—Outline of the theoretical treatment

1.1   General statement of the problem

The Principle

For the newcomer to the subject the term 'Numerical control of machine tools' is really rather misleading. One speaks, for example, of the mechanical, electrical, hydraulic, or pneumatic controls on machine tools, when referring to the physical means employed to influence or produce a desired pattern of movement within a machine. The uninitiated then inevitably tend to regard 'numerical control' as a further variant of these types of control, and as a matter which is concerned entirely with the internal workings of the machine. This is, however, by no means the case. The concept of numerical control ultimately extends far beyond the individual machine tool; it is much more a comprehensive form of organisation of internal data transmission (in a quite general sense) and data processing than an extension of the methods of machine control. It is certainly true that this concept exerts a major influence on the design of machine tools, and indeed part of the following discussions will deal exclusively with this aspect. Nevertheless, when considering the individual design problems one must never lose sight of the overall conception of a comprehensive flow of data and of the laws governing its structure. Also, the principles of numerical control are by no means limited to mechanical production methods in general and machine tools in particular, but they can also be applied with success to other areas of production. Consequently it is first necessary to determine the special features that arise in the practical application of numerical control techniques to production technology so as to enable the appropriate concepts, types of equipment, and method of combination to be established.

'Numerical' stands for 'by means of numbers'. The concept of the numerical control of machine tools, which was first mooted in the USA around 1949/50, is basically very simple: From the earliest times all dimensions of workpieces have been defined by numbers. Under certain circumstances it is possible to save both effort and expense if automatic machine tools are made in such a way that they can directly 'understand' and 'process'[1] our numerical language without human intervention.

Advantages of the concept

The advantages of constructing reliable machine tools which 'understand' numerical dimensions are:

1. All drilling and marking-out jigs, as well as master patterns, templates, sample workpieces, etc., can be eliminated, with the result that certain subsidiary workshops will no longer be required and storage space can be saved.
2. Numerical programmes can be changed more easily and quickly than can mechanical fixtures. The resultant reduction in setting-up time opens up many possibilities for automating small batch production runs.
3. This in turn may make it unnecessary to maintain large stocks of spare parts, especially for obsolescent equipment or where there is little demand.
4. Elimination of human errors in setting-up and of the effects of operator fatigue enables the average quality of the product to be improved and utilisation of the machines to be increased. At the same time there is a marked reduction in the mental concentration demanded of the operators.
5. Where there is a shortage of skilled labour it is relatively easy to go over to multi-machine operation or to shift working.
6. The organisation of the factory can be rationalised, and ultimately the entire production process can be centrally controlled.
7. In all cases where the shape of the workpiece can be described by mathematical equations, as is the case with many turbine blade profiles or hole patterns, it is possible to dispense with drawings entirely and instead to derive the profile directly from the numerical function and to feed the numbers into the machine. It is even possible to control the machine directly from experimental results that have been evaluated by means of a computer.
8. The use of remote-controlled servomechanisms equipped with powerful driving motors makes it possible to construct novel and automatic machining units which are free of all the limitations imposed by manual operation, and which are capable of machining large, complicated workpieces (e.g. aircraft wings, marine propellers, rocket components, dies, etc) in one piece, and often without rechucking.

The list is by no means complete, but even so it opens up some interesting possibilities.

---

[1] A more detailed explanation is given in the Appendix.

The idea of the numerical control of machine tools originated in the USA in connection with the production of large aircraft components which could no longer be machined by existing methods, i.e. the need for a large master pattern had to be eliminated. The fundamental studies that were undertaken at the MIT[2] soon showed, however, that the applications implicit in the concept were considerably more versatile than had at first been assumed. In view of these wider possibilities, considerable efforts have been made in industrial countries throughout the world since about 1955 to overcome the technical difficulties involved in implementing the concept.

Criteria the solutions are required to meet

The most important basic features in the design of machine tools should be:

Reliability, i.e. minimum down-time arising from failures;

Economy, i.e. a favourable ratio of capital cost to increase in productivity and/or reduction in production costs;

Accuracy, i.e. a constant level of product quality within specified tolerances.

All these three factors are to some extent interrelated. It is therefore desirable to explain them in somewhat greater detail, especially as applied to numerically controlled automatic machines. It must, however, be pointed out even at this stage that the economics of numerically controlled machines do not depend solely on the design of the machine and its controls, but are also affected to a very considerable extent by the overall organisation of the factory (cf. Chapter 12).

Reliability

All automatic machine tools, and especially those which are numerically controlled, are inevitably more complicated than simple, manually-controlled machines. Now there is a law which is universally applicable to the development of any automatic machine; it first became glaringly apparent in the construction of aircraft. This law relates increase in liability to failure with increase in complexity (11). If we denote the mean rate of failure (e.g. in parts per thousand) after a standard period of time (e.g. 1000 hours) of a critical component by p, then the reliability R of this component will be:

$$R = 1 - p/1000 \text{ (if } p \text{ is specified in parts per thousand)}$$

Two such components, considered in combination, then have an overall

---

[2] Massachusetts Institute of Technology, Cambridge, Mass., USA.

reliability of only:

$$R_{total} = R_{individual} \cdot R_{individual};$$

while if there are n components of equal significance to the functioning of the equipment the overall reliability will be:

$$R_{total} = R_{individual}^{n}$$

A numerical example will make this relationship clearer:
For an individual component reliability of

$$R_{individual} = 0.999 \; (p = 1\,{}^o\!/_{oo})$$

and with 1000 components of this type in a particular item of equipment, the overall reliability of the equipment will be only:

$$R_{total} = 0.999^{1000} = 0.368$$

i.e. the initial 99.9% has been reduced to 36.8%.

In very complicated automatic equipment, such as unmanned aircraft, this can lead to a high failure rate. In these cases we speak of a 'critical complexity' at which such systems can become virtually useless (11). In the circumstances under consideration here, however, it must be borne in mind that

a) not every component is of critical importance to the proper functioning of the equipment as a whole, and

b) automatic machine tools are fortunately not nearly as complicated as unmanned aircraft

The simple relationship

$$R_{total} = R_{individual}^{n}$$

should nevertheless be taken as a warning by all designers of automatic machines. It implies that:

1. the larger the number of components that are to be combined to form a single system, the higher the component reliability should be, and

2. the number of components appears as an exponent in the equations and must therefore be regarded as having a particularly critical effect.

Economic aspects

There are two major points that arise in connection with the economic aspect of numerical control:

a) When considering the capital cost, the old notion that the cost of a control system should not exceed a certain percentage of the cost of the machine tool proper must be discarded, since no control system (in the narrower sense of the term) is nowadays merely an appendage to the machine, but rather is combined with it to form an organic whole. The kinematics of the functional movements cannot, for example, be arbitrarily divorced from the machining of the workpiece.

b) Since the concept of numerical control extends far beyond the individual machine tool, and affects the organisation of the whole manufacturing plant (including auxiliary workshops, stores, etc.) any cost comparisons must be made on a much wider basis than is usual when considering the purchase of new machine tools (10).

The subject of economics in the sense of the various methods that are available for reducing the cost of producing work on numerically controlled machines is discussed in greater detail later.

Accuracy

This is difficult to define precisely even for non-automatic machine tools (8); accuracy can be precisely described only by a figure which, in actual fact, represents the 'inaccuracy' of a system. As suggested by the AEF(Committee for Units and Terms in Formulae) it is therefore advisable.to refrain entirely from using the term 'accuracy' in conjunction with numerical values, widely-used though this somewhat imprecise expression may be (16, 17). Instead the differences between the required and actual values of slide positions, which are the root causes of inaccuracy, can be expressed by means of terms such as 'deviation' and 'scatter', for which numerical values based on experimental results are available. With numerically-controlled machines the errors in the control system are superimposed on the mechanical errors in the machine tool in a very complicated manner; the inter-actions are considered in Chapter 7 in conjunction with a discussion of possible machine acceptance tests (17). For the present it may suffice to point out that there are no technical difficulties in producing numerically-controlled machine tools that are capable of working to within an accuracy of ± 0.002 mm or ± 0.0001 in, and that the question as to how far the search for accuracy is to be taken is in the end simply governed by economic considerations.

1.2   Basic possible solutions

The initial formulation of the problem is to construct a machine capable of understanding numbers. The dictionary definition of a number is 'a concept arising from the collection of a quantity of similar items'. If this definition is accepted, and it is borne in mind that in production technology the term 'number' relates primarily to the dimensions of the workpiece, it will first be necessary to find the similar items that can be collected together in quantity. Now these are the 'individual steps' or 'unit steps' of all the components of motion involved in shaping the workpiece. Following the numerical principle, it is therefore necessary to establish these unit steps. Taking the dimension 623.47 mm, for example, this can:

a) be resolved into 1 mm steps, in which case 623 or 624 steps will be obtained. In this case no intermediate values can be defined:
b) be resolved into 0.1 mm steps, in which case 6234 or 6235 steps will be obtained. The remaining 0.07 mm will have to be neglected;
c) be resolved into 0.01 mm steps, in which case 62347 steps will be obtained, which agrees precisely with the required dimension.

Resolution of this type into integral numbers of units is termed quantification (3, 12); while the representation of this quantified magnitude in the form of numbers written down in sequence to show the hundreds, tens, units, etc., involved is termed digitalisation.

The size of the unit step can be made as large or as small as desired, and as will be shown in more detail later, this plays a decisive role in the design of machine tools. Having made this point the first step has been taken towards solving our problem. So far as putting it into practice is concerned, there are two possibilities, namely the open-loop or closed-loop system of operation (12).

Open-loop system

Suppose Item 4 in Figure 1A) to be the slide of a machine with guides, feedscrew, etc., and Item 3 to be a pulse-controlled driving element which moves the slide, for example by means of a threaded spindle. Suppose this driving element is constructed in such a way that it is rotated through an angle $\Delta \varphi$ by each pulse (electrical, hydraulic, etc.), thereby moving the slide by means of the feedscrew through a short distance Δs. If this angular rotation is performed with a high degree of reliability and independently of the load imposed by Item 4 and the forces acting on it, the number of pulses will determine the total movement of the slide, and the pulse repetition frequency will determine the velocity of the slide. In practice this conception can be put into effect by means of suitable electro-magnetic clutches, stepping mechanisms, or even stepping motors. The same ideas can, of course, also be applied to turntables.

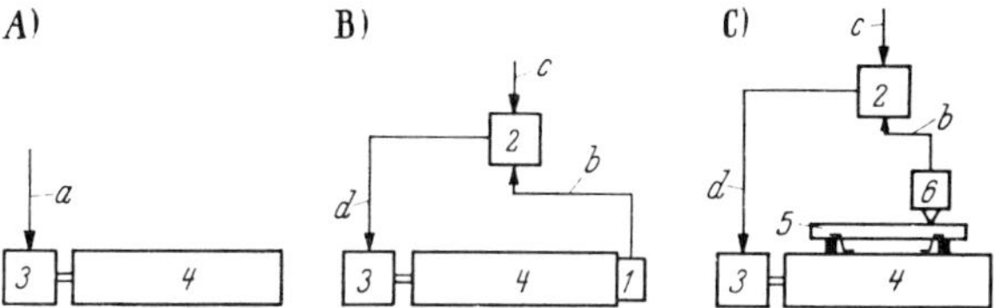

Fig. 1 **Three basic methods of applying input data to the numerical control of the motion of a slide along one axis of a machine.**

A) Slide displacement control (open-loop system). B) Slide displacement measurement and control (closed-loop system). C) Workpiece measurement and control (closed-loop system).

1. Displacement measuring system. 2. Comparator. 3. Drive system. 4. Machine slide. 5. Workpiece. 6. Workpiece measurement system (with sensor).

a. Pulse input for stepping motor. b. Actual-value feedback system (generally known as the controlled variable X in control technology). c. Desired value input (generally known as the command W in control technology). d. Positioning signal for drive 3 (generally known as the regulating variable Y in control technology).

In the present state of the art it is unusual to rely solely on the reliable functioning of power stepping motors of this type under all speeds and cutting forces likely to be encountered in practice, although this would be a very simple way of building a numerically-controlled machine tool. (Chapter 6).

Nowadays designers generally prefer to use a closed-loop system in which the actual position or actual displacement of a slide is monitored by some auxiliary measuring devices which are independent of the feed system. These are briefly discussed below.

Closed-loop system

In Figure 1B) Items 3 and 4 again represent the driving element and the machine slide. In addition, however, a displacement measuring device 1 is fitted to slide 4 to determine the instantaneous position or the distance travelled. This value b is compared in a device known as the comparator, 2, with the desired value c which is fed in from an external data unit. This comparator supplies a signal d to the drive 3, corresponding to the difference between the desired value and the actual value. This arrangement provides a closed-loop system:

$$1 - b - 2 - d - 3 - 4 - 1$$

The signal b is called the feedback. The external control of the closed loop is effected by feeding the desired value signal into the comparator. The desired value is also termed the command signal (always designated by W in control technology) (12).

Items 1 and 2 can assume many different forms in practice, and these are described in detail in the following chapters. It is, however, important to emphasise here that the comparator can operate in two entirely different ways. These are:

> Type S. The comparator generates a signal only when the actual value and the desired value are identical (coincidence). This signal can then be used, for example, to stop or change the speed of drive 3. The comparator does not exert any influence on the drive when the actual value differs from the desired value.
>
> Type R. The comparator generates a signal all the time the actual value differs from the desired value. The magnitude and sign of this signal can vary with the difference between the actual and the desired values. The drive will then be stopped, for example, only when the difference between the actual and the desired signals is zero.

If these two methods of operation of the comparator are examined more closely it will be found that they give rise to two different versions of the closed-loop system (3).

On-off type circuit and continuous-control circuit

Since in the case of a type S comparator a signal will be emitted only when the desired value and the actual value are the same, the number of output signals will be equal to the total number of points along the path at which control is to be exercised. From its very nature a comparator of this type cannot operate continuously, but only in discrete steps (3).

No matter how any individual comparator may be constructed, it operates in precisely the same way as a normal limit switch operated by mechanical cams where, again, only discrete signals are emitted at intervals along the path of of travel, these being frequently employed to switch off the motion. A closed-loop system of this type is therefore referred to as an on-off circuit. With the limit switch the desired value is selected by adjusting the mechanical cams, and coincidence is then achieved when the cam and the switch operator are in the same position so that the switch is actuated.

The displacement-dependent discrete output signals from a type S comparator can be used in exactly the same way as the signals from a limit switch, not only to switch off a process, but also, for example, to effect a change of speed along a displacement path (changeover from quick feed to normal cutting speed, and so on). The control systems fitted in lifts and elevators are a well-known example of this on-off circuit technique.

If the problem is to specify the positions of a slide in a machine tool in numerical form, the simplest comparators, of type D, will be pulse counters with preselector setting. The fact that in most cases these will have to take the form of electronic preselector counters is of no significance. What is important in this connection is the correct adoption of the basic idea of resolving the given number or dimension into the unit steps Δs. In the case of an open-loop system (control system with stepping motor) the pulses have to be provided by some external means and the stepping motor acts as an integrator which totals up the pulses in order to determine the distance travelled. When a preselector counter is used as the comparator, this can carry out the summation process; being an electronic device it is practically free from inertia, and can cope without difficulty with the high pulse reception rates that arise during quick-feed motions. The drive system can now rotate at a constant speed, and a stepping motor is not required. To explain this, consider once again the numerical example referred to above. Suppose the dimension 623.47 mm is split up into 62347 elementary steps Δs (displacement quanta) each 0.01 mm in length. Then the displacement measuring system, no matter how it is constructed in detail, must be such that it generates a signal for each 0.01 mm displacement of the slide. The preselector counter, which is a type S comparator, is then adjusted to the desired value of 62347 pulses, and the pulses emitted by the displacement measuring system are summed in the counter. When 62347 pulses have been passed to the counter, the slide will have been displaced through 623.47 mm. At this point the counter will produce a coincidence signal which will, for example, stop the motion of the slide, or change over from the quick-feed speed to the normal cutting speed, depending on the use that is made of the coincidence signal. It need hardly be mentioned that in practice matters are usually not quite so simple as indicated in this initial example.

However, such counters can be provided with two or more preselection mechanisms so that, just as with the cam switching technique, it is possible for several discrete coincidence signals, each controlled by the displacement of the slide, to be emitted.

The preselector counter thus provides a major and relatively simple means of accomplishing numerical control using a closed-loop system. The limitations and difficulties arising with this type of control will be discussed in detail in Chapter 3.

The characteristic feature of a type R comparator is that it emits a continuous signal whenever the desired value and the actual value are not the same. A continuous signal is therefore fed to the driving element, and this becomes zero only when there is no longer any difference between the desired and the actual values. As a result of the continuous comparison between the desired and the actual values the driving system is, in effect, under continuous supervision. If the circuit is arranged so that the positioning signal d (Figure 1B) controls the drive 3 in such a way that the difference between the desired value and the actual value is kept to zero at all times, the slide will automatically follow every change in the desired value. There is nothing intrinsically novel about this idea; we are dealing with a simple control loop (12) in which the command is represented by the desired value input. The new feature is that the input takes the form of the desired numerical value, and this may cause considerable design difficulties. Without going into detail at this point it may be mentioned that a closed-loop system having a continuous-control characteristic is much more complicated than one with an on-off characteristic. This difference is discussed at length in the important chapter dealing with the selection of suitable control systems for various types of machine tool on the basis that the outlay involved should not be higher than is absolutely necessary (cf. reliability and economics).

Slide displacement measurement control and workpiece measurement control

So far it has been tacitly assumed that controlling the movements of the slide will be sufficient to ensure that the correct dimensions of the workpiece are maintained. This is by no means correct. The slides and tables of a machine tool are basically only provided to form a means of securing the workpieces or the tools. The workpieces are shaped only as a result of the relative movement between them and the tools. In order to draw any conclusions as to the shape of the workpiece from the movement of the slide a whole series of assumptions must be made, of which only the three most important will be mentioned here:

1. The positions of the tool and the workpiece relative to the table or support must remain the same within sufficient accuracy when these parts are replaced by others.
2. The cutting edge of the tool must remain virtually unchanged over a sufficient period of time.
3. During the machining operation the elastic deformations of the tool, the workpiece, and the machine must not exceed certain values.

These requirements are by no means new and are not restricted to numerically controlled machine tools. The same problem arises with all cam-controlled automatic machines and copying machines. It assumes increased significance, however, owing to the possibilities opened up by numerical slide displacement control. At this point another associated concept is of considerable importance. It is an attractive idea not to measure the displacement of the slide at all, but instead to use the workpiece dimensions directly to provide a feedback signal for the comparator. This arrangement is shown schematically in Figure 1C). Compared with Figure 1B) the dispacement measuring system 1 has been replaced by the workpiece measuring system 6 as shown. The measured value is then processed in the same way as in Figure 1B). The whole problem thus becomes one of measurement rather than control.

It is, however, precisely in this measuring process that very considerable difficulties arise in practice; only the three most important will be mentioned here:

1. In most cases it is almost impossible to make an accurate measurement while machining is in progress. Large hot pieces of swarf, vibration, streams of coolant, etc. together with difficulties of access to the workpiece greatly increase the difficulties of measurement.

2. Very many workpieces have such complicated internal and external shapes that it would be necessary to use a large number of sensitive measuring instruments of various types in order to monitor the machining operation and the changes in the dimensions of the workpiece during the various stages of machining.

3. In order to exploit the principles of numerical control to the full, the measuring instruments would in many cases have to be remote-controlled (e.g. for scanning types of operation) and also be remote-reading.

In the present state of the art it is only in rare instances that all three of these difficulties can be overcome. The practical applications of control by workpiece monitoring are, therefore, at present for the most part restricted to cylindrical and surface grinders, and even then the coupling of such devices to a numerical data input system is still infrequently encountered (10). Further developments in this field depend entirely on the availability of suitable measuring equipment (cf. Chapter 5).

Inspection machines (measuring machines)

Matters are different if we dispense with direct measurement during the machining process, and instead automatically check the dimensions of the workpiece on a special inspection machine after it has been machined and removed from the machine tool (cf. Fig. 89). This measuring machine can then be designed solely for performing a wide range of measurements, since it is not affected in any way by the machining process. The dimensions that are obtained in this way can, of course, be compared with the desired values in suitable comparators, and the differences can be used to control the machine tool. In this case a certain time delay between the shaping and the measurement of the workpiece will have to be accepted. With many workpieces and with suitable processings of the measurements this arrangement may lead to technically and economically satisfactory solutions of the problem.

1.3   Summary of Sections 1.1 and 1.2

Summarising the basic considerations that have so far been put forward, a number of important pointers emerge as to the next stages that must be discussed:

1. Numerical control involves more than the control of a machine in the usual sense. Above all it extends far beyond the limited field of the machine tool into the overall production process; it is more a form of organisation rather than simply a control system for a machine. The concept therefore brings in its train a series of far-reaching technical and economic consequences.
2. There are three special criteria that must be observed when numerical control is to be adopted in practice; these are reliability, economy, and accuracy.

Basic design rules for *reliability* are:

> Minimum number of components,
> Maximum preliminary testing of components.

Two basic rules for economy are:

a) The cost of a numerical control system must not be assessed in terms of a percentage of the cost of the machine; the sole determining factor is the reduction in the overall production costs.
b) Cost comparisons must take into account the entire factory organisation (10). For *accuracy*, it is necessary to consider the machine and its control system as a whole. The recommendations of Chapter 7 should be borne in mind with a view to selecting the most economic control system.
3. Of the three possible basic methods, numerical control using measurement of the displacement of the slide in a closed-loop is the most practical solution in the present state of the art. In this connection a special study of the applicability of on-off circuits and of control circuits to the various types of machine tool is urgently required.
4. Workpiece inspection machines providing numerical values of the measurements are very promising so far as future developments are concerned, especially with the introduction of integrated data processing.

1.4   Analysis of the information flow in production technology

It will be convenient to pursue two aspects of this subject, limiting the discussion to the essential factors:

   a) What is the general nature of the information that is required to make a workpiece on a machine tool?

   b) What particular variants arise with the individual types of machine tool?

Workpieces are made on a machine tool in accordance with information supplied to it in some way by a human agency. This is true whether the machine is manually controlled by a skilled man in accordance with the information supplied on a drawing, or if it works automatically in accordance with a programme that is fed into it.

Quite generally all the information required for machining a workpiece can be termed 'working information' which is divided into two main groups depending on its origin:

1. Information relating to the shape of the workpiece, i.e. the specification of the path to be followed by the point, line, or surface forming the contact between the tool and the workpiece in the case of a cutting operation, or the design of the press tool, die, or punch in a forming operation (29). This is all information that has to be supplied by the designer and which at present is usually given on a drawing which forms the information store. This type of information is referred to below as 'dimensional information.'

2. Information as to the techniques by which this shape is to be produced, i.e. the required speeds (cutting speeds and feeds), choice of tools, breakdown into separate operations, their sequence, etc. These are for the most part factors which production engineers in the planning departments add to the drawings produced by the designer. This information is referred to below by the general term 'process information' (6, 18).

These two types of information represent the working information in what might be termed the raw state. In many cases, especially in the more highly organised undertakings, they are not suitable in this form (drawings, lists, work tickets etc.) for use as input data for machine tools; the information must first be 'worked up.' So as not to cloud the issue by going into too much detail, this information flow is shown in as general a manner as possible in Table 1.

The administrative aspects of this data processing operation need not be considered for the moment, only the technical features are of importance at present. The short list of examples that is given shows clearly that the operations are mainly ones of storage. All, or at least a part, of the working information is kept in fixed mechanical storage devices (e.g. drilling jigs, templates, etc) in such a way that no error can arise on the machine or so that the latter can even operate automatically. It may be pointed out in passing that this form of 'data processing' must be carried out in auxiliary workshops by highly-skilled personnel(10).

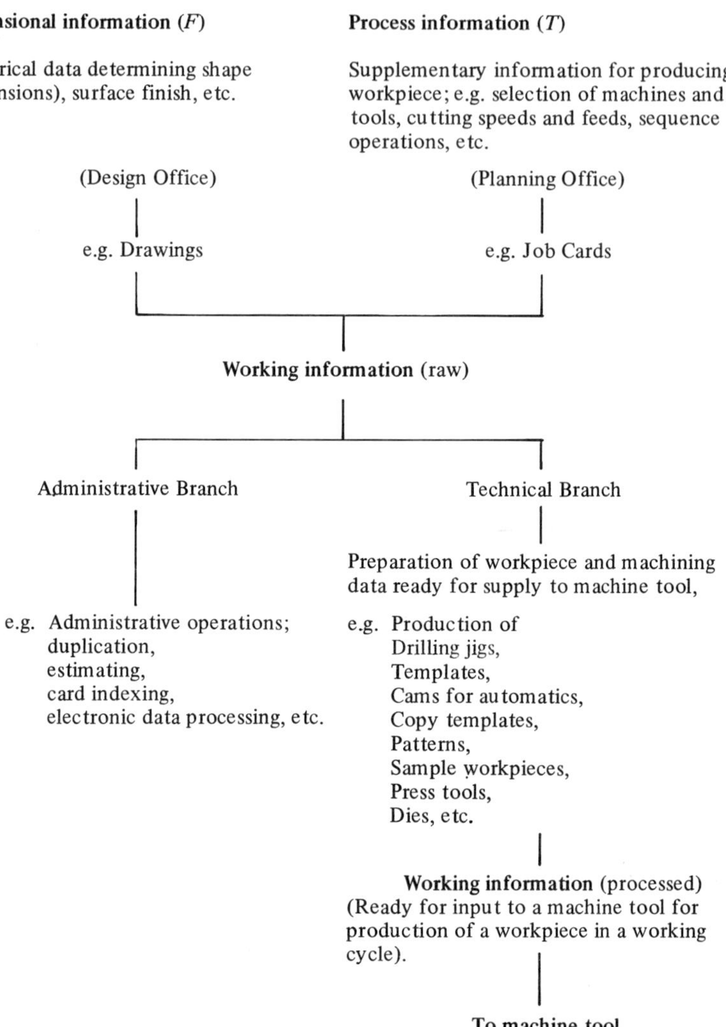

Table 1.   Arrangement and conventional processing of working information

Up to this point nothing that has been said will be unfamiliar to the practical engineer. Now, however, something entirely new will be added. Assume that a machine tool which can understand numerical dimensions has been produced in some form or other. In this case all these fixed mechanical storage devices will no longer be required. It is only necessary to process numbers and to store the results in suitable information carriers (e.g. punched tapes). This is more compact, and, if electronic office machinery is used, can be performed much more quickly than it is possible to make jigs and tools. The first step in the field of data processing has been taken, and it is now possible to bring the administrative branch shown in Table 1 into the planning operation. One of the most important problems to arise and which requires more close investigation, is the transmission and processing of information.

1.5 Rationalisation of the flow of information as a new engineering problem

The study of the optimum form of data transmission and processing forms part of information theory. The problem of dealing with energy effectively and economically having been mastered, it is now also necessary to learn how to handle information in a rational manner. The whole structure of the theory of information was built up mainly by Shannon (19) between 1942 and 1948. It would be outside the scope of this book to go into details at this point, but information theory forms the starting point for many further considerations. Anyone who wishes to delve more deeply into the problems of numerical control, which form a part of the general question of data processing is therefore strongly advised to supplement his reading of this book with a study of this new branch of science (1,2,6,23).

The most important step that has to be taken to enable the machine tool to be incorporated into a comprehensive information flow arrangement is that of regarding the machine itself as a data processing system. This may sound rather novel to many production and design engineers, but it is the only possible conclusion. It is only by extending the considerations of the flow of information to the machine tools themselves, by adopting and adapting the various concepts that have already become established for general data processing equipment, and by incorporating machine tools into the structure of the data processing systems that it is possible to obtain an overall view of the complicated relationships involved (18). This also implies that it will be necessary to introduce the methods of systems engineering into production engineering discussions; this point will be considered in greater detail in Chapter 2.

Information flow and coding theory

The next task can be formulated, quite generally, in the following terms:

Find the most satisfactory and efficient form for the transmission and processing of data within a given system (2, 6, 15).

The problem of conveying information in a form in which it can be read and processed by man is an ancient and important one, and has largely been solved. The problem of presenting imformation in such a way that it can also be read and processed by automatic machines, on the other hand, is more recent, although basically it involves nothing new. In principle one always uses a store of characters. In our culture, for example, in order to pass information from one person to another we employ:

> The 26 letters of the Roman alphabet a to z,
> The 10 Arabic numbers zero to nine,
> Various punctuation marks, such as .,!?(): etc.,
> Various mathematical symbols, such as = + −, <> etc.

Every written language and scientific discipline has its own 'alphabet', the word alphabet being used in its most general sense. Usage is the same in all cases; a list of symbols is adopted by common agreement, and a given message is denoted by a certain combination of characters drawn from this list. The basic stock of characters can be large or small.

Many written languages have a very large number of characters (e.g. the hieroglyphic languages). In the Western world 2 x 26 symbols (capital and lower case letters) prove, in general, sufficient to describe most things. And none of the information content is lost if the capital letters are dispensed with. As a first approximation, it can be said that for written communication something on the following lines would suffice:

1. A selection from 26 characters of the first type (e.g. Roman letters);
2. A selection from 10 characters of the second type (e.g. Arabic numbers);
3. A selection from about 8 characters of the third type (e.g. punctuation marks), termed 'operative symbols' below.

This would provide a basic store of 44 characters which would have to be learned. Any other characters would be superfluous, since even the most complex concepts can be expressed by a suitable combination of these symbols (20). In addition, the term 'word' may have a much more general meaning than that attributed to it in everyday conversation, so that it can signify any group of characters which, in combination, expresses a valid concept, since so far as automatic machines are concerned, it is the valid concepts which matter, and not pronounceability. The question that now arises is whether even this store of 44 characters may not still be too large; or in other words, what is the minimum number of basic characters with which it is possible to express a given item of information? The answer — which is important in practice — is two basic characters (1,20). A selection of one of two mutually exclusive conditions is termed a binary decision. Examples of this are the opposites:

> Yes − No
> True − False.

A binary decision can be made using very simple technical equipment. Any switch has only two states:

ON − OFF.

Consequently, a binary decision can be made with any kind of switch (mechanical, electro-mechanical, magnetic, electronic, pneumatic, hydraulic, etc). An elementary binary decision of this type is also called '1 bit'. It can be represented symbolically by:

$$1 - 0$$
$$\text{or } L - O$$

To avoid any confusion with the Arabic numbers 0 and 1 the two possible forms of the binary decision will be represented in this book by the capital letters

L and O.

The next step must be to find a relationship between this binary decision and any arbitrary number of units, i.e. the principles underlying the ordering of binary decisions must be found.

Every correlation between two groups of symbols is termed a 'code'. Many years ago Leibniz showed that calculations can be performed just as readily using a binary code (to base 2) as with the conventional decimal code (to base 10). The rules for forming these numbers may be compared by means of an example: The quantity three hundred and eighty two can be written in coded form as follows:

In the decimal system:

$$3 \cdot 10^2 + 8 \cdot 10^1 + 2 \cdot 10^0$$
$$3 \qquad 8 \qquad 2 \qquad \qquad \triangleq 382$$

In the pure binary system:

$$L \cdot 2^8 + O \cdot 2^7 + L \cdot 2^6 + L \cdot 2^5 + L \cdot 2^4 + L \cdot 2^3 + L \cdot 2^2 + L \cdot 2^1 + O \cdot 2^0$$
$$L \qquad O \qquad L \qquad L \qquad L \qquad L \qquad L \qquad L \qquad O$$
$$\triangleq \text{LOLLLLLLO}$$

In both cases the rule for forming numbers involves making use of the number of columns; From right to left these run in increasing powers of 10 or 2 respectively[3].

Another numerical example will show how a binary coding system rationalises the process of forming a number:

To obtain a physical representation of any five-digit decimal number (maximum 99999) in order to store or display it, etc., we require, if we use normal decimal coding, a matrix of 5 x 10 = 50 elements (relays, ferrite cores, lamps, etc), whereas for binary coding a maximum of only 17 elements is required. By a suitable choice of code (in this case by selecting the base) the number of components required has been reduced by almost 2/3, which is highly desirable in view of what has been said regarding reliability and economy. The two arrangements are compared in Figure 2, which shows how important the choice of a suitable coding system is on the dimensions of data processing equipment.

A)

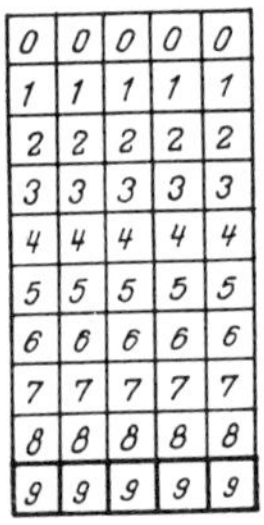

0	0	0	0	0
1	1	1	1	1
2	2	2	2	2
3	3	3	3	3
4	4	4	4	4
5	5	5	5	5
6	6	6	6	6
7	7	7	7	7
8	8	8	8	8
9	9	9	9	9

B)

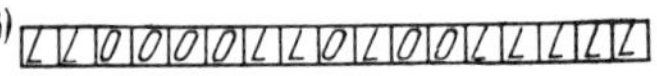

Fig. 2
Number of elements required to present a five-figure decimal number (maximum 99999)

A) in the decimal system (50 elements) B) in the binary system (maximum 17 elements)

Now a system to a base of 2 is by no means the only one for describing binary states, nor is it always the most suitable. It is beyond the scope of this book to go further into the details of coding theory, but, for future reference, the following two fundamental rules are of practical importance:

1. Reliability of operation may be impaired if the absolute minimum number of elements needed for representing a given amount of information is chosen, since a change in only one binary column will then change the information. A certain degree of 'redundancy' is essential.
2. It is impossible, in practice, to find a unit code which is equally well suited for all data transmission or data processing purposes.

---

[3]The expression for forming numbers by power series is:

$$Z = \sum_{\nu = 0}^{\nu = n} a_\nu \cdot b^\nu,$$

where Z is the value of the number, a_ν is the digit in the νth place, b is the base, and g is the number of the column (increasing from right to left).

Redundancy and the choice of a code

A simple example from everyday life can be used to explain the concept of redundancy (2, 6).

Mr. Smith wishes to send a telegram conveying the following message 'I shall arrive on Tuesday, 12 June, 1962, at 3.2 p.m. at Munich Main Station'. Here we have:

> 47 letters of the Roman alphabet (selected from 2 x 26 symbols)
> 9 Arabic numbers (selected from 10 symbols), and
> 20 operative signs (punctuation marks and spaces).

This amounts to a total of 76 symbols.

This message will cost something to send. Ignoring the postal rates for sending telegrams, which involve a minimum word charge, Mr. Smith, on hearing the cost of sending this message, will decide for reasons of economy to send a message somewhat as follows:

> 'arrive tuesday 12 june 62 15.22 munich main stn'

There has been no significant change to the content of the message; the recipient will know exactly what is meant, even if it is no longer good grammatical English.

For this message we need only:

> 30 letters (selected from 26 symbols)
> 8 digits (selected from 10 symbols)
> 9 operative signs.

So that a total of 47 symbols is required.

The cost of transmission is now only 62% of its initial value. Even if some of the symbols are mutilated in transmission the message will probably still be understood correctly. To reduce the number of characters still further, Mr. Smith might send 'arrive 12.6.62 15.2 mn. stn. mnch'.

The number of characters will now be only 34, or 45% of the original number. It is still possible to recognise the sense of the message, but there is now the danger that if by any chance one of the numbers is incorrectly transmitted a major error can be introduced. The figure 12 is no longer backed up by the additional word 'tuesday', while the 6 is not confirmed by the four-letter word 'june'. A limit in the possible reduction of the number of characters has been reached; any further reduction would greatly increase the likelihood of alterations arising in the message as it is transmitted. In other words, a certain excess over the minimum number of characters that is absolutely necessary, or a certain amount of redundancy, is necessary to prevent errors occurring. A code chosen on this basis is termed a 'redundant' code. This concept will be encountered repeatedly in the sections which follow. One point that may be made here, however, is that redundancy and economy are to some extent contradictory. The greater the redundancy in an item of information that is to be transmitted or processed, the more expensive its transmission or processing will become. Some acceptable compromise will always have to be found. Quite apart from this, if there is an undue degree of redundancy the test circuits required for its automatic evaluation will introduce a number of additional components which is incompatible with the basic rule regarding economy and reliability. In many cases it will prove impossible to investigate in detail the extent to which it is possible or desirable to make a code redundant; often experience and statistical information will be the only guide.

This leads to discussion of the second basic rule: The necessity of using different binary codes in different cases. Every binary code offers the advantage that it enables binary decisions to be made very simply with the aid of switches (of all types). It is not possible to discuss the many special codes here (6, 27); to enable the basic factors to be made clearly apparent, only five binary codes will be considered.

Other possible methods of coding and selection criteria will be discussed in conjunction with the practical applications which are described later: these include

> recordings of measurements
> programming (punched-tape coding, and similar)
> design of counters and storage devices, etc.

Pure binary code (Code A in Table 4)

The logic on which the binary system of numbers is based enables all calculations to be performed in an automatic computer that is kept as simple as possible. The binary code has the disadvantage, however, that it is very difficult, if not impossible, to read by the computer operator. When designing any automatic machine it must be assumed that the usual letters, decimal numbers, amd operative signs, together with combinations of these, will form the basis of the instructions which the operator feeds into the machine (Chapters 10 and 11). In some cases only a part of this store of characters will be used, to simplify the task of learning to operate the machine. Considering the numbers only, the conventional decimal system must always be taken as the starting point. If for economic reasons the automatic machine is to employ the binary system of numbers internally, the decimal data must be converted into the binary system at the input to the machine, and converted back into the decimal system at the output of the machine. Converters for transforming complete numbers from the decimal system into the binary system (e.g. for transforming 9 figure decimal numbers into the corresponding 30-place binary numbers) and vice versa are, however, relatively expensive. They are an economic proposition only if the installation in which they are to be used exceeds a certain size. Consequently, it is much simpler if, instead of converting the whole number, each digit is converted in turn, as is done using the codes C, D, and E which are described below (Table 4).

Before continuing, however, reference may be made to another disadvantage of the pure binary code. Each time there is a change to the next higher power of 2 there is a change from L to O in several columns simultaneously. This is particularly obvious in the change

> from 01111 for the decimal number 15
> to LOOOO for the decimal number 16 (Table 4, Code A)

If one wishes to portray a continuous process, e.g. the uniform motion of a machine slide, numerically in this way, there is a major unreliability of the reading at such a changeover point (Chapter 3). In practice it is impossible to ensure that all five columns change simultaneously and a wrong reading can be obtained for a certain short period of time. This disadvantage can be overcome if, for example, the code is constructed in such a way that only one column changes for every increment.

Gray code (reflected binary code)

As can be seen from Figure 3, in the Gray code (also known as the cyclically changed binary code) only one place changes at a time where there is a uniform progression, so that this code is suitable for converting continuous movements into numerical values. Unfortunately these numerical values are then of such a nature that they cannot be used in calculations. To enable the measured value to be processed, the Gray code has to be converted into another code which is suitable for computing, and this conversion is time-consuming. This limits the practical value of this code very severely. In many cases other methods are preferred for recording numerical values (Chapter 3 describes digital methods of measuring displacement), since it is then simpler to process the numerical values than when the Gray code is used. The advantage of being able to record the numerical values in a simple manner will have to be dispensed with in the majority of cases.

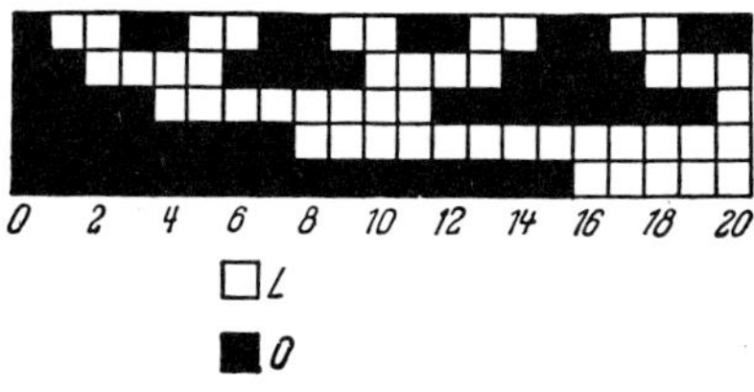

Fig. 3
Gray code, reflected binary code.

(Shown only for the decimal numbers from 0 to 20.)

Binary codes with error detection

In the form in which they have been presented both the binary codes that have just been described have another major disadvantage: They are not redundant, i.e. any alteration in a binary position will inevitably change the value of the number without any means of checking, so that in the simplest case a 'check number' or 'check bit' has to be added to ensure at least some degree of reliability in the automatic transmission and processing of numbers. Table 4 therefore shows as Code B a 6-bit binary code which in each case has a single check bit added to such a way as to give an odd number of L values in each line. Using a comparatively simple test circuit (6) it is then possible to check each individual binary number automatically for minor errors at least. In many cases this simple parity check is regarded as adequate; some practical applications of this very simple process are given in Chapter 9. A code of this type is also known as a 'single-check code' since it will not reveal the presence of two errors in a binary code. The simple parity check can, of course, be applied to binary numbers of any length; the six-place series of binary numbers with $2^6 = 64$ information bits and one check bit in each has been taken only as an example. In principle the check bit can be arranged in any location; in the example shown the check bit is located in the third column as specified for the EIA Code (Chapter 9), and is printed in bold type. There may also be cases where this simple test procedure is inadequate; since the recognition of an error in itself will, at most, cause the auotmatic machine to emit an error signal, and the correction will have to be made by an operator.

A method which is more satisfactory technically, but more complex, is the construction of 'self-correcting' codes in which the redundancy must be greater than one check bit. The general conditions for the detection of two errors or for the automatic correction of a single error were first established by Hamming in 1950 (14). The greater the number of check bits employed, i.e. the greater the redundancy, the more complicated the equipment required for the automatic checking and decoding of the codes. The amount of redundancy to be employed in coded numbers is therefore limited by economic considerations.

Code C in Table 4 is an example of a code designed by Hamming in which 4 information bits are accompanied by 3 check bits (14,15). In view of its basic importance the construction of this code and the manner in which it works will be briefly described. The four information bits are in binary code and are located in positions 3, 5, 6, 7 (reading from left to right). The check bits, which are printed in bold type, occupy positions 1, 2 and 4. It is now necessary to perform three parity checks to test each combination of holes. In the code illustrated the positions:

I	1, 3, 5, 7
II	2, 3, 6, 7
III	4, 5, 6, 7

must each be checked to see that there is an even number of L symbols in order to enable a wrong symbol in a seven-place binary number not only to be detected, but also to indicate the place where the wrong symbol must be 'reversed'. The principles of this process will be shown in an example:

The correct code for the decimal number 7 is

OOOLLLL

Due to a transmission error, or to a control tape with a hole that is not properly punched, or to a faulty contact in the tape reader (Chapter 9) the following code is entered into the automatic machine in error:

OOOLOLL

The three parity checks for even numbers of L referred to above then give:

I				1 . L
II				2 . L
III				3 . L,

which firstly indicate that the code is incorrect. If the convention that the presence of even numbers of L, as required, is indicated by O, while a departure from the required parity condition is indicated by L, the three parity checks give:

									I II III
the binary number   L   O   L or the decimal number 5.

This indicates that the error is located in the 5th column of the coded entry, and that there the symbol O must be replaced by an L, as was in fact originally intended. The error has been detected in the correct location and has been corrected.

It will be obvious that this process provides a very elegant solution, and that it enables the accuracy and reliability of data transmission to be considerably increased, but that at the same time it involves considerable technical effort. A code that is used in practice in conjunction with machines, and which, while of similar construction, is capable of higher outputs, and therefore is economically more advantageous, is described in Chapter 9. (25)

A common feature of all the binary codes that have been described up to the present is that they require a complete conversion of the numbers when they convert to the decimal system. The conversion process can be greatly simplified if, instead of the complete number being converted, conversion takes place digit by digit. In this case only the ten decimal numbers 0 to 9 have to be changed, and sometimes the same converter can be used for each decimal place. We thus arrive at the very widely used group of 'binary-coded decimal representations' (BCD representations). The position of the decimal place is maintained in all cases; the various codes differ only in the method of representing the digits 0 to 9 in binary form. Since it is always possible to represent any set of ten digits (0 to 9) with a minimum of four binary digits, the saving in the number of elements required will be 40% instead of 33% as achieved by the use of the pure binary code. This reduction in the savings that can be achieved is, however, more than offset by the savings in the decimal/binary and binary/decimal converters. A four-place binary number is also known as a tetrade.

Pure binary-coded decimal system (Code D in Table 4)

In this code each digit of a decimal number is converted separately into the pure binary form. For example, the number 962 is converted into the pure binary tetrades:

$$\begin{array}{ccc} 9 & 6 & 2 \\ \text{LOOL} & \text{OLLO} & \text{OOLO} \end{array}$$

Thus, calculations within each decimal place are carried out in binary form, whereas calculations between the decimal places are carried out in decimal form. Since only 10 of the 16 possible combinations of a tetrade are used there is even a certain degree of redundancy and error indication. This code is very suitable for many practical applications, but suffers from the disadvantage that the decimal zero must be represented by the binary number 0000, and is thus identical with the absence of information. Should a wire break or a dry joint be present (e.g. in a digit store) or a moving contact fail to operate correctly, it is possible for a decimal zero to be indicated where it should not exist. Clearly it would be desirable to provide some means of avoiding this contingency.

3-excess Code (or Stibitz code)

If advantage is taken of the fact that a binary tetrade can be used to represent not merely 10, but up to $2^4 = 10$ units (e.g. the decimal numbers 0 to 15), it is possible to select 10 suitable combinations from these 16. In the 3-excess code three of the possible combinations are omitted at each end of the pure binary series (Table 2).

The remaining binary numbers from OOLL to LLOO are allocated in sequence to the decimal numbers 0 to 9. This produces two advantages:

1. The zero has now become the unambiguous binary number OOLL, and is clearly distinguishable from the absence of information.
2. The numbers equally spaced on either side of a line of symmetry are inversely related, i.e. O and L are interchanged.

| Old relationship | | New relationship | |
decimal number	pure binary number	binary number	decimal number
0	OOOO		
1	OOOL		
2	OOLO		
3	OOLL	OOLL	0
4	OLOO	OLOO	1
5	OLOL	OLOL	2
6	OLLO	OLLO	3
7	OLLL	OLLL	4
			Line of symmetry
8	LOOO	LOOO	5
9	LOOL	LOOL	6
10	LOLO	LOLO	7
11	LOLL	LOLL	8
12	LLOO	LLOO	9
13	LLOL		
14	LLLO		
15	LLLL		

Table 2.   Construction of the 3-excess Code (Stibitz code).

Example: The figure 2 is symmetrically located with respect to 7, and the corresponding binary numbers are OLOL and LOLO.

A result of the latter feature is that the sum of two symmetrical decimal numbers is always 9, or in other words, the second number is always the complement of 9 of the first (e.g. $2 = 9 - 7$).

In addition it presents no difficulty whatever from a technical point of view to invert a binary number (e.g. by means of a changeover contact on a relay). With this code subtraction thus becomes very easy (6, 27), a feature which can prove extremely useful in designing comparators, for example. The redundancy is the same as for Code D in Table 4.

Decimal number	Binary number
0	LLOLO
1	OOLLL
2	OLOLL
3	LOOLL
4	OLLOL
5	LOLOL
6	LLOOL
7	LLLOO
8	OLLLO
9	LOLLO

Table 3.   Example of the construction of a $\binom{5}{3}$ Code

Decimal Number	Whole-number conversion			Digit-by-digit conversion Binary coded decimal system (BCD)		
	A	B	C	D		E
	Pure binary code with 5 information bits (Code without checks)	Pure binary code with 6 information bits and 1 check bit (simple error recognition code)	Pure binary code with 4 information bits and 3 check bits (simple error-correcting Hamming code)	Pure binary code for the ten digits		$\binom{5}{3}$ code for the ten decimal digits
				tens	units	
0	OOOOO	OOLOOOO	OOOOOOO	OOOO	OOOO	LLOLO
1	OOOOL	OOOOOOL	LLOLOOL	OOOO	OOOL	OOLLL
2	OOOLO	OOOOOLO	OLOLOLO	OOOO	OOLO	OLOLL
3	OOOLL	OOLOOLL	LOOOOLL	OOOO	OOLL	LOOLL
4	OOLOO	OOOOLOO	LOOLLOO	OOOO	OLOO	OLLOL
5	OOLOL	OOLOLOL	OLOOLOL	OOOO	OLOL	LOLOL
6	OOLLO	OOLOLLO	LLOOLLO	OOOO	OLLO	LLOOL
7	OOLLL	OOOOLLL	OOOLLLL	OOOO	OLLL	LLLOO
8	OLOOO	OOOLOOO	LLLOOOO	OOOO	LOOO	OLLLO
9	OLOOL	OOLLOOL	OOLLOOL	OOOO	LOOL	LOLLO
10	OLOLO	OOLLOLO	LOLLOLO	OOOL	OOOO	
11	OLOLL	OOOLOLL	OLLOOLL	OOOL	OOOL	
12	OLLOO	OOLLLOO	OLLLLOO	OOOL	OOLO	Continues to occur
13	OLLOL	OOOLLOL	LOLOLOL	OOOL	OOLL	in same form in all
14	OLLLO	OOOLLLO	OOLOLLO	OOOL	OLOO	decimal places
15	OLLLL	OOLLLLL	LLLLLLL	OOOL	OLOL	
16	LOOOO	OLOOOOO		OOOL	OLLO	
17	LOOOL	OLLOOOL		OOOL	OLLL	
18	LOOLO	OLLOOLO		OOOL	LOOO	
19	LOOLL	OLOOOLL		OOOL	LOOL	
20	LOLOO	OLLOLOO		OOLO	OOOO	
21	LOLOL	OLOOLOL		OOLO	OOOL	
22	LOLLO	OLOOLLO		OOLO	OOLO	
23	LOLLL	OLLOLLL		OOLO	OOLL	
24	LLOOO	OLLLOOO		OOLO	OLOO	

25	LLOOL	OLOLOOL		OOLO OLOL	
26	LLOLO	OLOLOLO		OOLO OLLO	
27	LLOLL	OLLLOLL		OOLO OLLL	
28	LLLOO	OLOLLOO		OOLO LOOO	
29	LLLOL	OLLLLOL		OOLO LOOL	
30	LLLLO	OLLLLLO		OOLL OOOO	
31	LLLLL	OLOLLLL		OOLL OOOL	
32		LOOOOOO		OOLL OOLO	
33		LOLOOOL		OOLL OOLL	
34		LOLOOLO		OOLL OLOO	
35		LOOOOLL		OOLL OLOL	
36		LOLOLOO		OOLL OLLO	
37		LOOOLOL		OOLL OLLL	
38		LOOOLLO		OOLL LOOO	
39		LOLOLLL		OOLL LOOL	
40 to 59		etc.		etc.	
60		LLLLLOO		OLLO OOOO	
61		LLOLLOL		OLLO OOOL	
62		LLOLLLO		OLLO OOLO	
63		LLLLLLL		OLLO OOLL	
Advantages	Good for calculations	Good for calculations, slightly redundant, (check bit in third place printed bold)	Good for error-free transmission, heavily redundant (check bits printed bold)	Good for calculations, easily converted to decimal system and vice versa	Redundant, easily stored
Disadvantages	Not redundant, awkward to convert to decimal system and vice versa	Only *one* error can be detected, awkward to convert to decimal system	Complicated to check and awkward to convert to decimal system	Limited redundancy, decimal zero not unambiguous (corresponds to absence of a character)	Unsuitable for calculations

Table 4. Comparison of 5 binary Codes and their properties

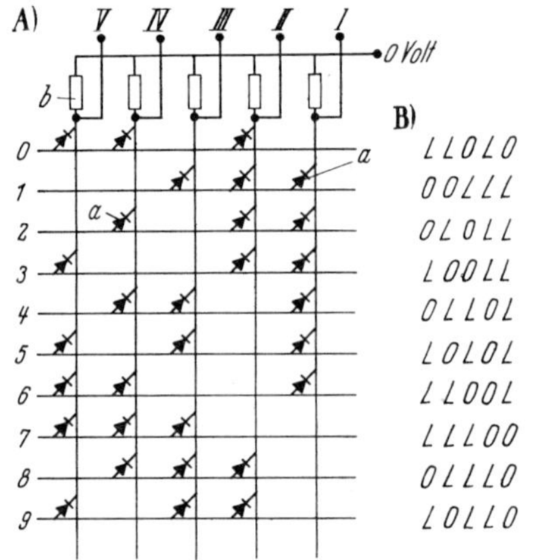

Fig.4
Example of a diode matrix for the code conversion of decimal numbers to binary numbers; the form of the signal on the binary side can, for example, be defined by 0 = 0 volts, L = + 10 volts.

A) Matrix construction. B) The binary numbers that correspond to the decimal for this matrix construction.

a Semi-conductor diodes.
b Resistances. 0 . . 9 Decimal input.
I . . V Binary output.

If a greater degree of security is to be achieved for a BCD code it will no longer suffice to use a tetrade for the 10 digits. An example of a simple code with a relatively high redundancy will therefore now be given.

The $\frac{5}{3}$ Code (in words: the three-out-of-five code, code E in Table 4)

Several groupings are possible with this code, but only one will be described here.

With five binary places there are 2^5 possible combinations of L and O. If, however, we require three L's to occur in each number there will be precisely ten possible combinations. These can be allocated to the ten decimal digits, thus providing a numerical code which is very simple and which can easily be checked (Table 4); because a simple test circuit can be arranged to indicate an error if there is one L too few or too many (Chapter 9). Unfortunately this code again suffers from the disadvantage that it cannot be used for calculations. Against this it is very suitable for monitoring the storage and transmission of numbers, and can therefore be used in number stores, data transmission devices, etc.

The five binary codes that have been described are compared in Table 4.

Code conversion

As mentioned above, codes D and E can easily be derived from the decimal system. As an example, a simple diode matrix for such a conversion is given for code E (Figure 4). It consists of ten transverse wires and five longitudinal wires, insulated from each other, but connected together at the junctions corresponding to the coding by means of semi-conductor diodes, for example. If a positive (+) 10 volt supply is applied to the wire corresponding to the decimal number 3, a current will flow in wires I, II, and V where it can be stored and monitored as a binary number. Only one matrix of this kind will be required if the digits forming the decimal number are fed into it in sequence (in the same way as they are written).

1.6   Summary of Sections 1.4 and 1.5

If the various points regarding coding that have been discussed are summarised, it is possible to draw the following conclusions:

1. Dimensional information and process information form the basis for the flow of data required for production. The system used for processing and transmitting this data must be considered from both the technical and the economic points of view, use being made of information theory.
2. The choice of code is closely linked with the economic aspects and the operational reliability of an installation.
3. There is no one 'best' code for all purposes. It must therefore be accepted that the code will have to change several times within a large system. The codes employed in a system should thus be capable of being converted without difficulty.
4. The meeting point of the human operator and the automatic machine (15, 29) must always be studied in detail, since this is where interaction occurs between standard practice and learning and familiarisation problems on the one hand and the rationalisation of the information flow required for the proper operation of the plant on the other hand (Chapters 9, 10, and 11).

Sections 1.4 and 1.5 provide an initial outline review of the problems of information flow outside the machine tool and of some of the most important possible solutions. It will still be necessary to consider some of the problems of information flow within the machine tool.

2 Machine tool as a link in the chain of data-processing systems— The fundamentals of slide displacement measurement

2.1 The machine tool as a data-processing system

Internal and external data processing

It seems appropriate to regard the machine tool from here on as a data processing system, so that it can be integrated into a larger information system. No matter what form the data processing chain takes from the initial design of the workpiece to the finish-machined, or even inspected, component, it is possible to divide it into two clearly separate parts:

1. Production and preparation of the working information from the stage where the form of the product is first specified (e.g. in a drawing, a table of dimensions, or a mathematical equation). To this is added the processing information up to the point where a single data carrier (such as a punched tape) is produced which can be used for the direct control of the machine tool. All these operations can be grouped together as the external data processing system, of which the operations listed in Table 2 (Chapter 1) will by this definition all form a part.
2. Processing of the working information inside the machine tool — logically described as internal data processing. It always begins with the introduction, by hand or by a data carrier, of the working information into the machine. It thus embraces the entire control of the machine in the narrower sense of the term.

Merely subdividing the operations in this way gives no indication as to whether they are performed automatically, and if so to what extent; it merely distinguishes between two fields of data processing which complement each other and indeed interact, but which by their nature are entirely different in form. The production of drilling jigs or of templates in a tool room, for example, would be part of the external data processing system by this definition, since this basically merely provides a means for storing the information given on the drawings (shape indications and dimensions) in rigid mechanical fixtures. The action of following a template in a machine and the setting up of a copying system of some sort would, on the other hand, be classed as internal data processing (6).

If a closer look is taken at the machine tool with its internal data processing system, it is possible to subdivide the input working data again according to the methods used to process it within the machine. It can be shown that two widely different types of data processing are involved. A large part of the working information is used directly to control the energy flow within the machine by means of simple switching operations. Another part, however, is used only after it has undergone a more or less extensive data processing operation (such as computation) to control the movements of the machine slides, i.e. the kinematics of generating the workpiece shaping. It is convenient to describe these two types of information as 'switching information' and 'displacement information'. The distinction between these two is of sufficient importance to be considered in greater detail:

Switching operation

Every designer of push-button control systems is familiar with the idea of interlocks. The procedure is always the same:

Pressing a button energises a circuit and at the same time applies the interlock, which normally acts as a safety device. A simple example from machine tool practice may make this process clearer:

Suppose the headstock of a lathe is equipped with an electro-magnetic clutch drive providing six speeds. The following safety measures may be specified:

1. The supply pressure for the compressed air operated collet or chuck must not drop below a minimum value.
2. The collet or chuck must be closed.
3. The motor for the cutting fluid pump must be running.
4. Only one clutch may be engaged at any time.

The circuit must then be designed in such a way that when, for example, button 3 is pressed (Input of information 'ENGAGE Clutch 3'), clutch 3 will not, in fact, engage until all the four conditions given above are satisfied. This is basically the same technique as that used for interlocking points and signals in railway engineering.

Basically this problem has nothing to do with numerical control; ultimately, however, the machines that can understand dimensions will be entirely remote-controlled, so that inevitably the question of logical linkages will arise in the design of the circuits. In simple cases problems of this type can be solved by an experienced maintenance electrician. In more complicated cases, especially where the number of components employed is to be kept to a minimum, empirical methods soon become inadequate. Those engaged on this work are therefore strongly advised to make themselves familiar with switching theory and its associated mathematics, switching algebra (4, 6), since the later chapters of this book assume a basic knowledge of these subjects.

At this point it is, however, important to point out that a further subdivision of the switching information is necessary. The case of the manually-operated push button is clear; this is concerned with direct switching information. The same process also takes place if we feed in the switching information (e.g. ENGAGE Clutch 3) by means of a particular combination of holes on a punched tape. But there is another class of switching operation which acts indirectly rather than directly on the motive power. Like direct information, this information is fed in from outside, but it is then held in a buffer store. From there it determines the pattern of the flow of signals within the machine which form the switching information, and channels them to the points at which the drive is controlled. (28) Let us consider another practical example from the machine tool field:

Suppose the longitudinal slide of a lathe generates four discrete signals along the length of its travel, e.g. by means of limit switches and cams. These signals do not originate from outside the machine, but are produced in a definite sequence by the machine as it operates, in this case being dependent on the displacement of the slide. Unless the machine is permanently wired up to perform one specific function it is, however, necessary to determine externally the use that is to be made of these signals. Assume that the signals have the following meanings:

> Signal 1: ENGAGE main spindle speed 4,
> Signal 2: ENGAGE longitudinal feed 5,
> Signal 3: ENGAGE main spindle speed 2,
> Signal 4: ENGAGE longitudinal feed 3,

and that these instructions are programmed externally to form part of the switching information input (e.g. by a crossbar distributor). The signals themselves then emanate from inside the machine and at different times (hence the buffer store), are defined externally, and then, like any other item of switching information, act either directly or through a logical linkage, or an energy controller. This type of switching information will be referred to as indirect information. The simplest method of carrying out such signal processing in practice is to make use of manually set plug boards, jacks and switchboards, etc. It is, however, necessary to obtain a clear picture of the logical relationships underlying these techniques at this stage, since otherwise great difficulty will be encountered later on for example when drawing up complicated programs for tape-controlled machines. In principle the many ideas and methods used are well known, but they seem much more abstract when remote programming is used than when the machine is controlled directly.

This concludes the discussion of this aspect of the switching information. It should, however, be pointed out once more that switching information, although it may in some cases comprise a considerable proportion of the working information, has nothing to do with the actual numerical concept of a machine that is capable of understanding numbers, but is primarily the way in which the Technological Information (cf. Table 2) is processed mechanically. The way in which this processing is performed is basically the same for all machine tools.

Displacement information

The situation is completely different as regards the displacement information which controls the slide movements that determine the shape imparted to the workpiece; this is derived from the shape data (cf. Table 2). So far as the numerical concept is concerned, the numerical dimensions represent the core of the displacement information and the numerical control relates only to the displacement information. With non-numerical processes the displacement information is supplied to the internal data processing system by other means, such as tracers which scan templates, patterns, cams, etc. (10, 24, 29). The displacement information is stored in these devices during the external data processing operation, and is then 'recalled' during the internal data processing operations by means of suitable devices (e.g. copying systems).

Clearly no machine slide can be moved by numerical values alone. A more or less comprehensive signal processing operation must always be employed. An initial discussion of this was given in Chapter 1, and it was established that at the present time the most convenient way of doing this is by measurement control of the slide displacement. Figure 5 shows the four basic units of Figure 1B) extended in accordance with what has been stated up to the present, and in particular a central input position *f* for the working information has been added. No mention has been made up to the present as to how the working information *a* (supplied from the external data processing system) is fed into the input unit *f*. This may be done either manually or by means of a data carrier (cf. Chapter 9). It will merely be pointed out here that it is one of the important functions of the input unit *f*, especially where it automatically reads information off a data carrier, both to decode this

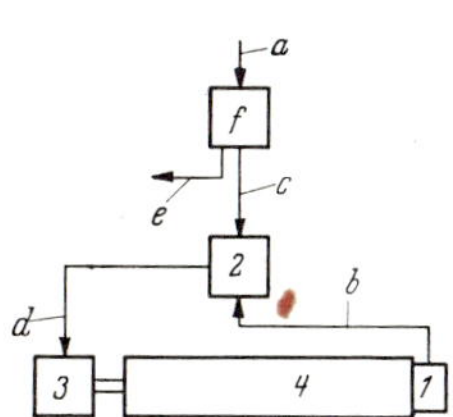

a Working information (supplied by external data processing system), *b* actual value feedback, *c* (identical with command *W* in control technology), *d* positioning signal for drive 3 (identical with regulating variable *Y* in control technology), *e* switching information, *f* input point for working information, (e.g. punched tape reader, control console, etc) which also establishes the machine-specific type of signal for further processing of the displacement information (see Chapter 9).

Fig. 5   **Internal data processing for displacement along a single axis using a closed-loop system**

information into switching and displacement information, and also to establish the machine-specific type of signal (e.g. electrical or pneumatic signals).

Classification of machine tools by type of control

Machine tools can be divided into two broad groups, according to the method by which their slides are controlled:

1. Machines in which there is no functional relationship between the movements along the different coordinates. This group includes machines such as coordinate drilling machines, jig borers, boring and milling machines, turret lathes, turret presses, spot welders, metal-forming machines, lathes and milling machines on which only operations parallel to an axis are performed, etc.
2. Machines in which there is a functional relationship between the movements along the different coordinates. This group includes profiling machines, such as profile milling machines, profile turning machines, profile flame-cutting machines, etc.

Differentiating the machines in this way, i.e. according to the functional relationship between the movements, is important from both the technical and the economic viewpoints. It may be pointed out here that to maintain a functional relationship accurately by means of numerical control methods can even nowadays be a very expensive matter, so that numerical control systems of the second type (with functional relationships) can cost twice as much and more as numerical control systems of the first type (without functional relationships). (10). This is the main reason why initially (until about 1965) the main emphasis was on the development of control systems without functional relationships, only the introduction of integrated circuits (cf. Chapters, 3, 8 and 12) enabling more satisfactory cost ratios to be obtained (37, 38).

Positioning control and straight-line control

Strictly speaking it is necessary to make a further distinction between two sub-groups when dealing with control systems without functional relationships.

a) The tool is never in contact with the workpiece while the slide is in motion, and the control merely moves the slide to individual discrete coordinate points. The most typical example of this class of machines is the coordinate drilling machine. This system is known as positioning control.

b) The tool is in contact with the workpiece while the slide is moved along one of the coordinates; the movements are thus always parallel to the axes of the machine and extend from a specified starting point to a specified finishing point. Since here again there is no functional relationship between the individual directions of motion the use of this type of control system in conjunction with cartesian coordinates (e.g. milling machines with traverse tables) only enables rectangular shapes to be machined, or in the case of lathes, only cylindrical workpieces with perpendicular shoulders can be produced. Since machining always takes place along a straight path, this type of control is known as straight-line control. In most cases, however, these two types of control can be considered under one heading.

Continuous path control

Positioning and straight-line control systems (without functional relationships) thus differ from the control systems employed in profiling machines (with functional relationships) in which the tool is guided along a path of any desired shape. The latter type of control is known as a continuous path control system.

These terms that have been introduced for the grouping of information and of machines are summarised and briefly defined in Table 5. For the remainder of this general survey of the position, however, it will be sufficient to consider the switching information as a whole without sub-dividing into direct and indirect categories, and also to ignore the differences between positioning control and straight-line control systems. The most important feature which both these types of control system share is the absence of any functional relationship. Since, as has been shown, this functional relationship plays such a decisive role, from here on all discussions of fundamental principles must involve at least two axes.

Internal data processing for movements without functional relationships

Figure 6 shows the information flow [1] for a two-axis control system of this type. Not only is there duplication of parts 1 to 4 but the input units *e* and/or *f* are additional items. The theory and practice of input units is discussed in detail in Chapter 9, but it will be convenient to deal with some of the basic features of these devices here in order to enable an overall view to be obtained.

---

[1] Strictly speaking one should speak of signal flows in this context, since the actual carriers of information are termed signals; as the later discussions deal with information flow in general, however, the distinction may be ignored here.

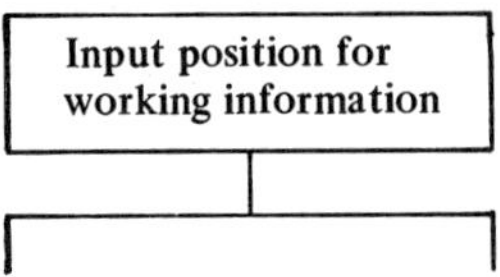

Switching information

All working information which directly affects a flow of energy by simple switching operations and their logical linkages within the machine (mathematical treatment by circuit technology methods). The information can be subdivided into direct switching information and indirect switching information. Both include, for example,

1. Switching of main spindle speeds, feed speeds, etc.

2. Recovery of information from memory stores, such as correct indexing of turret heads, switching tool-changing systems on and off in conjunction with tool magazines (see Chapter 7) etc.

3. Switching of auxiliary movements, such as chucking and releasing workpieces, starting and stopping automatic workpiece loading equipment, switching measuring instruments on and off, controlling coolant flow, etc.

Important equipment:

Push-button switches, relays, contactors, electromagnetic clutches, hydraulic and pneumatic valves, solenoids, magnetic chucks, combinations of switching transistors [39], etc.

Displacement information

All working information which requires a greater or lesser degree of processing before it can intervene in the running of the machine and thereby determine the path to be followed by the point of contact between the tool and the workpiece (mathematical treatment primarily by control technology methods). For example, by digital programming, control of follow-up circuits, control of interpolators, etc.

In this connection it is necessary to distinguish between two groups of machine tools:

Machine tools with positioning and straight-line control systems

Characteristics:

a) Without functional relationships between the coordinates,

b) Often relatively little displacement information, relatively large amount of switching information (manual input of all working information therefore possible).

c) During changes of position there is often no tool in contact with the workpiece; if a tool is in contact with the workpiece, however, only straight-line movement along an axis is possible.

Examples:

Boring and drilling machines, jig borers, turret lathes, turret presses, spot welding machines, lathes for cylindrical workpieces, milling machines for flat surfaces, forming machines, etc.

Machine tools with continuous path controls

Characteristics:

a) With functional relationships between the coordinates,

b) Large amount of displacement information, relatively little switching information (therefore displacement information must be fed in by means of a data carrier).

c) Tool continuously in contact with workpiece while movement takes place along a path of any length and shape.

Examples:

Profile milling machines, some types of lathe (e.g. profiling lathes), flamecutting machines, etc.

Table 5.   Summary of processes and examples of internal data processing.

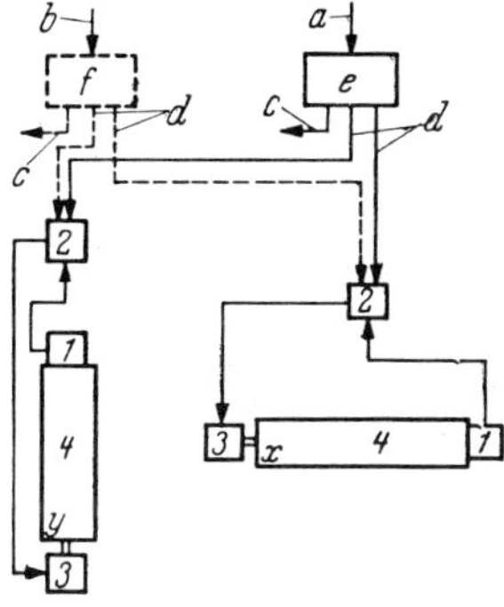

Fig. 6
Information flow in a control system without functional relationships between the motions

a Working information stored in data carriers (e.g. punched tape), *b* working information stored in drawings, instruction sheets, job cards, etc. *c* switching information, *d* displacement information, *e* reader for data carriers, *f* manual input for working information *b*, *1* displacement measuring systems, *2* comparator, *3* drives, *4* machine slide.

As will be shown later, the number of items of working information that have to be fed in and the rate at which they must be supplied is not high with many machines equipped with positioning control or straight-line control. In many instances manual input of switching information by means of push buttons, levers, hand wheels, etc. in the usual way will be satisfactory; while numerical displacement information (coordinates) is fed in by decade switches and similar items. In Figure 6 the symbol *f* indicates the point at which this information is fed in manually. The automatic input of the coordinates and the absence of drilling jigs, etc., often enable appreciable savings to be achieved even with semi-automatic operations of this type.

If the whole cycle of operations is to be made automatic and readily reproducible, all the necessary working information can be stored in a stack of punched cards, a punched tape, or a magnetic tape (Chapter 9) which acts as a data carrier, and the information can be fed into the machine automatically by means of the reader. In many cases it may be desirable to provide a free choice between the insertion of the working information either manually or by means of the data carrier, to provide the flexibility needed for handling different batch sizes. In this case the two input units will be arranged in parallel, as shown in Figure 6. For fully-automatic operation all the working information (both switching and displacement information) will then be stored in the data carrier that has been selected, and this arrangement then completes the actual machine-tool control system. The production of the data carrier, i.e. the programming of the automatic machine, forms a part of the external data processing system. Initially (about 1956-1960) it was thought that programming and the production of the data carrier e.g. punching the punched tape, would have to be entrusted to the skilled machine operator. As further experience was gained it was found that the production of the punched tape should, with very few exceptions (cf. Chapter 9), be performed as part of the external data processing.

The absence of a functional relationship has also a considerable effect on the design of parts 1 to 4. Even the first cursory inspection reveals a number of special features:

Parts 1 and 2: It is not essential to have a continuously operating control loop to regulate the movements of the slide; it is quite sufficient to have an on-off circuit enabling the speed to be changed or the drive to be switched off when a predetermined position has been reached (similar to a lift control system). Continuous comparisons need not be made in the comparator (Part 2); all that is necessary is a type S comparator (cf. Chapter 1). Obviously the task can also be performed by a continous-control circuit, but in view of the basic rules regarding reliability and cost which were mentioned earlier careful consideration must be given to the need or otherwise of equipping a machine with a continuous-path control-circuit.

Part 3: With positioning control and straight-line control, stepped drives can be used, running at various selected speeds (e.g. through gearing and clutches. Because of the large differences that exist between rapid traverse and inching speeds, the speed range must often be wide (1 : 2000 or more). The steps between the fixed speeds can, of course, be large, depending on the type of machine. Of course, an infinitely-variable drive can also be used if this is thought desirable for some other reason, but it is not necessary purely for the purpose of numerical control (cf. Chapter 6).

Part 4: In view of the low inching speeds, no difficulties are likely to occur with stick-slip phenomena where position or straight-line control systems are used (cf. Chapter 7). Depending on the chosen displacement measuring system and the other control devices, play between the leadscrew and its nut may not be critical. In many cases, for example, the slide will, for operational reasons, always approach its final position in the same direction so that there is no need for a backlash free nut to be employed (cf. Chapter 6).

Internal data processing for movements which are functionally related

Conditions are entirely different where continuous path control is employed. In order to enable a workpiece of a particular shape to be machined with sufficient accuracy there must not only be a functional relationship between the coordinate inputs, but the displacement information density must be very high. If the accuracy of a profile is to be maintained to within 0.01 mm, the required coordinate values would, strictly speaking, have to be programmed and fed in for every 0.01 mm. This could not be attained with the means conventionally available. One method which is practically feasible for a two-axis continuous-path control is shown in Figure 7. Here the displacement information is supplied in relatively coarse steps, i.e. with a rather wide spacing between coordinate values, and, by means of a punched tape, fed into the reader e, together with the switching information. Here the switching information is separated out and handled in the same way as for the position and straight line control systems; it need not be discussed further at this point. The displacement information is fed into an interpolator or curve generator i which has two main functions to perform:

1. It must either interpolate between the fixed data points so as to produce a straight line, parabola, cylindrical arc, or similar curve, or it must generate a particular type of curve (e.g. a circular arc).
2. It must resolve the resultant feed rate, i.e. the tangential velocity vector, into the component velocities of the individual machine slides, and it must monitor the component velocities to ensure that the functional relationship is maintained.

Without going into details at this stage it should, nevertheless, be pointed out that this equipment always takes the form of a special digital or analogue computer (for details see Chapter 8). The interpolator then guides the motions steadily and centrally along each axis, which explains why it is also known as a 'director'. It is also now possible to draw some conclusions regarding the requirements Parts 1 to 4 must fulfil when continuous path control is being used.

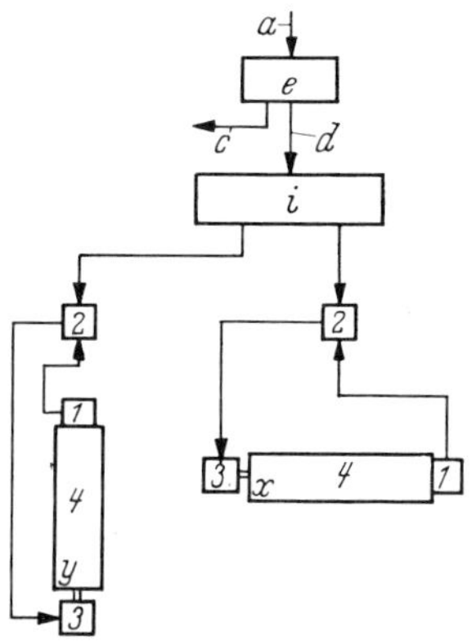

Fig 7
Information flow in a control system with a functional relationship between the movements

a working information stored in data carrier, *c* switching information, *d* displacement information, *e* reader for data carrier, *i* interpolator or curve generator, *1* displacement measuring system, *2* comparator. *3* drives, *4* machine slide.

Parts 1 and 2: The displacement measuring system and the comparator must now work continuously so that they can follow the constantly changing commands; this means that the closed-loop must take the form of a continuous control circuit, and using an on-off type of circuit is no longer possible. In order to understand and design such systems a thorough knowledge of control technology is essential (3).

Part 3: The drives must be infinitely variable from maximum speed in one direction to a maximum speed in the opposite direction, and they must be powerful enough to ensure that the control circuit possesses a good response despite the relatively large masses and frictional forces in part 4.

Part 4: Since reversals are required to preserve the proper functional relationships between the movements, the lead-screws and nuts must be free from backlash. In order to avoid excessive demands on the control circuit, the frictional forces between the leadscrews and nuts and at the guides must be kept low, and, in particular, at a uniform level.

If the requirements for the internal data processing of positioning and straight-line control systems on the one hand, and continuous path control systems on the other, are compared even this short survey will indicate clearly the reasons for the large difference in cost. It will be obvious that this is largely due to the interpolator used with the continuous path control system.

2.2   The incorporation of the machine tool into a larger information structure

Discussion of the system and development of model systems

Finally there is the question of incorporating the various types of numerical control, i.e. position, straight-line, or continuous path control, into the comprehensive system employed for the plant as a whole. In practice there are very many possible combinations that could be employed (10). To avoid raising too many difficulties at this stage, the discussion will initially be limited to a small number of typical systems. Experience has shown that it will be sufficient at this stage to restrict the discussion to the three systems illustrated in Figure 3. Positioning control and straight-line control systems can be considered together; but continuous path control can conveniently be divided into two arrangements. In all cases the broken line $k \ldots l$ represents the boundary between external and internal data processing.

Figure 8A), shows the complete information flow system for positioning control and straight-line control. As has already been discussed, the working information comprises the dimensional information and the processing information (Table 2). This can be supplied to the manual input unit n from a conventional, non-automatic data processing unit e (dotted line) as described in Chapter 1. Alternatively the working information can be listed in a schedule and arranged by the programmer c in such a form (manuscript d) that an assistant can transfer it by means of the coder f to the data carrier g that is to be employed (cf. Chapter 10). In practice this generally takes the form of punched tape. The subsequent treatment within the internal data processing system has already been discussed.

Figures 8B) and 8C) show two systems for continuous path control. The main difference between them is the location of the interpolator i. The internal data processing system employed in Fig. 8B) has already been discussed. In practice the only convenient data carrier is a punched tape. The external data processing system is relatively simple. The dimensional information and the processing information are brought together in the programmer in much the same way as for either positioning or straight-line control. Within the programmer the amount of processing and computing that is required will depend on the shape of the workpiece. Assume that the sketch of the workpiece represents a steel plate in which a recess of a shape as shown is to be machined by means of an end-mill. The programmer then has to determine a number of datum points on the path to be followed by the

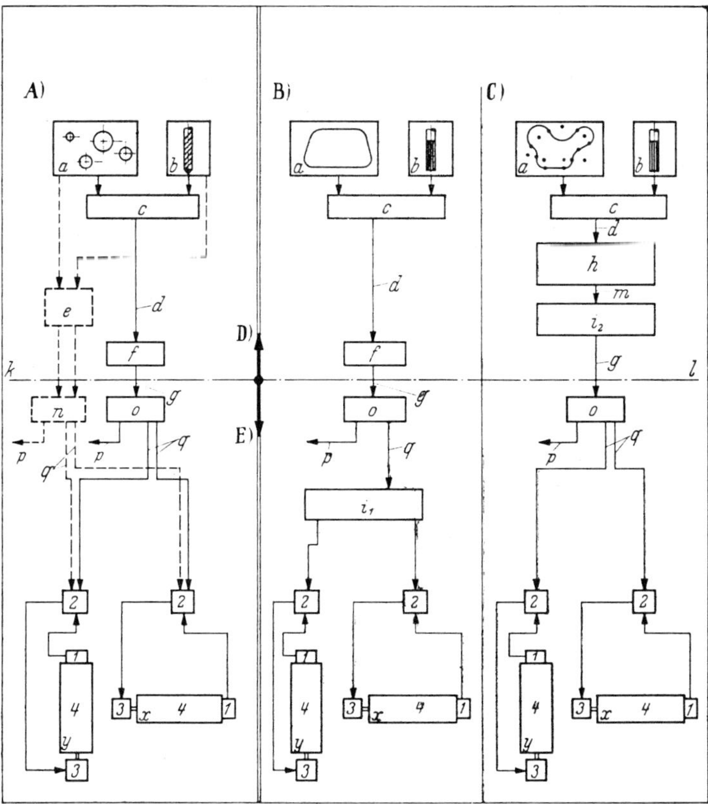

Fig. 8　**Three characteristic data-processing systems for machining operations**

A) System for positioning and straight-line control (no functional relationship between the movements along the individual axes). B) System for continuous path control with internal interpolator. C) System for contouring control with external interpolator. D) External data processing, E) Internal data processing *a* dimensional information, *b* processing information, *c* programming point, *d* program manuscript, *e* conventional external data processing (tool room, pattern stores, etc), *f* coding point for data carrier, *h* universal electronic computer (can also be used for administrative work), i_1 internal interpolator, i_2 external interpolator, k - 1 boundary between internal and external data processing, *m* intermediate data carrier, *n* manual input for working information, *o* reader for data carrier, *p* switching information, *q* displacement information, *x* X-axis, *y* Y-axis, 1 displacement measuring systems, 2 comparators, 3 drives, 4 machine slides.

cutter, basing his calculations on the shape of the workpiece and the diameter of the milling cutter, and these have to be entered in the schedule in the correct sequence. The tape *g* can then be punched at the coding station *f*. This concludes the external data processing operation. It will be clear even from this simple example that the amount of work involved in the external data processing depends to a large extent on the shape of the workpiece. With complex workpieces it may be well worth while to make use of electronic data processing equipment; to avoid complicating the issue this possibility will not be further discussed here (cf. Chapter 11). Figure 8C) shows a system of completely different layout. Here a comprehensive electronic data processing system is assumed to be available. Under these circumstances the workpiece, as shown in the sketch of the dimensional information, need only be described by means of a few parameters, such as the coordinates of centre points, points of inflection, radii of curves, slopes, equations for envelope curves, etc. The electronic data processing unit can then calculate the path of the end-mill, making allowance for its diameter (cf. Chapter 11). In many cases it may be desirable, for administrative and economic reasons, to calculate the complete interpolated path of the milling cutter in two steps:

> 1st Step: Work out the positions of a number of fairly widely-spaced points along the path from known geometrical data on a universal computer (either company owned or at a computer bureau) which is also used for other technical and administrative tasks.
> 2nd Step: Interpolate between these datum points by means of a separate interpolator, which is employed solely for the external data processing of the machine tools which it serves.

The advantage of this is that the problem is subdivided into a relatively simple general operation followed by a specialised operation. This subdivision is shown in Figure 8C) in which the two data-processing units are linked via the intermediate data carrier *m*.

In the case of Figure 8C) it is therefore possible to speak of an external interpolator, and in the case of Figure 8B) of an internal interpolator.

The removal of the interpolator from the internal to the external data-processing operation has far-reaching consequences, of which only a few can be mentioned here. They are:

1. The interpolated displacement information becomes so extensive that it can no longer be stored economically on a punched tape; instead a magnetic tape must be used as the data carrier (cf. Chapter 9).
2. All the electronic data-processing equipment is concentrated in the external data-processing system, where it can carry out the necessary operations at high speed without taking into consideration the machining conditions, etc. This means that equipment of this kind can supply control tapes to up to about 50 machine tools (40).

3. The cost of the control equipment located on the machine tool itself is considerably reduced, since the machine no longer has its own interpolator. The signals from the magnetic tape are now used directly as the input to the follow-up control loop circuits of the machine tool.

4. It is a perfectly practicable proposition to select the data-processing unit i so that is it capable of controlling several machine tools directly without the intermediary of a data carrier which is always a possible source of faults (on-line working, Chapters 9 and 12).

It is easy to see that the system shown in Figure 8C) requires extensive automation of the entire plant with all that this entails. In general this stage has not yet become common. At present the major development effort, at least in the non-military area, is therefore concentrated on the systems shown in Figure 8A) and 8B). A wide range of variations and combinations of the different systems is possible, especially as far as external data processing is concerned. It is nevertheless quite reasonable to limit the number of different systems to the three described, since these form the basis of most conceivable arrangements (10).

2.3   Summary of Sections 2.1 and 2.2

If the main conclusions reached in these two sections are summarised, the following picture emerges as a basis for the subsequent treatment of the subject:

1. To enable machine tools to be incorporated in large data flow systems it is convenient to introduce a number of new concepts:
Internal and external data processing
Switching and displacement information
External and internal interpolators.
2. When assessing control systems and developing appropriate models, it is very important to pay attention to the functional relationships between the movements of the slides of the machine tools. In numerical control systems it is necessary to distinguish between:
Positioning control systems }
Straight-line control systems } without functional relationships
Continuous path control systems with functional relationships.
3. Three characteristic systems suffice for an initial survey of the problem. All variations which are known at present can be derived from these three systems.
4. Until about 1967 the main emphasis of the development work was on

positioning control and straight-cut control systems; while today it is on contouring control systems (10).

5. For a more detailed study of the various problems it is advisable to acquire a basic knowledge of information theory, logic algebra (switching circuit techniques) (1,4,6,22,23) and control technology (3).

This summary concludes the broad outline of the study of systems. One important fact emerges from this overall study, namely that a change in one component of the chain leads inevitably to changes in other components. No program, can, for instance, be drawn up for the external data processing operation without some information having been given on the design of the machine, and vive versa (cf. Chapters 9 and 10). The components of the internal data processing system are also interdependent. Consequently the elements forming the four-part basic form (machine slides, drive system, displacement measuring system, comparator) cannot be considered separately beyond a certain point, but must be regarded as a functional whole. Only when the system is viewed in this way is it possible to define the most important features on which the design should be based. Before, however, discussing in detail in the following seven chapters, the individual components that form the internal data processing system, it seems advisable to consider some measurement techniques in machine tools and the processing of the results. This is appropriate after it has now been established that control based on the measurement of slides displacement is the primary factor to be considered.

2.4   The principles of displacement measurement

There are fundamentally two different methods for determining, processing, and displaying measured values.

Analogue and digital methods of operation

A quantity that is to be measured can be expressed relatively easily as a proportional analogue of another physical quantity. An instrument used for this purpose is generally called a measuring transducer. Consider the ordinary mercury thermometer as a simple example. Man has no sense organs which enable him to determine temperatures accurately, so that the temperature that is to be measured has to be transformed into a proportional elongation of a column of mercury. For a given temperature range, it is therefore possible to represent the value that is to be determined in a way which can easily be appreciated by visual observation; but so far this does not include any simple and objective means of describing the termperature to another observer. To do this, a scale must be mounted behind the mercury column, this scale being calibrated in some agreed manner. It is normal to read off and if necessary process the result in degrees Celsius or degrees Fahrenheit. In general terms this means that from a certain point onward it is useful to change from the range of a length variation to representation by discrete numerical values. This is done so much as a matter of course in everyday life that the fundamental change in the method of operation is no longer appreciated. The process can now be taken a step further and the chain of analogue transformations can be extended. The temperature can, for example, also be determined by means of a thermocouple which transforms it into an analogue electrical potential. This potential can then be applied to a moving-coil voltmeter, where it is converted by the effects of electro-magnetism into a corresponding angle of rotation. If this is connected to a pen recorder and this in turn is arranged to record on a moving paper strip, all changes of temperature with time will be converted into analogous curves. If these curves are to be evaluated, however, it will be convenient to print a grid on the paper and to calibrate this with numerical values.

Even though a considerable amount of time and effort is spent in an analogue conversion, in the end the results have to be presented in the conventional way by figures, in order to enable the information to be stored or further processed. Essentially, the same process takes place for all measuring operations (voltages, currents, pressures, velocities, etc.), and the displacements of the machine slides form no exception. The result of these considerations can be expressed quite generally as follows:

A more or less extensive chain of analogue transformation stages and processing operations can occur between a physical measurement and the numerical representation of the measured values (required, for example, for calculations or for conveying information from one person to another).

The basic rules regarding reliability and economy suggest that it would be desirable to keep this chain as short as possible, but there is also a further point to consider. Any analogue transformation can be carried out only with a certain degree of accuracy, and this is equally true of the analogue processing of measurements. Increasing the length of the chain of analogue conversions and processing operations increases the inaccuracy. Once a numerical value is obtained for a measurement, however, it can be processed further without loss of accuracy, which will depend only on the number of significant figures in the initial numerical value. This is another reason why it is desirable to keep the number of analogue transformation processes to a minimum. At the same time economic considerations oppose this argument, since, in general, analogue processing of measurements is much cheaper than numerical (digital) processing. In many cases it will therefore be necessary to take the trouble and work out the most suitable points at which to change from one principle to the other (cf. Chapters 4 and 5).

Since numbers are composed of elements consisting of figures coded in any desired way, and a representation in the form of figures is also known as a digital representation, the term 'digital technique' has become widely used for this more recent form of data processing. The old-established analogue/digital transformation, of which the mercury thermometer is an example, is always in essence the transformation from infinitely small steps to the finite steps represented by real numbers. Basically this type of measurement transformation (e.g. by means of the calibrated scale of the thermometer) is one of the functions that is most commonly used in human mental processes; and it is only when this function is to be performed by automatic equipment that it is necessary to become conscious of it at all.

The reverse process, that of digital/analogue transformation is of more recent origin, and in most cases is less easy to realise in practice. This subject is therefore discussed in some detail in Chapter 4. The effects of this approach on the numerical measurement and control of slide displacement is shown in Figure 9.

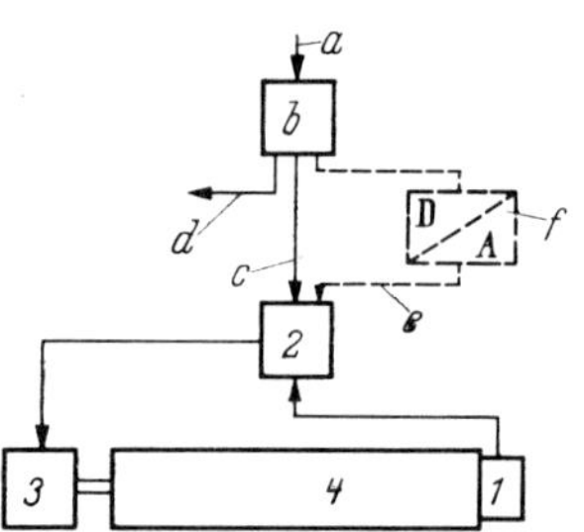

Fig 9
Elements for numerical data input in conjunction with digital and analogue displacement measurement (for one axis)

a Working information, *b* input point for working information (manually or by data carrier), *c* numerical (digital) displacement information (desired values), *d* switching information, *e* analogue displacement information (desired values), *f* digital/analogue converter.

No matter whether the input point *b* is designed for the working information to be fed in manually or by means of a data carrier, the entire numerical concept requires the working information *a* to be in numerical form. If the displacement measuring system 1 is designed to measure the values directly in digital form (acting as a local analogue/digital converter), the comparator 2 must also work in digital form, and the numerical displacement information *c* can then be the desired values that are fed directly into the comparator. If, however, for some reason an analogue method of making measurements is chosen for item 1 (e.g. the use of an electrical potential to represent the position of a slide), then item 2 can also be allowed to operate as an analogue system (e.g. as an electrical bridge circuit). The desired value *e* must then also be available as an analogue of the same physical form (e.g. a voltage). A suitable digital/analogue converter must be located between the numerical input point *b* and the analogue comparator 2. The conversion from an analogue to a digital mode which is necessary to comply with the numerical concept has therefore merely been shifted; it is essential that it takes place at some point within the system. Here again very careful study will be necessary in each individual case to determine which of the two methods is the more economical and reliable. The decision will depend largely on the nature and quality of the components that are available.

In many cases the digital method of displacement measurement may involve the least amount of equipment. A study of the various possibilities leads to a further important subdivision.

Incremental and absolute methods of measurement

As has already been mentioned in the first chapter, the simplest method of measuring displacement by digital means is to divide the distance traversed into equal elements Δs (distance quanta). The process is shown schematically in Figure 10A). Suppose the scale b measures the distance travelled by the slide a up to the limit of its movement. Suppose also that the scale is graduated uniformly in steps of Δs and that it is scanned by a device g in such a way that as each mark is passed an electrical pulse is generated. There are many possible methods of doing this (e.g. electro-mechanical, photo-electric, and magnetic scanning); they are described in Chapter 3 and will not be considered at this point. The only important fact to note here is that the pulses obtained in this way can be added up in a counter (shown in Fig. 10 as

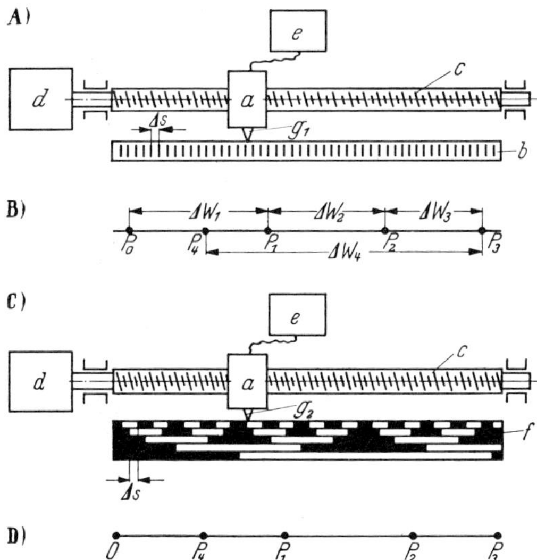

Fig. 10 **Comparison of incremental (relative) and absolute measurement systems**

A) Incremental process, B) control points along path, C) absolute process, D) control points. a machine slide, b incremental scale, c leadscrew, d drive, e comparator, f absolute scale (shown in Gray-code in example), g scale scanner, Δs distance element, distance quantum, or digital programme step, ΔW distance increment.

the comparator *e*), and the counter reading is then an accurate indication of the distance travelled by the slide. This system will basically measure displacements only, and to obtain the actual position of the slide it is necessary to add the individual displacements. In Figure 10B) the process is shown schematically in another form. The datum point is the arbitrarily chosen point at which the movement starts, at the left-hand limit of the movement of the slide *a*; when the slide is in this position, the counter is also set at zero. If *n* is the number of displacement quanta through which the slide moves, the positions of the individual points P_1 to P_ν are defined by:

$$0 + n_1 \cdot \Delta s \qquad \text{for } P_1$$
$$0 + n_1 \cdot \Delta s + n_2 \cdot \Delta s \qquad \text{for } P_2$$
$$0 + n_1 \cdot \Delta s + n_2 \cdot \Delta s + n_3 \cdot \Delta s \qquad \text{for } P_3$$

and so on, to

$$0 + n_1 \cdot \Delta s + n_2 \cdot \Delta s + n_3 \cdot \Delta s + \ldots + n_\nu \cdot \Delta s \quad \text{for } P_\nu$$

The scale scanning system and the counter can be so arranged that the direction of motion is also determined, and in this case the individual increments are automatically added or subtracted (e.g. from point P_3 to point P_4). It nevertheless remains true that each position is determined solely by the displacement from its predecessor. The process is therefore known as the 'Incremental Process' or occasionally as the 'Relative Process'. As in all such additive systems there is the danger that an error in the scanning or counting operations or the introduction of spurious pulses, etc., will result in the whole chain of measurements breaking down so that all the subsequent positional readings will be incorrect.

From the technical point of view it is possible to overcome these difficulties to a large extent by the use of suitable auxiliary devices; these devices, however, represent additional expense and make this intrinsically simple process much more complex (cf. Chapter 5). Of all the possible arrangements, the pattern of movement which involves the lowest outlay is that in which the machine slide always moves directly from a fixed datum position to the required position and then back to the datum position. As shown in Figure 10B) the pattern of movement is then

$$O \ldots P_1, O \ldots P_2, O \ldots P_3, O \ldots P_4, \text{etc}$$

This type of movement pattern is encountered, for example, in drilling spindles, the slides of turret lathes, etc. With all these single-stroke motions the mode of operation is such that there cannot be any cumulative error. The situation becomes more critical when setting up coordinated tables. Here the sequence

$$O - P_1 - P_2 - P_3 - P_4 - \ldots - P_\nu$$

and the associated cumulative measurements can conceal certain possibilities of error which become larger the larger the number of points that are to be controlled in succession without a return to the datum position. In addition, if there is a power failure or similar break-down it is impossible to reproduce the current slide position; instead it is necessary to reset the slide and the counter to their zero positions and to start the count again. On closer examination the incremental process, which seems so simple, will be found to possess certain complexities, and it can be used without modification only for certain patterns of movement.

In many cases the absolute measurement of the position, which is reproducible at any time, will be preferred. As was indicated in the first chapter, however, this means that the unknown number of displacement quanta Δs has to be encoded directly at the measuring point so that it can be represented in an unambiguous numerical system. The procedure is shown schematically in Figure 10C). In this case there is a coded scale on the bar f. To enable an operator to read off the position it is natural to choose the decimal code, and this is, in fact, adopted on manually-adjustable coordinate tables. As was shown in the first chapter however, the decimal scale is uneconomical for automatic data reading and processing, and it is better to make use of the binary system. In the diagram the Gray-code has been adopted; as has already been mentioned, this suffers from a number of disadvantages and is little used, instead the pure binary code is employed in conjunction with special scanning methods (cf. Chapter 3.) In any event, every position of the slide is unambiguously defined, and can be reproduced whenever necessary. On the other hand the equipment has become considerably more complex. The bar now carries a large number of scales (corresponding to the number of binary places) instead of just the one, and the scanner g_2 must be designed to cover all these scales; while the comparator e has to process complete numbers and not just separate pulses (as is the case with counters (cf. Chapter 3)). The absolute measuring process is therefore usually more expensive than the incremental process, but even so there are many cases where no other process can be employed because of reliability considerations. Moreover, it is definitely possible to reach the wrong decision if a choice is based solely on cost and complexity (10).

A difficulty which is common to both processes is that of reliably resolving continuous movements into the very small displacement elements Δs with the high accuracy of 0.01 mm and better which is usual for machine tools. It is these difficulties, and the costs in solving them, which in many cases justify the adoption, at least at the present state of the art of digital techniques, of the analogue measurement of displacement and subsequent conversion to digital form (cf. Chapters 4 and 5), possibly in a separate unit located away from the machine tool, such as a control cabinet.

Direct and indirect measurement of displacement

In conclusion, the various possible locations of the measuring point on the machine tool will be considered. Figure 11 shows three of the possible methods of measuring the displacement of a slide, although there are, of course, others. In A) a translatory measuring system (in this example an incremental scale with graduations) is coupled directly to the longitudinal motion of the machine slide. The measurement of the displacement is therefore direct, and is not affected by the accuracy of the leadscrew or of the forces acting on it. The measuring system also does not depend on the nature of the drive, and the slide could equally well be moved by a piston and cylinder as by a screwed shaft. It is always desirable to aim for a direct measurement system of this type, but unfortunately most linear measuring systems are still relatively expensive (cf. Chapters 3, 4 and 5). In B) the measuring system is attached to the leadscrew. The displacement or the position is thus measured indirectly by the angular rotation of the leadscrew. Consequently any errors in the leadscrew, deflections due to the forces acting on it, etc., will affect the readings. At the same time this arrangement offers the advantage that:

a) Rotary measuring systems are often simpler and less expensive to produce than linear systems,
b) the pitch of the leadscrew produces a degree of magnification which simplifies the problem of scanning very small elements of displacement Δs, especially with digital measuring systems.

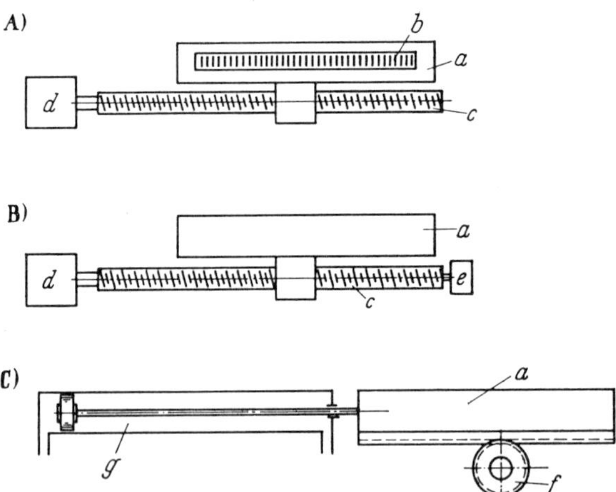

Fig 11   Three basic arrangements for the measuring unit.

A) Measuring unit on machine slide (linear system, also known as direct measurement). B) Measuring unit on leadscrew. C) Measuring unit on machine slide (rotary system).

a Machine slide, *b* Linear displacement measuring system. *c* Leadscrew. *d* Rotary drive. *e* Rotary measuring system on leadscrew. *f* Rotary measuring system with rack and pinion. *g* Linear drive system.

This example shows very clearly how closely the mechanical component (the leadscrew) is related to the electrical component (the displacement measuring system). Leadscrews and nuts that are sufficiently accurate, stiff, and operating at low friction (cf. Chapter 7) greatly facilitate the recording of the measurements and hence the construction of the control system.

Case C) represents a compromise between Solution A) and solution B). As in B) a rotary measuring system is adopted, but the conversion from longitudinal movement into rotary movement is effected with practically no loss of energy. The forces resisting the feed motion cannot introduce errors into the readings. On the other hand, any wear that may take place, and also the innate accuracy of the measuring point, will affect the results. The example shown consists of a precision rack and pinion. It could equally well take the form of a measuring screw that does not transmit any forces, or a similar device. In any event the performance of this movement converter is to a very large extent affected not only by the accuracy with which it is made, but also by its design and by the choice of materials (wear).

2.5 Summary of the principles underlying the measurement of displacement

Briefly summarising this very general discussion of the principles underlying the measurement of displacement, a number of important pointers emerge for detailed discussion in the next chapter.

1. It is necessary to distinguish between:
   a) The method of carrying out the measurements: digital or analogue,
   b) The nature of the measuring process: incremental or absolute,
   c) The location of the measuring position: direct on the machine slide or indirect on the leadscrew or a movement converter.
2. The choice of the method for carrying out the measurments often depends on calculating the cost, and is basically independent of the nature of the machine and of the movements which it performs (cf. Chapter 5).
3. The choice of an incremental or an absolute measuring process depends very largely on the task to be performed by the machine slide.
4. The choice of location for the measuring position depends on the conditions prevailing on the machine tool. It is important to divide the considerations of displacement measuring systems into those for linear and rotary systems.

II   Internal Data Processing

3   Digital displacement measuring systems and comparators

Digital measuring methods offer three fundamental advantages when applied to numerical control systems for machine tools, viz:

1. The measured value can be transmitted directly to a numerical data processing unit. This means that once the measurement is available in digital form, the accuracy of its further processing depends only on the number of significant figures it contains.
2. The data can be processed numerically, right from the moment when it is originally produced (e.g. from the time when the drawings are made) to the end of the process (even to the inspection machine), without the use of any linear measurement converter, digital/analogue converter, etc. It is therefore possible to minimise the outlay of equipment.
3. It is possible to provide a numerical value for the actual position of the slide, using very simple means. In many instances this position indication can itself prove of great assistance in rationalising the work even if it is not extended to form part of a numerical control system. (10)

It has already been mentioned (Chapter 1) that these theoretical considerations cannot always be realised in practice. There are many and varied reasons why this should be so. To establish a firm foundation on which to base a critical comparison with other displacement measuring methods, the only satisfactory course is to consider, point by point, the principal components and assemblies that are available. Only the special features of digital measuring techniques as applied to machine tool control systems, will be discussed here; there is already a considerable amount of material published which considers the subject in greater detail (e.g. 24, 41).

Frequency range

One of the most important preliminary decisions that has to be made when selecting suitable components for digital processing of machine tool data is the maximum pulse-rate of the displacement quanta to be processed. This depends on the velocity of the machine slide and the magnitude of the displacement elements Δs. Typical conditions which may occur with machine tools are shown graphically in Fig. 12. To reduce idle times, rapid-traverse speeds of 10 to 12 m/min are by no means uncommon, so this range must be included in the considerations. In the majority of practical instances it is, therefore, necessary to use electronic components and circuits that are suitable for frequencies of 20-30 kHz and above. These frequencies are well within the capabilities of modern semi-conductors. In practice, it is often better to design a system on the pulse rise-time (switching time) instead of the frequency. It should also be borne in mind that digital measuring scales, no matter how they are made, become very costly when graduated to closer than 0.01 mm because of the high precision required in their manufacture and installation; hence, where such extremely small displacement quanta have to be used (such as for jig borers) use is either made of some other method (Chapters 4 and 5), or an electronic dividing circuit (such as that shown in Fig. 26), is used. It will probably be sufficient to allow for a switching time of

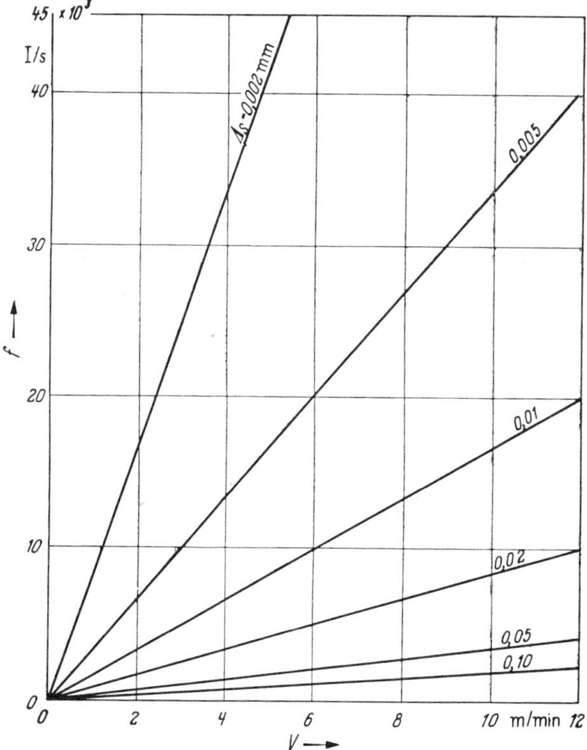

Fig. 12 Pulse frequencies f of digital displacement-measuring systems, as a function of slide speeds V (m/min) and of displacement elements Δs (mm) as encountered in practice.

about 10 μ when designing the circuits. Pneumatic control systems form an exception to this: they are discussed separately in Chapters 6 and 9.

Another factor that affects the design and selection of components is the necessity of ensuring reliable operation—even for very slow movements of the slide (inching speeds).

It is often necessary to ensure that the scanner operates satisfactorily when the slide is stationary. As will be shown later, the requirement for a lower frequency limit of zero Hz also introduces a number of special design features.

There are numerous ways in which it is technically possible to create digital displacement-measuring systems and comparators. Some basic design possibilities have already been mentioned in Chapter 2. These are:

1. Incremental and absolute measuring processes,
2. Displacement-measuring systems using rotary and linear movements,
3. Groups S and R types of comparator designs.

It is also necessary to distinguish between at least two physically different methods of scanning the scales in displacement-measuring systems whilst there are also at least three different methods by which comparators can be created, even when considering only the use of semi-conductors.

Listing the various possibilities in this way can cause considerable confusion to the newcomer in this field, due to the many available designs and their possible combinations. An attempt has been made to summarise the most important arrangements and their characteristic features in Table 6; the details will be discussed later. The reader who merely wishes to gain a quick grasp of the essentials, however, may find it sufficient to study this table carefully and to read the summaries 3.1.5 and 3.2.5; the illustrations referred to should also be studied.

3.1   The digital-incremental process and its associated equipment

The characteristic feature of the incremental process is that the displacement elements Δs are summed in a counter which is external to the displacement-measuring unit itself. It is a fundamental requirement for this process, to ensure every precaution is taken to eliminate possible counting errors. In the displacement-measuring unit, the displacement quanta are transformed into electrical pulses which are routed via a connecting cable to an electronic counter which must be easily capable of dealing with a counting rate of at least 30 000 pulses. Experience shows that practical difficulties occur not so much in the electronic counter itself, (which, if properly designed as a self-contained unit is very reliable in operation) but, almost exclusively, in the pulse generating system; that is to say, in the displacement-measuring unit and the connecting cable.

3.1 Digital-incremental process				3.2 Digital-absolute process			
Displacement measurement				Position measurement			
	Type of scanning	Remarks	Fig. No.		Type of scanning	Remarks	Fig. No.
3.1.1. Rotary systems	Photoelectric	Gratings on glass discs, scanned by photo-diodes and elements	16, 17, 18	3.2.1 Rotary systems	Photoelectric	Coded glass discs with photodiodes, V-scanning	47, 48, 49
	Magnetic	Toothed discs of soft-iron with induction coils supplied with the carrier frequency. Discs incorporating small permanent magnets, and used in conjunction with Hall-effect generators.	20 19		Magnetic	Coded drums with photodiodes, V-scanning Similar in principle to 3.1.1; scarcely used.	50
3.1.2 Linear systems	Photoelectric	Glass scales with photodiodes (transmitted light) $\Delta_S \geqslant 0.008$ mm/0.0004 in Polished steel scales (reflected light) $\Delta_S \geqslant 0.04$ mm/0.002 in. Further sub-divisions can be achieved in both examples by adoption of suitable electronic circuitry. Usually used with Hall-effect generators. Only for very coarse scales.	21, 22, 23, 24, 28 25, 29, 26 Chapter 7	3.2.2	Photoelectric	Binary-coded scales with photodiodes and V-scanning. (Currently costly for lengths above 1000 mm).	44, 45, 46
	Magnetic				Magnetic	Only for very coarse scales usually in conjunction with Hall-effect generators.	Chapter 7

3.1.5 Types of comparator for digital-incremental process (counters)			3.2.3 Types of comparator for digital-absolute process			
Semi-conductors used	Type S comparators for positioning and straight-line control systems, forward and backward counter with pre-set	Type R comparators for continuous path control systems (forward and backward counter) in conjunction with logic elements and digital/analogue converter.	Semi-conductors employed	Types of comparator for positioning and straight-line control systems		Comparator Type R_2 for continuous path control systems (numerical proportional regulator)
				Type S without direction sensing (on-off) circuit.	Type R with direction sensing (numerical three-point regulator)	
Transistors, integrated circuits.	Fig. 31, 33, 34, 35 Fig. 36, 37	Figure 43	Transistors Diodes integrated circuits	Fig. 51 A + B Fig. 52	Fig. 51 C + D Fig. 53	(6, 54)

Table 6.   **Summary of various types of displacement-measuring systems and comparators.**

There are mainly two possible sources of error in these measurements:

1. Vibration of the machine tool, caused by heavy cuts and by large tools rotating at high speed. These can be transmitted to the displacement-measuring unit and can generate additional pulses. If under adverse unfavourable conditions the amplitude of the vibration exceeds about 3 to 4 μm, such spurious pulses can be generated in the measuring system and be registered by the counter, even though the slide is stationary.

2. Modern machine tools often incorporate extensive electrical power systems incorporating numerous relays. There are also frequently a large number of inductive loads supplied with direct current (electro-magnetic clutches, solenoid-operated valves, etc., with one side of the supply frequently earthed to the body of the machine), and these produce very undesirable switching effects, despite the low voltage (24 V) on which they operate. As a result machine tool circuits carry many spurious voltages, and inductive or capacitative coupling can cause these to have very adverse effects on precision measurements involving electronic instruments.

Precautions against spurious pulses

Two methods are available for countering spurious pulses of the first type (vibration), and these can of course be combined:

   a) Mechanical method

      The machine tool and the measuring instruments are made as stiff as possible and, in addition, the displacement-measuring unit is arranged so that any vibrations that do occur will have minimal effect upon it.

   b) Electrical method

      The displacement-measuring unit is designed so that it is directional in its response. No vibratory motion can then produce a spurious increment. One basic solution for this is shown in Figure 13.

Here, the uncoded binary scale c, which has divisions $\tau = \Delta s$ (corresponding to 360° electrical) is scanned by two scanners a and b which are a distance d apart. The nature of the scale and the method of scanning is immaterial (cf. Figures 16, 17, 18, 19, 20, 21, 22, 23, 24, 28, 29). The distance d, however, must satisfy the following condition:

$$d \approx n \cdot \frac{\tau}{2} \pm \frac{\tau}{4} \text{corresponding to } n \cdot \pi \pm \frac{\pi}{2}$$

    where $n = 0,1,2,3,4,5 \ldots k$

Pulses e and f are therefore always mutually displaced by 90° electrical. If these two pulse trains are applied to a discriminating circuit, as shown in

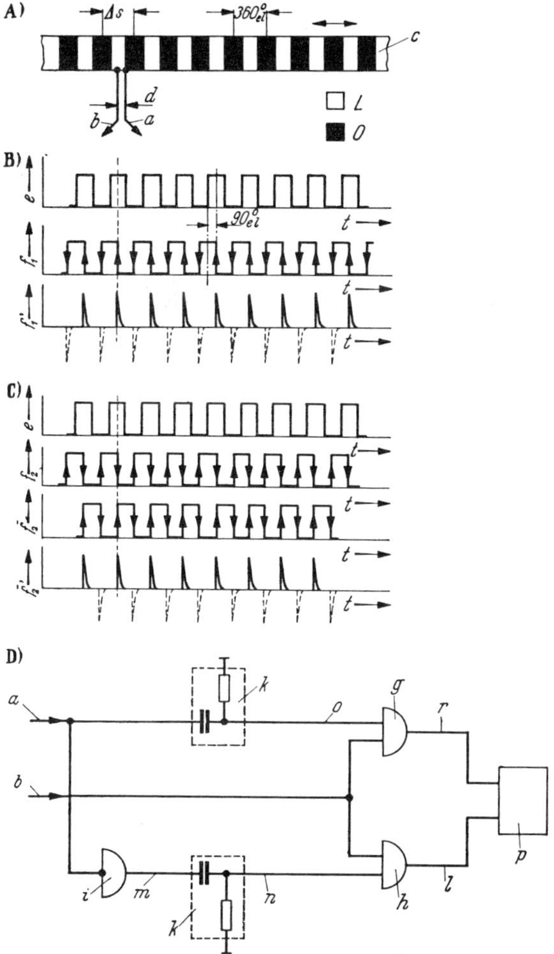

Fig. 13　Method of operation of an electronic direction discriminator.

A) Diagrammatic arrangement. B) Waveforms, scale c moving to right. C) Waveforms, scale c moving to left. D) Logic diagram of a discriminator circuit.

a Scanner 1 for signal f. b Scanner 2 for signal e. c Uncoded binary scale. d Distance between scanner 1 and scanner 2. e Waveforms of signal on scanner 2. f_1 Waveform of signal on scanner 1 when scale moves to right. f_1' Signal at point o (Figure 13D)). This is the derivative with respect to time of signal f_1. f_2 Waveform of signal at scanner 2 when scale c moves to left. $\overline{f_2}$ Inverse of signal f_2 at point m. f_2' Signal at point n. This is the derivative with respect to time of signal $\overline{f_2}$. g h AND logic elements. i Inverter. k Differentiating blocks. l Signal output for movement of scale c to left. m, n, o Signal transmission lines. p Forward and backward counter (cf. Figure 34). r Signal output for movement of scale c to right. Δs Displacement element.

Figure 13D), pulses will be obtained at its outlets r or l depending on the direction of motion. If the scale c is moving to the left, the pulses will be emitted by outlet l, while if the scale c is moving to the right, the pulses will be emitted by outlet r. Direction discriminating circuits of this type nowadays employ semi-conductors, (transistors and diodes) (24, 41). The method of operation can be seen from Figure 13B). This will now be described in somewhat greater detail since it can serve as an example of the methods used in digital techniques. Assume that the scale c comprises a number of segments of the same size, which are alternately passing and blocking current, and that the brushes a and b, which are apart, are very narrow. Then as b moves along c the train of signals $e = f(t)$ will appear at brush b. At the same time the train of signals $f_1 = f(t)$ or $f_2 = f(t)$, $90°$ electrical out-of-phase with the first signals, will appear at the brush a, the direction in which the voltage or current is flowing varying with the direction of movement of the scale relative to the brushes. The same effect can be achieved when a photoelectric, magnetic, or other scanner is used. The directional arrows on the diagram clearly show the basis of the system. The relationship between the phase direction of the electrical current and the direction of motion of the slide can be utilised in a discriminator circuit. Here it is necessary to differentiate the signal train $f_1 = f(t)$, with respect to time. This can be achieved by arranging a capacitor and a resistance in series as shown in Figure 13D)[1].

Corresponding to the sharp rise of the signal train $f_1 = f(t)$, transient pulses of the function $f_1' = f(t)$ will occur at point o, their polarity depending on the direction of current flow (see arrow) of function $f_1 = f(t)$. If the two signals $e = f(t)$ are applied to an AND circuit (cf. Appendix 2), a pulse will occur at point r only if e and f' are in the same direction, or in this example, if scale c is moving to the right whilst the brushes remain stationary. If the scale moves to the left the pulses of the function $f_1' = f(t)$ are reversed, while the variation of $e = f(t)$ remains unchanged. Under these circumstances the effects of the AND circuit is to suppress any signal at the point r.

The reverse effect is required at point l. To achieve this it is first necessary to reverse the direction of the original signal. This is achieved by an inverter circuit i (termed an inverter for short, cf. Appendix 2). The original function $f_2 = f(t)$ at point a is then inverted into the form $\overline{f_2} = f(t)$ at point m.

If this signal is again differentiated by an R-C circuit, pulses of the function $\overline{f_2}'$ will occur at point n, and these, together with the signal train

---

[1] The charging current of a capacitor C is proportional to the derivative with respect to time of the charge voltage e:

$$i = C \cdot \frac{de}{dt}.$$

The voltage e_r across the resistance r is, by Ohm's Law, proportional to the charging current $i(e_r = i.r)$. The voltage at point o is therefore proportional to the derivative of the original voltage e.

$e = f(t)$ are again applied to an AND circuit h. This enables pulses to appear at point l only if $e = f(t)$ and f_2' are in the same direction. This will be the case only if scale c is moving to the left. Pulses corresponding to the displacement quanta will thus appear at either l or r, depending on the direction of movement, and these can be summed by means of a forward and backward counter p. This arrangement offers two advantages:

Firstly no permanent errors can be introduced by any vibration, since each addition is immediately followed by a subtraction. Secondly, all increments due to movement of the scale, both quanta Δs and any short displacements, are added or subtracted depending on the direction of movement. The displacements can take place in either direction in any sequence, and this circuit will provide a constant positional check (cf. Figure 42) or it can be used for forming an R type comparator (cf. Figure 43).

The situation is different when spurious pulses of the second type (induced spurious pulses) must be eliminated since these are always purely electrical phenomena. Nevertheless there are again various possible solutions, two of these being briefly as follows:

   a) Suppression of the spurious pulses at source.
     If this is attempted, careful suppression will be needed throughout the installation. It will be necessary to avoid inductive effects by embodying the usual circuit devices, such as local surge voltage suppression on all switched inductors, spark-suppression at all switch contacts; strict separation of the power and control circuit wiring from the measuring circuit wiring; independent earthing of all electronic components avoiding the body of the machine; careful shielding of the measuring circuit wiring, with the screening insulated from the body of the machine, etc.
   b) Elimination of spurious pulses by an evaluation circuit. This can be achieved, for example, by giving the measuring pulses a characteristic 'signature', which differentiates them from the spurious pulses directly at the pulse source, i.e. at the displacement measuring unit. One possible method of achieving this is shown schematically in Figure 14 (42). c is, as in Figure 13, a grating with the two scanners a and b. The spacing of these two scanners d is, however, determined by a different principle.

Here, the requirement is that

$$d = n \cdot \tau \pm \frac{\tau}{2}, \text{ corresponding to } n \cdot 2\pi \pm \tau$$

where, as before, $n = 0,1,2,3,4 \ldots k$.

The measuring impulses are now $180°$ electrical apart, and not $90°$. Let i be the transmission line between the scanners a and b and the

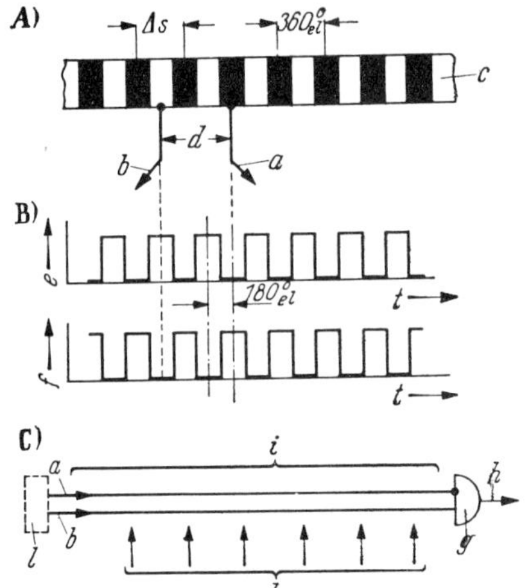

Fig. 14
Method of operation of a system for marking measuring pulses.

A) Diagrammatic arrangement. B) Waveforms. C) Logic diagram of a comparison circuit.
a Scanner 1. *b* Scanner 2. *c* Uncoded binary scale (strip scale). *d* Distance between scanners 1 and 2. *e* Waveforms of signal on scanner 2. *f* Waveforms of signal on scanner 1. *g* EXCLUSIVE OR logic element. *h* Pulses at counter. Δ*s* Displacement element. *i* Transmission line. *l* Pulse generator (displacement-measuring system). *k* Spurious pulses.

counter input. At the latter there is a logic element *g* to perform the EXCLUSIVE OR function (cf. Appendix 2) (6). This logic element operates in such a manner that pulses are passed to the counter *h* only if the two input pulses are of opposite polarity. The inverse measuring pulses *e* and *f* are thus allowed through and combine to form the output pulses *h*; all spurious pulses, on the other hand, are of the same polarity along both wires of the transmission line and so are also of the same polarity on arrival at the element *g* where they are therefore blocked. This comparatively simple device will effectively separate the majority of the spurious pulses from the measuring pulses even when the wiring system is not so well arranged, and when the machines are not completely suppressed.

Obviously, these methods for eliminating the effects of vibration and electrical interference can be applied either individually or in combination, so that (at some additional cost and effort) it is possible to make the incremental process completely reliable (cf. also Figure 29). It will be self-evident that the measuring equipment must also be properly protected from dust, swarf, damp, etc.

Zero-point displacement and location

Reference has already been made (Chapter 2) to the important feature of the incremental process that the datum (which is the same as the point at which counting begins) can be selected anywhere along the length of the scale and, hence, at any position within the working range of the machine tool. This simple means for locating the datum in any desired position can be a major advantage of an incremental control system. It is, nevertheless, often desirable to locate the datum in some fixed position, depending on the type of machine, or to arrange for it to be variable only within certain limits. The situation is straightforward for machines that perform a single-stroke operation (e.g. movement of drilling spindles or of turret lathe slides); here, the datum should coincide approximately with the stationary position of the moving part of the machine when no operation is in progress. In the simplest example one could fit a cam in a fixed position, so that by operating a limit switch it always produces the counting pulse at the same point of the stroke. It should be noted, however, that:

there is a tendency for the contacts of every mechanically-operated switch to rebound. The electronic counter will therefore be subjected to a series of pulses, all of which it tends to count. There are various ways of overcoming this difficulty. One, for example is to use a "normally-closed" contact instead of a "normally-open" contact, i.e. the counting pulses are short-circuited before the count begins, and the circuit is opened when counting starts. This will eliminate the risk of contact bounce interfering with the count. Provided certain safety precautions are taken, this is a very simple method of obtaining an accurate switch-on point. An electronic method of selecting the datum is given by the arrangement of the scale as shown in Figure 23.

In every instance where advantage is to be taken of the simple, free, choice of datum position, care must be taken to ensure that equipment is available on the machine which will enable the initial position of the workpiece, relative to the tool, to be determined with sufficient accuracy to enable the counter (used as a comparator) to be 'set-to-zero' at this point, i.e. when the datum is fixed on the machine slide, this must be provided with an index mark to enable the particular tool in use to be calibrated (cf. Chapter 7). When the datum is determined relating to the workpiece, the workpiece datum that is fixed by the program (cf. Chapter 10) must be defined with sufficient accuracy to enable the operator to readily recognise when the tool and workpiece zero positions coincide, so that he can set the counter to zero when the slide is in this position. In the latter instance, incorrect positioning due to faulty chucking, or clamping, of the workpiece can be easily eliminated.

3.1.1 *Rotary displacement measuring systems for the incremental process*

The advantages and disadvantages of rotary displacement measuring systems have already been discussed in very general terms in Chapter 2. When the displacement is resolved into digital displacement elements Δs, the reduction ratio involved in converting a rotation into a longitudinal movement (e.g. the pitch of a leadscrew) can offer the advantage of increasing the spacing of the scale graduations, thus simplifying the scanning process.

Because of their tendency to faulty operation, electro-mechanical methods using springs or brushes for scanning are rarely used nowadays.

The photoelectric scanner offers the advantage that there is no mechanical contact between the scale and the scanner, but it requires a number of components in combination, and some of these are fragile. As mentioned earlier, the photoelectric scanners must be capable of working at frequencies of at least 20 kHz, i.e. they must be capable of detecting switching times of 20 μs or less. Virtually only germanium or silicon photodiodes and elements can be used for this purpose. The sizes of these semiconductor components and the scale graduations required will eventually determine the design details of the photoelectric scanning head. In order to give an indication of their size, a number of the semi-conductor elements currently used are illustrated in Figure 15. The effective area of the more widely used photodiodes and elements ranges from about 1 to 150 mm^2. In machine-shop conditions it is perhaps not normally advisable to take full advantage of the degree of component miniaturisation now possible, since this can often lead to a loss in reliability. Safety considerations of this type have an effect on the design features of scanners. To within a certain minimal magnitude of Δs it is possible to select the graduations such that the width of the slot a is about the same as the effective photodiode width (Figure 16A). For small values of Δs it is necessary to use an auxiliary grating with the same subdivisions (Figure 16C). The light passing through a number of narrow slits is then summed by the collecting lens, and sufficient light is obtained to operate the photodiode e effectively. Extensive investigations have been made into the geometrical and optical conditions of photoelectric scanning systems (42). The light sources currently used are mainly low-voltage incandescent lamps

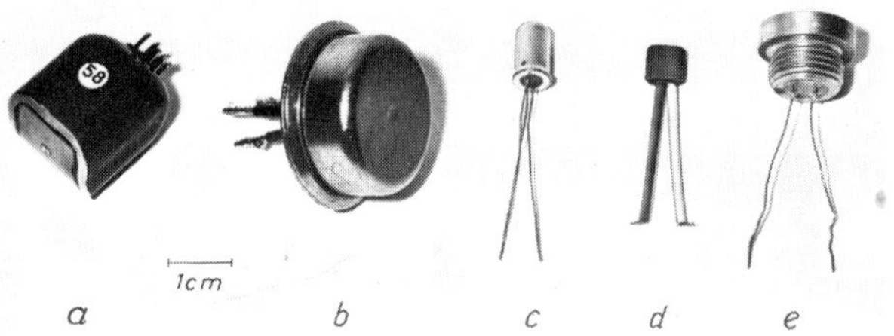

Fig. 15 Selection and size comparison of some important electronic semiconductor components. The trend towards the use of small components is clear (6, 32, 37) (cf. Figures 36, 37).

a Pick-up head for the speed-independent application and scanning of magnetic markers. This is a combination of a small electro-magnet (for application) and a Hall-effect generator (for speed-independent scanning). *b* Transistor for 600 mA collector current. *c* Transistor for 50 mA collector current. *d* Germanium photodiode (light-sensitive surface about 1 mm^2). *e* Silicon photoelectric cell (light-sensitive surface about 1.5 mm^2).

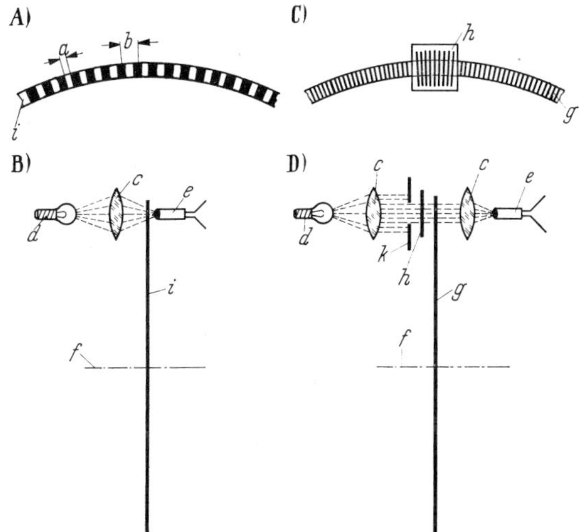

Fig. 16 **Diagrammatic arrangement of rotary displacement measuring systems using photoelectric scanning.**

A)and B) Arrangement without auxiliary grating (A) Plan, B) Side view). C) and D) Arrangement with auxiliary grating (C) Plan, D) Side View).

a Light slit (1.5 mm minimum), *b* coarse graduation $\tau \approx 2a$, *c* lenses, *d* light sources, *e* photodiodes, *f* axes of rotation of graduated discs. *g* disc with fine divisions (spacing up to about 0.02 mm), *h* auxiliary grating (same graduations as *g*), *i* disc graduated with widely-spaced light and dark segments (for transmitted light), *k* diaphragm.

(with thick filaments), which are usually underloaded for two reasons:

a) Lamp life is increased: a life of 2000 hours or more is common.
b) The red component of the light is larger, and hence it has a greater effect on the photodiodes which are red-sensitive.

Usually, the direct currents produced by the photodiodes and elements are so small, and the distances to the counter are so large, that it is advisable to provide at least one small, transistorised, amplifier stage directly after the photocell. If the methods that have been described for ensuring reliable optical scanning and electrical transmission of displacement measuring pulses are all adopted, and if it is borne in mind that these delicate components have to be assembled into robust units which are easily exchanged in case of defective items, the basic principle for the design and manufacture of these displacement measuring systems will be largely self-evident. Figures 17, 18, 22, and 23 show examples of photoelectric measuring systems.

Inductive (electro-magnetic) scanning also offers the advantage of dispensing with contacts. Here the problems are, however, very different. The

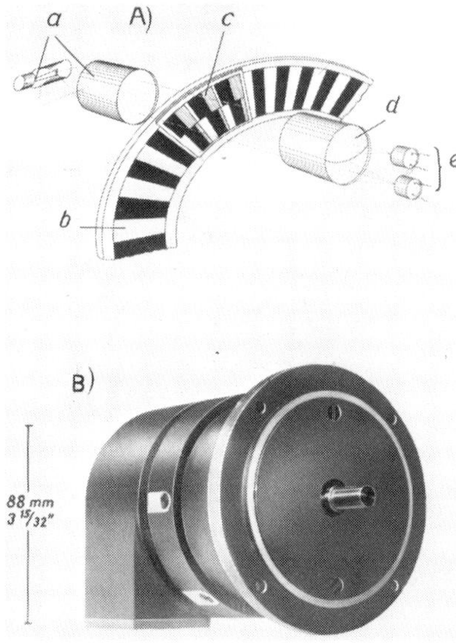

Fig. 17
Example of a photoelectric incremental displacement-measuring system for rotary motion and direction-sensitive scanning (auxiliary grating system as shown in Figure 16C), 1000 pulses/rev).

A) Principle of operation, B) Complete unit.

a Light source, *b* Circular grating *c* Auxiliary grating with two mutually displaced sectors (displacement about 90° electrical), *d* Objective lens, *e* Photodiode.

requirement that the displacement quanta must be capable of being converted into electrical pulses, with equal facility over the whole frequency range up to 20kHz, introduces two boundary conditions:

1. At very slow speeds the rate at which the lines of force are intersected is so low that the voltages induced cannot be evaluated with any degree of confidence [2].
2. At high speeds (rapid traverse motions) the voltage increases appreciably as $\frac{d\phi}{dt}$ increases.

The designer must therefore attempt to overcome these drawbacks, which are due to the conditions described by Faraday's Law. There are two basic means by which this may be done:

a) Using Hall-generators. These are components made of compound semiconductor materials (such as indium arsenide) which have a strongly marked Hall-effect. In this context a thin semiconductor

---

[2] From Faraday's Law the induced voltage is:

$$e = -N \frac{d\phi}{dt}$$

where N is the number of turns on the coil, ϕ is the magnetic flux enclosed by the coil, and t is the time.

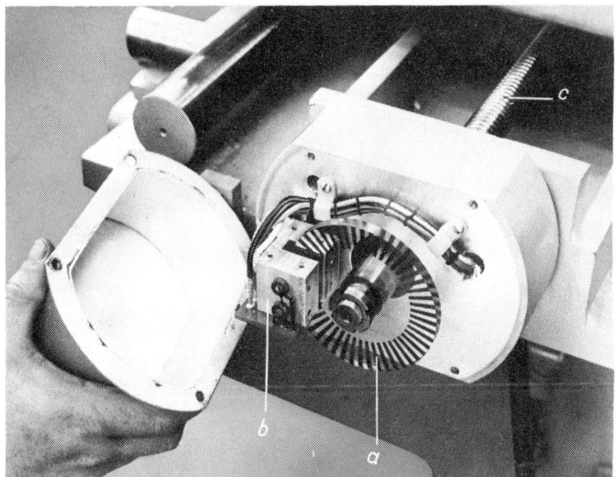

Fig. 18 Example of a photoelectric incremental displacement measuring system for rotary motion with direction-sensitive scanning (without auxiliary grating as shown in Figure 16A) fitted to the feed spindle of a drilling machine, (see also machine illustrated in Figure 179, photo courtesy of *Pratt & Whitney*).

a Graduated disc. *b* photoelectric scanning head, *c* feed spindle.

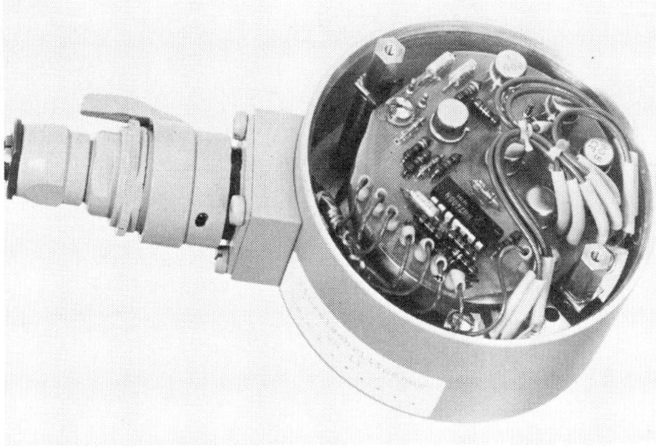

Fig. 19 Example of a rotary-type direction-sensitive incremental displacement measuring system with permanent magnets, Hall generators, and evaluation circuit. (Photo courtesy of *Siemens*).

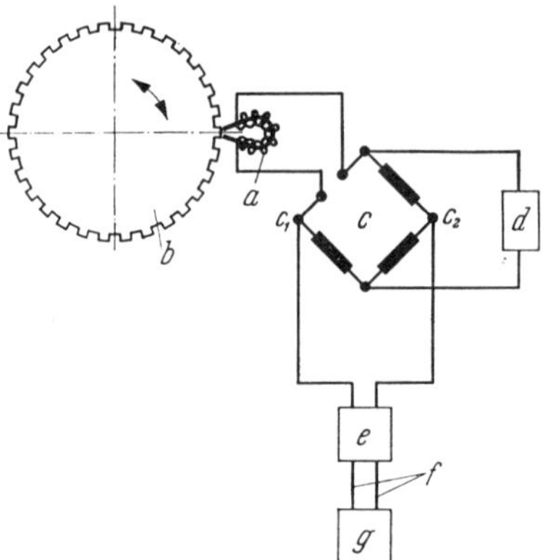

Fig. 20
Diagrammatic representation of inductive scanning with a toothed-wheel using the carrier-frequency system

a Coil with slotted iron core, *b* Soft-iron toothed wheel, *c* Inductance bridge (carrier-frequency signals at points c_1 and c_2'), *d* Oscillator generating sinusoidal voltages at about 100-200 kHz (carrier-frequency), *e* Demodulator, *f* Signal leads to counter *g*, *g* Counter (in this example, counts in one direction only).

plate will, under certain circumstances generate a voltage that can readily be measured when it is placed in a magnetic field (34). The output depends on the field strength, and not on its variation with time. In the range under consideration it is also independent of frequency and, with this type of element, the same voltage amplitudes are attained at both inching and rapid traverse speeds. A rotor fitted with small permanent magnets together with one or more Hall probes (depending on the additional scanning requirements) constitute a very simple rotary measuring system. One of the many possible types of Hall probe is illustrated in Figure 15; whilst Figure 19 shows an example of an angular step transmitter using this technique.

b) Using a carrier frequency. A basic circuit is shown in Figure 20. The teeth of a soft-iron disc *b* (e.g. an unhardened gear wheel with no remanent magnetism) pass across a coil with a slotted iron core *a* similar to the magnetic pick-up head of a tape recorder). The air gap between *a* and *b* will vary according to the position of the teeth, and this causes the inductance of coil *a* to change. This coil forms part of an inductance measuring bridge *c* which is supplied with sinusoidal voltages by an oscillator *d* at a carrier-frequency of about 100 to 200 kHz. This relatively high carrier-frequency is necessary to ensure that it is sufficiently remote from the fundamental (about 20 kHz) of the maximum scanning frequency. The fluctuations in the inductance of *a* are shown at points c_1 and c_2 as variations in the amplitudes of the carrier-frequency voltage generated in *d*. The carrier-frequency is separated in the demodulator *e* and only the signals that are produced by variations in inductance of *a* remain in the conductors *f*. Their amplitude is virtually independent of the speed of rotation of gearwheel *b*, and they can be fed to the counter *g*.

3.1.2   Linear displacement measuring system for the incremental process

It is not practicable to use electromechanical scanning (e.g. by brushes and segments) with the small displacement elements that occur in machine tool practice. Inductive scanning has also, as yet, not been widely used, although it appears to be quite suitable for accurate, but widely-spaced, graduations of 10 mm or 1 inch (cf. Chapter 5). The main emphasis in the development and application of uncoded linear scales has been for photoelectric scanning. Various designs are currently in use, incorporating both transparent and opaque scales (reflected-light methods). The simplest devices use glass scales about 6 to 15 mm in thickness, and varying from 30 to 60 mm in width, depending on the method of scanning that is used. Glass scales of this type are currently made in lengths of up to 1200 mm, and with graduations down to a minimum spacing of 0.004 mm. Where longer scales are required, it is possible to arrange several sections, each of say 1000 mm in length, butted end-to-end.

It is necessary to distinguish between two basic scanning methods, the transmitted-light and the reflected-light method. Figure 21 shows a

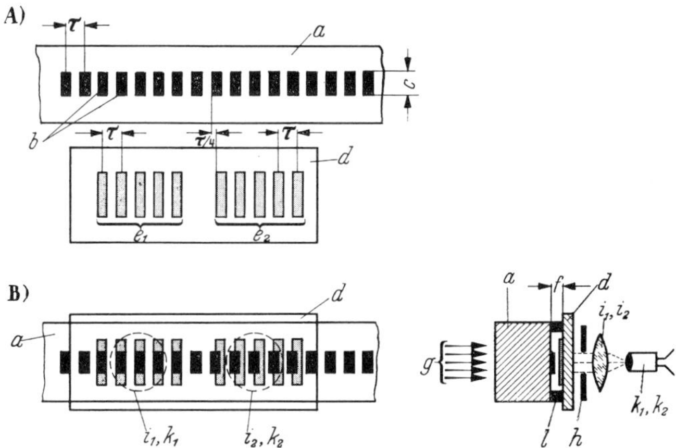

Fig. 21   **Principle of a photoelectric direction sensitive linear grating scanning system using short graduations**

A) Grating and auxiliary grating, drawn side-by-side. B) Grating and auxiliary grating drawn superimposed.

a Rectangular glass bar (8 x 25 mm to 12 x 30 mm). b Chemically or physically applied light and dark pattern (grating). c Length of graduations (2 to 3 mm). d Base for auxiliary grating (thin glass plate). e_1 and e_2 Two groups of light and dark strips (displaced by 90° electrical = $\tau/4$). f Spacing between grating b and auxiliary grating e. g Parallel light. h Diaphragm. i_1 and i_2 Condensing lenses. k_1 and k_2 Photodiodes. l Spacer strips between a and b (about 5 μm to 20 μm, depending on width of graduation marks. Spacing of graduations (in this case equal to Δs).

Fig. 22 Example of a photoelectric grating scanner using transmitted light techniques with electronic discriminator unit to give output signals depending on the direction of movement (Photo courtesy of *Heidenhain*).

direction-sensitive scanning arrangement for transmitted light using two similarly graduated auxiliary scales, e_1 and e_2 which are displaced relating to each other in the direction of movement. With this arrangement the lengths of the graduations c can be kept relatively short (about 2 to 3 mm), which means that the cost can be reduced. Figure 22 shows an example of a photoelectric grating scanner. With these auxiliary grating systems it is important to accurately maintain the distance f between the main grating b and the auxiliary grating e over the entire path length. This distance should be less than the spacing of the graduations (42). In practice this condition can be met most easily by vacuum deposition of spacer strips l on the scale a and holding the plate d against the scale by spring pressure. Today users of systems of this type (machine-tool designers) need not consider such details, since the units are commercially available and ready for fitting (cf. Figure 22). Particularly (in view of the problems that will be discussed in Section 3.1.3) it is often advisable to use four photodiodes for scanning. This will, of course, increase the space requirements. The current trend towards the miniaturisation of electronic components (6, 35) has led to many attempts to minimize space requirements. One possibility is shown in Figure 23. Here use is made of concentrated rays of light by using fibre-optics as light guide. (43, 44, 45). As a result the photoelectric elements can be located closer together, and the heat from the light source is kept away from the scale. In addition, modern components have enabled the size of the electronic discriminator to be reduced (6, 30, 37, 38) compared with earlier versions.

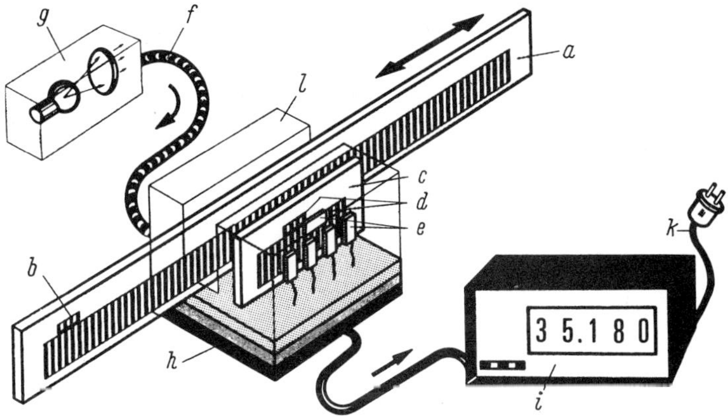

Fig. 23 Example of the application of fibre-optics to the construction of a compact linear displacement measuring system

a Digital-incremental graduated scale. *b* Auxiliary scale for zero pulse. *c* Scanning plate. *d* Scanning fields (auxiliary grating). *e* Photoelements (miniaturised). *f* fibre-optics light guides. *g* Lamp casing. *h* Electronic discriminator circuit. *i* Counter. *k* Mains supply. *l* Transmitter casing (Principle: Heidenhain LID 2).

Figure 24 also shows a method of arranging more than two scanning systems, side-by-side, in a relatively simple manner (in this example up to four photodiodes can be fitted, each displaced 90° with respect to the next); here the distance *l* between the gratings is not so critical as in the preceding example (6, 40). It is a requirement for this method, however, that the length *c* of the graduations is comparatively large (about 10 to 12 mm). The auxiliary grating *e* has the same graduations as the main grating *b*. Rotation of the auxiliary grating through an angle δ produces a moiré pattern, the width of which will depend on the angle δ and the pitch of the graduations τ. Depending on the direction of the movement of *d*, this moiré pattern will move either upward or downward. By suitable choice of the angle of rotation it is possible to magnify appreciably the pitch of the graduations to $m = \tau'$, and so to match the scale to the space requirements of the photodiodes, etc. In general, a scanning system of this type is equipped with up to four photodiodes displaced at intervals of 90° electrical, thereby increasing the reliability of operation and enabling all types of discriminator circuit to be accommodated. This is not the place to go into complete detail; the machine shown in Figure 89 is equipped with a measuring system of this type (40). A similar magnification, from τ to τ', is also obtained if the auxiliary gratings e_1 or e_2, shown in Figure 21, are provided with graduations slightly different in pitch from those on the main grating *b*. The progressive shift in the markings has a similar effect to a vernier scale, and also produces alternate light and dark fields spaced relatively far apart (42).

The transmitted-light processes described offer the advantage of good optical resolution, but have the disadvantage that they require glass scales (easily broken, attract dirt under the conditions prevailing in a machine shop, coefficients of thermal expansion different from those of the machine parts, etc), and are frequently by no means easy to install. The reflected-light

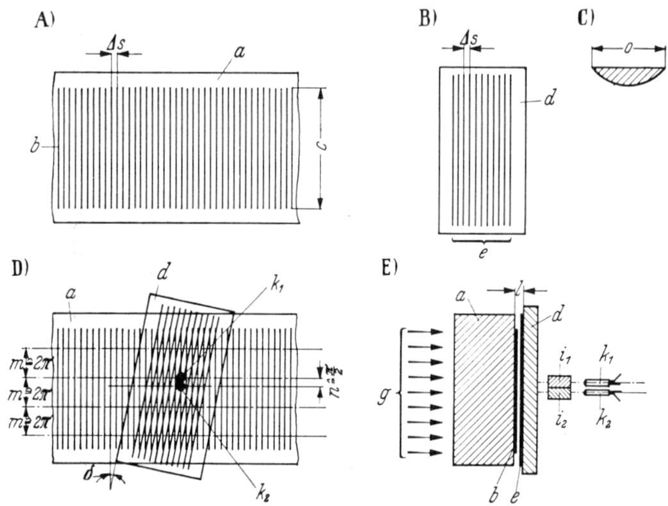

Fig. 24 Principle of a photo-electric grating scanner with graduation magnification.

A) Part of a long graduated scale. B) Short auxiliary scale, C) Side view of a cylindrical lens plate *i*. D) Grating with auxiliary grating rotated through δ and superimposed. E) Side view of D).

a Glass scale, *b* Grating with relatively long graduations (about 12 mm). *c* Length of graduations. *d* Glass plate with auxiliary grating. *e* Auxiliary grating. *g* Parallel incident light. i_1 i_2 Cylindrical lenses. $k_1 k_2$ Photodiodes. *l* spacing between scale and auxiliary scale. *m* Spacing of graduations τ' increased by moiré effect. *n* Distance between the two photodiodes k_1 and k_2. $\tau = \Delta s$ original spacing of scale graduations.
Note: In practice, the grating lines must have a thickness of about ½τ.

Fig. 25
Principle of operation of the reflected-light method.

a polished reflecting steel scale (moves linearly) with etched graduations. *b* transparent auxiliary grating. *c* Lens system for concentrating light beam. *d* Light source. *e* Condenser lenses. *f* Four photocells for discrimination of signals as shown in Figure 26.

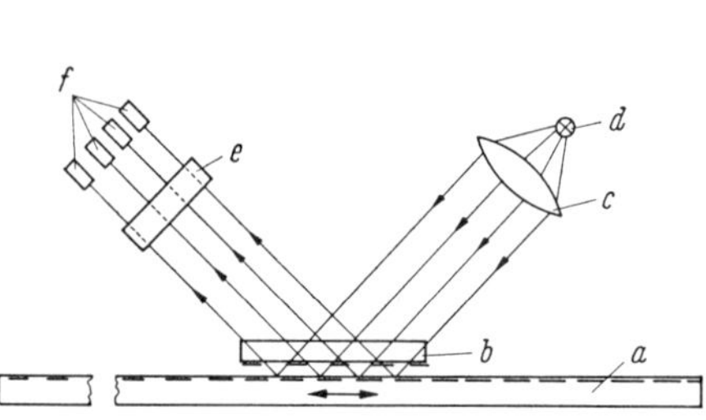

methods that have been developed in recent years often do not suffer from these disadvantages, since most of them use polished stainless-steel scales. The principle on which these measuring processes work is very simple, and is shown diagrammatically in Figure 25. The long graduated scale frequently comprises a steel tape bonded to a rigid steel bar. The graduations must be spaced further apart when using the reflected-light method than for the transmitted-light method so as to ensure reliable operation. The closest spacing of graduations currently in use is about 0.04 mm (compared with 0.008 mm and less for the transmitted-light method). A method is described in Section 3.1.4. in which the graduations on the steel scale can be spaced as far apart as 0.64 mm.

3.1.3 Subdivision of scales by means of electronic circuits

Some of the principles underlying discriminators have already been indicated in Figures 13 and 14. With linear scanning of a scale there is a further reason for multiple scanning. For economic reasons it is desirable to avoid making the initial scale graduations too fine (Figures 21, 23, and 24). Efforts should,

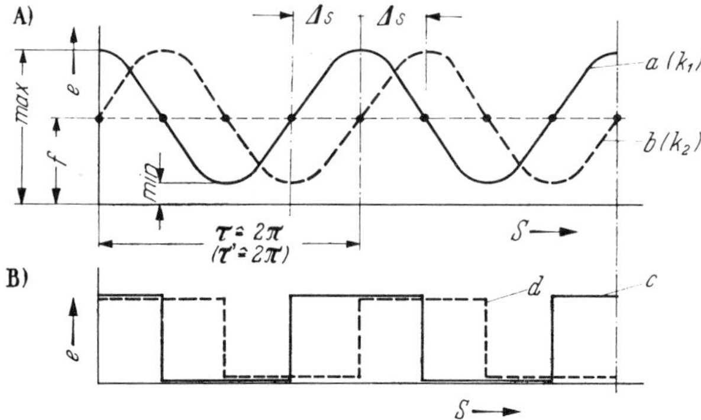

Fig. 26 **Signal waveform (photodiode voltage) and subdivision of scale by means of electronic circuits.**

A) Original voltage waveform (approximately sinusoidal). B) Converted waveform.

a Voltage waveform, $e = f(s)$ of photodiode k_1 (see Figures 21 and 24). b Voltage waveform of photodiode k_2, c Sinusoidal waveform a converted into a square wave (e.g. by means of a Schmitt trigger). d Sinusoidal waveform b converted into a square wave. e Photodiode voltage. f Switching voltage for Schmitt trigger. s Direction of relative motion of gratings. Δs displacement element that is to be processed ($\Delta s = \tau/4$ or $\tau'/4$). For τ see Figure 21 and for τ' see Figure 24.

Note: For simplicity one triggering voltage f only is shown. In fact, because of hysteresis of the bistable circuit, two triggering voltages will be present (one for making and one for breaking the circuit).

therefore, be made to obtain displacement quanta which are sufficiently small even with relatively coarse scale divisions. It is, for example, possible to subdivide the spacing of a scale by a factor of 2 or even 4 using purely electrical methods. This is explained in Figure 26 which shows, in idealised form, the signal waveform of two photodiodes displaced by 90° electrical. The photodiode voltage fluctuates as the relative positions of the gratings vary, and hence the amount of light that is passed, between a maximum and a minimum value. As a first approximation the waveform may be taken as sinusoidal. The use of a voltage-dependent electronic flip-flop circuit (e.g. a Schmitt trigger) enables each cycle of this wave form to be converted into a square wave (6, 41). If the triggering level f (Figure 26A) is correctly chosen it is, theoretically, possible to obtain four equally-spaced voltage rises or falls per scale division (τ or $\tau' = 2\pi$) (Figure 26B). These voltage changes are differentiated (as indicated in Figure 13) and in this way it is possible to obtain up to four times as many counting pulses as there are graduations on the scale. In practice, the effort required to calibrate this arrangement (and the auxiliary equipment required to ensure that precise and reliable division by four is obtained) is likely to be considerable, so that this method is costly compared to other methods (cf. Chapters 4 and 5). If pulse-doubling only is to be effected, however, and if advantage is taken of the convenient calibration of the photodiodes by means of the moiré pattern technique, the subdivision of the scale by means of electronic circuits can lead to an overall design simplification. The inspection machine described in Chapter 5 (Figure 89), for example, is equipped with a displacement measuring system of the type shown in Figure 24, in which four photodiodes scan the moiré pattern at intervals of 0°, 90°, 180°, and 270° electrical. Two scanning positions would theoretically be sufficient to determine the direction of movement, but for reliable pulse doubling and to suppress spurious pulses it is also advisable to scan the inverse signals and to apply them to a discriminator circuit (41, 46).

3.1.4   Methods for simplifying linear scales

As has already been mentioned, glass scales graduated at intervals of 0.01 mm are comparatively expensive; also, the transmitted-light method can prove unreliable when using very fine gratings under the adverse conditions in a workshop. Many attempts have therefore been made to develop different types of scale. In each instance the object has been:

a) To increase the spacing of the graduations, and so to reduce costs, and/or

b) To use the reflected light method.

To supplement what has already been discussed, two further incremental processes that use optical scanning will be briefly described.

Multiprismat Method (Zeiss-Hensoldt) (26)

This method makes use of the laws of reflection that relate to a pair of mirrors arranged at 90°. The basic principle is shown diagrammatically in Figure 27. If the mirror assembly is displaced by the distance x, a beam of light will be displaced by $2x$ owing to the double reflection that it undergoes. If an incremental grating pattern is reflected, the mirror image will move twice as fast as the mirror assembly itself. If a large number of accurate, right-angle mirror assemblies are located on a bar on a machine slide, a short grating pattern with graduations spaced at 2τ will suffice to give the impression in the mirror image of a continuous grating with graduations spaced at τ. The length of the transmission grating need amount to only about three mirror assembly spacings to give the impression of a continuous grating equal in length to the bar carrying the mirror assemblies. This optical device achieves two objects:

1. Only a short, and therefore inexpensive, graduated scale is required.
2. The graduations on the scale can be spaced twice as far apart as the displacement element it is required to measure.

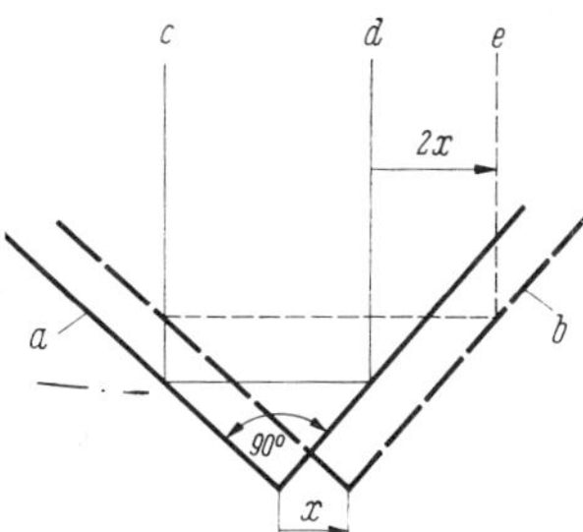

Fig. 27
Reflection at mirrors at 90°

a Initial position of mirror assembly.
b Mirror assembly shifted distance x to the right. *c* Incident beam of light. *d* Reflected beam of light when mirrors are in position *a*. *e* Reflected beam of light when mirrors are in position *b* (Shifted distance $2x$ to the right).

To realise this simple principle in practice requires a considerable amount of optical equipment, however, and the strip of right-angle mirrors must be made with great accuracy. The main path of the light rays is shown in Figure 28; this diagrammatic arrangement gives a good indication of the number of lenses, prisms, etc., required. In principle, it makes no difference which grating is reflected in the Multiprismat bar, and one is not restricted to a single incremental scale. It would also be possible to use a binary-code arrangement (as shown in Table 4, possibly even with a check-bit) as the transmission grid, and so virtually to produce an absolute linear scale. Whether the accuracy with which the prism scales can be made in practice would be sufficient for this, and if so, whether there would be a steep rise in their cost, cannot be answered merely on the strength of theoretical considerations. In practice, only incremental scales have been illustrated in use with the Multiprismat arrangement. On the other hand, the short

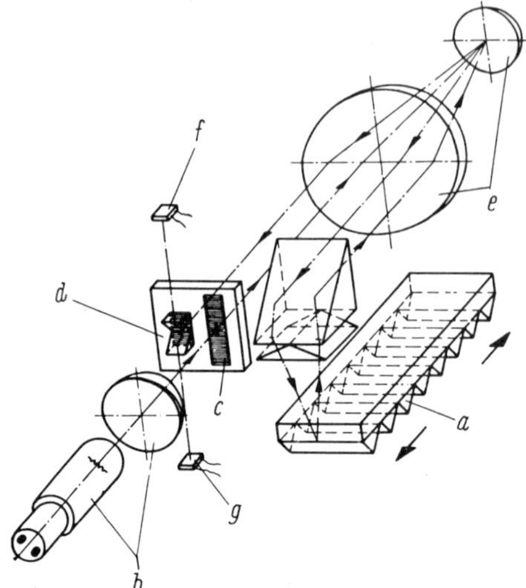

Fig. 28
Diagrammatic illustration of complete path followed by light beam in the *Multiprismat* **(26)**

a Multiprismat bar (length equal to maximum movement of machine slide). *b* Light source for transmission grating. *c* Transmission grating *d* Reception grating with prism to resolve beam into two rays which are directed at the two photoelectric cells *f e* Mirror and objective lens. *f* Two photoelectric cells for direction-sensitive scanning as shown in Fig. 13 (*Zeiss-Hensoldt* principle).

transmission scales can easily be changed, so that it is simple to change over from the metric to the inch system of measurement.

The Polygon Mirror Process (Philips) (47)

This process could also have been included for various reasons in the group of special measuring systems that is described in Chapter 5. Since it is, however, also often used in conjunction with a suitable discriminator circuit to provide a digital read-out (see Figure 41) it will be explained in principle at this point. It also illustrates a number of ideas that are of importance for practical applications.

The measuring system consists of three items:

a) The measuring scale
b) A photoelectric scanner
c) An electronic discriminating and matching circuit which is separately arranged.

These three main components can be arranged in various ways, depending on whether the measuring system is to be used solely for visual read-outs or whether it is to be used for a complete numerical control system.

The measuring scale comprises steel bars to which glass scales with graduations spaced at intervals of $\tau = 0.64$ mm are bonded, these scales being scanned by reflected light. The scales are 250.24 mm long, and can be combined to produce any required length (Figure 30).

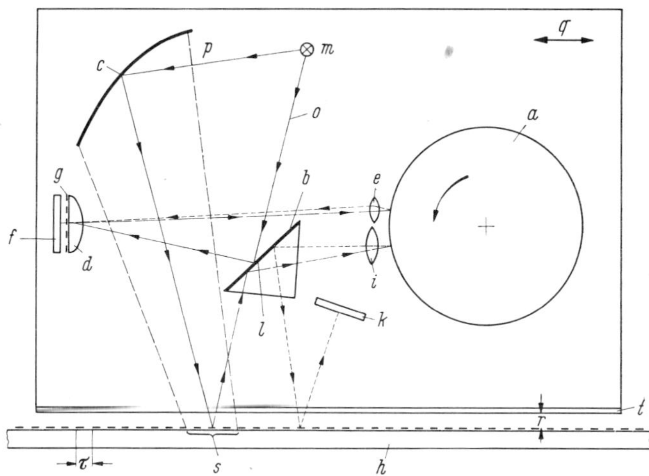

Fig. 29 **Illustration showing the principles of the** *Philips Polygon Mirror Process.*

a Polygonal mirror with 240 plane mirrors arranged on its periphery, driven by a small synchronous motor at $n_O = 375$ rev/min. *b* Reflecting external surface of prism. *c* Elliptical mirror. *d* Condensing lens (cylindrical lens) with grating *g*. *e* Condensing lens 1 in front of polygonal mirror. *f* Large-area silicon photoelectric cell (Figure 15). *g* Grating with marks spaced at $\tau = 0.64$ mm, partially reflecting and partially transparent. *h* Reflecting measuring bar with grating with marks spaced at $\tau = 0.64$ mm (length equal to maximum displacement of machine slide). *i* Condensing lens 2 in front of polygonal mirror. *k* Large-area silicon photoelectric cell (similar to item *f*). *l* Reflecting internal surface of prism. *m* Light source. *o* Reference light ray. *p* Measuring light ray. *q* Casing of measuring unit. *r* Air gap between casing *q* and measuring bar *h*. *s* Length of measuring bar on which measuring ray impinges. *t* Glass pane to cover sensitive optical system. τ Spacing of marks on measuring bar (0.64 mm).

Note: In practice, lenses *c* and *i* are placed side-by-side, so that the images they form lie on the same strip of mirror. They are drawn one above the other to make it easier to visualise the arrangement.

The most interesting part of the system is the photoelectric scanning head. The simplified arrangement shown in Figure 29 illustrates the main principles. The most important element is the drum-shaped polygonal mirror *a* which comprises some 240 plane mirrors arranged around its periphery; this drum is driven at $n_O = 375$ rev/min by a synchronous motor. Two rays of light, *o* and *p*, are emitted by the lamp *m*, and these are subsequently formed into strips of light by means of the mirrors and gratings. For simplicity, the light rays are represented by single lines with directional arrows. Ray *o* impinges on the mirrored surface *b* of the prism, and from there it is reflected through the condensing lens *d* (cylindrical lens) and projected onto a grating *g*: this grating is provided with strips that are alternately opaque reflecting and transparent and which are arranged at the same spacing of 0.64 mm as the marks on the measuring bar. A large-area silicon photoelectric cell f is arranged behind the grating and part of the light falls directly on this. The other part of the light beam is reflected from the grating g and so takes the form of the grating (shown by a dashed line in the drawing). This reflected light beam passes by the prism and impinges on the condensing lens e which projects an image of the grating on the rotating polygonal mirror. As a result a uniformly moving picture of the grating is produced. By a suitable arrangement of the optical system this 'moving grating' (shown by the dotted line) is reflected back onto the original grating g. This produces a sinusoidally varying illumination of the photoelectric-cell which then produces a constant alternating voltage having a frequency of 7.5 kHz; this acts as a reference voltage and frequency which is independent of the movement of the measuring bar h and the light reflected from it.

Beam p passes first to the elliptical mirror c where it is reflected in such a manner that a strip of light of definite width and length s falls upon the measuring bar. The mirror dimensions c are such that the distance between the scanner and the measuring bar is large and not particularly critical; it amounts to 15 ± 0.5 mm. The air gap is large enough to enable the whole system to be cleaned without difficulty, especially since the casing of the scanning head q is provided with a glass window t to prevent dirt entering. The beam of light is reflected from the measuring bar in the form of a grating (shown by the dashed line) and this is projected onto the internal mirrored surface of the prism l. From there the beam passes to the condensing lens i which is located *beside* the first condensing lens e, and then falls upon the rotating polygonal mirror which transforms it into another moving grating. The width of the polygonal drum a is such that there is no interference between the measured signal and the reference signal even though both are reflected by the same strip of mirror. A suitable optical system causes the second moving grating to be reflected once more from the measuring bar and then to be projected onto the second photoelectric-cell k.

If the measuring bar (or the scanning head) is stationary, an alternating voltage of the same frequency and (if the dimensions of the system are appropriate) the same magnitude as that on the first photoelectric-cell f will be produced by the second photoelectric-cell k. If, however, there is relative longitudinal movement between the measuring bar and the scanner q, not only will the phase relationship of the alternating voltages change, but also the frequency of the measuring alternating voltage; and this will become greater or less than 7.5 kHz depending on the direction of movement, the magnitude of the frequency change depending on the speed at which the machine slide moves. The electronic discriminating circuit connected to the instrument determines the direction and magnitude of the displacement. It is arranged so that a displacement of 0.32 mm causes $2^5 = 32$ counting pulses to be emitted by the circuit, so that 0.01 mm can be indicated on a counter for every pulse. Since the measuring light beam p has to cover the distance to the measuring bar twice, whilst the reference light beam o does not, a difference in the voltage amplitudes of the measuring and reference signals at the two photoelectric-cells f and k will indicate that the measuring bar needs cleaning. It is easy to arrange a pilot-light to indicate this and to cause it to be switched on by a monitoring circuit.

Another advantage of this whole arrangement is a.c. signals are emitted, which can easily be amplified, and a definite and unambiguous signal is given even when the machine slide is stationary. Furthermore, spurious pulses will have no adverse effect, especially since narrowband pass filters limit the frequency band of voltages emitted by the measuring photoelectric-cell k to such an extent that while it is perfectly capable of detecting slide speeds of up to a maximum of 13.5 m/min, higher frequencies are suppressed. The speed of 13.5 m/min is the fastest rapid-traverse for slide movements. Figure 30 shows an example of this measuring system fitted to a machine tool.

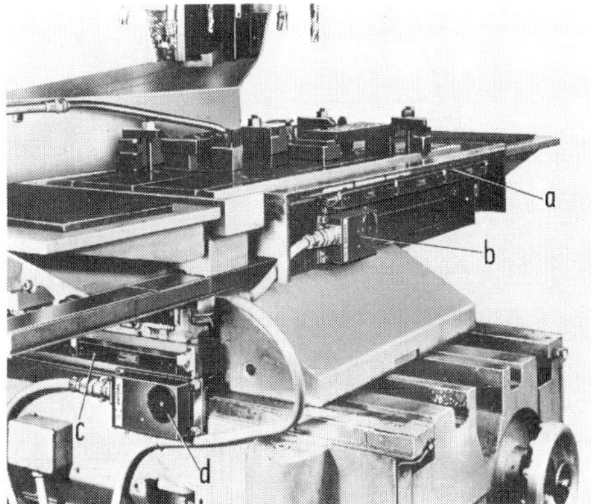

Fig. 30
Example of practical application of measuring system shown in Figure 29.

a Measuring bar for the X-axis *b* Measuring unit for the X-axis
c Measuring bar for the Y-axis *d* Measuring unit for the Y-axis.
(Photo courtesy of *Philips*)

3.1.5 *Pre-selector counters as comparators for the incremental process*

In any discussion of the main types of design, it is desirable to consider the two functions 'preselection' and 'counting' separately to some extent. As has already been explained, the initial design conditions are as follows:

a) Pre-selection (input of the desired value) should be capable of being performed both by manual setting and by means of a data carrier (cf. Chapter 9).

b) Counting frequency can be up to $20 \cdot 10^3$ pulses/s, or the switching time should be 20 μs or less.

The following important decision will have to be made when selecting the components that are to be used, the choice depending on the particular application:

Is it necessary to display the desired value and/or the actual value to the operator, or not? In the present stage of development many users of NC machines prefer to avoid having the actual values of the positions displayed. For an automatic machine that does not require an operator, there is no point in providing a visible display of the actual values; it was not provided on the

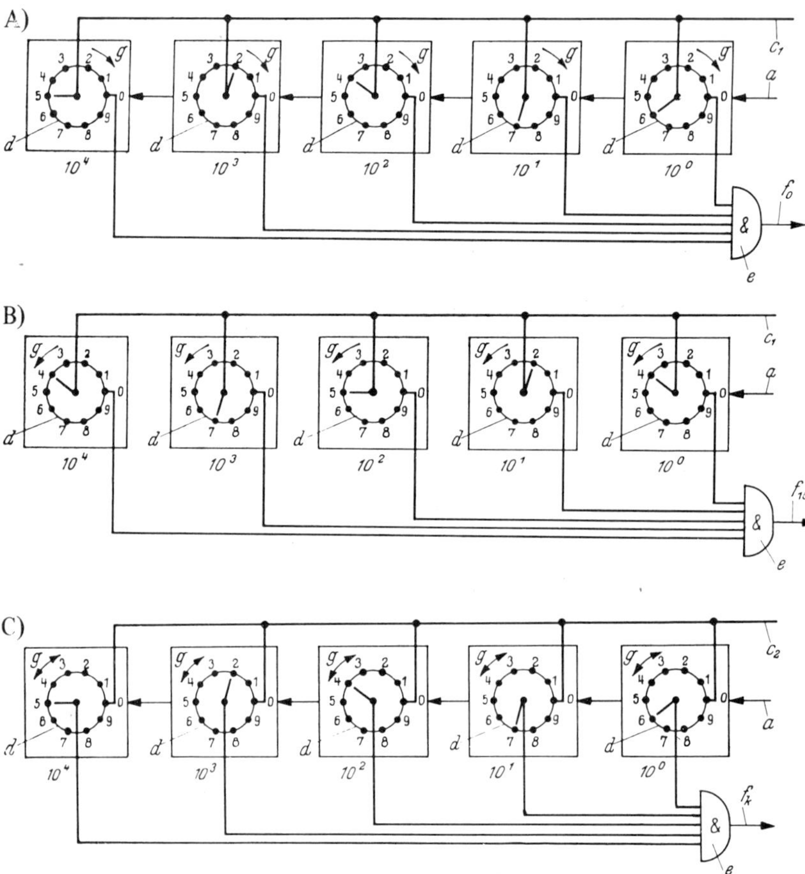

Fig. 31   Pre-selection principles in electronic counters

A) Subtraction of the pulses from the pre-selected number (example shown 52476) down to zero with coincidence check at 00000. B) Addition of pulses to complement of number (example :   100 000 - 52 476 = 47 524   or 99 999 - 52 476 = 47 523) to make up to 100 000 or 99 999, with coincidence check at 00 000 or 99 999. C) Summation of pulses from zero up to preselected number (52 476) and coincidence check at preselected number.

a Input for counting pulses. c_1 Signal to transmit (store) switch positions to counter circuits. c_2 Signal for setting counter to 00 000. d Counter circuits. e AND circuit. f_0, f_{10} and f_k Output signals at coincidence. g Direction of counting.

multi-spindle automatics, which have been widely used for a long time. There are instances, however, in which it is desirable to work merely with a numerical read-out of the slide positions, rather than using a completely automatic numerical control. The correct solution for any particular example will probably lie somewhere between these two extremes (10). When discussing the different forms of comparator design a brief reference will therefore be made to the possible ways in which a numerical display can be arranged.

Pre-selection

Three approaches are possible to the problem of obtaining an output signal from a counter when a given number has been reached. These are illustrated in Figure 31; at this point the construction or method of operation of the counter is immaterial.

Example A). The desired number (e.g. 52 476) is set on a series of decade switches and is passed to the counter circuits d by a pulse on the line c_1 to start the count[3].

The counting pulses are fed in at a and are subtracted from the pre-selected number, so that the count proceeds backwards towards 00000.

When all the decades are at the 0 position, this zero coincidence can be converted in an AND circuit e into an output signal f_O.

Example B) Here, the initial count that is set up on the counters is not the desired number but its complement (either to 100 000 or to 99 999). For the example (52 476) this would be 47 524 for the complement to produce 100 000. The count can now be made forwards until the counter reading is 100 000; this then produces the coincidence signal f_{10} when 52 476 pulses have been passed to the point a.

Example C). The switches are not used to set up the desired number but to detect it. The counter is set to zero by a pulse on line c_2. The counting pulses are again fed in at a, it being assumed that a forward count is made (g moves anti-clockwise). When the pre-selected number is reached, signals will appear on switches d and these can again be combined in an AND circuit to form a single output signal f_k (coincidence signal). All three of these possible methods of pre-selecting the number are used in practice, but the current trend appears to favour Example C).

In the majority of applications in the machine tool field, however, at least two methods of pre-selection along a path that is to be measured are desirable, as discussed in Chapter 1. The choice is then either to connect up

---

[3] In the simplest case it is possible to visualise a five-digit mechanical counter on which the pre-selected number is set up manually digit-by-digit. If the counter is then turned backwards by its shaft until the number 00 000 appears, the exact number of pulses, rotations, etc. that have been preselected will be obtained.

two counting circuits in parallel (one for each pre-selection), or, by a suitable coupling of the output signal f, the pre-selection pulses c and a brief interruption of the input a to the counter, to use the same counting circuit all the time.

In Example C) the circuitry will probably be slightly more elaborate but, on the other hand, the direction of counting makes little difference, and any required number of tappings (pre-selection) can be made without affecting the speed of counting. It is therefore probable that this method of pre-selection will be the most satisfactory in most cases because of its ready adaptability to the requirements of machine tools.

The nature of the switching systems has not yet been discussed. For manual setting stepped-decade rotary switches or tabulator keys (Chapter 9) are the most satisfactory. Relay stores are sometimes used for remote setting using punched-tape readers, since the reading speed of these used on machine tools is well within the working range of the relays (Chapter 9). Also, where relays are used it is easy to add a desired value indicator (cf. Figure 38). In the majority of instances, however, electronic 'stores' are used (e.g. transistorised flip-flop circuits as shown in Figure 33).

The counting process

It is outside the scope of this book to go into the very numerous designs of counters and their associated circuits which have been developed in recent years. And it is necessary to restrict the discussion to the fundamentals required for an understanding of numerical control. Many publications exist on this subject. They can be consulted for further details (e.g. 6, 24, 41). Because machine tool control systems encompass a range of up to at least 20 000 pulses/s it seems desirable, in the first instance, to note some of the development trends for the electronic components that are used for this purpose.

It is unnecessary today to discuss the 'cold cathode' tube (48), which was still of considerable importance when the first edition of this book was published (1962/63). It has since been found that these do not withstand the adverse conditions to which they are subjected in a machine shop so well as modern semiconductors (greater space required, more sensitive to shock, shorter life, etc). Magnetic-core counters are also obsolete, mainly because of the difficulty of fitting them with visual indicators (49). Present-day counter designs reflect the rapid developments that have been made in the field of semiconductors. They range from those using individually-wired components (35, 50), by way of printed circuits (36) (cf. Figure 35), to integrated circuits (30, 37, 38) (cf. Figures 36, 37). It seems advisable to make a brief survey of the various types that are available, and to illustrate the trends which the development of these devices are following. Some simple examples will indicate the user requirements associated with NC machines.

Transistors and their application to pre-selector counters

To introduce the layman and the newcomer to this field, Figure 32 gives a diagrammatic illustration of a transistor together with some of the basic concepts. Without going into the theoretical background of semiconductor physics (50, 51) it will be sufficient for the present to establish the following points:

  a) The most important transistor materials are germanium and silicon.
  b) A relatively large collector current i_c can be controlled by a very small base current i_b. Depending on the type of transistor and the circuit in which it is used it is possible to obtain a current gain of i_c/i_b = 70 to 500 with current models.
  c) The switching time that elapses between a change in i_b and the consequent change in i_c is of the order of 10^{-8} to 10^{-4} seconds, depending on the type of transistor and the circuit.
  d) The most important application of a transistor to the control of machine tools is its use as a contactless electronic switching element which operates with no detectable wear or fatigue.
  e) The practical difficulties which are occasionally still encountered lie less in the large-scale production of extremely reliable pre-tested components (whose characteristics are maintained within close tolerances) but rather in defective wiring (e.g. cold joints) or in unsuitable ambient temperatures (above 40-50°C) for germanium transistors.

Of the innumerable possible combinations of transistors, only one basic circuit, which is probably the most important for this particular application, will be considered; since this is important for an understanding of many of the items of electronic equipment that are used in conjunction with numerical control systems for machine tools. The basic circuit diagram is

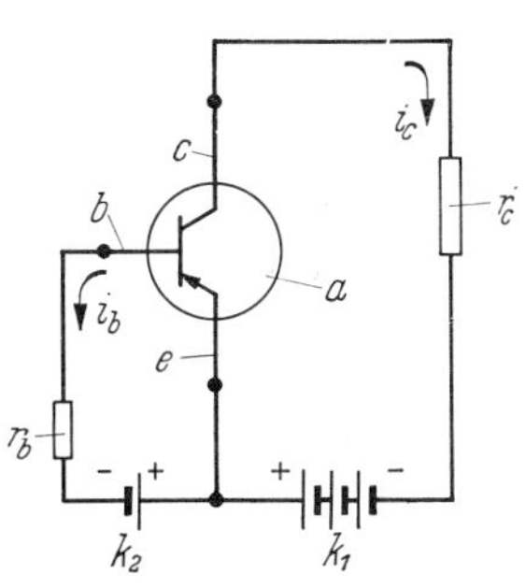

Fig. 32
Theoretical circuit showing the operation of a transistor (pnp) (also Figure 15) in an emitter circuit

a Transistor container. *b* Base and connection. *c* Collector and connection. *e* Emitter and connection. i_b Base current. i_c Collector current. k_1 Voltage supply for emitter-collector circuit (always a d.c. source). k_2 Voltage supply for the emitter-base circuit (can also be an alternating or pulse voltage; the important feature is that the collector current i_c will flow only when the potential applied to *b* is negative with respect to that on *e*).

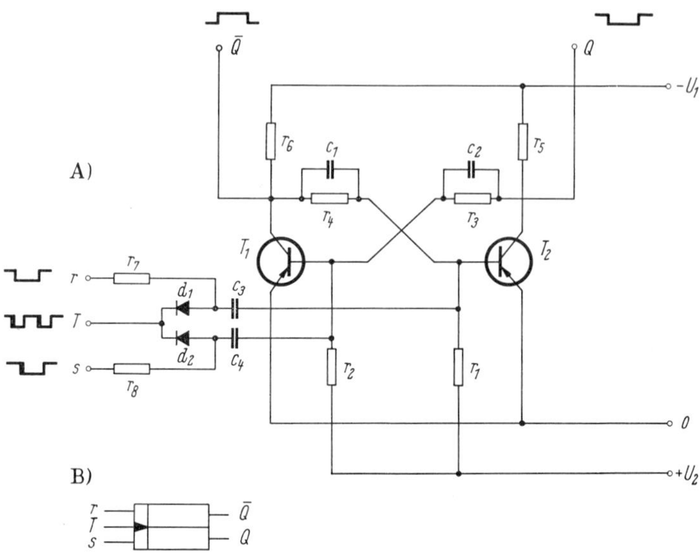

Fig. 33 Example of a bistable transistorised flip-flop circuit (pnp transistors)

A) Basic circuit diagram. B) Symbol (DIN 44700, Sheet 14).

T Input for cycle pulses (the presence of the diodes and differential elements results in only the negative flanks of the pulses being effective). *s* Input for 1st static signal (*L* signals on *s* and negative pulse edges on *T* produce *L* signals at output *Q*. *r* Input for 2nd static signal (*L* signals on *r* and negative pulse edges on *T* produce *L* signals at output *Q̄*. *Q*, *Q̄* Flip flop outputs (output voltages are square-wave and of opposite polarity). c_1 c_2 Negative feedback capacitors to increase slope of pulse edges of output signals. c_3 c_4 Capacitors to differentiate input pulses. *d* Semi-conductor diodes. *r* Resistances. T_1 T_2 Similar pnp transistors. U_1 U_2 d.c. supplies

shown in Figure 33. The essential feature of this combination of transistors (flip-flop) is that it acts as a switch with two conditions of equilibrium:

Condition 1: Transistor 1 is passing current, transistor 2 is cut-off.
Condition 2: Transistor 2 is conducting, transistor 1 is cut-off.

There is no other stable condition, so that this arrangement can be used for producing and storing binary signals. The circuit is completely symmetrical using the same components in both halves. The method of operation is briefly as follows:

Assume that a larger collector current i_c is being conducted by transistor T_1 than by transistor T_2. There will then be a slightly larger voltage drop across resistance r_1 than across resistance r_6. This causes a reduction in the voltage applied to the base of transistor T_2 through the potential divider r_4 r_7

with the result that this transistor will conduct even less collector current compared with T_1 than it did before. This, in turn, causes a further reduction in the voltage drop across r_6, and the negative base voltage of T_1 rises with the result that the collector current in T_1 becomes higher still. The whole process takes place very rapidly, and a stable condition will not be reached until transistor T_1 only is conducting. It could equally well have been transistor T_2 which first conducts the full current. The important point is that only one of the two transistors can conduct at any one time, and the other is cut-off, with no stable intermediate condition.

It is a comparatively simple matter to 'reverse' a contactless non-wearing switch of this type, i.e. to cause the transistor which has been cut-off to conduct and, at the same time, to cut-of the transistor that has been conducting, simply by applying a pulse at the appropriate point of the circuit to upset the stable equilibrium condition momentarily. Assume that initially T_1 is conducting and T_2 is cut-off. If a negative pulse is applied to the base of T_2 this transistor will conduct, the voltage drop across r_6 will fall. This reduces the collector current of T_1, the voltage drop across r_1 is reduced, and the negative voltage applied to the base of T_2 is increased further. The negative triggering pulse is thus further amplified, and within a very short time T_2 will have become fully conducting and T_1 will have cut-off. If a negative pulse is then applied to T_1 the system reverts to the original condition and T_1 becomes fully conducting. Precisely the same effect can also be achieved if alternate negative and positive pulses are applied to the base of one of the transistors only. The nature of the transistor operating characteristics and the effect of the coupling capacitors c_1 and c_2 also causes the circuit to be reversed if negative pulses are applied to points a_1 and a_2, coupled through the semi-conductor diodes, or to point f. This brief description shows how important this very reliable and robust, bistable, flip-flop circuit is for data processing systems. Its main features can be summarised as follows:

1. A binary state can be stored or erased by applying pulses to the base electrodes of T_1 and T_2.
2. Two input pulses of the same polarity applied to points a_1 and a_2 (dynamic triggering, dynamic input) will produce a square-wave voltage at output b.
3. If this square-wave curve is differentiated (cf. also Figure 13) a pulse having the same polarity as the two input pulses will be obtained on one of the edges. One output pulse is then obtained for each pair of input pulses. A stage of this type can thus be used for halving the pulse frequencies.
4. By connecting several flip-flop circuits in series in such a manner that the differentiated output signal from each stage is used to trigger the following stage a very simple counter for the pure binary system can be built up.

For the practical construction of counters, the basic remarks on codes given in Chapter 1 are, of course, relevant. From these the following points emerge:

a) If the counter is designed to operate on the pure binary system, the minimum number of counting elements will be required, but it cannot be used without some auxiliary devices for the pre-selection or display of decimal numbers.

b) If ten flip-flop circuits are joined together in such a manner that only one pulse is displaced in each case, and if such a chain is connected to a decade ring counter (41, 52), the counting circuits will, on the one hand, require additional components, but, on the other hand, the equipment needed for the pre-selection and display (if required) of

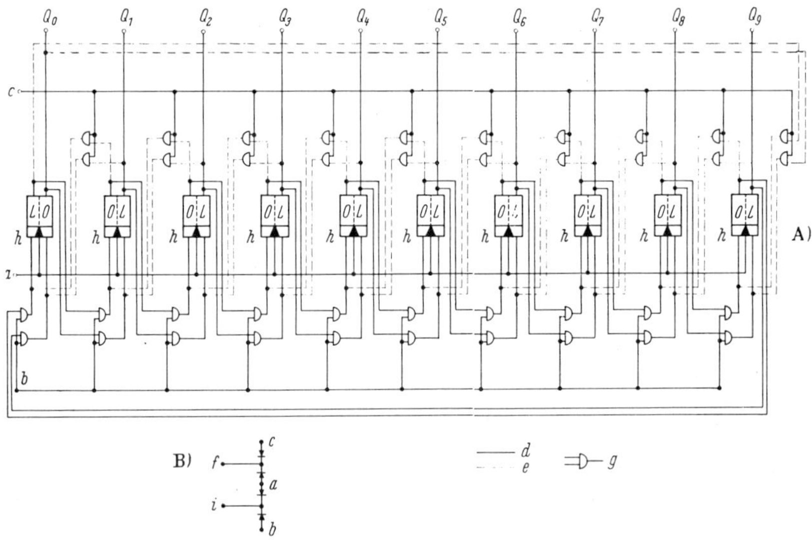

Fig.34   **Basic switching logic diagram for a decimal ring counter (6, 41).**

A) 10-bit sliding store connected up to form ring counter. B) Input circuit for processing forward and backward counting pulses (cf. Figure 13).

a Counting pulse input (dynamic input for cycle pulses, corresponds to input T in Figure 33). b Input point for forward counting direction (negative potential at b, zero potential at c). c Input point for backward counting direction (negative potential at e, zero potential at b). d Signal flows for forward counting direction. e Signal flows for backward counting direction (dashed lines). g Symbol for AND logic. h flip-flop stages as Figure 33. Q_0 to Q_9 connections for numerical display (signal Q in Figure 33).

Note: If inputs a and b are connected together via decoupling diodes, *one* input pulse will be obtained for forward counts for connection to the direction discriminator as shown in Figure 13. If the inputs a and c are similarly joined, the connection for the discriminator for backward counting will be obtained (Section B).

decimal numbers will be reduced. Figure 34 shows the basic circuit of a decade ring counter (6, 41, 52).

c) The circuits used in the semi-conductor counters (6, 41) at present commercially available, fall for the most part between these two extremes. An attempt is always made to keep the overall outlay for the functions 'counting, pre-selection, and display' to a minimum by suitably coding the counting stages. As shown in Figure 34, a counter suitable for both forward and backward counting can be produced comparatively simply by suitable coupling of the bistable flip-flop stages. As negative pulses are applied at points a and b, condition L will shift from left to right; applying pulses to points a and c will shift condition L from right to left.

The relative sizes of typical conventional electronic components are illustrated in Figure 15. The trend, however, is very definitely towards further miniaturisation, the use of printed circuits (6, 36) that carry the individual components (Figure 35), and also micro-miniaturisation using integrated circuits. Whilst the arrangement shown in Figure 35 enables a density of about five components per cubic centimetre to be achieved (6), with integrated circuits it is nowadays possible to arrange functional elements at a density of about 1000 cm^2 (37, 38). Figure 36 is a highly-magnified view of one of these monolithic circuits which is made from a single Si crystal (30, 32, 37, 38).

The current trend towards the adoption of integrated circuits of monolithic construction has two major consequences:

a) The smaller dimensions result in an appreciable reduction in the switching times (currently about 10^{-9} seconds).

b) The equipment is becoming much smaller and lighter, and the cost-performance ratio is constantly being improved.

The effects of this new semi-conductor technique are reflected most strongly in the design of computers (cf. Chapter 11); integrated circuits are a feature of what are termed third generation computers. With numerical control systems fullest advantages of this development are at present being taken in the construction of internal interpolators (cf. Chapter 8). Less advantage can be gained from the characteristics of integrated circuits in the construction of comparators for the incremental process; that is, for counters, since the associated components (such as counter-display tubes, relays for coupling with the other parts of the control system, etc) are naturally slower in operation and have larger dimensions (cf. Figures 37 and 38). In addition, the very short switching times have an adverse effect when it comes to designing reliable units incorporating integrated circuits. This is because these circuits are much more readily affected by transient stray pulses than are more conventional transistor combinations with their slower switching speeds, even when these are used with printed circuits. Numerical control systems for

Fig. 35 Example of a control system using plug-in printed circuit boards (6, 36).

a Interchangeable digital element (counter decade) comprising individual components mounted on a plastic panel carrying a printed circuit; maintenance is simplified (Photo courtesy of *Hughes/Cobelda/Burckhardt & Weber*).

machine tools must, inevitably, work close to power circuits which can emit considerable stray pulses. Under these conditions integrated circuits have to be designed with great care, and special attention must be paid to their being matched correctly to the input and output circuits and to providing adequate screening.

So far as the display of numerical information is concerned, it must be borne in mind that transistors generally operate at low voltages, so that the choice of rapid-switching, reliable, and low-loss luminous indicators is narrow. Very often, additional amplifying stages have to be provided to enable the visual display to be read conveniently and easily. It is therefore worth while to give careful consideration to whether a numerical display is really required on a machine tool (10).

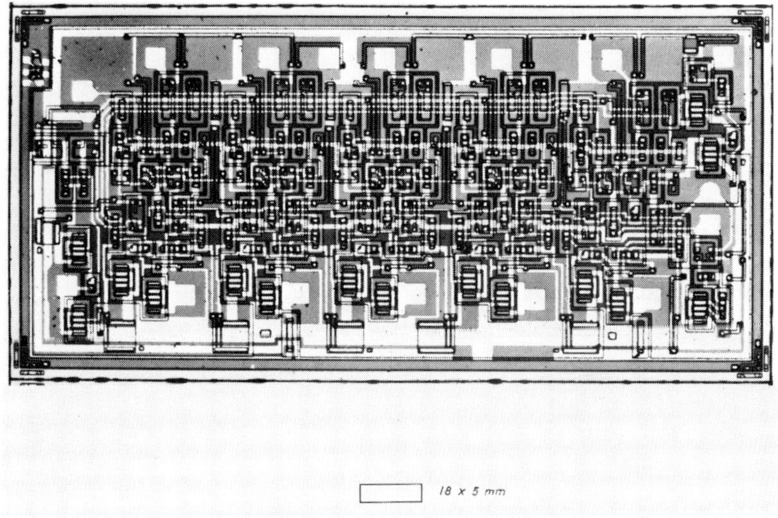

Fig. 36 **Example of an integrated circuit for a 5-bit sliding register produced on a single Si-crystal. Magnification 160 x. (Chapters 8 and 9).**

Note: The true size of this element with five flip-flops corresponding to Figure 33 is shown at the bottom of the illustration. (Photo courtesy of *Texas Instruments*).

In this connection, there is also another point to consider. There are two forms of numerical display that predominate in modern installations (41):

a) Neon tubes in which there are ten cathodes, made of wire bent to the shapes of the figures, and one anode (Figure 39). If a voltage is applied to each cathode in turn, the cathode that is passing current will glow, and the relevant number can be read off with ease. The disadvantage of these tubes is their relatively high operating voltage (about 100 V) and their limited size.

b) The use of floodlight display tubes which make use of the ability of curved glass (e.g. Plexiglas) to channel a beam of light. The individual numbers are engraved on the Plexiglas sheets, so that when a low-voltage lamp lights up the observer sees only one of the engraved numbers (41). The advantage of this arrangement is the use of a low voltage compatible with transducers and the possibility of using large figures (up to 150 mm and more in height).

Figures 41 and 42 show a method of displaying numbers, that is currently widely employed. No provision is made for pre-selection of the numbers or for obtaining signals for automatic control of the machine, and the instrument is used solely for the digital read-out of the slide position. A direction-sensitive displacement measuring system (photoelectric, either rotary or linear) produces the counting impulses for the forward and backward counter shown in Figure 41 (transistorised model). Displays of the type shown in Figure 40 are fitted in the casing where they are readily visible, and the unit as a whole can be mounted in some convenient position on the machine. Figure 42 shows units of this type installed on a lathe, where the direct indication of length and diameter enables considerable time to be saved since it is no longer necessary to stop and check the dimensions (10). Another application, this time in conjunction with an analogue measuring system, is shown in Chapter 5 (Figure 90).

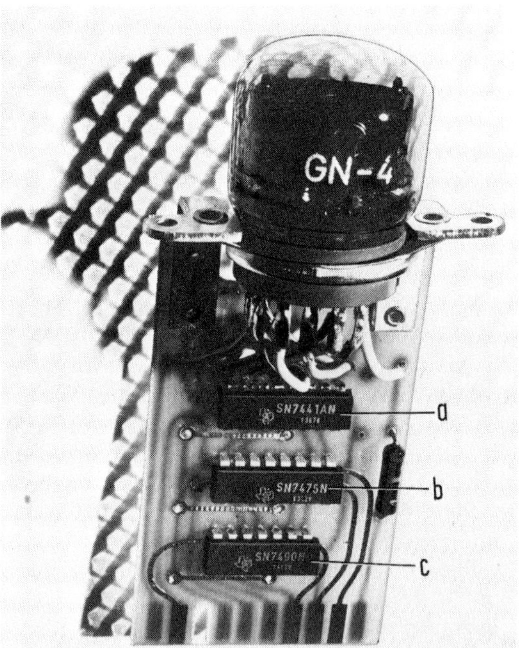

Fig. 37 **Example of a single decade decimal counter of integrated circuit techniques with construction neon tubes for storable display of numbers (similar to Figure 39).**

a Decimal counter (in BCD code). *b* Four flip-flops as intermediate store. *c* BCD/decimal converter with outputs for operation of gas-filled tubes (Photo by courtesy of *Texas Instruments*).

A)

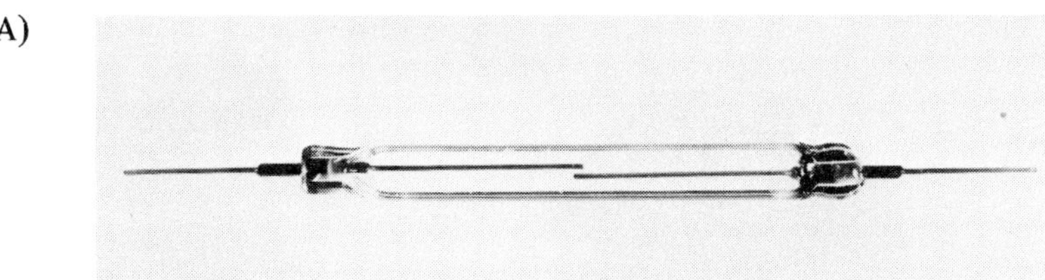

B)

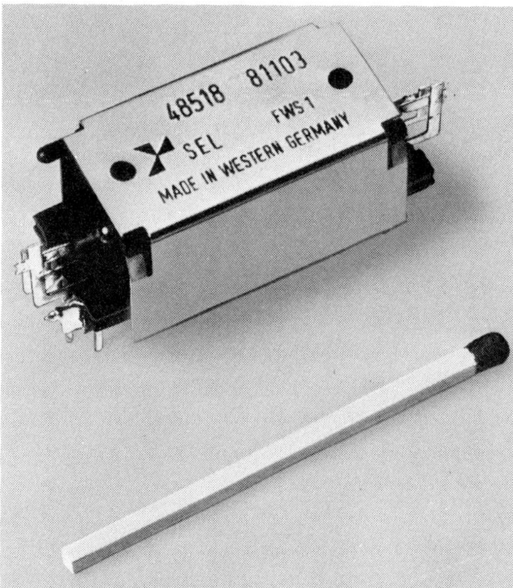

Fig. 38
Example of a gas-shielded relay for very high switching frequencies *(Reed Relays, Herkon Relays,* and similar trade names)

A) Sealed contact tube without surrounding coil, B) Complete relay unit with match to indicate size.
(Photo courtesy of *ITT/SEL*).

So far the use of transistorised counters in on-off circuits has been considered, this being the form (S type comparators) in which they have, for the most part, been used up to the present. If they are to be utilized as R type comparators, continuous comparison will be necessary, and so the counters must be in the form of forward and backward counters, used with certain auxiliary equipment. Details of the circuitry are beyond the scope of this book; it falls within the field of digital-control circuit design (54). Here, reference will merely be made to two new problems that arise out of the need for a continuous comparison of the pulses.

1. Both the desired value pulses and the actual value pulses must be able to indicate the direction of motion; to do this four information inputs must be presented and processed. The block diagram of an extended counter device of this type is shown in Figure 43 (54). The desired

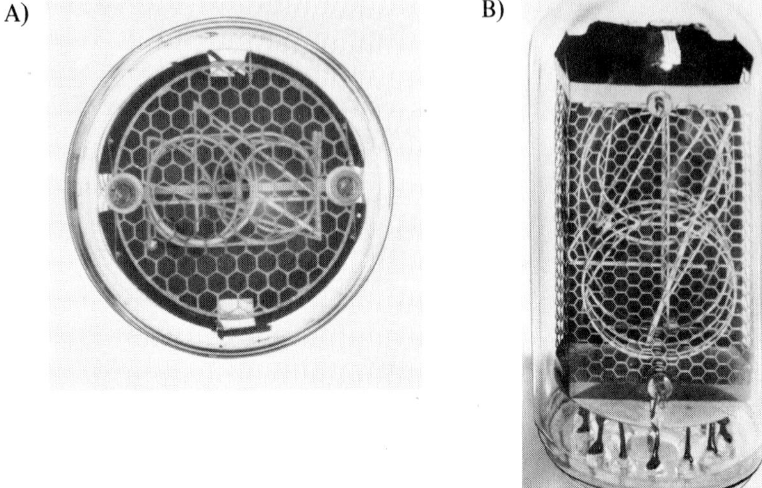

Fig. 39 Examples of incandescent tubes with decade numerical display

A) Small model in which the cathode can be read from end of the neon tube. B) Large model in which the cathode is read from one side of the valve. (Photo courtesy of *Valvo*)

value pulses a and b must first be processed in logic units e and h, together with the actual value pulses c and d, before they are passed to the binary counter as counting pulses i and k which are of a type that can be dealt with by this counter. The method of operation is briefly as follows:

Assume, in the first instance, that unit h is not provided, and that g represents the forward pulses required for the counter, and f the backward pulses. If now a desired forward pulse b is fed in, the counter is moved on one step by means of the OR circuit e and the line g. If this is followed by an actual forward pulse, the counter will be moved back one step by means of the other OR circuit and the line f. The counter reading will therefore be unchanged. The same applies if a desired backward pulse is set off against an actual backward pulse. If the quantity and direction of the actual pulses is the same as the quantity and direction of the desired pulses, there will be no change in the readings of the counter. If, on the other hand, the numbers of pulses vary, the forward and backward counter will act as a differential indicator (continuous comparator) and the deviation will

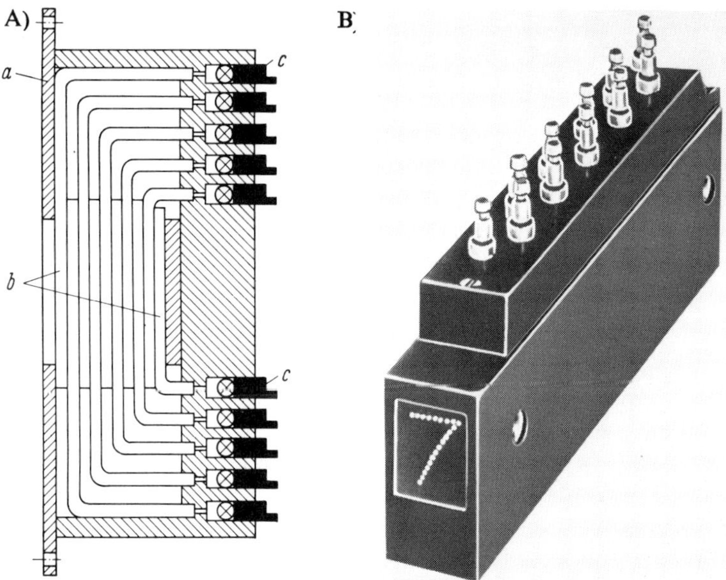

Fig. 40 **Example of floodlight display unit**

A) Section showing principle of operation. B) External view (Photo courtesy of *TWK-Electronik*).

a Front panel. *b* 10 Plexiglas panels with engraved numbers capable of being illuminated from the side. *c* Interchangeable miniature bulb with terminal tags.

Fig. 41
Example of a six-digit transistorised decade counter using floodlight counter units (Photo courtesy of *Philips*)

be shown in numerical form *l* on the counter. The counter will, however, work correctly only if the counting pulses *f* and *g* do not occur at the same instant. In practice, this cannot be guaranteed, since the desired value and actual value pulses stem from quite different sources. The coincidence gate *h* must therefore be included to prevent any simultaneous pulses *f* and *g* from reaching the counter. The counter can then easily cope with pulses *i* and *k*.

2. To operate a controller, the adjustable variable Y must in most cases be produced in a suitable analogue form (e.g. as current for controlling a d.c. motor, an electro-hydraulic valve, or the like; cf. Chapter 6). It is therefore usually necessary to place a digital/analogue converter m after the counter (6). This is one of the examples referred to in Chapter 2, in which the digital technique is retained as long as possible (the deviation is then a precise small binary number), and the continuous mode is not adopted until just before the actuator control point (e.g. in the feed of a machine tool slide). In Figure 43 it is assumed that the remainder of the signal processing p (amplification, control-loop stabilisation, etc) is effected in analogue form. The final output is the regulating variable Y which is used to control the relevant actuator.

Detailed calculations and thorough tests will have to be performed to determine whether this construction is more reliable and less costly than other possible solutions (cf. Section 3.2 of this Chapter and Chapters 4 and 5)

Fig. 42 Example of the use of a digital read-out on a manually-operated lathe

a Diameter reading. *b* Length reading (Photo courtesy of *Philips*).

Note: There is no automatic comparison of the desired and actual values; this is made by the operator. As indicated in Chapter 1, Figure 1, the dimensions of the workpiece are not measured directly, but indirectly by measuring the slide displacement. This can be allowed for after a tool change by calibration (setting counter to zero at a certain point); cf. also Chapter 7.

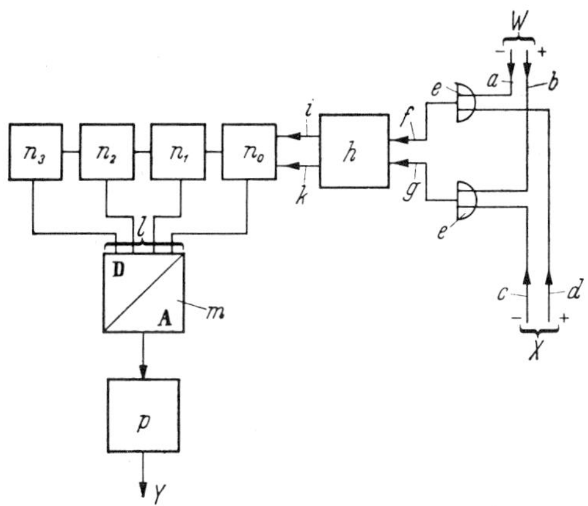

Fig. 43 Logic diagram of a comparator for a digital-incremental control circuit

a Desired value backward pulses. *b* Desired value forward pulses. *c* Actual value backward pulses. *d* Actual value forward pulses. *e* OR elements. *f* Backward counting pulses (unordered). *g* Forward counting pulses (unordered). *h* Coincidence gate. *i* Backward counting pulses (ordered). *k* Forward counting pulses (ordered). *l* Deviation $(X \cdot W)$ in numerical (digital) form. *m* Digital/analogue converter. n_0 to n_3 Binary stages of a forward and backward counter. *p* Additional control circuit components (amplifiers, differentiating stages, etc.). *W* Desired values (Commands emanating from an internal interpolator or a magnetic tape, cf. Chapters 8 and 9). *X* Actual values, produced by a direction-sensitive displacement measuring system. *Y* Regulating variable (for controlling the drive system, cf. Chapter 6).

taking into account the basic considerations for the use of the incremental process for machine tools (cf. Chapter 2). As is explained in greater detail in Chapters 8 and 9, the process may well at present be limited to contouring control systems with digital internal interpolators.

Finally, one design condition that applies to all pre-selector counter designs that have been mentioned: the selectivity of counters for machine tool control systems must be at least $20 \cdot 10^3$ pulses/s, that is, the duration of the primary output pulse will be:

$$t_{ia} \approx \frac{1}{2 \times 20\,000} = 25 \; \mu s \text{ and possibly even less.}$$

Almost always, however, the counter output signal is used for controlling components which have a certain mass, and hence a certain inertia. Experience shows that the reaction times of control circuits that contain large relays, contactors, solenoid valves, etc., are of the order of 10 to 100 ms. To enable these counters to be used as widely as possible it is, therefore, advisable:

a) to extend the primary electronic output signal, which is of short duration, to 100 ms or more, or even to store it, directly in the counter, and

b) to make the counter output voltage-free, i.e. to conclude the internal counter circuit with the coil of a small relay, and only to lead out the neutral relay contact connections. Figure 38 shows a relay design with hermetically sealed contacts. Relays of this type can nowadays be made small enough to be soldered onto printed circuit boards.

3.1.6   *Summary of conclusions - incremental process*

The following are the main points which have arisen relating to the various posssible applications of the incremental process to numerically-controlled machine tools:

1. The incremental process is a displacement measuring process; the instantaneous position of a machine slide can be obtained only by summing all the increments. The precautions that have to be taken to protect the system against spurious pulses are of vital importance in machine tool applications. Adequate protection can be provided only by careful design and manufacture of the components. One of the main advantages of this process over its competitors is, however, its greater simplicity. But whether this will be retained under these circumstances cannot be determined without taking into account the points raised in Chapters 4 and 5.

2. The components and units used must be designed for switching times of 20 μs and less.

3. Datum point determination on the machine tool is very simple, although, where mechanical switches are used to initiate counting, it is essential to ensure that they are bounce-free in operation.

4. Owing to the high velocities encountered in rapid-traverse motions, contactless scanning should (as far as possible) be used for the displacement measuring systems. Magnetic scanners appear to offer certain advantages for rotating systems (high reliability, freedom from maintenance, etc). With linear systems only photoelectric scanning is likely to prove effective at present.

5. For a low-cost installation the need for providing a numerical display of the actual values must be carefully considered. It is, however, probable that for lathes and large machine tools for example, digital read-out will become a standard feature because of the simplicity with which it enables measurements to be made while the machine is running (reduction of idle-time to check dimensions). The operator then performs the functions of the actual/desired value comparator (10).

6. The most suitable form of construction for counters and the most appropriate choice of components should take into account the reliability and cost. It is possible to form three categories based on the machine-slide movements and the method of their suitable control.

Group I: On-off circuits with no change in the direction of counting (e.g. for single-stroke motions).

Group II: On-off circuits with a change in the direction of counting (e.g. co-ordinate table settings).

Group III: Continuous-control circuits with constant changes in direction and continuously-monitored slide positions (digitally-controlled synchronous operation).

7. Since the use of continuous control circuits is essential only for continuous path control systems, this is the only application in which Group III will be encountered. The question of the suitability of digital-incremental control circuits in conjunction with internal interpolators and magnetic tape control systems will be further discussed in Chapters 8 and 9.
8. The considerations set out in 6 above assume that an individually optimum solution is being sought. If the manufacturing problems are approached from the point of view of the maker of the control system, it will be clear that a unit-construction system using standardised building-blocks will be preferred; even though in some individual cases this will lead to a more complex installation. From this point of view, complete transistorisation of all the building-blocks is the best solution, and this also offers the user advantages in standardisation and ease of maintenance. With the current state-of-the-art in the design and use of semiconductor devices, using standard components and units to the greatest possible extent will support the tendency for the prices of electronic equipment to fall and also avoid excessive specialisation. The trend towards the more widespread use of integrated circuits will encourage the still wider use of standardised items.

3.2   The digital-absolute process and its associated equipment

As has already been mentioned in Chapter 1, the decimal system of counting is costly when used for automatic measurement and data processing; for purely economic reasons, it is unlikely to be adopted. The following discussion is therefore largely restricted to binary systems.

The incremental measuring scales that have been described so far are inscribed with unmarked graduations, and in the simplest case these are spaced at intervals equal to the digital displacement element Δs. This will in no way be changed if, in addition to a grating of this type with the fundamental subdivision of $2^0 \times \Delta s$, gratings with the binary subdivisions $2^1 \times \Delta s$, $2^2 \times \Delta s$, $2^3 \times \Delta s$, $2^4 \times \Delta s$, and so on, are used. All the points that have previously been made about the scanning of scales (frequency range, use of auxiliary gratings, etc) will, therefore, still be valid, provided that each track is considered separately. Depending on the total length that is to be measured twelve to twenty incremental scales are now placed side-by-side, the divisions being in binary steps so that the scanning operation is simplified for higher orders of numbers. The overall resolution and the critical operating conditions depend, however, solely on the value chosen for Δs, or in other words:

a) The choice of the digital step Δs, which is entirely arbitrary, exerts a larger influence on costs than it does where incremental measuring scales are used.
b) The outlay on the scanning system increases, as a first approximation, with the number of tracks used in an absolute measuring system. This means that the cost of scanning increases with the length to be measured.

There is one further difficulty. As was mentioned in Chapter 1, when scales graduated in a pure binary system are used, read-out errors can occur at all those points on the scale where several binary digits change simultaneously as the changeover is made from one number to the next (cf. Chapter 1, Table 4). This difficulty can be overcome by using the Gray-code, for example, instead of the pure binary-code (cf. Chapter 1). This is, however, difficult to use for processing the measured values. Many attempts have, therefore, been made to avoid the read-out ambiguities by other methods. Only one of the many proposals will be discussed at this point.

Scanning using switched scanning zones (zone-scanning, V-scanning)

At first glance this process appears rather complicated, but it offers so many advantages in manufacture and operation that it has been widely adopted. In its original form, it is suitable only for pure binary-coded measuring scales (not for binary-codes in general). To understand the method of operation it is necessary to visualise a pure binary number sequence. Several binary numbers are listed, one below the other, in Table 7.

If, firstly, only two right-hand columns are considered (the first column for 2^0 and the second column for 2^1), it will be seen that when 0 is in the first column, the adjoining binary digit in the second column appears twice in

```
OOOO  |
OOOL  |
OOLO  |
OOLL  |
OLOO  |
OLOL  ↓  Increasing binary numbers
OLLO
OLLL
LOOO
LOOL
LOLO
←——— Place numbers
```

Table 7.   Structure of binary numbers.

succession with rising binary numbers. On the other hand, when L is in the first column, the adjoining binary digit in the second column appears twice in succession with decreasing binary numbers. The same applies to the other columns of the binary numbers. Use can be made of this property of the numbers to ensure a correct read-out. Figure 44A) is a section of a binary-coded measuring scale. The actual method of scanning employed (electro-mechanical, photoelectric, magnetic, etc.) is immaterial so far as the following discussions are concerned. If, instead of a single row of scanners along the line b - b, a double scanner system (as shown in Figure 44B)) is fitted, and if the spacing from the line of symmetry b - b is as shown, the following switching rule will be obtained for the scanning system:

1. If an O is read on any track (binary column) then the scanner of Series II (in the direction of increasing binary numbers) must be switched on for the next higher track.
2. If an L is read on any track, then the scanner of Series I (in the direction of decreasing binary numbers) must be switched on for the next higher track.

This rule is explained in Figure 44C) for two scanning positions b_1 and b_2. It can be seen clearly that this procedure displaces the scanning operation into the unambiguous zone. The electronic switching devices that are needed with this procedure will, of course, add to the cost and complication of the installation, but this seems justified in view of the gain in reliability of read-out that is achieved. This V-scanning system also offers an advantage in the manufacture of the coded measuring systems in that the accuracy with which the higher tracks are divided can be lower the higher the track concerned.

Only in recent years has it proved possible to make relatively inexpensive binary-coded linear scales. Figures 45 and 46 show a glass scale of this type with a scanning head which also contains the logic circuit for the V-scanning. Scales of this type are nowadays made with up to eighteen binary tracks and

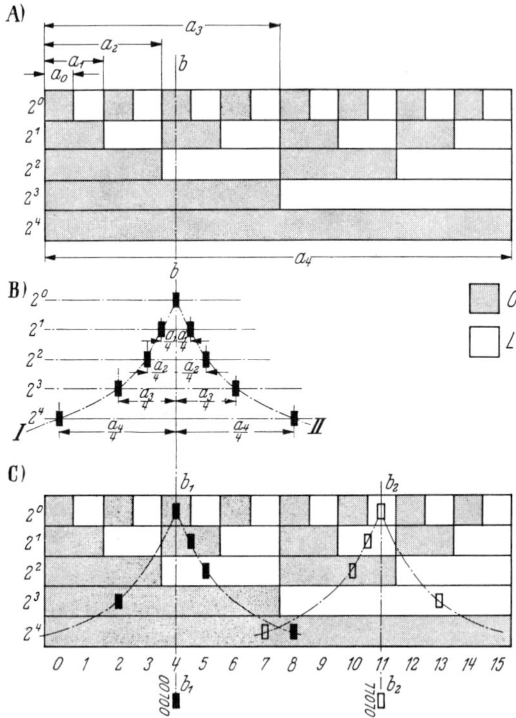

Fig. 44 **Illustration of the principles for *V*-scanning (zone-scanning) of a binary-coded scale**

A) Binary-coded scale. B) Arrangement of scanners. C) Scale *A* and scanner system *B* superimposed at two points (b_1 and b_2) (only the scanners that are switched on are shown).

a_0 to a_4 Field widths of scale (½ x division spacing) for individual tracks. b, b_1 and b_2 Scanning positions (line of symmetry between I and II). I First scanning system. II Second scanning system.

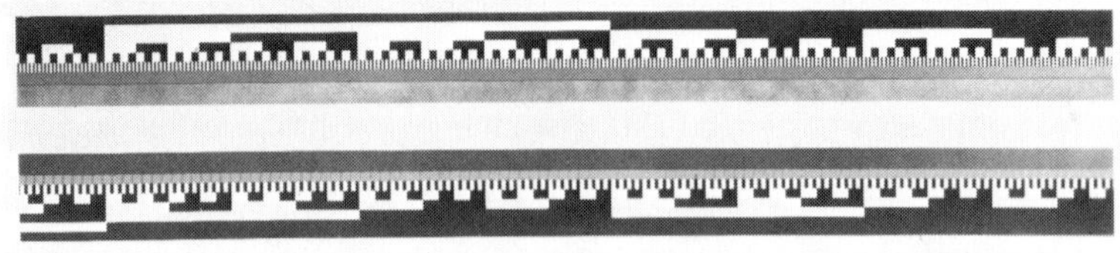

Fig. 45 **Example of an 18-track binary-coded scale (Photo courtesy of *Heidenhain*)**

Note: The finest tracks are located in the centre of the scale; these are not reproduced by the printing process.

A)

B)

Fig. 46
A) Example of a binary-coded linear scale with light source and photoelectric scanning head. The discriminator electronic circuit (together with the circuits for V-scanning, cf. Figure 43) are located in a separate unit which can also accommodate the binary-decimal converter if this is required. (Photo courtesy of *Heidenhain*).

B) Example of an absolute scale fitted to a machine tool (Photo courtesy of *Heidenhain*).

a scale length of about 1310 mm. Using electronic devices of the type shown in Figure 26, it is possible to obtain a resolution capacity of as little as 2.5 μm. The finest track is then located in the centre of the scale (Figure 45). Ideal though binary-coded scales may be technically they are not widely used on large NC machines, even today, because of the high cost of this type of displacement measuring system (10). The coded displacement measuring systems that are most widely used are those that incorporate rotary coded scales. In these, coded discs are coupled by means of gears of suitable ratios. The use of V-scanning greatly reduces the accuracy required of these gear trains, so that their cost is reduced—which will help in offsetting the expense of the extra circuitry referred to above.

The scanning of rotating coded discs, however, introduces further problems. The pure binary system has the disadvantage that it is relatively complicated to transform into the decimal system. It is only practicable to use with relatively large installations (e.g. in conjunction with electronic computers or interpolators). In most instances a mixed decimal-binary system of coding will be preferred (cf. Chapter 1). Each disc (decimal place) can then carry a binary code and the individual discs can be coupled by gearing with a reduction ratio of 10:1. The V-scanning system that has been described is, however, suitable only for the pure binary system.

It is therefore necessary to use binary codes which are compiled in such a manner that they can be used in conjunction with zone scanning, using the simplest possible circuit. One possible method is indicated in Figure 47. The basis on which the code is compiled is shown in Table 8 (6).

Decimal code		Gray-Code	Examples of a tetrade modification of the Gray-Code
0	(0)	OOOO	OOOO
1	(1)	OOOL	OOOL
2	(2)	OOLL	OOLL
3	(3)	OOLO	OOLO
4	(4)	OLLO	OLLO
5	–	OLLL	–
6	–	OLOL	–
7	–	OLOO	–
			——— Line of symmetry
8	–	LLOO	–
9	–	LLOL	–
10	–	LLLL	–
11	(5)	LLLO	LLLO
12	(6)	LOLO	LOLO
13	(7)	LOLL	LOLL
14	(8)	LOOL	LOOL
15	(9)	LOOO	LOOO

Table 8. Rules for forming a modified Gray-code for binary tetrades

The modified Gray-code obtained in this manner has three characteristic features:

a) As with the Gray-code, only one binary place is changed at a time; in this case, however, the range 0 to 9 completes a cycle.
b) The first five binary numbers have 0 in the fourth digit (from the right), while the next five numbers have L in the fourth place.
c) The remaining three binary digits are mirror images of each other about a line of symmetry.

These features enable a very simple zone scanning circuit to be achieved (Figure 47).

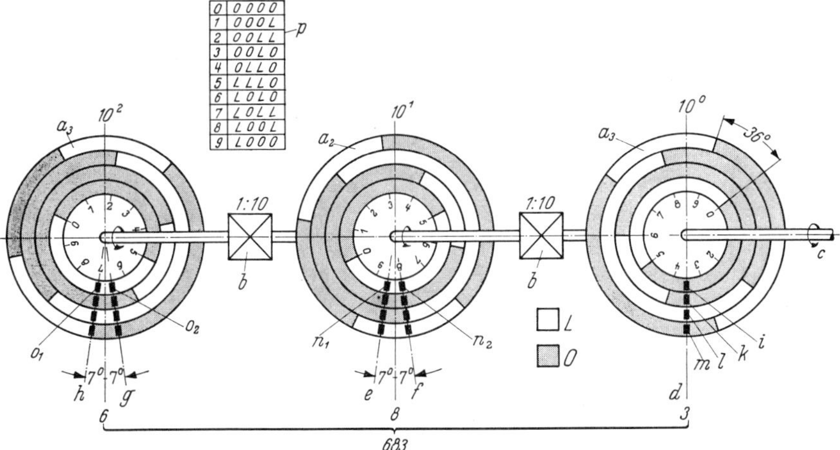

Fig. 47 **The zonal scanning of binary-coded discs.**

a_1 Coded disc for units. a_2 Coded disc for tens. a_3 Coded disc for hundreds. b 1 : 10 gearing (high precision not required because of zone scanning). c Drive. d Axis for the single row of scanners for a_1. e, f Axes for the two rows of scanners for a_2. g, h Axes for the two rows of scanners for a_3. i, k, l, m Signal wires from the four scanners (brushes, photodiodes, etc.) of disc a_1. n_1, n_2 Signal wires from the inner scanners of a_2. o_1 o_2 Signal wires from the inner scanners of a_3 (the other signal wires have been omitted in the interests of clarity). p Code table (disc coding reading from inside outwards).

On the units disc only one row of scanners i, k, l, m is required, located along the zero line (line of symmetry) d, since the nature of the code prevents any possibility of ambiguity at the changeover points. The width of the field is $360°/10 = 36°$. The tens disc is provided with two rows of scanners e and f, which are offset about 7° on either side of the zero line (line of symmetry)[4].

The hundreds disc is of similar construction, as are all further discs for higher orders. The rules for switching the rows of scanners *e - f, h - g*, and so on are, then:

1. If an 0 is detected on the inner track of a disc, the scanner row corresponding to higher numbers (in this case the row offset against the direction of rotation) must be switched on on the next disc (for a higher decimal place).
2. If an L is detected on the inner track of a disc, the scanner row corresponding to smaller numbers must be switched on on the next disc.

When all the discs have been scanned in accordance with this rule, the row of scanners in use will always lie in the unambiguous area. It is therefore unnecessary to use high-precision gearing. The inner disc track also forms the switching track for the changeover to the next series of scanners.

The binary code described has the advantage that the changeover arrangement is very simple, but against this it has the disadvantage that it cannot be used for calculations. This exerts a considerable influence on the design of the comparator (cf. Figures 51, 52, and 53).

The need for the rule established in Chapter 1, that the designer must compare the various possibilities with great care, is well illustrated here. It may prove necessary to fit a small code converter between the measuring system and the comparator. The example shown in Fig. 50 is equipped with a different binary code from that described above. It would, however, be beyond the scope of this chapter to describe all the possible variations (6); a description of the basic problems and the principles by means of which they are solved will have to suffice.

The method adopted for the scanning operation could theoretically assume as many different forms as are used for the scanning of incremental measuring scales.

Electro-mechanical scanning using brushes seems a rather risky choice for use with machine-tool controls, even where a rotary system is being used, owing to the large number of scanning points and the high rubbing speeds, (rapid-traverse motions and high resolution); although a number of attempts have been made in the past to use this type of equipment (55).

Electromagnetic scanning is also likely to prove complicated and costly, again, owing to the large number of scanning points; although it is quite possible that low-cost equipment of this type (perhaps based on Hall-effect generators) may be introduced at some future date.

---

[4] In theory, the offset could be $36/4 = 9°$, but because of the dimensions of the scanners it is advisable to choose a slightly smaller angle. This does not affect the operation of the system in any way.

Up to the present time, however, the only system that has achieved any practical importance is that using photoelectric scanning. The examples to be described are therefore all of this type.

3.2.1 Rotary systems for the numerical measurement of position

Two code discs carrying pure binary codes and designed for photoelectric scanning are shown in Figure 48; these discs are mounted on a single shaft. They and the following units are coupled by means of reduction gearing (ratio $1 : 16^2$). This form of construction is, however, used only in conjunction with computers (e.g. internal interpolators, cf. Chapter 8), since the cost of a converter for transforming the readings into a decimal code is relatively unimportant (6, 56). Figure 49 shows a more recent type in which the use of very small components enables the information to be produced with the aid of a single code disc (which has, however, many tracks).

The photoelectric system incorporating V-scanning shown in Figure 50 has no discs, but blackened drums provided with slits arranged in the code configuration. The light source is located between the six drums (coded masks); photo-conductive transistors are arranged inside the drums and act as photoelectric scanners. This results in a very compact form of construction. The 'units' drum is provided with a cyclic code (similar to the code shown in Figure 47) and a tetrade scanning system; all the other masks carry a 3-excess code (cf. Chapter 1, Table 4) and have V-scanning. The equipment is intended for use primarily in conjunction with computing operations (derivation of differences), for which the 3-excess code is particularly suitable (6).

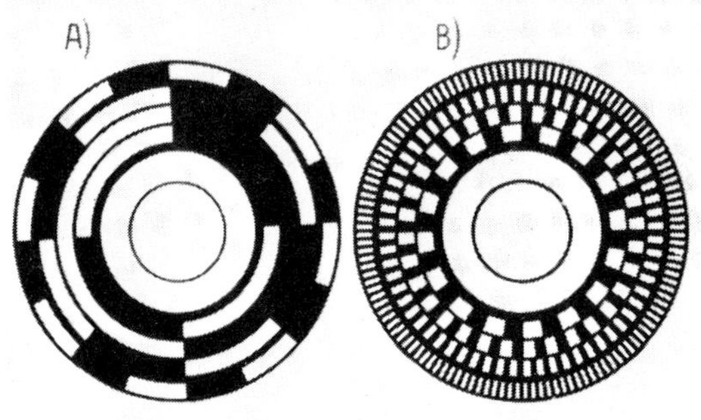

Fig. 48 Discs with pure binary coding (for practical reasons the discs have the same diameter and rotate at the same speed. Photo courtesy of *AEG*).

A) Track sequence (from outside inwards) 16, 8, 2, 4.
B) Track sequence (from outside inwards) 256, 128, 64, 32.

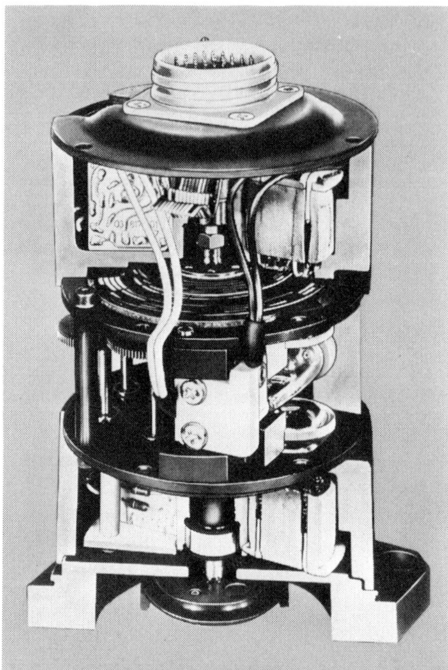

Fig. 49
Angle coder with integ-
rated circuits and
miniature photoelectric
cells. Several decades are
grouped together on a
single code disc. Scanning
follows the principles
shown in Figure 47.
Signal amplification takes
place in the angle coder,
but the V-logic is located
in the machine control
system to enable advan-
tage to be taken of the IC
technique for other
purposes (cf. Chapter 8).
(Photo courtesy of *AEG*)

Fig. 50. Example of a rotary position measuring unit for six decimal places with photoelectric scanning (Photo courtesy of *Güttinger*).

a Drive shaft. *b* Connection for signal wiring. *c* Drum coded by light slots (the photoelectric scanners are located inside the drum, while the common light source is in the centre of the unit). *d* Electronic changeover switches for *V*-scanning.

References to other forms of rotary measuring systems using coded discs will be found in the literature (6), so that it will be unnecessary to go beyond the basic features of the subject at this point.

3.2.2 Linear systems for the numerical measurement of position

Inevitably many attempts have been made to avoid the need for using a rotating system and, instead, to determine the position of a machine slide directly by means of coded linear scales. For the reasons that have already been mentioned (small displacement elements Δs, large number of tracks, precautions needed to ensure accurate scanning, etc.) the cost of producing coded linear scales is very high. The cost of a scale will obviously increase considerably with increasing length (large number of tracks and calibration difficulties). For economic reasons, this method will therefore initially be restricted to short scales. It is quite conceivable that equipment for numerically controlled measuring of workpieces (cf. Chapter 1) could use such code scales. Longer measuring scales are currently obtainable (cf. Figure 4b), but little used owing to the high cost of manufacture (10).

3.2.3 Possible designs of numerical comparator for the absolute process

At the present time, only transistorised equipment, or more recently integrated circuits, find any practical application. Even if the discussion were to be limited to these components it would go beyond the scope of this book to deal with complete circuit layouts. To enable the basic phenomena to be understood it will be sufficient to describe two of the most important types and their logic circuit (54).

Depending on the required output signals, it is convenient to distinguish between three different arrangements. The three possible output functions are illustrated in Figure 51. If, as recommended by the German Standard DIN 19226 (12) for control technology, the desired value is designated W and the actual value X, the difference $W - X$ will represent the output signal of the comparator in digital form.

In Example A) of Figure 51, the comparator is of such simple construction that it can only determine whether two numbers are identical, i.e. it can only check that $W = X$ for each digit in the number. Only if this is so will the output signal L be produced. This simple design corresponds exactly to the usual S type of comparator employed in on-off circuits as described in Chapter 1, and which is also employed in a pre-selection counter. As will be shown, the difference as compared with the pre-selection counter is merely that coincidence is established not only in one track, but also in all the tracks of a coded binary scale. The type of binary coding employed is immaterial, since the co-incidence condition $W = X$ is established for each track separately. This co-incidence check can also be made relatively easily for several numbers along a displacement distance; the process corresponds to the use of several pre-selected numbers for effecting speed changes.

In Example B) of Figure 51 a further step has been taken. The comparator must now also decide whether $W > X$ or $W < X$, and must deliver direction-indicating digital positional signals p_r and n_r, the magnitudes of

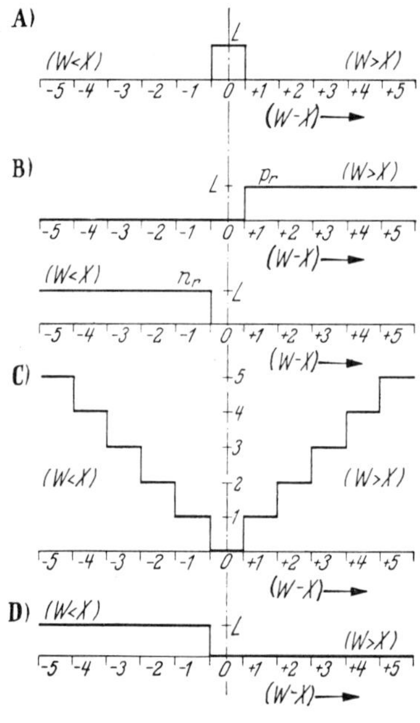

Fig. 51

Comparison of the overall decisions (output signals) for three different types of numerical comparator (54, 56)

A) Basic diagram of the output signal of a comparator of the type shown in Figure 52 (simple co-incidence check device for on-off circuits). B) Basic diagram of the output signals p_r *and* n_r of a comparator of the type shown in Figure 53 (three-point controller for non-continuous regulation). C) Basic diagram of the output signals of a comparator with computer characteristics (proportional controller for continuous regulation), numerical absolute value of the deviation. D) Decision regarding direction made by computing system (sign shows deviation).

which depend on the value of L to two separate output lines. When co-incidence occurs both output lines will carry the signal 0. To some extent this arrangement thus corresponds to the general type or R comparator that is used for continuous-control circuits (described in Chapter 1). In control theory, this can be described as a three-point controller (non-continuous regulation). As will be shown, the cost of this arrangement is appreciably higher than for Example A). In addition, it is no longer possible to choose any binary code arbitrarily; the code chosen must be suitable for determining differences. Otherwise a code converter will have to be installed between the displacement measuring system and the comparator - which again increases the cost.

In Example C) of Figure 51, the number of elements required in the comparator has been increased still further, and the equipment is designed as a computer for determining differences. It thus acts as a continuously-operating proportional controller. Now the value $W - X$, distinguished by magnitude and by sign, appears as the output signal. The choice of the code is then subjected to much more stringent requirements (performance of computing operations) than was the case for examples A) and B). In general, either the pure binary code or the 3-excess code will be used. Logic circuits will be given for Examples A) and B). The construction of comparators for Case C) and the understanding of their design requires a good basic knowledge of the construction of electronic computers; possible logic circuits for this case will therefore not be given (6, 54).

Numerical comparator with output signal of undetermined direction

As has already been mentioned, the co-incidence check is performed for each binary digit separately, so that any binary code can be chosen. Any code that appears satisfactory for the scanning operation can therefore be used in the comparator, and it is only necessary to make sure that this same code is used for the input of the desired values. If required, code conversion can be performed between the input equipment and the comparator using relatively simple means. The logic circuit for four binary digits is shown in Figure 52. The logic elements a and c act as AND elements, which produce the L decision only if all the input signals are L. This makes the basic mode of operation of the comparator clear: two AND elements are first used to establish at each binary digit whether $W = X = L$ or whether the inverse value $\overline{W} = \overline{X} = L$ i.e. a check is made whether in every case $W = X$. If both W and X have the value L, the logical decision of the upper AND element will be L, while if W and X both have the value 0, then $\overline{W}$ will have the value L and $\overline{X}$ will have the value L, so that the logical decision of the lower AND element will again be L. In both cases the affirmative signal L will appear at the output of the elements a. For further processing it makes no difference whether the co-incidence is due to two L values or to two 0 values. The OR element b collects the two cases to form a decision X_{wv} ($v = 0, 1, 3, .. n$).If these logic decisions are carried out in the same way for all the binary digits, it is only necessary to feed all the partial decisions Xwv to an AND element c to obtain the final decision "co-incidence achieved in all binary digits." This

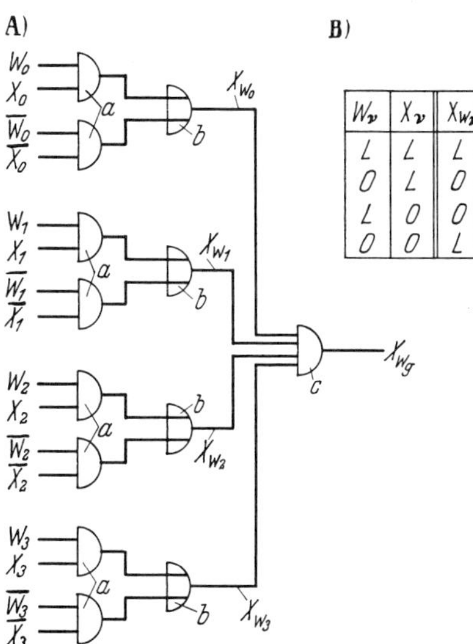

W_v	X_v	X_{Wv}
L	L	L
0	L	0
L	0	0
0	0	L

Fig. 52
Diagram showing the principle of operation of a numerical comparator with an output signal which is independent of direction (Type S comparator) at co-incidence and the use of a general binary system

A) Logic circuit for four binary digits. B) Binary decision table. a and c AND elements. b OR elements. W_0, W_1, W_2, W_3 Desired values in the individual binary digits. $\overline{W}_0$, $\overline{W}_1$, $\overline{W}_2$, $\overline{W}_3$ Inverse (negative) desired values. X_0, X_1, X_2, X_3 Actual values in the individual binary digits. $\overline{X}_0$, $\overline{X}_1$, $\overline{X}_2$, $\overline{X}_3$ Inverse actual values. X_{W0}, X_{W1}, X_{W2}, X_{W3} Evaluated results for the individual binary digits. X_{Wg} Final result (output signal, coincidence signal, cf. Figure 51A).

can only be L (cf. Figure 51A); for all other relationships between X and W in any binary digit the final answer will always be X_w overall $= 0$. The logic problem is therefore solved.

The logic decision elements can take many different forms, such as relays, semiconductors, hydraulic or pneumatic elements (cf. Chapter 9), etc, so that electrical methods are not essential. In practice, however, only contactless, quick-acting and non-wearing semiconductors are used. It is, of course, possible to use the 4-element decision $X_{w\ overall}$ given in the example to determine when co-incidence is achieved in all the decimal places, when binary-decimal coding is used, and then to obtain the overall co-incidence signal by adding an AND element. As can be seen, the construction of a co-incidence monitor of this type, which does not indicate direction, is comparatively simple. The decision as to the direction in which a machine slide must move to achieve co-incidence between the desired and the actual values cannot, however, be provided by a simple device of this type, and some other device is needed for this purpose. Its selection will depend on the remainder of the machine control system and on the extent to which it is automated (10). It is, however, appropriate to look into the cost and complication involved if the decision as to direction is to be made by the comparator itself.

Numerical comparator with two direction-dependent output signals

Figure 53D) shows a logic circuit for four binary digits (columns) (54). To clarify the apparently complex relationships the partial logic decisions are shown in coded form in diagrams A(, B), and C).

In the first place, four basic decisions have to be made in each binary column for:

$$W_\nu\ = X_\nu\ = L$$
$$W_\nu\ = X_\nu\ = O$$
$$W_\nu > X_\nu$$
$$W_\nu < X_\nu$$

This task can be performed by two AND elements (Fig. 53A). An original value and an inversion is applied to each AND element. The four possibilities are listed in a table in the sequence given above. The logical decisions made by the AND elements a are shown in the same sequence to the right of the elements. As can be seen, the first part of the problem is basically already solved:

If $W = X = L$ or $W = X = O$, then the decision is O.
If $W > X$ ($W - X$ positive), then the decision L appears at the
output of the upper AND element.
If $W < X$ ($W - X$ negative), then the decision L appears at the
output of the lower AND element.

For subsequent processing operations it is better to use the inverse values of these output decisions, rather than the decisions themselves. For this reason an inverter is included after the AND element. The result of this is given at the outputs p and n. It must however be borne in mind that the numbers being investigated have several decimal places. It is therefore quite possible that in the lower places W is always less than X, and that W is greater than X only in a higher place. For the number as a whole W is then greater than X. The decisions relating to the individual binary digits are then of no value; they must be coupled together in such a manner that the decisions for the higher binary digits influence those for the lower binary digits. One basic method of doing this is shown in Figure 53B for the νth binary digit. The requirement here is for a code to be used in which the next higher binary digit also has a higher value. Assume that the pure binary code is used in

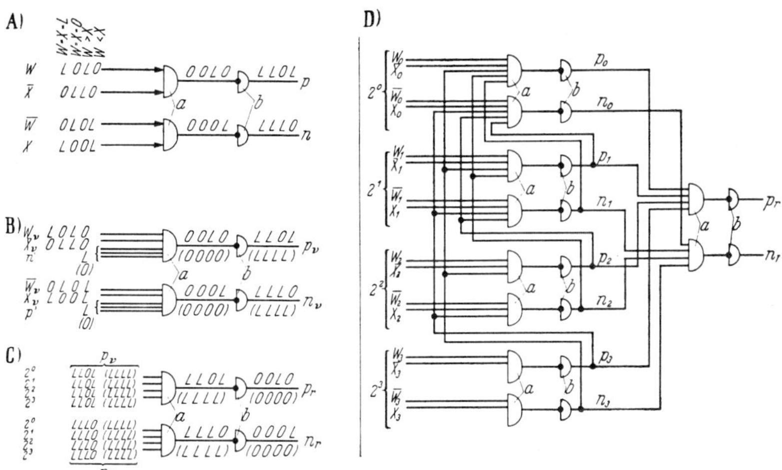

Fig. 53 **Diagram showing the principle of operation of a numerical comparator with two direction-dependent output signals of constant amplitude (Type R comparator used as a three-point regulator) using the pure binary system**

A) Logic circuit and summary of input conditions for a binary digit ignoring the results in the other binary digits. B) Logic circuit for a binary digit of νth order taking account of the results of all higher binary digits. C) Logic circuit for deriving output decisions p_r and n_r from four binary digits. D) Complete logic circuit for four binary digits (combination of A, B, and C).

a AND elements. b Inverter. n Logical decision for the case $W < X$ (W-X negative). p Logical decision for when $W > X$ (W - X positive). n_ν p_ν Decisions at νth binary digit. n', p' All results for binary digit higher than νth place. n_r and p_r Final decisions (output signals) for all binary digits (cf. also Figure 51B). W Desired values. $\overline{W}$ Inverse desired values. X Actual values. $\overline{X}$ Inverse actual values.

Note: The elements a and b can be combined and indicated by a single symbol; this has not been done here for the sake of clarity.

which the values of the digits increase in powers of 2. In the νth binary digit the input to the AND elements will then not only be the basic input information W, X and $\overline{W}$, $\overline{X}$ (as already explained in Part A of the Figure), but also the output decisions of all the binary digits higher than the νth. Let these be represented by p' and n'. They can also be only L or 0. Considering first the upper AND element (Part B of the Figure), this will give a decision L only if all higher binary places have produced the decision L in the outputs of n'; in all other instances it will give the decision 0. If then an 0 appears in one of the higher places at the n output, the output decision of the upper AND element is always 0 (shown in parentheses in the diagram). In other words, only if all the binary digits higher than the νth have signalled L at the outputs n will the original decision L remain in force for $W > X$; otherwise the decision will always be 0. A similar condition exists at the lower AND element (Part B of the Figure). Here the summarised decisions p' of the higher binary digits assist in forming the decision. The result is as follows: Only if an L decision has appeared at the p output of all higher binary places for the case $W < X$ will the original decision L be maintained for the νth place. In all other cases 0 will appear.

The third logical step is shown in Figure 53C). Here the decisions in the four binary digits (of this example) are combined in two further AND elements and are inverted. The resulting decisions p_r and n_r are then:

If $W = X = \text{L}$, then $p_r = \text{O}$ and $n_r = \text{O}$

If $W = X = \text{O}$, then $p_r = \text{O}$ and $n_r = \text{O}$

i.e. the result O will be obtained at both outputs if a co-incidence obtains.

If $W > X$ ($W - X$ positive), then $p_r = \text{L}$ and $n_r = \text{O}$

If $W < X$ ($W - X$ negative), then $p_r = \text{O}$ and $n_r = \text{L}$

If there is no-co-incidence, therefore, L will appear at one of the two outputs. This type of output function behaviour is represented graphically in Figure 51B). The problem has therefore again been solved.

Figure 53D) shows the complete logic system for the four binary columns. The system as a whole functions as a three-point regulator and therefore belongs logically to the R type of comparator which is suitable for incorporation in a control-loop circuit (cf. Chapter 1). Semiconductor logic elements are again used almost exclusively for this application.

The extra complication in the introduction of a simple direction-dependent decision can be visualised from Figs. 52A) and 53D), even though the actual comparisons of the complexity require the use of circuit algebra. If the arrangement is to be taken a step further to enable the output values to be made proportional to the numerical difference (corresponding to Figures 51C) and D), the complexity increases appreciably, so that it is necessary to consider carefully whether this method of producing a direction-dependent numerical proportional positioning signal is economically worth while. This decision must take the following factors into account:

1. alternative solutions available for the same problem
2. the type of machinery on which a digital-absolute position control system would be technically and economically superior to all other systems.

The answer to these questions must be deferred to some chapters later in this book, since it is necessary to view the problem as a whole before arriving at any valid results.

3.2.4 Zero-point shift

In conclusion, reference can be made to one feature of the digital-absolute process. The machines are now, in a sense, 'calibrated' with a certain scale. The position of the absolute scale is fixed with respect to the machine, in the same way as normal precision scales with microscope readers are used on manually-operated jig borers. If the position of the datum point is to be shifted for any reason (i.e. for setting up), this is not as easy to do as with the incremental process. It is either necessary to displace the whole scale (e.g. by disconnecting a coupling between the drive shaft and the rotary measuring system) or to feed the appropriate signals into the comparator to add or subtract a selected figure to all measured values (coordinate transformation) (6, 10, 57). On the other hand, shifting the datum point by this numerical computation method has the advantage that factors such as corrections for the diameters or the widths of milling cutters can be fed into the machine directly in numerical form when it is being set up. In practice, this feature can be utilized in a milling machine equipped with a straight-line control system, for example, so that when a cutter is reground the new diameter is set up manually on a decade switch (for example on a control console as shown in the figure) and the same punched tape (containing the co-ordinate information for the old cutter) can still be used. The small computer will automatically make the necessary corrections. The costs of computers equipped with integrated circuits have been reduced by the adoption of standardisation and miniaturisation, and they are being ever more widely used for the numerical correction of tool dimensions (cf. Chapters 8 and 9).

3.2.5 Summary of discussion of absolute process

Summarising the description of the current state of the digital-absolute position measuring systems and the associated comparators, the picture that emerges is briefly as follows:

1. The present state-of-the-art is such that no difficulty is encountered in the production of coded linear measuring scales and scanners (as a development of the uncoded scales used for the incremental process)

without undue technical complication or cost; the cost of the scales is, however, such that at present the use of the process is restricted to the smaller sizes of plant.

2. At the present time only rotating position measuring systems using photoelectric scanners are a practical proposition. Among the points that should be borne in mind are:

   a)   the code chosen must be suitable for further numerical processing in the comparator, interpolator, etc.

   b)   reliable and unambiguous detection of the numbers is necessary; this is achieved most commonly by V-scanning.

3. In practice, only semiconductors, possibly in the form of integrated circuits, are used for comparators. It is necessary to distinguish between three types of comparator, mainly for economic reasons:

   a)   simple, numerical co-incidence monitors which do not specify the direction (for use with straightforward on-off circuits, or for stepped switching).

   b)   Numerical three-point controllers

   c)   Numerical proportional controllers.

   The number of digital elements and the cost of the circuits will increase from the simple co-incidence monitor up to the numerical proportional comparator.

4. Compared with the digital incremental measuring system there is a steep increase in the number of components required and the complexity of the circuits, so that the designer must give careful thought both to the size of the machine (Maximum displacements) and to the minimum requirements governing the kinematics of the slides. (Choice of the digital displacement element Δs, decisions regarding the need for on-off circuits with stepped switching, non-continuous and continuous control circuits, etc.). For bringing a heavy machine slide to rest it may prove necessary to provide 4 or 5 preliminary stopping signals (cf. Chapter 7), so that the cost differences between the S type and R type comparators disappear.

5. The expense of providing scanners and the comparator requirements does not, in general, make it possible to provide additional tracks for a redundant measuring system code.

6. If a zero shift is required this will prove expensive compared with the incremental process, but it may make the task of operating a numerically-controlled machine tool much easier. The trend towards the use of standardised integrated circuits will probably cause the disappearance of this initial cost disadvantage of the absolute system (10).

4   Analogue displacement measuring systems and comparators

Compared with digital systems, analogue displacement measuring systems and comparators possess two features of particular note:

1. There are no finite quanta - because of the principles on which these devices operate. All processes are continuous in nature i.e. variations take place by infinitely small steps. If measurable finite or widely scattered steps occur, they originate in the equipment; keeping these steps small is one of the major objectives in the design and production of the equipment.

2. In general, the quantity that is to be measured (in the examples under consideration a linear displacement or an angle) is converted into some other analogous physical quantity which is more convenient to measure and to data process than the original quantity (cf. Chapter 2).

4.1   Principles of measuring analogue displacements or angles

The basic relationships that arise from these two features are illustrated in Figure 54. Part A) of this figure shows the simplest case. The two annular resistances a (a_1 and a_2) and b in this example made from wound thin wire. They convert the angles ϑ into analogue resistance values. Obviously there is a minimum limit to the thickness of the wire that can be used, whilst the sliders c and d also have a finite width. As the sliders are moved, the resistances will change in small, but quite noticeable, steps. These small steps, which govern the resolution capacity of a measured-value converter, are an inevitable accompaniment of the technical methods employed, but must not be allowed to disguise the fact that the conversion process is continuous in nature. If a d.c. voltage is applied to the two annular resistances, the angular positions ϑ are converted into analogue d.c. voltages which can easily be further processed. If the two angles ϑ_1 and ϑ_2 are equal, there will be no potential

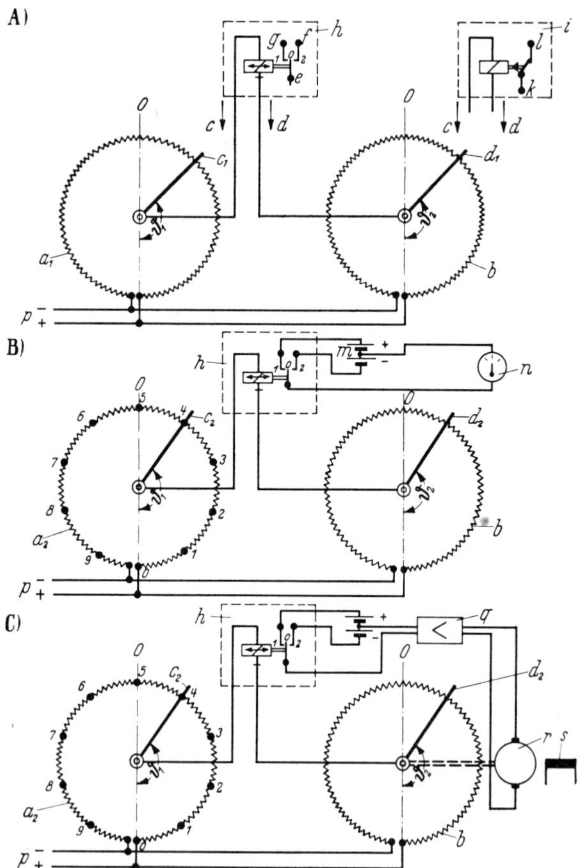

Fig. 54 **Circuit diagrams of an analogue displacement or angle measuring and data processing system**

A) Continuous input of desired value. B) Digital (numerical) desired value input for one decimal digit. C) Digital desired value input with automatic analogue follow-up control loop circuit.

a_1 Wire-wound annular resistance. a_2 Resistance with uniform steps and numbered contacts. b Wire-wound resistance. c_1 Sliding contact, manually rotatable, acts as desired value input. c_2 Sliding contact, manually rotatable in steps (with mechanical indexing device) as digital desired value input. d Rotatable sliding contacts tapping off actual value. e Armature of polarised relay h. f Working contact for position 2 of armature e. g Working contact for position 1 of armature e. h Polarised relay. i Non-polarised simple relay. k-l Normally-closed contacts on relay i. m Auxiliary voltage supply for indicating and evaluating measured values. n Moving-coil instrument to indicate measured values. o Geometrical zero axis. p d.c. voltage source for bridge circuit. q d.c. amplifier (e.g. amplifying rotary machine). r Armature of d.c. motor. s Constant excitation of d.c. motor. ϑ_1, ϑ_2 Angle of rotation.

difference between the tapping points c and d; if they are not equal, there will be either a positive or a negative potential difference. This can be detected in two different ways. If the sign of the difference is of no importance it will suffice to feed the voltage between sliders c and d to a simple unpolarised relay i. The contacts joining the signal wires k and l will then be closed as long as there is no voltage applied to the relay and hence no potential difference between c and d. If a voltage is applied to the relay, which occurs if angles ϑ_1 and ϑ_2 are not the same, the contacts joining k and l will be opened, no matter whether $\vartheta_1 > \vartheta_2$ or $\vartheta_1 < \vartheta_2$. This simple arrangement thus provides an S type comparator (Chapter 1). If the unpolarised relay i is replaced by a polarised relay h a simple means is provided for determining the sign of the difference $\vartheta_1 - \vartheta_2$; the relay armature e will occupy the zero position only when $\vartheta_1 = \vartheta_2$. This arrangement provides - in terms of control technology - a three-point controller, and is one of the simplest comparators R type (Chapter 1). The relays i and h will, of course, have a certain response sensitivity, i.e. the voltage between sliders c and d must exceed a certain threshold value before the relays operate. In addition to the resolution capacity of the measuring system, there is thus the response sensitivity of the analogue comparator which has a minimum value below which it is technically not possible to go.

In practice, with machine tools the angular displacements are usually of less importance than the linear displacements. If a rotary displacement measuring system of the type described above is used, a third source of error is added, namely the manufacturing inaccuracy, elastic deformation, and mechanical backlash in the members which convert the longitudinal movement into a rotation.

All these three sources of error:
> the resolution capacity of the measuring system;
> the response sensitivity of the comparator; and
> inaccuracies in the mechanical components forming the movement converters

act in the same direction. The desired continuity in the operation of the measuring and comparison processes will be noticeably interrupted unless a major effort is made to keep all the errors as small as possible. This simple example thus illustrates one of the features of all analogue processes. The basic simplicity of the method is limited by costs; as higher degrees of accuracy are required the costs for the necessary precision components rise rapidly (cf. Fig. 93).

And there is a further design requirement. The original problem was to determine the angle of rotation, or the movement of the associated slide in numerical terms. A simple solution for this is indicated in Fig. 54B). Let the total resistance a_2 of the desired value transducer be divided into ten equal parts, the tapping points being taken to the numbered contacts 0 ... 9, and let the slider c_2 be fitted with an indexing mechanism such that only the ten contact positions provided can be selected. The variation of resistance or voltage at the desired-value transducer can thus take place only in equal integral steps.

The moving-coil instrument n, which acts in conjunction with the relay h and the auxiliary voltage supply m will indicate only the values +, 0, and −. Value 0 is shown only if the angles ϑ_1 and ϑ_2 are equal; in all other cases the instrument shows either + or − depending on the sign of the difference. The magnitude of ϑ_2 is still adjusted manually. The process, which was initially continuous or nearly so, has now become quantified; the stepped resistance a_2 and the numbered stepping switch c_2 are probably the simplest form of a digital/analogue converter. In one form or another a device of this type is to be found in every analogue measuring system used with a numerically controlled machine tool.

In Figure 54C) the adjustment of ϑ_2 is no longer manual, but by an electro-mechanical actuator (in this case, for example, a remote-excited d.c. motor). It is obvious that a suitable amplifier q must be located between the precision relay h and the armature r of the d.c. motor. This completes the automatic analogue follow-up control loop with digital (numerical) desired-value input. Unfortunately simple components of this type are usually not suitable for use in engineering machine shop conditions. In particular, to ensure reliability no slide wires or contacts that are subject to wear should be employed. In practice, therefore, the analogue systems that are most widely employed make use of contact-less inductive methods for measuring angular and linear displacement.

4.2   The synchro as an inductive angle measuring device

Two basic forms of synchro are shown in Figure 55. Type A) has a non-symmetrical air gap and is used for accurate transmission of angular movements at low torques (e.g. remote indication of turntable or instrument settings). In Type B) the constant air gap between the stator and the rotor, together with the uniform arrangement of the windings in slots (similar to a slip-ring motor) results in symmetrical multi-phase voltages being transmitted between the stator and the rotor in proportion to the angle of rotation. Only Type B) is suitable for use in machine tool control systems, and the following remarks will be concerned with this design.

The uniform arrangement of stator and rotor windings in slots enables a wide range of combinations of windings to be used. Some of the more common arrangements are shown in Figure 56. The resolution capacity for the conversion of mechanical rotation into a voltage depends largely on the construction of the iron core, the fineness of the slots, and the nature of the winding. Currently, precision synchros are made with a resolution capacity of 3 to 10 minutes of arc. For various reasons the frequency of the alternating

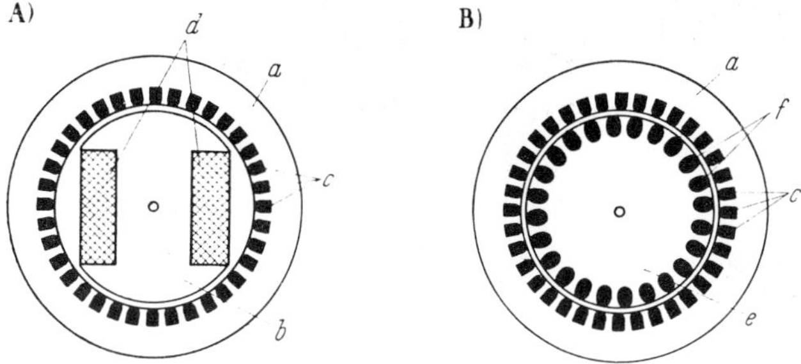

Fig. 55 Section through two types of synchro

A) Type with non-uniform air gap for transmitting low torques. B) Type with uniform air gap for measuring purposes.

a Stator. *b* Double-T armature. *c* Stator slots to accommodate windings. *d* Armature windings. *e* Slotted rotor. *f* Rotor slots to accommodate windings.

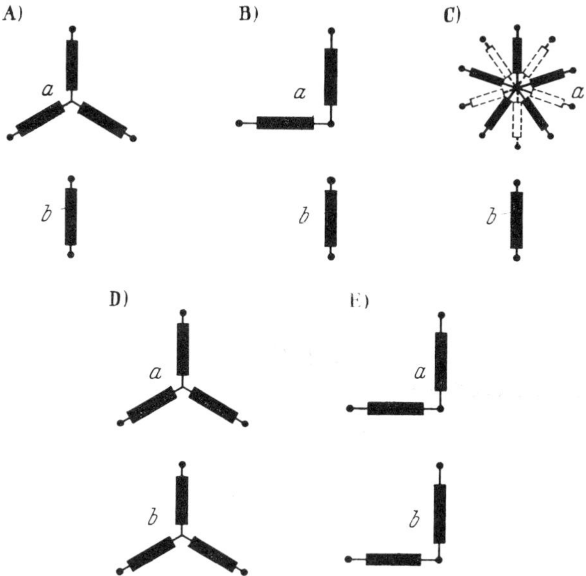

Fig. 56 Common arrangements of windings in synchros.

A) Single-phase – three-phase combination. B) Single-phase – two-phase combination. C) Single-phase – five-(ten-)phase combination (decimally divided connections). D) Three-phase – three-phase combination (differential synchro). E) Two-phase – two-phase combination (differential synchro).

a Stator windings. *b* Rotor windings.

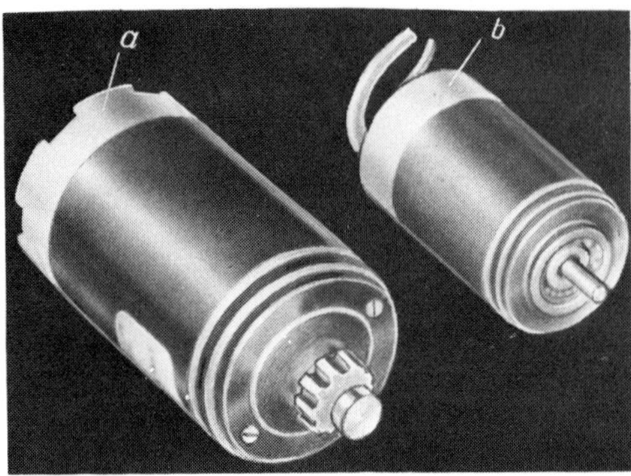

Fig. 57
Synchros for various frequencies

a Synchro for 50 Hz (larger unit). *b* Synchro for 400 Hz (smaller unit).

current that is supplied to the synchro is also important. A 50 Hz synchro is compared with a 400 Hz unit in Figure 57. Using a higher frequency enables the dimensions of the unit to be appreciably reduced; in particular, the rotor can be made with a very low moment of inertia. This can be of advantage where the accelerations and decelerations are high and the synchro is used in conjunction with sensitive mechanical transmission components (e.g. high-precision rack-and-pinion units). On the other hand, the provision of a separate 400 Hz supply on a machine tool for measuring purposes results in additional cost.

The method of operation of a synchro is shown diagrammatically in Figure 58 which shows a single-phase wound stator combined with a single-phase wound rotor. An alternating voltage

$$e_a = E_1 \cdot \sin \omega t$$

is applied to the stator winding *a*

where $\omega = 2\pi f$ is the angular frequency corresponding to the frequency f of the measuring circuit. This alternating voltage sets up a magnetic flux in the synchro, which induces a voltage e_b of the same frequency in the rotor winding. This voltage, apart from its sinusoidal variation with time (sin ωt) and the ratio of the numbers of turns in the windings, will, however, depend on the effective linkage of the rotor windings with the stator flux. If, for example, the rotor winding is turned slowly through an angle ϑ, the amplitude of the secondary voltage will vary in accordance with the law:

$$e_b = (E_2 \sin \omega t) \cos \vartheta$$

where $E_2 = C_1 \cdot E_1$ is determined solely by the constant ratio between the turns of the primary and secondary windings and by E_1 (ignoring voltage drops).

The fluctuation with time (sin ωt) is superimposed on the angular displacement cos ϑ. At the nodes *c* (Figure 58C) there is a phase shift of

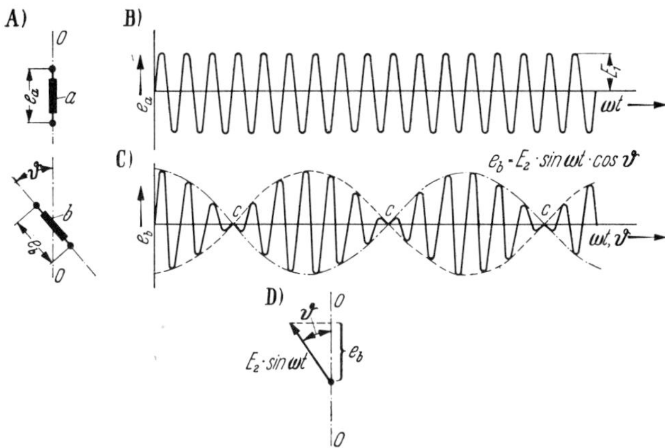

Fig. 58 Method of operation of a synchro used for measuring (measurement of angles)

A) Single-phase winding combination (solely to illustrate the principle). B) Primary voltage waveform. C) Variation of the secondary voltage with time and angle of rotation. D) Vectorial representation of the secondary voltage.

a Primary winding. *b* Secondary winding. *c* Node positions for secondary voltage with phase shift. e_a instantaneous values of primary voltage. e_b Instantaneous values of secondary voltage. E_1 Maximum value of primary voltage. E_2 Maximum value of secondary voltage (E_2 is determined from E_1 by means of the numbers of turns on the windings *a* and *b* just as for a transformer). *o* Spatial zero axis. *t* Time in seconds. ω Angular frequency (= $2\pi f$). ϑ Angle of rotation of winding *a* relative to winding *b*

$180°$. If the variation with time of the voltage ($e_a = E_1 \sin \omega t$) is assumed to be uniform, the variation with angular position can be shown in simplified form vectorially as illustrated in Fig. 58D). The functions of the stator and the rotor are, of course, interchangeable.

The relationships become clearer if the r.m.s. value $E_1/\sqrt{2} = E_a^*$ is substituted for the voltage e_a, which fluctuates with time, and similarly the r.m.s. value $E_2/\sqrt{2} = E_b^*$ for the secondary voltage, which also fluctuates with time. The maximum voltage $E_b^*{}_{max}$ across the secondary will then occur when the angle $\vartheta = 0$, and this can be treated vectorially.

Figure 59 illustrates what occurs frequently in practice where there is a single-phase primary winding *a* and a two-phase secondary winding *b* - *c*. Since the two identical secondary windings *b* and *c* are displaced relative to each other by $90°$ electrical ($\hat{=} \pi/2$) the sine and cosine components of the voltage $E_2^* = E_b^*{}_{max} = E_c^*{}_{max}$ can be tapped-off *b* or *c* respectively. With an arrangement of this type the angle is resolved into its sine and cosine components, which explains the use of the term 'resolver' for this type of device.

Conversely a symmetrical two-phase system can be applied to the windings *b* and *c*, the magnetic fields of which combine geometrically to form a

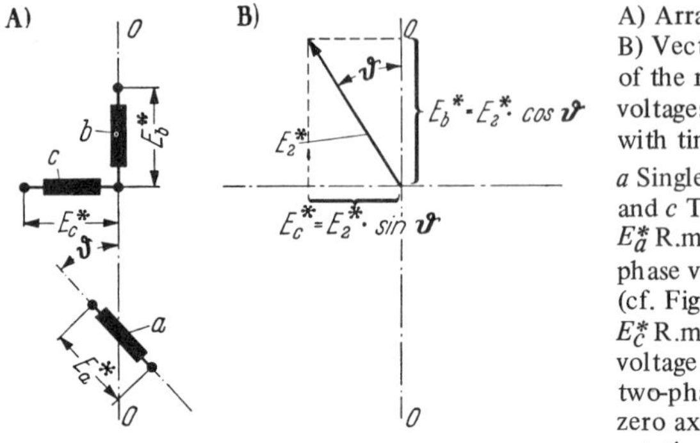

A) Arrangement of windings.
B) Vectorial representation of the r.m.s. values of the voltages which fluctuate with time.

a Single-phase winding. *b* and *c* Two-phase windings. E_a^* R.m.s. value of the single-phase voltage ($E_a^* = E_1 / \sqrt{2}$) (cf. Figure 58). E_b^*, E_c^* R.m.s. values of the voltage components of the two-phase system. *o* Spatial zero axis. ϑ Angle of rotation.

Fig. 59 Method of operation of a single-phase — two-phase winding combination in a synchro (used as a resolver).

resultant field which will in turn produce a single-phase voltage in winding *a*. Considerable use is made of this arrangement in combination with digital/analogue converters; the details will be discussed when converters of this type are described (cf. Section 4.6).

The simplest arrangement of the resolution and combination of components is shown in Figure 60, where two resolvers are coupled together. If the voltage E_a^* is applied to the single-phase winding *a*, the sine and cosine components in the windings *b* and *c* are transmitted to the windings *d* and *f* of the second system. A magnetic field will therefore be established in the stator of the latter and occupying the same spatial position as the field in the primary system (with respect to the same zero position of the axes). If the

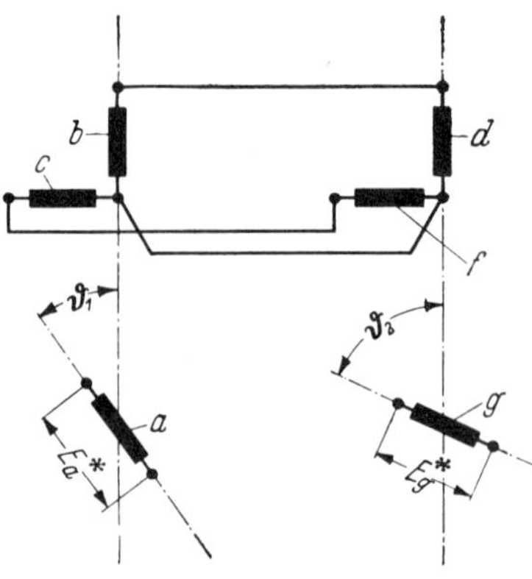

Fig. 60
Coupling of two synchros for synchronous operation (transmitter and receiver system for two-phase arrangement.

a Single-phase winding of transmitter system. *b* and *c* Two-phase windings of transmitter system. *d* and *f* Two-phase windings of receiver system. E_a^* R.m.s. value of single-phase receiver voltage. ϑ_1, ϑ_2 Angles of rotation.

rotor with the winding g of the second system is turned through the angle ϑ_2, a voltage E_g^* such that

$$E_g^* = E_2^* \cos (\vartheta_1 - \vartheta_2).$$

will be generated in it.

The voltage E_g^* thus reaches its maximum value E_2^* when $\vartheta_1 = \vartheta_2$, and it is zero when $(\vartheta_1 - \vartheta_2) = \pm \pi/2$. As will be shown later, such a system can be used very conveniently to build up a follow-on control circuit (similar to Figure 54C) if a phase-sensitive comparator (Figure 82) is included. It is the suitability for use in the synchronisation of rotary movements that accounts for the use of the term 'synchro' to describe these units. It should be borne in mind that the angle function is ambiguous. In practice, only the range $(\vartheta_1 - \vartheta_2) = 120$ to $150°$ can be used for control (i.e. for the provision of an unambiguous feedback signal) (cf. also Figure 1B).

The same effect as that just described can also be obtained by using a three-phase stator winding (corresponding to Figure 56A) in place of a two-phase winding. The more uniform distribution of the windings offers certain advantages. Another advantage is that it is easy to operate several systems in a three-phase rotating field mode; one such possibility is shown in Figure 61. Here three synchros are connected in parallel to a three-phase supply on the primary side. If the windings are suitably designed, a rotating

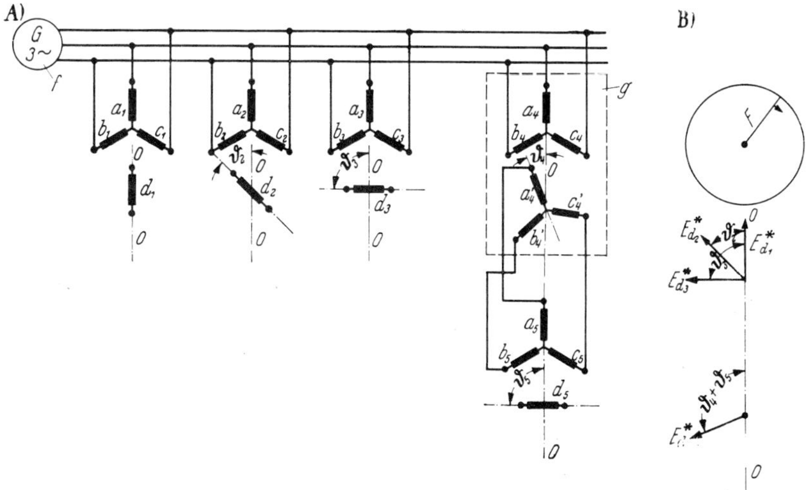

Fig. 61　Three-phase supply of several synchros (rotating field operation)

　　$a, b\ c$ Three-phase windings of synchros. d Single-phase windings of synchros.
E_d^* R.m.s. values of constant a.c. voltages induced by rotating field F_d.
F_d Rotating field in the synchros. g Three-phase differential synchro. o Spatial
zero axis. ϑ Angle of rotation.

field F having a very low harmonic content will be set up in all the synchros. This causes alternating voltages to be induced in the single-phase rotor windings, equal in amplitude but differing in phase according to the angular position ϑ relative to the common zero position 0 - 0. In theory, any number of synchros can be connected in this way, all of which will be synchronised by means of the common three-phase supply. In other words, the movements of any number of machine slides can easily be referred to a common spatial axis (zero point) and controlled simultaneously. If differential synchros (as illustrated diagrammatically in Figure 56D) are located in advance of the measuring systems in the various axes, another major practical advantage is obtained in that the second three-phase winding, which is arranged so that it can be turned within the continuously rotating field, provides a datum-point shift facility. A differential synchro of this type is shown at g in Figure 61, and is followed by a normal synchro with the original mechanical angle of rotation ϑ_5. If the differential synchro is rotated by ϑ_4 from its zero position in a direction opposite to that in which the field rotates, a voltage that is shifted in phase by the angle $(\vartheta_4 + \vartheta_5)$ will then appear at the output of winding d_5. An arrangement of this type makes it a simple matter to carry out tool adjustments, datum-point corrections, etc.

It will be clear that a three-phase measuring and control system of this type, which establishes the reference axis for all slide movements, has to meet stringent requirements of symmetry and freedom from harmonics.

Simple, inexpensive, and reliable though synchros of all types may be, they nevertheless suffer from two basic disadvantages:

1. Their analogue representation of the angle of rotation is expressed unambiguously only over a relatively small range.

2. They are primarily capable of dealing with rotational movements only since they are rotary measuring systems. In most instances, however, it is ncessary to deal with the rectilinear motions of slides in machine tool control systems.

The resolution capacity of a modern precision synchro is about 3 to 10 minutes of arc, instruments having a value of 3 minutes of arc being rather costly. If the rectilinear movement of a slide is to be represented in analogue form unambiguously and to an accuracy of 0.01 mm, in theory the maximum displacement that can be covered by one complete revolution of the synchro is about

$$\frac{360 \cdot 60}{3} \cdot 0.01 \approx 72 \text{ mm}$$

In practice, the displacement that can be dealt with is less than this, for various reasons:

a) It is not advisable to work down to the lowest limit of the electrical resolution capacity, because the tolerances are not close enough to ensure that this can be maintained at all times [1]. It is better to work on a value of at least 10 minutes of arc.

b) Errors are introduced every time a rectilinear movement is converted into a rotary movement. As mentioned in Chapter 2, the only conversion systems that are feasible are screwed spindles (e.g. the leadscrew itself) and rack and pinion units. Both of these have the effect of increasing, to some extent, the minimum unambiguous step that can be dealt with.

c) It is very difficult to cover, by simple means, the full revolution of a synchro corresponding to an electrical angle of $360°$ without obtaining ambiguous readings. The maximum angle that can be used in practice amounts to less than $180°$ and, in many cases, angles of 110 to $150°$ are adopted to ensure that reliable results are obtained.

The result of these three factors is that a single synchro is unable to cope unambiguously and reliably with a rectilinear displacement of more than 4 to 6 mm (10 mm at the most). The displacements encountered in practice are, however, much greater than this. There are three basic methods of overcoming this difficulty:

1. The individual synchros and their associated comparators can be used solely for synchronising the rotations, i.e. the permissible deviation in designing a follow-up control loop must be restricted to the range of not more than $150°$ as mentioned above. If the measured deviation exceeds this amount the 'electrical shaft' will 'break' and there will no longer be an unambiguous relationship between the two speeds or angular displacements. In many numerically controlled machine tools the command (desired-angle input) is produced by an electronic computer (cf. Chapters 5 and 9).

2. Each time the demodulated sinusoidal oscillation (cos component in Figure 58C) passes through a node this can be transformed into a counting pulse which is applied to a counter. If the circuit is suitably arranged it is even possible to use one of the forward and backward counters described in Chapter 3. This produces a mixed digital and analogue system in which the comparatively large digital step $\pi \triangleq 180°$ electrical is finely divided into analogue elements (cf. Chapter 5).

3. In most instances, several synchros are coupled together by precision gearing to enable large angles of rotation or large displacements to be

---

[1] If very close tolerances have to be maintained and the synchro is required to have a resolution capacity of less than 5 minutes of arc it will become very costly.

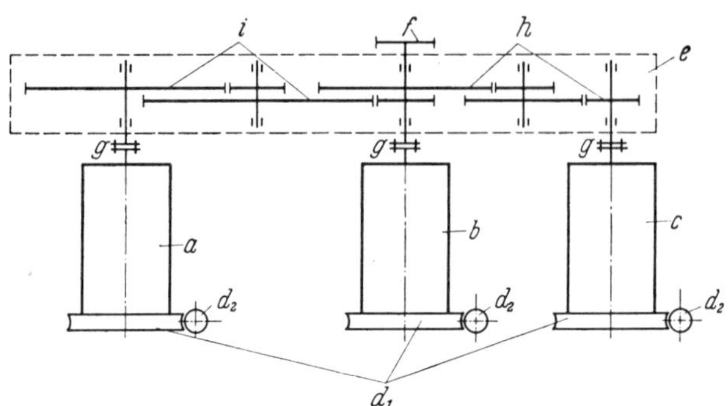

Fig. 62 **Diagram illustrating the mechanical coupling of three high-precision synchros for analogue representation of a displacement of 1000 mm with a resolution capacity of 0.01 mm.**

a Synchro for coarse range. *B* Synchro for medium range. *c* Synchro for fine range. d_1 and d_2 Worms and wormwheels (mounted on synchro casings) for adjusting zero readings, to simplify installation. *e* Precision reduction gear with minimum play. *f* Pinion (engages in a rack of pitch $5/\pi$. *g* Couplings, free from play. *h* 10 : 1 reduction gear.

Note: Calibration *d* is provided solely to simplify installation. Zero point displacements for the machine as a whole are nowadays almost without exception performed digitally (cf. Section 4) using small computers (cf Chapt. 8)

represented. The basic principles of a coupling system of this type, using three synchros for the analogue/digital measurement of displacement with a resolution of 0.01 mm over a total length of 1000 mm, are shown in Figure 62. Figure 63 shows a practical example fitted to a milling machine. To position the moving pad to a given

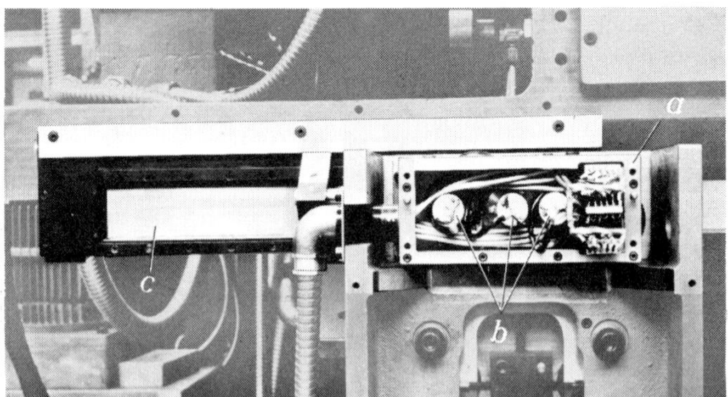

Fig. 63 **Example of an analogue measuring system with three synchros. (Photo courtesy** *of Fritz Werner GmbH* **).**

a Gearbox (Part *e* in Figure 62). *b* Three synchros. *c* Flexible cover (steel tape) to protect rack from dirt.

desired value it is necessary to use an electrical switching mechanism - which switches from one measuring system to the other as the actual and desired values gradually become closer until in the final stages only the fine control system is working. It is not appropriate to consider the details of the circuits here, but the process is represented diagrammatically in Figure 64 (58).

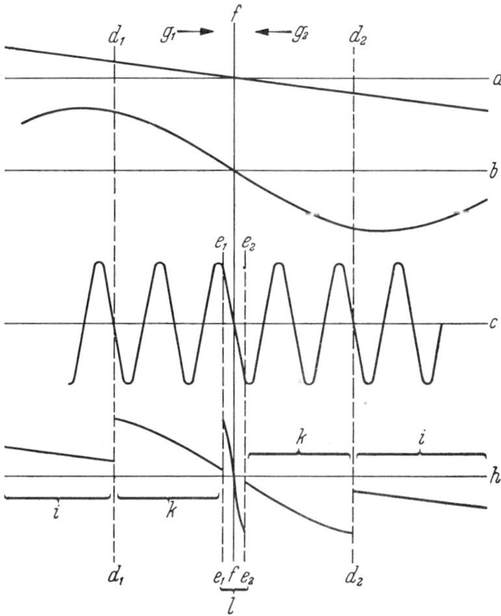

Fig. 64 Diagram illustrating the principles of measuring range switching when using synchros and an Inductosyn (58)

a Coarse system. *b* Medium system. *c* Fine system (or could be finely divided linear Inductosyn). d_1 and d_2 Voltage-dependent switching points for changeover from coarse to medium system. e_1 and e_2 Voltage-dependent switching points for changeover from medium to fine system. *f* Desired value. g_1 and g_2 Directions of approach to the desired value. *h* Voltage waveforms in the individual systems. *i* Range covered by coarse system. *k* Range covered by medium system. *l* Range covered by fine system.

The datum-point corrections (cf. Figure 61) that have been described for individual synchros using differential synchros can, of course, also be adopted where several synchros are coupled. In this instance, however, several differential synchros will have to be coupled by precision gearing, the number depending on the range over which the datum-point correction is to be performed. Because of the expense this will usually be limited to coupling two differential synchros, and the resultant restriction of the possible range of adjustment is then accepted unless it is decided to perform the zero-shift operation entirely in the digital range (cf. Chapter 9).

The production and installation of precision gearing and racks and pinions which are free from backlash can be very expensive. Systems of this type will, therefore, become the more attractive for incorporation in a machine tool the lesser the need for high overall accuracy of the machine, and the smaller the machine.

Many attempts have been made to develop analogue-linear displacement measuring systems, to overcome the difficulties in the conversion of rectilinear motions to rotations. Of the many proposals that have been made, only two will be discussed in further detail at this stage (55, 60, 61). Mention is made of a number of other methods in Chapter 5.

4.3   The Linear Inductosyn[2]

In its basic form the linear Inductosyn consists of an iron-less resolver which has been unrolled into a plane. It has two parts, the scale and the slider, which move relative to each other. Both are made of non-magnetic material (glass, ceramics, non-magnetic steel, et al); the current-carrying windings, which produce the magnetic fields, are made of copper in the form of a printed circuit (36). the coefficient of thermal expansion of the carrier material is matched to that of the machine tool. The basic principle is shown diagrammatically in Figure 65. The printed rectangular windings a of the scale have the poles spaced at intervals of τ_p and are made to the following dimensions to suit the measuring system that is to be used:

$$\tau_p = 0.1 \text{ inch for inch system}$$
$$\tau_p = 2 \text{ mm for metric system.}$$

The scale is also made in two standard lengths:

    10 inches for inch system
    200 mm for metric system.

Longer scales are produced by placing a number of standard lengths end-to-end. Suitable arrangement of the slider windings results in an average reading being obtained, so that any errors are compensated.
The scale represents the single-phase wound portion of a four-pole resolver (cf. Figure 59), although there is only one turn per pole. The slider carries two groups of windings c_1 and c_2 which are displaced relative to each other by

$$g = n \,.\, 2\,\tau_p \pm \frac{\tau_p}{2}$$

---

[2] Registered trade name of a component marketed by Inductosyn Corporation, formerly Farrand Control Inc., New York. (59)

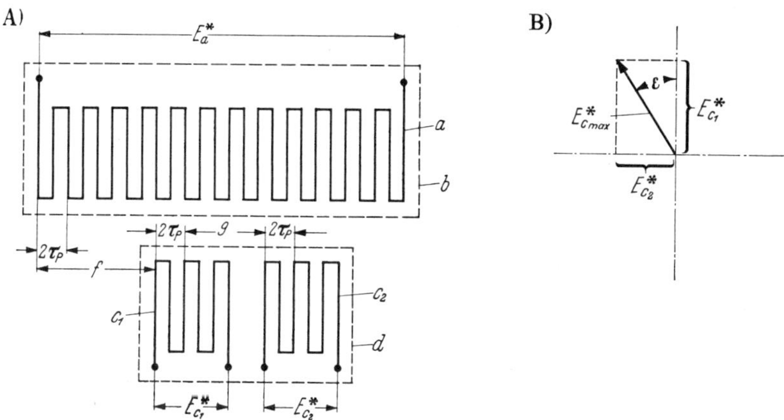

Fig. 65 Diagram illustrating the principle of the linear Inductosyn with finely-spaced poles.

A) Layout of printed circuits containing windings. B) Vectorial representation of the voltages.

a Windings for single-phase system (fixed scale). b Supporting plate for scale. c_1 and c_2 Windings for Two-phase system. E_a^* R.m.s. value of single-phase voltage. $E_{c_1}^*$ and $E_{c_2}^*$ Voltages of two-phase system displaced by 90° electrical $\triangleq \pi/2$ relative to each other. E_{max}^* r.m.s. value of a slider voltage. f Longitudinal displacement of slide relative to scale zero. g Longitudinal separation of the two-phase windings within the slider. ϵ Effective electrical angle. τp Pole spacing.

where $\tau_p/2 \triangleq \pi/2 \triangleq 90°$ electrical and n is an integer. When assembled for operation the slider is located above the scale with an air gap of $\delta \approx 0.05$ to 0.15 mm between them. The air gap δ itself is not critical, but it should be maintained constant over the full length of the scale (which may extend to several standard scale lengths). To meet this requirement the individual scale lengths must be carefully aligned. If an alternating voltage having an r.m.s. value of E_a^* is applied to the scale windings a, magnetic fields which fluctuate with time will be set up between the straight elements of the windings, and these will also link with the windings on the slider. Since the base of the scale and the slider are made of non-magnetic materials the angular frequency $\omega = 2\pi f$ must be fairly high to obtain measurable voltages (E_a^*, E_{c1}^*, and E_{c2}^*) with the small numbers of turns that are available. In practice frequencies of between 1 and 20 kHz are used, with about 2 kHz being a common value. The reason for this will be explained later. The construction of the slider is such that induced secondary voltages E_{c1}^* and E_{c2}^* are displaced in phase by $\pi/2 \triangleq 90°$ electrical, as with the rotating resolver. If the position of the slider $f = n\,2\tau_p + h$ (where n is any integer with respect to the spatial zero point and $0 < h < 2\tau_p$) is changed because of a longitudinal displacement, a new electrical cycle begins as every second pole position is passed. The effective electrical displacement angle is then

$$\epsilon = \frac{h}{2\tau_p} \cdot 360°.$$

The same vector diagram as is shown in Figure 59 for the rotary resolver can, therefore, be drawn for the periodic longitudinal displacement component $h \triangleq \epsilon$, with the exception that the angle of rotation δ is replaced by the equivalent angle ϵ. Depending on the pole spacing employed, one electrical half-cycle (180° electrical) will be passed through for every 0.1 inch or 2 mm of linear displacement.

Exceedingly simple though the principle of this device may be, considerable difficulties were encountered until a few years ago in actually making the linear Inductosyn as a reliable measuring device. The dimensions of the individual winding elements and the pole spacing must be maintained with great accuracy; suitable design of the windings c_1 and c_2 on the slider enables any unavoidable inaccuracies in the spacing to be largely compensated. Furthermore, the slider windings are arranged in such a manner that an approximately sinusoidal field distribution is achieved. An example of a scale and slider is shown in Figure 66. As with a rotary resolver, a two-phase supply can also be applied to the slider of the linear Inductosyn; the voltage E_a^* on the scale windings a will then be the geometrical sum (ignoring scatter and ohmic losses). This arrangement is used in particular with the digital/analogue converters that are described later.

It is first necessary, however, to draw attention to a further development of the linear Inductosyn. Owing to the small pole spacing τ_p, the linear Inductosyn must usually be combined with a medium and coarse measuring system. These two coarser systems can, as indicated in Figure 62, use rotary resolvers, which may be driven off the leadscrew; the medium measuring system need only be accurate enough to detect a longitudinal displacement of 0.7 to 0.8 mm, and to be capable of switching over reliably to the linear Inductosyn within this range (cf. Figure 64).

The combination of a measuring system containing iron (rotary resolver) and one that does not contain iron (linear Inductosyn) on one machine makes it necessary to arrive at a compromise so far as the choice of frequency for the measuring voltage is concerned. The iron-less linear Inductosyn will work best with a high frequency (10 kHz to 30 kHz). This is much too high for the rotary resolver, which contains iron, and which is often designed for a frequency of 400 Hz. It is undesirable to use two frequencies for the measuring voltages, so a compromise of between 1.2 kHz and 2 kHz is frequently adopted in practice. Figure 67 illustrates a linear Inductosyn fitted to a machine.

In view of these diffficulties steps have recently been taken to extend the effective range of the linear Inductosyn. The principles adopted, which are illustrated in Figure 68, are relatively simple. The scale windings a are no

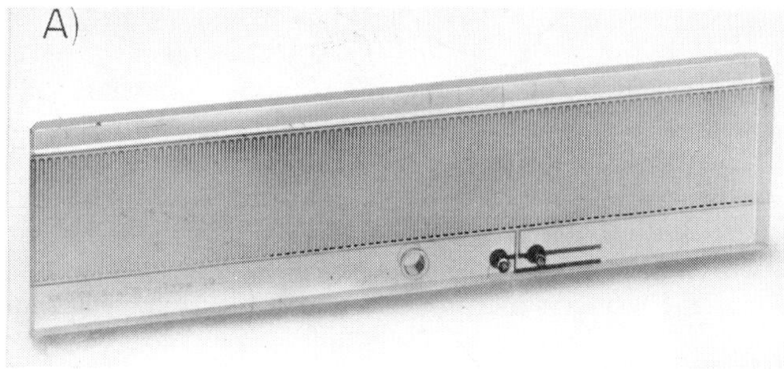

Fig. 66
Examples of a linear Induc-
tosyn with closely spaced
poles

A) Scale B) Slider.

Fig. 67 Example of a linear Inductosyn fitted to a machine (Photo courtesy of *Siemens*)
a Scale (fitted to machine bed). *b* Slider (fitted to machine slide). *c* Flexible
cover (steel tape).
Note: Front cover plate has been removed to show interior.

longer arranged perpendicular to the direction of movement but instead are inclined at an angle to suit the required spacing. The slider windings c_1 and c_2 are arranged precisely in line with the direction of movement, and intersect the scale windings at a definite angle. If an alternating voltage E_a^* is applied to the scale windings a, alternating magnetic fields are set up between the turns of the winding. As indicated in the diagram, unlike poles are arranged adjacent to each other. Considering first position g of the slider windings c_1 and c_2, it will be seen that windings c_1 are cut by equally balanced opposing fields, so that the resultant secondary voltage E_{c1}^* is zero. The windings c_2, on the other hand, lie entirely within the influence of exciting fields of the same polarity, so that E_{c2}^* has a maximum value. If the slider is displaced by a distance d to position h, E_{c1}^* becomes a maximum and E_{c2}^* will become zero.

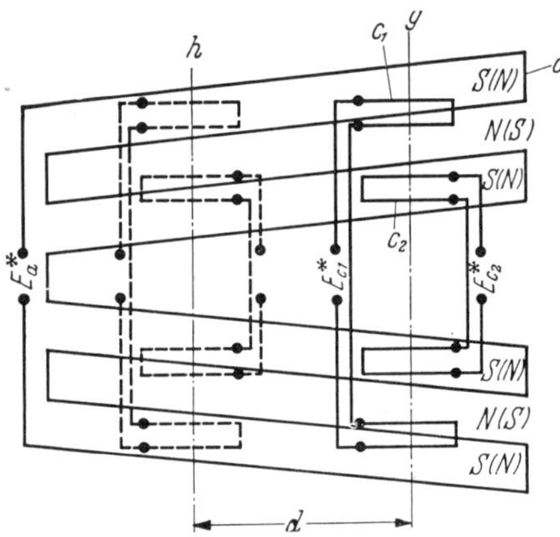

Fig. 68

Principle of the linear Inductosyn with widely spaced poles

a Scale windings (stationary). c_1 and c_2 Slider windings (can be moved longitudinally). d Example of longitudinal displacement of the slider. E_a^* R.m.s. value of the single-phase voltage. E_{c1}^* and E_{c2}^* Voltages of the two-phase system, mutually displaced by $\pi/2$. *N* North magnetic pole (viewed from above) during positive half-cycle of the voltage E_a^*. (N) North magnetic pole during negative half-cycle of E_a^*. S and (S) Corresponding south poles.

Three points are apparent from a study of this theoretical diagram:

a) The voltages E_{c1}^* and E_{c2}^* are clearly mutually displaced by 90° electrical. If care is taken to ensure that the slider windings are designed in such a manner (e.g. several parallel wires) that when the slider is moved longitudinally from position g to position h the voltage variations are at least approximately sinusoidal, the action will again be that of a resolver, but with a large pole spacing.

b) The pole spacing depends on the angle of inclination of the scale windings to the direction of displacement.

c) The magnetic coupling between the scale windings and the slider windings is weaker than in the simple Inductosyn described above. As a result, operating frequencies that are higher than previously must be

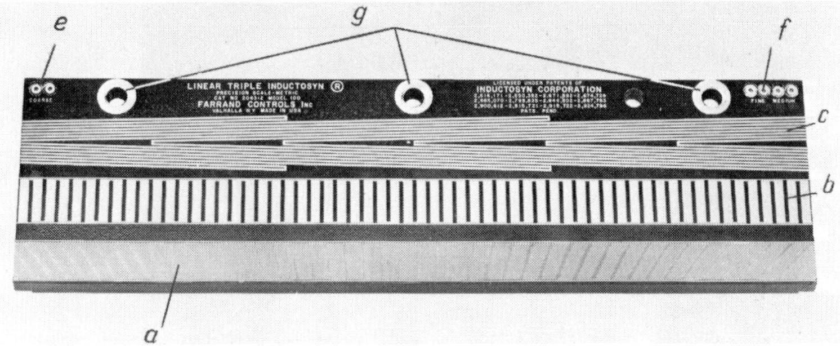

Fig. 69 **Scale for a linear Inductosyn with closely and widely spaced poles. (Triple phase unit. Photo courtesy of** *Farrand/Siemens*)

a Windings with close pole spacing (corresponding to Figure 66A). *b* Windings with medium pole spacing. *c* Windings with wide pole spacing. *e* Terminal sockets for interconnecting several scale panels and for tapping-off the single-phase voltage of the coarse measuring system. *f* Terminal sockets for the medium and fine measuring systems. *g* Mounting holes.

adopted ($\geqslant 10$ kHz) if the resulting secondary voltages are to be capable of being reliably evaluated. Since it is now no longer necessary to use also rotary resolvers, a single higher-frequency supply can be used in the machine.

Figure 69 shows a scale with three different pole spacings. The fine and medium elements employ the same principle, but for the poles that are widely separated the inclination is chosen so that the poles are spaced 100 mm apart. There are no changes in the coarse, medium, and fine switching process when a triple scale of this type is employed. Also, when the larger pole spacings are employed two-phase voltages can be fed to the slider (Figure 70) in the same manner as with the rotary resolver (cf. Figure 59) and the simple linear Inductosyn (Figure 65), and the geometric sum of the voltages is then tapped-off the scale. This proves of advantage when working with digital/analogue converters.

The above remarks relate to the linear Inductosyn for translatory movements. The same principle can, however, also be employed for rotary movements for which the rotary Inductosyn is available (Figure 71). This is older than the linear Inductosyn, since it has long been used in the USA in the aeronautical field (e.g. for controlling radar sets). In this context it was developed to a high stage of precision, designed so that it could be produced economically, and then modified into the linear Inductosyn, which is of less interest to the aeronautical industry. It was with the introduction of numerically controlled machine tools between 1950 and 1956 that a worldwide interest developed in the linear form described (59). The licensors, the *Inductosyn Corporation* state that with the normal type of rotary

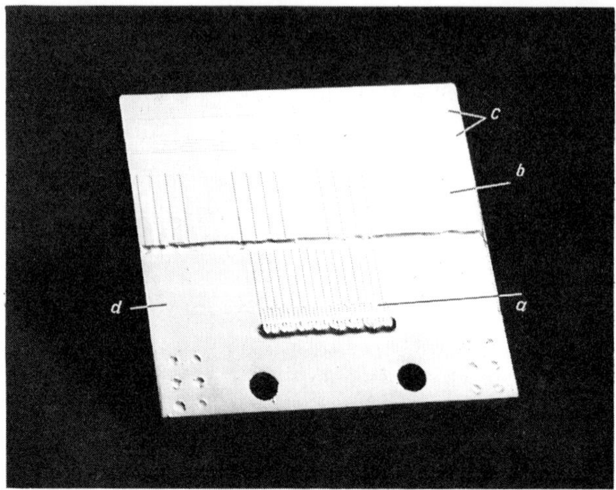

Fig. 70 Slider for triple Inductosyn scale shown in Figure 69; the windings are shielded from electrostatic fields by metal foil, and so cannot be clearly seen. (Photo courtesy of *Farrand/Siemens* **).**

a Two-phase windings with close pole spacing. *b* Two-phase windings with medium pole spacing. *c* Two-phase windings with wide pole spacing. *d* Screening foil.

**Fig. 71
Annular scale of a rotary
Inductosyn (360 poles);
resolution capacity approx.
3 seconds of arc. (Photo
courtesy of** *Olivetti* **).**

Inductosyn it is possible to achieve a resolution capacity of 1 second of arc, and that this can even be improved on in special versions. The normal resolution capacity of a linear Inductosyn is said to be 1.3 μm whilst, in special cases, it is possible to achieve smaller values than this, although the costs will then rise.

With the increasing use that is being made of numerically-controlled turntables and machining centres with pivot axes (cf. Chapter 7), precision methods of measuring rotary movement are assuming more and more importance in the machine tool field (cf. Figure 91, and also the alternative solution in Figure 74). Whether it is always necessary to use the high accuracy of the rotary Inductosyn (or the rotary Accupin), however, or whether the less expensive synchro would prove adequate, is an economic question that will have to be worked out for each individual application.

4.4   The Accupin[3]

Of the many alternative solutions for analogue displacement measuring systems that were proposed in the early days of numerical control systems (55, 60, 61), only comparatively few have actually been put to practical use and survived to the present day. Of these, Accupin will be described, primarily for instructive purposes.

With the Inductosyn, the accuracy and reliability is to a greater extent dependent on the mastery of the more photo-chemically oriented technology of printed circuits (36), whereas in the manufacture of the Accupin, very high demands are made on the mechanical technology (62, 63). The basic principle is very simple, and is illustrated in Figure 72. Coils d with the laminated iron cores a are excited by a 200 - 1500 Hz alternating current. A precision scale (Figure 73A) consisting of magnetic nickel-steel pins which are normally spaced at 0.1 in (2 mm) moves between the cores a and the appropriate return paths for the magnetic flux. The pins c have a diameter which is slightly smaller than 0.1 in (2 mm) to ensure that a uniform and stable space is maintained between the cylinders.

The pin diameters are all equal; they must be made to a tolerance of 0.00003 in (0.75 μm). These magnetic pins are inserted with a high degree of accuracy either in the grooves of a straight metal bar (Figure 73A) or in a cylindrical drum (Figure 74), made of non-magnetic steel. The straight metal bars are supplied in lengths of 6 and 10 in (152 mm and 254 mm) and can be combined to produce any required length. The cylindrical drums for the rotary Accupin measuring system (Figure 74) have a circumference of 36 in (914 mm) on which 360 pins are arranged at intervals of 0.1 in (2.5 mm), so that the geometrical relationships in the magnetically effective air gap are the same for the linear Accupin and the rotary Accupin.

The extreme positions for the coil inductance due to the variation in the air gap are drawn in Figures 72A) and B). The geometry of the coil cores is chosen with reference to the pin diameters c and the pin spacing e such that the variation in inductance that results from displacement of the scale is

---

[3]   Registered trade name of a device introduced by the General Electric Company/USA.

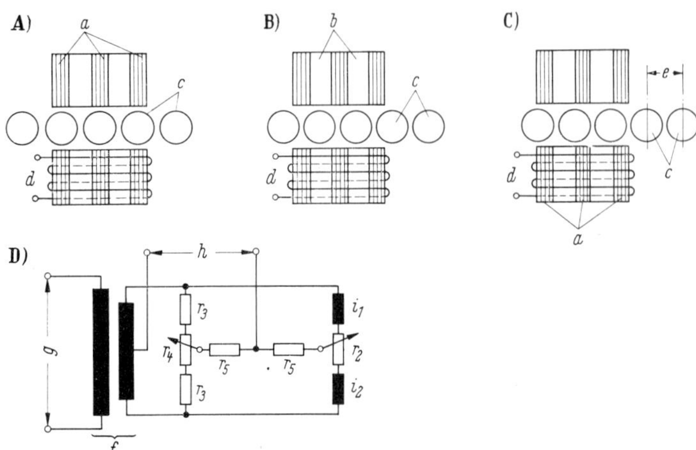

Fig. 72 Schematic diagrams showing the principle of operation of the linear Accupin.

A) Position of the measuring head when inductance is a maximum. B) Position of the measuring head when inductance is a minimum. C) Position of measuring head when inductance has a mean value. D) Schematic diagram of the discriminator circuit and means of adjustment.

a Three-part laminated solenoid cores. b Non-magnetic intermediate layers. c Ferro-magnetic pins (embedded in non-magnetic steel). d Induction coils. e Pin centres spaced at 0.1 in (or 2 mm). f Transformer with centre tapping. g A.C. supply (200 to 1500 Hz). h Signal voltage (output voltage) of inductance bridge. i_1 First inductance (e.g. in maximum position A). i_2 Second inductance (if i_1 is for maximum position A, i_2 must be for minimum position B). r_2, r_4 Infinitely-variable ohmic balancing resistances. r_3, r_5 Bridge elements with constant ohmic resistance.

Note: A second similar system is employed in which the two coils and solenoid cores are displaced spatially by 90 or 270° electrical with respect to those of the first system, whilst the supply is 90° electrical out-of-phase with g. The two supply voltages can be drawn, for example, from a resolver as shown in Figure 56B; to make adjustment easier (elimination of mechanical faults) it is even possible to incorporate a differential resolver, as shown in Figure 56E, in the circuit. At the same time the bridge voltages h can be added vectorially to provide a single measuring voltage. The complete measuring head then contains at least four inductances spaced at 0, 90, 180 and 270° electrical (cf. Figure 73B).

virtually sinusoidal. To overcome geometrical errors that arise during manufacture, not only are the coil cores made in three parts (Figure 72A) and 73B)), but pairs of cores located some distance apart are connected up in a bridge circuit as shown in Figure 72D) in such a manner that electrical calibration by potentiometers is possible. If the scale is displaced relative to the measuring head not only will a carrier-frequency alternating current corresponding to the supply voltage g appear at point h in Figure 72D), but this current will be modulated by an amount dependent on the scale

A)

B)

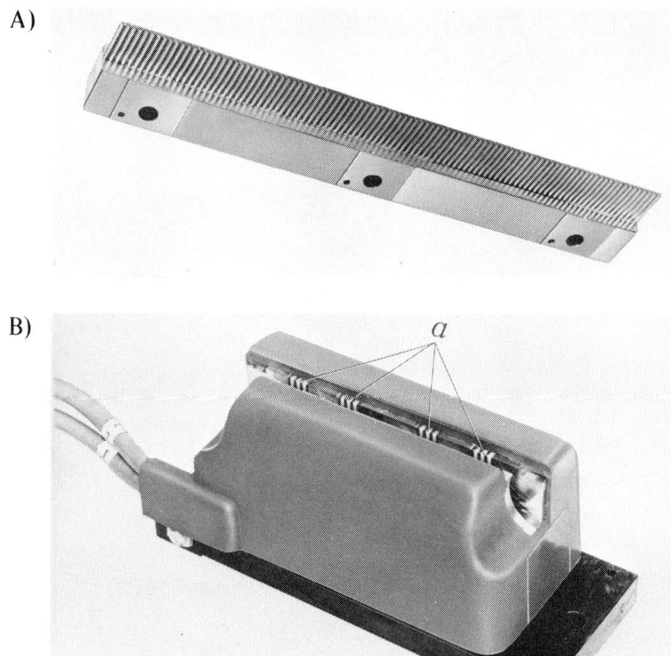

Fig. 73 Scale and measuring head of a linear Accupin

A) Scale with 0.1 in (0.2 mm) spacing; the scales are made in standard lengths of 6 and 10 in (152 and 254 mm) which can be joined together. B) Measuring head with 4 three-part inductances; the effective electrical spatial displacement of the inductances from each other is 90° (cf. Figure 72).

a Inductance with three-part iron cores (cf. Figure 72); the cores can also carry several coils. (Photo courtesy of *General Electric*)

displacement. The conditions are therefore the same as illustrated in Figure 58 for the synchro. In addition there are two further solenoid cores arranged in the measuring head (Figure 73B) which are displaced 90° from the first. In this manner a direction-sensitive two-phase system is obtained, and so the conditions are basically similar to those obtaining in the resolver (Figure 59) or the Inductosyn (Figure 65). The resolution capacity of the Accupin that is used for numerical control purposes amounts to 0.0001 in and so is similar to that of the linear Inductosyn. With the Inductosyn, the electrical outputs from the Accupin measuring head and the corresponding bridge circuit correspond to those of a two-phase synchro (Figure 59) with one revolution for a displacement of 0.1 in (2 mm). To enable them to be used internationally both the Inductosyn and the Accupin are also made for use with metric units. Figure 74 shows a rotary Accupin.

Fig. 74
Example of a rotary Accupin fitted to a turntable

a Rotary Accupin with 360 pins spaced at 0.1 in (2 mm) intervals.
b Corresponding measuring heads. (Photo courtesy of *General Electric*)

4.5   Visual indication of analogue displacement measurements

As has been mentioned in Chapter 3, there are a number of applications of digital precision indicating methods, which are quite independent of numerical control. It seemed desirable, therefore, to produce attachments for the analogue measuring systems, which have many advantages, so that these too would provide a means of indicating positions numerically. It will be obvious that this was not as easy to achieve as with a digital measuring system. In many instances, however, the expense is economically justifiable, and so a brief description of one possible application will be given. Figure 75 shows the principles of operation of this equipment.

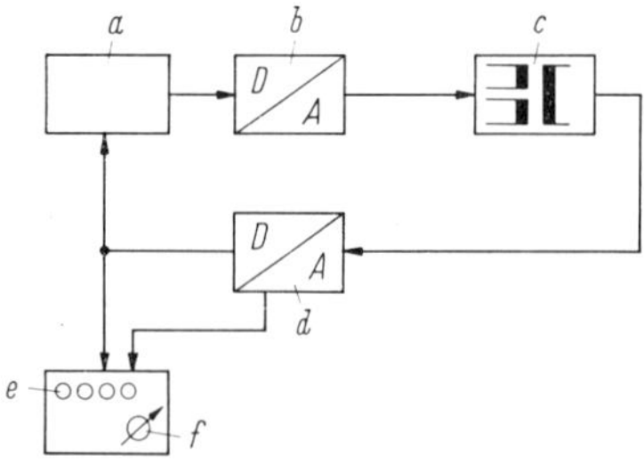

Fig. 75
Principles of operation of a visual numerical indication system combined with an analogue measuring system (e.g. Inductosyn)

a Counter (without indicator) for a maximum of 3-4 decades. *b* Digital/analogue converter. *c* Single Inductosyn with 0.1 in spacing (or 2 mm spacing) *d* Analogue/digital converter. *e* Counter with digital position indicator (e.g. in steps of 0.01 mm). *f* Indicator unit for analogue values (calibrated, for example, in 0.001 mm).

The displacement measuring system works on a cyclic-absolute process. Absolute analogue measurements are made within 0.1 in (2 mm) intervals.

Assume that the signal flow for the positioning process is described in one axis. Group *a* represents a digital counter having a maximum capacity of three or four decades (corresponding to a 2 mm interval and a resolution of 0.01 or 0.001 mm). The initial reading of the counter *a* can be any arbitrary value (e.g. 0.00); this is transformed by the digital/analogue converter into an analogue pair of voltages, 90° electrical out-of-phase (similar to Figure 80). These two voltages are applied to the rotor winding of a single Inductosyn. Depending on the relationship between the initial arbitrary reading of the counter *a* and the instantaneous relative positions of the rotor and the Inductosyn scale winding, an analogue error signal, which may be either positive or negative, will be produced in the scale winding. This is transformed into a digital value by the analogue digital converter *d*. This digital value is applied to both the digital counter *a* and the indicator counter *e* and is added or subtracted in both counters, depending on its sign. The resultant value in the digital counter *a* will correspond to the relative position of the rotor and the Inductosyn scale within an interval of 2 mm, and the error signal from the scale winding will have the value zero.

If the rotor-scale position changes, a new error voltage will be produced, and this will again be transformed into a digital value and be applied to counters *a* and *e* where it will act in accordance with its sign. No matter what

Fig. 76 Application of a numerical read-out to a large machine tool with a central measuring and control station.

a Numerical read-out (Photo courtesy of *Olivetti/Schiess*).

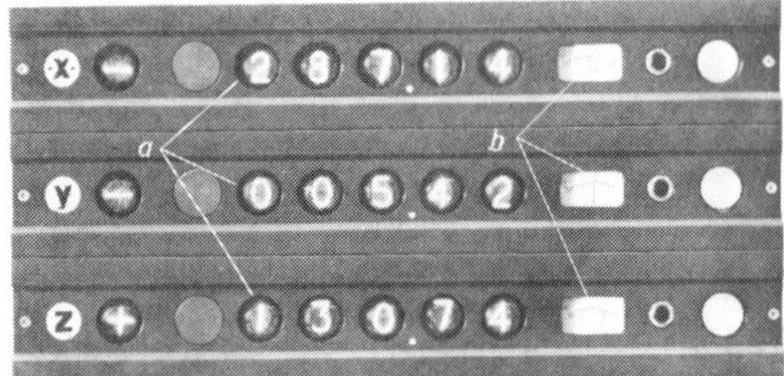

Fig. 77 **Example of a numerical read-out (to 0.01 mm) used in conjunction with an analogue precision scale.**

a Numerical read-out using indicator tubes. *b* Analogue indicator for the range of 0.01 to 0.002 mm (Photo courtesy of *Olivetti*).

the number of 2 mm intervals that have been covered, the indicator counter *e* will always show the total displacement in numerical form.

The analogue indicator *f* will also measure every voltage error that lies below the resolution capacity of the A/D converter, and so improve the electrical resolution of the measuring method.

Figure 76, like Figure 42, shows a typical application for a numerical position read-out. On large machines - in this example a large vertical lathe - it is advisable to concentrate all the readings that have to be monitored at one central control point. If the numerical position read-out is also located at the central control point it is possible under certain conditions (cf. Chapter 7) to follow, as they occur, all the changes in dimensions due to the machining process. The task of making measurements on the component, which can occupy a great deal of time, is thus eliminated. Depending on the accuracy required it is even possible, as indicated in Figures 75 and 77, to use the advantages of the analogue method of measurement - for example, by making use of the Inductosyn with its high resolution capacity - by showing values of less than 0.01 mm not digitally, but by means of a voltmeter which is calibrated directly in micrometres. It is thus economically quite sound to combine the digital and analogue measuring techniques. A further example of this is the measuring machine shown in Figure 90.

4.6   Digital/analogue converters (D/A converters)

The digital displacement measuring devices described in Chapter 3 are signal-processing devices which act as analogue/digital converters; the constant changes in displacement that inevitably occur are immediately digitalised. With analogue (continuous) displacement measurement and signal processing, the reverse path must be followed, as has already been mentioned in Chapter 2. To meet the basic requirement, that displacement information be fed-in in numerical form, each analogue measuring system and comparator must be provided with a device which transforms the digital (numerical) desired-values into analogue desired-values. With numerical machine control systems, mainly the electrical signals, the magnitude and phase relationships must be altered digitally. Tapped transformers of all types are (6) the principal components that are used for achieving this. They must always be very carefully wound and must have a low no-load current and low iron core losses. Apart from mentioning the principles, only two practical examples will be discussed in this chapter.

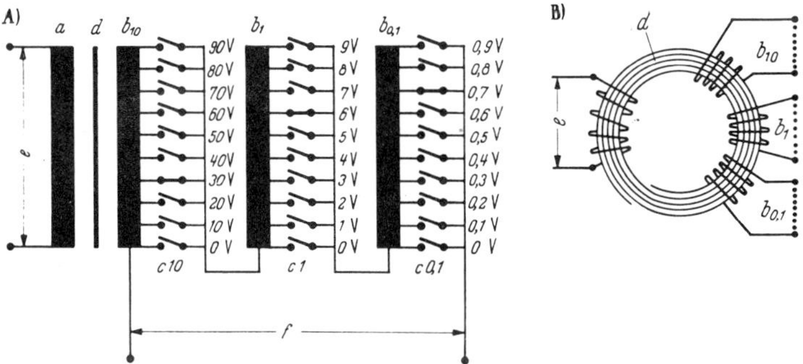

Fig. 78 Principles of operation of a voltage-defining digital/analogue converter.

A) Diagrammatic arrangement of tapped transformer. B) Construction of tapped transformer.

a Primary winding. b_{10} Secondary winding for max. 90 volts with 10 tappings (to produce decimal places 10^1). b_1 Secondary winding for max. 9 volts with 10 tappings (to produce decimal digits 10^0). $b_{0,1}$ Secondary winding for max. 0.9 volts with 10 tappings (to produce decimal digits 10^{-1}). c Switch systems (manually or remote operated). d Annular iron core with windings. e Constant a.c. input. f Output a.c. (in this case 36.7 volts to represent the number 367).

One particularly simple example is outlined in Figure 78. An annular core d made of special material having low iron core losses carries a primary winding a and three secondary windings b, each of which has 10 tappings with the total number of windings arranged in decade steps. The secondary voltages can be tapped-off by a suitable arrangement of switches, although care must be taken to ensure that only one switch can be closed at a time in each decade (e.g. a manually-operated stepping switch with ten positions, cf. Chapter 9), interlocked relays, or even solid-state switches (6).

The three secondary windings are connected in series as shown so that the voltages tapped-off from each decade are additive. With the switches in the positions shown, the number 367 has been converted into the analogue

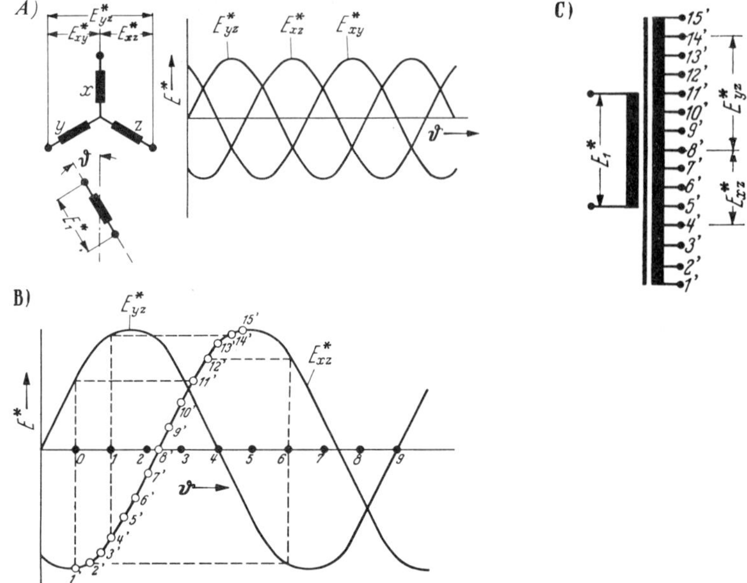

Fig. 79 **Diagrammatic arrangement of principle of operation for a simple angle-defining digital/analogue converter for a three-phase synchro.**

A) Voltage relationships in a three-phase synchro. B) Determination of transformer tappings for an angular step of $36° \triangleq 1$ decimal digit. C) Diagrammatic representation of a transformer. E^* R.m.s. values of alternating voltages, ϑ angle of rotation.

36.7volt alternating current. Two essential features of this type of equipment are apparent from this simple example:

a) The switches c must be very reliable in operation and must have a low contact resitance. Where relays are used, hermetically-sealed types filled with inert gas and equipped with gold-plated contacts are often employed.
b) The number of decades that can be used has an upper limit governed by the voltage (high voltages (several hundred) are not used), and a lower limit governed by the inevitable voltage drops at the contacts (0.001 volts, for example, can hardly be controlled reliably without using excessively costly equipment).

These boundary conditions govern the range of operation of D/A converters that is technically and economically possible and, hence, also the working range of numerically-controlled machine tools that use an absolute analogue method of measurement. The numerical range that is feasible in practice amounts to about 5 or 6 decades.

Some easing of the situation can be provided by making use of phase angles (i.e. voltage relationships) rather than of r.m.s. values. Figure 79 shows the circuit of a transformer transmitter for use with a three-phase synchro as the receiver. The component voltages in each symmetrical three-pulse system, which has a low harmonic content, always add up to zero.

Then for example,

$$E_{xy}^* = E_{yz}^* - E_{xz}^* .$$

This means that only two voltages are required to feed a symmetrical three-phase synchro, since they are sufficient to define the three-phase system. If the number of windings on the secondary of a tapped transformer (Figure 79C) is varied in accordance with a sine law, a three-phase desired-value input can be simulated very simply, and a stepped (digitalised) phase shift is obtained (65). The principle of determining the number of windings for the secondary of the tapped transformer can be derived from Figure 79B). The full $360°$ rotation of the synchro (assumed to be of the two-pole type) is divided into ten sections, so that each step corresponds to an angular change of $36°$. The symmetry of the sine function ensures that the same values of voltage recur time and again, so that it is sufficient to provide for 15 transformer tappings distributed over the unambiguous range of $180°$ electrical.

Table 9 shows the correlation between the variables:

Number	Simulated angle of rotation	Transformer tapping No. for E_{yz}^*	for E_{xz}^*
0	0	11′	1′
1	36°	14′	4′
2	72°	14′	7′
3	108°	11′	10′
4	144°	8′	13′
5	180°	5′	15′
6	216°	2′	12′
7	252°	2′	9′
8	288°	5′	6′
9	324°	8′	3′

Table 9. Correlation between transformer tappings and angle of rotation ϑ from Figure 79

If the terminal phase z is connected to the mid-point tap 8′ on the transformer, this will behave electrically as a digitally-stepped synchro transmitter (angular step $36°$). The single-phase output voltage of the receiver

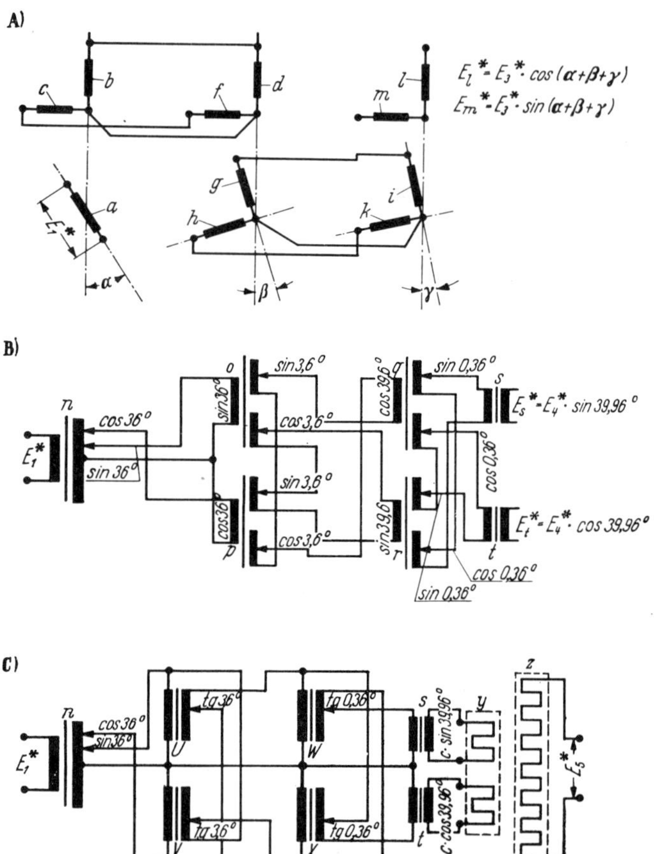

Fig. 80 **Principles of operation of an angle-defining digital/analogue converter for two-phase operation (59).**

A) Functioning of the system illustrated by means of a resolver system that performs the same operations. B) Circuit with tapped transformers for three decimal digits derived from the resolver circuit. C) A simpler circuit than that shown in B).

a Single-phase input winding. *b, c, d, f, g, h i, k* Two-phase intermediate windings. *l* and *m* Two-phase output windings. *n* Input tapped transformer. *o, p, q, r* Intermediate tapped transformers. *s* and *t* Output transformers. *u, v, w, x* Simplified intermediate transformers. *y* Two-phase slider of a linear Inductosyn. *z* Single-phase scale of a linear Inductosyn. E_1^* R.m.s. value of the single-phase input a.c. E_3^* and E_4^* R.m.s. values of the maximum two-phase voltages (magnitudes of these depend on the ratios of the numbers of turns). E_5^* R.m.s. value of the single-phase output a.c. of the Inductosyn (can be compared in phase with E_1^* of Figure 72)

The decimal number represented by the values of angles in Parts B) and C) of the figure is 111 ($36^\circ + 3.6^\circ + 0.36^\circ$); with this arrangement 360° electrical can be subdivided into 1000 parts.

synchro then directly reflects the difference between the desired and the actual values, as in Figure 60, and can be used as the basis for a follow-up regulator circuit.

The example considered above illustrates the behaviour of a single synchro with a relatively large angular step. The cost increases considerably if fine adjustments are required. Figure 80 shows the basic circuit of a three-digit D/A converter for two-phase feed-in of desired values to a linear Inductosyn (58, 59). To understand the transformer circuit, it is convenient to start with three two-phase synchros connected in series, which, according to Figures 56 E), and 61A), are suitable for use as differential synchros for the summation of angles. Consider that the three systems are connected by means of three sets of gearing with reduction ratios arranged in increasing powers of 10. System I takes the form of a rotary resolver and turns through the angle α. System II is arranged as a two-phase differential synchro, and owing to the gear reduction ratio this rotates only $\beta = \alpha/10$. System III is similar to System II, and rotates through $\gamma = \beta/10 = \alpha/100$. As has already been deduced from Figure 61, the sum of all the angles, resolved into sine and cosine components, is produced at the outlet windings l and m. If system I rotates $360°/10=36°$, the factors at the output will be $\cos 39.96°$ and $\sin 39.96°$; the angular subdivision is thus already very finely stepped. If the synchros are replaced by precision tapped transformers with the numbers of turns in the secondary windings stepped to represent angle functions (similar to Figure 79) the arrangement illustrated in Figure 80B) will be obtained, and an electrical angle can again be simulated by means of stationary components. The law for the formation of the chain is derived by the following arguments:

In general

$$\sin (\alpha + \beta) = \sin\alpha \, \cos\beta + \cos\alpha \, \sin\beta$$
$$\cos (\alpha + \beta) = \cos\alpha \, \cos\beta + \sin\alpha \, \sin\beta.$$

If the input transformer n is supplied with an alternating current having a r.m.s. voltage of E_1^*, the values that can be tapped off the secondary side, as in the three-phase system shown in Figure 69, are

$$E_1^* \sin\alpha \text{ and } E_1^* \cos\alpha.$$

For simplicity, only one example, $\alpha = 36°$, together with its angular functions, has been shown in the drawing. The two transformers o and p which follow are wound and connected so that the values at their outputs are $E_1^* \sin (\alpha + \beta)$ and $E_1^* \cos (\alpha + \beta)$. In the example the sum of the angles is $39.6°$. By a repeated application of the theorem of addition, the next two tapped transformers produce the functions:

$$\sin (\alpha + \beta + \gamma) \text{ and } \cos (\alpha + \beta + \gamma).$$

Taking the ratios of the output transformers s and t into account, the voltages available at the output of the chain will be

$$E_5^* = E_4^* \sin (\alpha + \beta + \gamma) \text{ and}$$
$$E_t^* = E_4^* \cos (\alpha + \beta + \gamma)$$

In the example, the angle γ is assumed to be $0.36°$ $(=\alpha/100)$.

Much simpler windings and circuit arrangements can be used if the ratios of the sines and cosines to each other, i.e. the tangents, are used rather than the sines and cosines themselves[4].

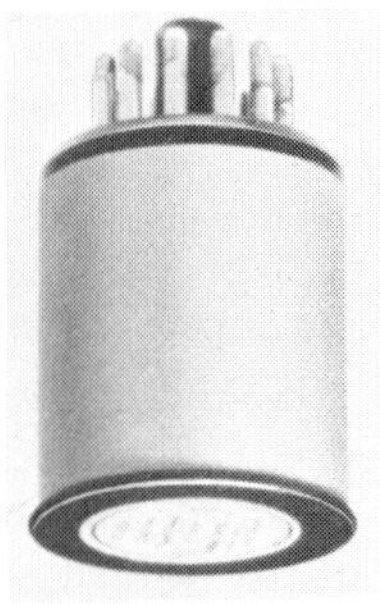

Fig. 81
Example of a medium-frequency tapped transformer (approx. 1 to 20 kHz) for use in a D/A converter (e.g. transformers u, v, w, x **in Figure 70C). Height of transformer about 50 mm. (Photo courtesy of** *Farrand/Siemens*)

If suitable output transformers are chosen (cf. Figure 81) it is possible to apply the two sine and cosine components to the two-phase slider of a linear Inductosyn. The voltage E_5^* will then be produced at the output of the Inductosyn scale, and if the transmission chain is suitably designed the phase angle of this voltage will bear some fixed relationship to the phase angle of the input voltage E_1^*. A follow-up control loop can then easily be set-up.

4.7   Analogue type comparators

One great advantage of using analogue desired and actual values is that the comparators for control circuits (R type comparators) can also be very simple. This is particularly so if only moderate sensitivity is required. The

---

[4] The simpler relationships are given by

$$\frac{\sin (\alpha + \beta)}{\cos \beta} = \sin \alpha + \cos\alpha \, \tan \beta$$

$$\frac{\cos (\alpha + \beta)}{\cos \beta} = \cos \alpha - \sin\alpha \, \tan \beta$$

polarised relay has already been mentioned as one very simple device in Figure 54, although this is suitable only for on-off operation. A relay of this type can be used to explain the concept of sensitivity very clearly. It is represented by the minimum voltage that is required to cause the relay armature to be moved reliably to one side or the other. In practice, polarised relays are not popular in machine tool control systems because of their sensitive contacts. Modern semiconductor techniques have resulted in a large number of solid-state devices so sensitive that they can provide virtually continuously operating comparator circuits. Of the many possibilities (3, 58, 64), only two simple examples will be mentioned at this point; they are illustrated in Figure 82.

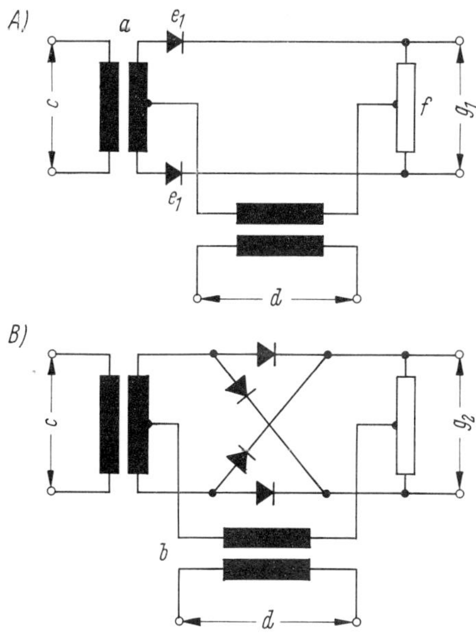

Fig. 82
Examples of circuits for phase-sensitive rectifiers (3)

A) Half-wave rectification.
B) Full-wave rectification.

a Input transformer for the comparison voltage (constant phase angle). *b* Input transformer for the phase-displaced voltage. *c* Voltage of fixed phase angle (e.g. voltage E_a^* in Figure 60). *d* Voltage of variable phase angle (e.g. voltage E_g^* in Figure 60). *e* Semiconductor diodes. *f* Output resistance with central tapping. g_1 Output signal (d.c. voltage from half-wave rectification). g_2 Output signal (d.c. voltage from full-wave rectification). i_1 and i_2 Component currents.

In the majority of cases a comparison of the phase angles of two alternating currents is preferred to a comparison of the amplitudes.
The simplest solution is then to consider only the two limiting cases:

Phase difference zero, or

Phase difference equal to $\pi/2$.

The circuits shown in Figure 82 consist basically only of transformers, semiconductor diodes, and ohmic resistances *f*, i.e. of a small variety of components which have a long life and are very reliable. In both circuits the d.c. voltage at the output g will be zero when the phase difference between the input voltages c and d is $\pi/2$. To visualise the operation of the circuit it is advisable to start from the voltage relationships existing in the synchros shown in Figures 58 and 59. Assume that the transmitter voltage E_a^* is to be compared with the receiver voltage E_g^* (Figure 60). If ϑ_1 is constant and ϑ_2 is varied, the voltage E_g^* will pass through the complete range of values shown in Figure 58C), whilst the voltage E_a^* remains constant. The nodes c are of special interest, because the phase step previously mentioned occurs at these. If a constant reference voltage E_a^* (as in Figure 58) is applied to position c of a comparator of the type shown in Figure 82A), and a variable voltage E_g^* is applied to point d, the following will occur:

No voltage will appear at the point d, if the rotors of the two synchros are perpendicular to each other. The centre tapping in the secondary winding of transformer a causes the reference voltage to be split into two opposing voltages ($180°$ electrical out-of-phase), and the two semiconductor diodes each perform a half-wave rectification of their positive half-cycles. The load resistance f is also symmetrically divided; the first positive half-wave current i_1, flows downwards through the top half, whilst the second positive semiconductor current i_2 passes upwards through the lower half. There will be no net voltage across the output g, since the mean values cancel each other out. If the rotors of the synchros are not perpendicular to each other, the phase reversal at the nodes c (Figure 58) will cause either i_1 or i_2 to increase, depending on the direction of the displacement. This upsets the symmetry of the voltage drops across the resistance f, and the mean values no longer cancel each other out. The d.c. signal at the output g is, within limits, proportional to the difference in the angles $(\vartheta_1 - \vartheta_2)$ and also indicates the direction of the deviation. The requirements for an R type comparator are thus fulfilled. The sensitivity of the system depends primarily on:

a) The quality of the transformation (iron core losses, no-load current, leakage, etc).
b) The performance of the semiconductor diodes, e, and
c) The fact that only half-wave rectification is employed, i.e. that only every other half-cycle of the alternating current is used for the signal.

The circuit shown in Figure 82B) is an improvement on this. The four semiconductor diodes provide full-wave rectification, which enables better use to be made of the alternating voltages for producing the signal. On the other hand, with this circuit the characteristics of the semiconductor diodes must be very closely matched and must remain very constant.

Depending on the care with which the components are selected, these circuits can be used to detect angular differences of $\Delta\vartheta \approx 0.8$ to $1.5°$. Better results are obtainable with different circuits; the sensitivity that can be achieved with analogue comparators is probably about $0.01°$. The number of high-precision components required will then, of course, increase, but it does not become as high as for digital comparators.

4.8   Summarising remarks on the analogue process

The following summarises what has been said regarding analogue displacement measuring systems and comparators:

1. The only measuring systems that are suitable for use on machine tools are those in which there is no metallic contact and here inductive devices predominate.
2. The range over which a single device (e.g. a synchro) can undertake absolute measurements is relatively small; larger angles and displacements require a cyclical repetition of measurements.
3. Linear inductive displacement measuring systems can be regarded as four-pole synchros, and must be treated electrically as such.
4. In view of the cyclic repetition of the readings, there are three ways in which inductive analogue measuring systems can be used as displacement indicators:
   a)   Close coupling of a single system with a continuously varying reference value by means of a carefully designed follow-up control loop circuit, briefly known as 'Synchronous operation'. If this principle is to be adopted for machine tools, an electronic computer will be required.
   b)   Extension of the measuring range for analogue-absolute displacement measurement by employing mechanical coupling elements of high precision between a number of individual systems.
   c)   A third possibility is to combine digital and analogue methods; this will be discussed in Chapter 5.
5. It is usual to divide the measurements into three systems (coarse, medium, and fine). The resolution of such an arrangement will, in practice, be about 1 part in 10^5.
6. Synchros form rotary displacement measuring systems that are of very simple construction, comparatively both robust and inexpensive.
7. Higher frequencies (400 to 2000 Hz) result in compact electrical components and low moments of inertia in synchros. From about 800 Hz upwards they also enable analogue systems to be used in combination with precision systems for linear displacement measurement (e.g. Inductosyn, Accupin, etc).
8. Comparators for all types of application (on-off circuits, continuous and non-continuous control circuits) are of simple construction; it is also easy to produce preliminary signals (e.g. for changes of speed).
9. The basic task of combining a continuous displacement, angle, or position statement with a numerical value is transferred from the measuring system itself into a separate digital/analogue converter (cf. Chapter 2), and in this respect the analogue systems differ from the

digital-absolute systems. In many instances the technical difficulties and the cost centres are merely transferred elsewhere.

10. Zero-point displacement and adjustment is easily performed in an analogue system by means of differential synchros. It is necessary to note the difference, as indicated in Point 4, between synchronous operation (1 synchro) and absolute measurement (several synchros coupled together by precision gearing). In current installations the zero-point displacement is, however, transferred into the digital field by using small computers. (cf. Chapter 8).

11. An actual value indication can be provided but this is more complicated than with a digital system.

Before it is possible to make a final comparison with digital systems, however, certain special arrangements, which are to be described in the next chapter, must be considered; these do not fit readily within the strict classification of digital and analogue systems as used for the methods of measuring displacments that have been considered up to now.

5   Special problems of numerical measurement technology as applied to machines — The measuring systems compared

Three methods of measuring slide displacements which do not fall neatly into the groups that have previously been discussed will be considered in this chapter, together with two recent methods for measuring the workpiece dimensions and using them for control purposes. This will enable some conclusions to be drawn concerning the possibility of applying the various measuring systems to numerically controlled machine tools in closed-loop systems as illustrated in Fig. 1B).

5.1   Two-part measuring systems

In the earlier chapters reference has been made on a number of occasions to the important and large group of machines which is equipped with positioning and straight-line control systems. In most instances these merely require on-off circuits so that the numerical control system can be based on simple trip dog techniques.

The basic concept underlying all these two-part systems is that they avoid the use of expensive linear precision measurement systems (digital or analogue) for measuring large displacements. This principle has been put into practice in a number of designs, which differ in detail 55, 60, 61). One possible method is illustrated in Fig. 83. The long machine slide, 4', is provided with an incremental scale which is accurately divided into widely-spaced steps. These can, as indicated in the example, consist of individual blocks a made of a ferro-magnetic material. They can be spaced at intervals of 1 cm, 1 inch, or 1 dm; the width of the blocks can be about half the length of the intervals or less. The blocks are scanned by an a.c. excited magnet system b, which may be described as a differential transformer with a variable air gap. The electromagnetic system is clearly shown in Fig. 83B). If an a.c. voltage of E_1^* (r.m.s. value) is applied to the central coil h, a magnetic flux will be produced the two halves of which tend to form a closed-loop through the outer legs of the core. If one of the blocks is located precisely symmetrically in front of the three-legged core, the two partial fluxes will be equal. Since the two secondary coils i_1 and i_2 are connected in series with

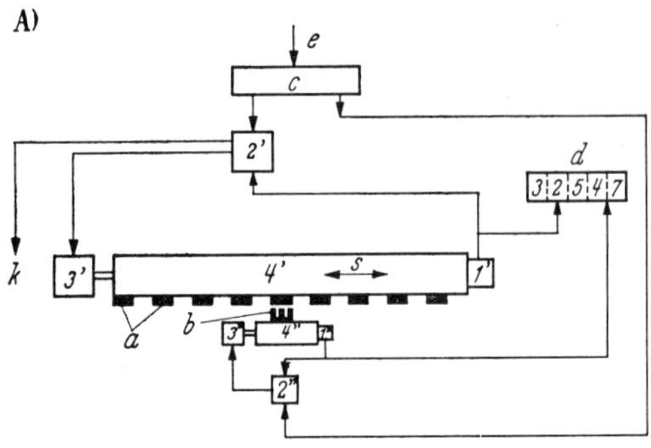

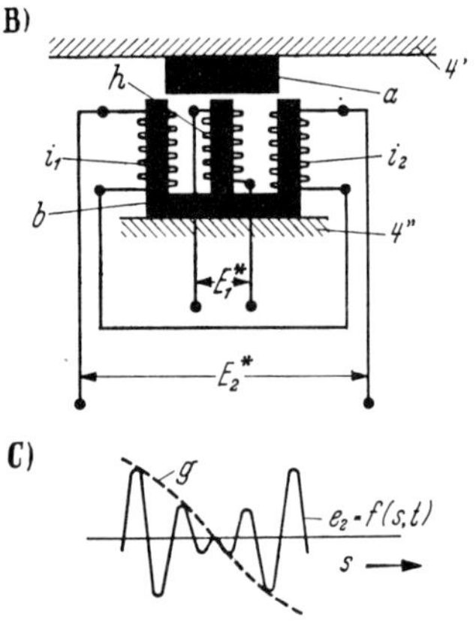

Fig. 83 Principles of operation of a two-part displacement measuring system (cf. also Fig. 122)

A) Signal flows through the two circuits. B) Construction of a zero-point indicator. C) Variation of signal voltage in a zero-point indicator.

a Ferro-magnetic block located on a precision scale with widely-spaced (1 cm, 1 inch, 1 dm) intervals. *b* Inductive differential scanner. *c* Input of numerical information (desired value input). *d* Numerical actual value indicator. *e* Desired values (in numerical form) E_1^* R.m.s. value of a constant a.c. voltage. E_2^* R.m.s. value of the signal voltage. e_2 Variation of signal voltage. *g* Phase-sensitive rectified signal voltage. *h* Primary coil. i_1, i_2 Secondary coils. *k* Preliminary signal to activate zero-point indicator. *s* Slide displacement. *t* Time. *1'* Displacement measuring system for large displacements. *2'* Comparator for this. *3'* Drive system for machine-tool table. *4* Main slide (machine-tool table). *1''* Displacement measuring system for small displacements (1 cm, 1 inch, or 1 dm). *2''* Corresponding comparator. *3''* Drive system for auxiliary slide. *4''* Measuring slide (auxiliary slide).

reverse polarities, the secondary voltage E_2^* will be zero when the two partial fluxes are the same. If the block a is not precisely central, the two partial fluxes, and hence the two partial voltages, will be unequal, and the resultant voltage E_2^* will not be zero. As it passes through the zero position phase reversal will occur, since the voltage difference changes sign. The variation of the voltage with changes in the position of the block about its central location is sketched in Fig. 83C). The behaviour of this sytem is thus very similar to that of synchros (cf. Fig. 58). There are now two possibilities:

1. The alternating voltage E_2^* is rectified by means of a simple bridge circuit (Graetz circuit), and only the point at which the voltage becomes zero is used for data processing and machine tool control.
   Here, the measuring device acts like a limit-switch working in conjunction with a trip dog. Automatic correction is not possible once the slide has passed the required point. (Type S comparator, pure on-off circuit),
2. If a more sophisticated control system is required, the alternating voltage E_2^* can be rectified in a phase-sensitive bridge as shown in Fig. 82. Here, the positive and negative deviations from the zero position can also be used, at least over a certain limited range (Type R comparator with limited working range). It is possible to correct for a slight overshoot of the zero position, provided that the drive system is of suitable design (cf. Chapter 6).

The way in which the two slides, 4' and 4", work together is very simple. The displacement measuring system, 1', for the large slide can be of relatively inaccurate and simple design. Its resolution capacity need only be about the length of the blocks a. There is also a free choice of the measuring system that is to be used and of the way in which it is attached to the slide; in practice measurements are always made at the lead screw, and any preliminary signals that may be required for speed changes, etc, will be derived from this measuring system. The coarse steps used in this system suggest the possibility of employing a simple, rotary, digital absolute measuring system (for example, as shown in Fig. 47). The small measuring slide, 4", carries no mechanical load whatever and has to traverse only short distances (approximately one interval of the coarse scale). The auxiliary measuring system, 1", can therefore be attached to a short precision leadscrew; here again a rotary, digital, absolute measuring system offers certain advantages. In practice, one can arrange for numerical values down to 1 cm to be allotted to the measuring system, 1', and the smaller values to system, 1" (see also Section 6.6). If system, 1', is designed for three decades and system, 1", also for three decades, a relatively large working range (six decades) or high precision (jig borers) can be achieved at relatively little cost. If digital-absolute systems are used, the method has the advantage of enabling the actual value to be displayed in absolute dimensions in a simple manner, and this is often very convenient, especially with large machines (e.g. heavy boring and milling machines). It is, of course, also possible to use synchro systems.

One of the special features of two-part measuring systems is that by using a separate setting motor and precision lead screw for the low values it is possible to achieve a comparatively high rate of changing the desired-value settings. With large machines this is probably of no practical importance, since for mechanical reasons (large components with long machining times, during which the new settings can be pre-selected on the small measuring slide, 4", without any difficulty) on heavy machine tools the desired values do not change very rapidly.

The accuracy with which the movement is stopped (with and without directional sensitivity) depends, in the main, only on the small measuring slide system, the construction of the differential transformer, and the coarse scale of the blocks. The stopping accuracy can be of the order of 2 to 5 μm; the process has been used in the USA for many years under the name "Electrolimit System" for non-numerically controlled jig borers (66). The methods of operation of these two-part measuring systems will probably prove of general interest for application to large machine tools with numerical positioning and straight-line control systems.

5.2   Electro-Optical System for Synchronous Operation

Figure 84 illustrates a method which, for a number of reasons (67) is of special interest. In this method a relatively coarsely divided incremental scale (grid) is scanned by the incident light process. This has the advantage that, as in Fig. 29, a comparatively robust steel scale which is provided with fairly inexpensive graduations can be used. The grid scale is not scanned digitally, but the constant beam of light emitted by the light source d is transformed by a rotating spiral into an approximately sinusoidal form. The pulsating beam of light is reflected from the polished steel scale, and passes to a photodiode e which converts the light pulsations into an alternating current. The glass disc with the spiral b is driven by a small synchronous motor c which must run smoothly and free from torsional oscillations. If the scale a (which is secured to the machine slide) remains stationary, the frequency of the light beam fluctuations, and hence those of the alternating current, will be f_o = 104 Hz for the proposed dimensions. This is calculated from the constant speed of the motor and the number of turns of the spiral. If scale a moves to the right, the frequency increases, whilst if it moves to the left, the frequency decreases. Small displacements only alter the phase angle of the measuring voltage which is maintained by the synchronous motor at a certain ratio of the reference voltage (e.g. the supply voltage of the synchronous motor). This arrangement thus enables the same effects as for the synchro illustrated in Fig. 58 to be achieved photoelectrically. The system can

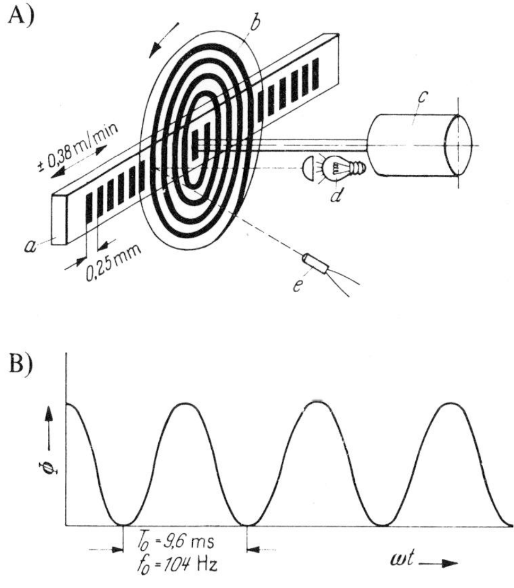

Fig. 84
Electro-optical scale scanning

A) Schematic diagram of operation. B) Fluctuation of the light-generated current.

a Highly-polished metal scale with relatively coarse graduations for incident-light scanning.
b Transparent rotating disc with spiral (same pitch as interval of graduations on scale).
c Synchronous motor to drive glass disc. *d* Light source. *e* Photocell.
t Time in seconds.
Φ Current generated by the light at photocell.
ω Angular frequency of light-generated current fluctuations.

therefore be used like a synchro with, f_o = 104 Hz for the construction of a follow-on control loop circuit; it cannot be used for the measurement of displacements over long distances. There is an obvious similarity with the method illustrated and described in Fig. 29, with the exception that the signal is processed in a completely different manner.

Superimposed scanning of this type offers two advantages:

1. The linear motion is transformed into a rotary motion with no direct contact being required and with no wear taking place.
2. The frequency is relatively low. It is therefore suitable, as will be shown in Chapter 9, for use with magnetic tape information carriers. In fact, up to the present the system has been used only in conjunction with external interpolators and magnetic-tape controlled machines[1].

5.3 Digital follow-on control circuit with a single synchro.
(Bendix system)

In Chapter 4 (Figs. 55 to 61) the synchro was described as an extremely simple and reliable element for use in analogue systems, and an appropriate

[1] Messrs. Ferranti, Edinburgh.

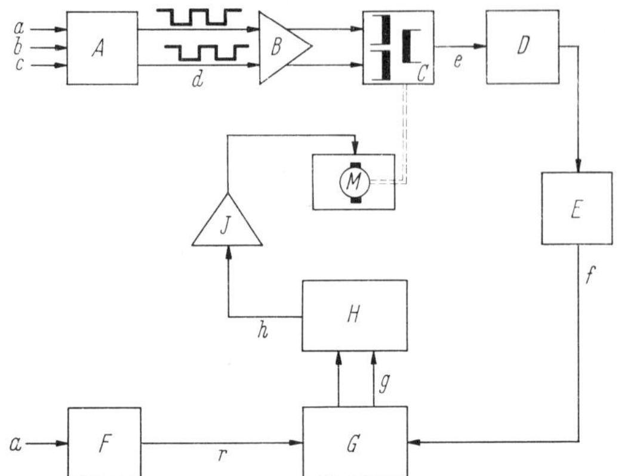

Fig. 85 Principle of operation of a phase-analogous slide displacement control system using digital methods

A) Frequency reducer 1000 : 1 (for principle, cf. Fig. 33). B) Amplifier (a c.). C) Resolver (coupled with machine slide and drive). D) Signal converter (sinosoidal wave to square wave). E) Frequency reducer 8 : 1. F) Frequency reducer 8000 : 1. G) Phase comparison logic. H) Summation logic with rectifier and filter. J) Amplifier (d.c.) M) Drive (d.c. motor or electro-hydraulic valve, cf. Chapter 6). *a* Square-wave voltage 3 MHz (initial frequency, clock frequency). *b* Directional signal for *c* *c* Traverse impulse. *d* Square-wave voltage, approx. 3 kHz phase displaced by 90°. *e* Single-phase output voltage of resolver (cf. Fig. 59). *f* Feed-back frequency. *g* Output signals from phase comparison logic. *h* D.c. control signal for drive. *r* Reference frequency (375 Hz).

analogue phase comparator was shown in Fig. 82. An indication was also given in Fig. 61 of means by which it was possible to control several machine axes simultaneously using analogue techniques. With the more widespread introduction of digital computers and of pulse techniques (cf. Chapters 8 and 11) using integrated circuits, it appeared desirable to apply the advantages of the rapid and accurate discrimination circuits that are possible with digital techniques to synchros, which have been widely adopted in machine control systems.

A phase-analogue slide displacement control system using digital methods will therefore be briefly described with the aid of Figs. 85 and 86 (104). A square-wave voltage produced by a signal generator and with a high initial frequency or clock frequency of, for example, 3 MHz, is reduced to a reference frequency *r* of 375 Hz, which is made available to all the axes that have to be controlled, by means of a group of frequency reducers F^2. The cycle time of the frequency that is processed in the digital unit *h* is thus 2.7 ms, so that with a scanning ratio of 1 : 1 the pulse duration is 1.35 ms. The

---

[2] The principles of frequency reduction have already been discussed in Chapter 3 with reference to Fig. 3.

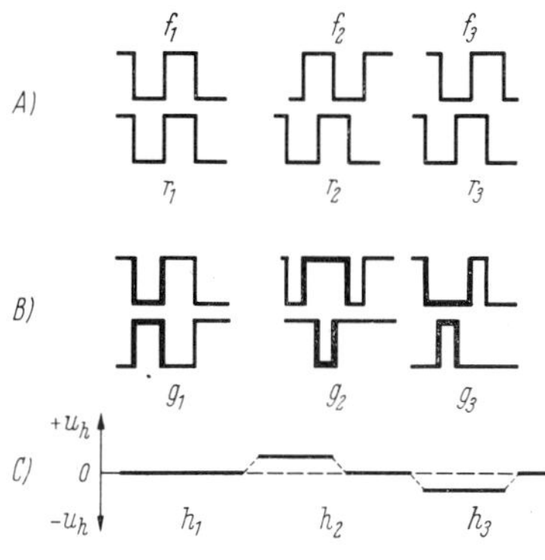

Fig. 86
Signals and signal processing in phase-analogous follow-on control circuit

A) Frequency inputs to unit G.
B) Signal inputs to unit H
C) D.c. outputs from unit H

f_1 Feedback frequency in phase with reference frequency. f_2 Feedback frequency leading by 90°.
f_3 Feedback frequency lagging by 90°. g_1, g_2, g_3 Corresponding signals from phase comparison logic G. h_1, h_2, h_3 Corresponding control signal for drive M. $r_1 = r_2 = r_3$ Reference frequency (375 Hz).

same initial frequency a is also applied to the frequency reducers A for each axis, where it is reduced at the ratio 1000 : 1. A square-wave voltage of 3 kHz is thus available initially in each axis for signal processing in units B, C, and D. Unit A also acts as a frequency splitter, so that at the output, two square-wave voltages d having the same frequency but 90° out of phase are available. Using digital techniques this double function of 'frequency reduction and phase displacement by $\pi/2$' can easily be achieved.

The two square-wave voltages d which are out of phase, are amplified in the subsequent amplifier B. The electrical conditions at the resolver are now the same as described in Figs. 59 and 65 in the discussion on analogue techniques. Depending on the position of the single-phase rotor, a variable-phase single-phase voltage e is produced which has an essentially sinusoidal wave form having the basic frequency of 3 kHz. To enable digital technique methods to be employed for subsequent processing, the sinusoidal signal voltage e has to be transformed back into a square-wave voltage in a signal converter, stage D, with no change in the phase angle. If unit D is followed by a digital frequency reduction stage E having a ratio of 8 : 1 (3 flip-flops of type shown in Fig. 33), a square-wave signal voltage f of the basic frequency 375 Hz, i.e. of the same magnitude as the reference frequency r will be obtained. These two differ only by their phase angle, which depends on the position of the rotor of resolver C. The digital phase comparison processes are shown in Fig. 86. The two square-wave voltages f and r are applied to unit G for digital phase comparison. The duration of the impulse is 1.35 ms; this represents a very long time for present day integrated circuit components (cf. Fig. 34 and the associated description). Figure 86A) shows the three extreme cases of phase differences between f and r. The logic circuits in unit G act in such a manner that, as shown in Fig. 86B), the output signals g appear with different impulse ratios and signs[3].

---

[3] for the most part inverters and NAND circuits are employed.

The impulse outputs g are now summed in unit H, rectified, and smoothed, so that the d.c. signal h can vary infinitely between positive and negative maximum values, depending on the phase relationship between f and r. There is thus again an analogous d.c. setting signal which, after suitable amplification in J (cf. Chapter 6) can be used for controlling an electrical or hydraulic drive M; the decision as to the type of drive to employ is discussed in detail in Chapter 6. The drive M moves the machine slide and hence also the resolver C until the phase angle of the signal e coincides with that of the reference signal r. It will be obvious that the use of digital techniques with their extremely high-speed components imposes very severe requirements on the mechanical design of the drives and on their stabilisation (cf. Chapter 6). These requirements are slightly modified by the possibility of summing the values which is afforded by unit H.

Every frequency splitter A which is allocated to a machine axis has superimposed on the reference frequency a, of, for example, 3 MHz, a lower 'traversing frequency' c which also takes the form of a square-wave voltage. Signal b indicates the direction of travel. It is thus possible to increase or decrease the frequency of the output signals d; as a result the machine slide - located between M and C - will move more or less rapidly forwards or backwards. Basically, it is of no importance how the 'traversing frequencies' are produced; they may be provided by a simple pulse generator (for straight-line control systems) or by an internal interpolator for continuous-path control systems (cf. Chapter 8), from a magnetic tape (cf. Chapter 9) or from an on-line computer (cf. Chapters 11 and 12). In principle, this control system can thus be used for every type of machine kinematics (positioning, straight-line, or continuous-path control) (105, 106, 107). The system is so dimensioned that there is a choice of two values for a traverse impulse (displacement increment); these are Value $1 = 0.002$ mm or Value $2 = 0.01$ mm. Depending on the value chosen the number of traverse impulses applied will thus determine the displacement per axis, whilst the impulse frequency will determine the speed of displacement along each axis. The functional similarity of this method with stepping motors (cf. Chapter 6) is obvious.

5.4   Numerical Workpiece Measurement Control Systems

In the first chapter, when discussing Figure 1, reference was made to the fact that there are two ways in which the principle of a closed-loop circuit can be realised. These are:

Numerical slide displacement measurement control, and
Numerical workpiece measurement control.

It is far easier to determine the slide displacements than to measure the dimensions of the workpiece during the course of the machining process, desirable though this would be, and in view of their more practical nature the methods of measuring slide displacement, which are widely employed, were therefore considered in detail in Chapters 3 and 4. It is only in recent years that machine tools have become available in which it is possible to obtain signals representing the workpiece dimensions directly in a form which lends itself to numerical processing. In view of the difficulties — access to the measuring points is usually poor, the presence of hot swarf and large chips, complicated internal and external shapes of the workpieces, etc — the application of this method is even nowadays confined to the control of grinding processes and, in particular, on external cylindrical grinders. Since it seems probable that in the relatively near future (10) such technological problems as are encountered in making measurements (cutting forces, vibrations, temperatures, torques, etc) will be overcome by the advances that are being made in automation (cf. Chapters 8, 11, 12), some mention will therefore be made here of the work that is being done to measure geometrical data directly on the workpiece. To simplify matters only one practical example will be discussed.

Figure 87 shows a system for measuring the external diameters of a number of workpieces, either continuously or on a sampling basis, on cylindrical grinding machines. In Fig. 88 the principles underlying the incorporation of a workpiece measurement system in the control system for a

Fig. 87 Example of a numerical workpiece measurement control system. Diameter tolerance attainable IT 4.

a Workpiece measuring device based on the Inductosyn (MULTI-FINITRON multi-position control unit with measuring range up to 170 mm diameter). *b* Workpiece with various diameters. *c* Grinding wheel. (Photo courtesy of *Fortuna*)

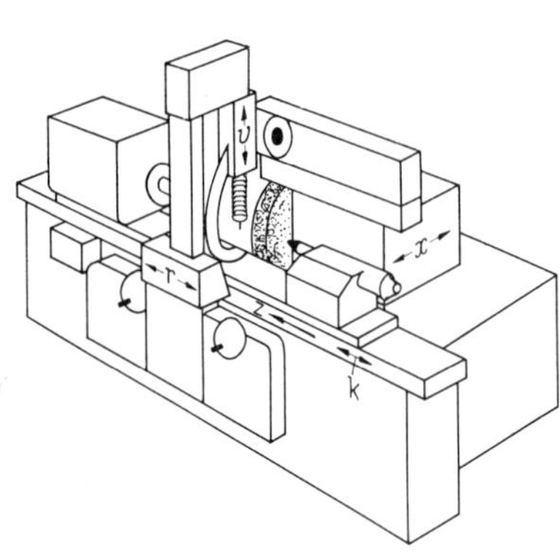

Fig. 88
Principle of operation of a cylindrical grinding machine with numerical workpiece measurement control - possibilities for slide control.

x Grinding slide position (e.g. by gearing and synchros as shown in Fig. 62). z Positioning of workpiece table (e.g. by gearing and synchros as shown in Fig. 62). k Hydraulic table traverse for longitudinal grinding. v Measurement and control of various workpiece diameters to a high degree of precision without machine re-setting. r Manual or automatic longitudinal displacement of measuring head - about 100 mm (Illustration courtesy of *Fortuna*)

cylindrical grinding machine are illustrated. The special features of the precision grinding process (rapid wear of the grinding wheels; production of smooth surfaces and the presence of small, cool chips; number of measurements to be taken simultaneously usually limited) led to the decision to attempt a solution to this difficult problem for this particular application. It must be borne in mind, however, that the various NC functions must be carefully considered in entirety. For this reason a complete machine is shown in Fig. 88, although the mechanical components will not be discussed in detail until Chapter 7. The diagram shows five degrees of freedom — x, z, k, r, v — which, in theory, should be monitored and controlled by NC techniques. Whether it is, in fact, practical to do this, and if so whether it is economical, cannot be decided until extensive calculations have been completed. It should also be borne in mind that the tolerance of IT 4 that is quoted (10 μm for external diameters of 80 to 120 mm) can be maintained only if the machine conditions are carefully controlled (temperature stabilisation of bearings, of the hydraulic fluid, of the grinding fluid and its filtration, remote-controlled balancing of the grinding wheel, etc); the hourly cost for a precision machine of this type is therefore inevitably high, and for economic reasons the automation should be limited. In particular, the same measuring system should be used in the measurement and control unit a (Fig. 87) as for the remaining NC part of the machine; a change of system would inevitably cause difficulties at the interfaces and would create additional costs. Figure 87 also shows a typical workpiece for an NC machine of this type (108). Similar approaches might well be feasible nowadays for surface grinders, but whether they would be economically justifiable under mass-production conditions is quite another question. This situation can, however, change radically as progress is made in measuring technology and as various machining processes become increasingly integrated with the assistance of large data processing installations (cf. Chapters 11 and 12). It is quite possible that the ease with which a production machine can be integrated in a highly organised production information and materials flow system will assume greater importance in the future than it does at the present moment, so that the criteria that are now used for assessing the acceptability of capital costs will no longer be valid (10). Further discussion on these basic points will be made again in Chapters 8 and 11.

5.5   Numerical Inspection Machines

The logical consequence of the difficulties of measuring workpiece dimensions while the workpiece is on the machine tool is to remove the important and difficult measuring problems entirely from the machine tool and to transfer them to a separate measuring station that is automated to a greater or lesser extent. The relatively simplest case is a numerical inspection machine of the general type illustrated in Figs. 89, 90, and 91. The most important point is that the workpiece dimensions obtained on the machines should be available in *numerical* form for further processing. Whether one satisfies oneself with a simple 2-D or 3-D visual display, and manual

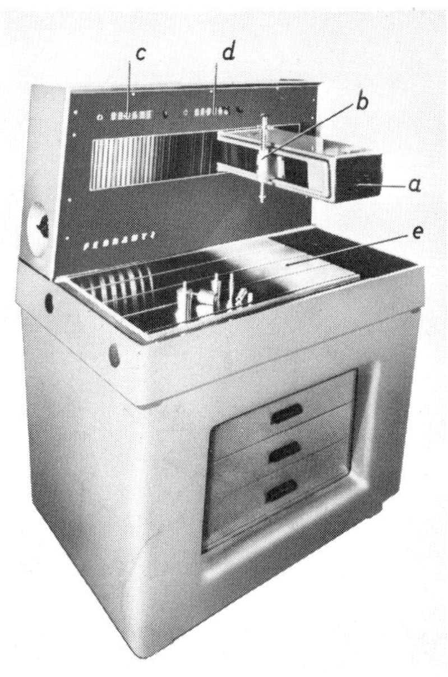

Fig. 89
Example of a 2-D measuring machine

a Arm movable in longitudinal direction (x-axis) that carries the measuring system for the transverse slide (y-axis). *b* Probe (without measuring system) on transverse slide. *c, d* Numerical displays as shown in Fig. 31A. *e* Mounting (clamping) surface for two-dimensional workpiece.

Note: The measuring systems incorporated in this machine are of the type shown in Fig. 24. (Photo courtesy of *Ferranti*)

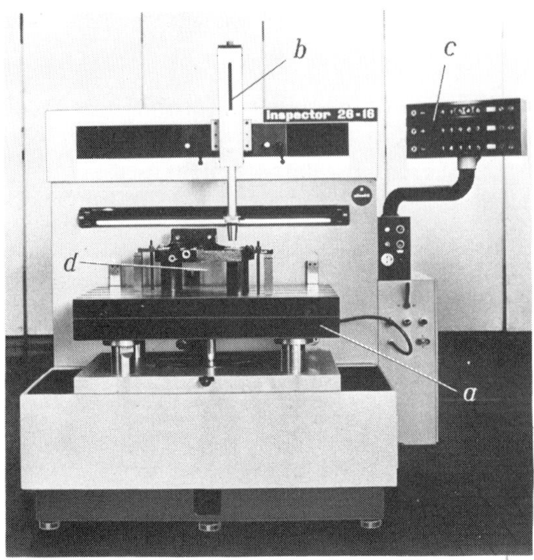

Fig. 90   Example of a 3-D measuring machine

> *a* Mounting table with ability to make measurements in the *x-y* plane. *b* Probe with measuring system for the *z*-axis. *c* Numerical display of dimensions (cf. Fig. 77). *d* Workpiece to be measured.
>
> Note: The measuring system corresponds to that shown in Fig. 75; comparison of the measured dimensions with those shown on the drawings is performed by the operator. (Photo courtesy of *Olivetti*).

Fig. 91   A version of the machine shown in Fig. 90 used for measuring angles and lengths (radii).

> *a* Rotatable chucking head with rotary Inductosyn as shown in Fig. 71. *b* Distance (diameter) probe with linear Inductosyn as in Fig. 66. Item being inspected is a camshaft.

comparison with the data given on the drawings, whether one arranges for the actual dimensions to be calculated by a small computer and automatically listed by means of a recorder (Fig. 92), or whether one uses the measured values in a large feed-back loop in a central processor will depend on the overall organisation adopted for the factory (cf. Chapters 11 and 12). In any event, this is a decision which cannot be based solely on the conditions prevailing at a single machine tool, but can only be considered for the system as a whole, where system in this context refers to the production plant in its entirety (10).

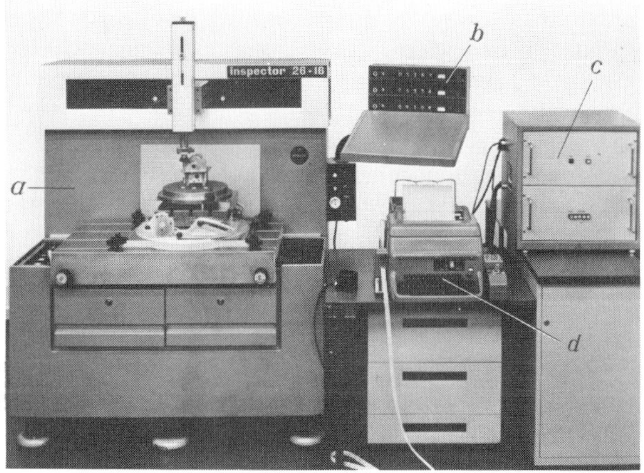

Fig. 92 Measuring machine with automatic evaluation of results (comparison of actual and desired values, classification of components, calculation of average values, etc.) and recorder for listing results.

a Measuring machine as shown in Fig. 90. *b* Numerical display of dimensions. *c* Electronic evaluation equipment. *d* Recorder producing printed lists. (Photo courtesy of *Olivetti, Italy* and *Siemens/R.I.D.*)

5.6 Comparison of numerical measuring systems used in conjunction with machine tools

Chapters 3 and 4, as well as Sections 5.1 to 5.3 of Chapter 5, are of importance to the design and layout of numerically controlled machine tools, and so are directly related to Chapters 6 to 9, which follow. Sections 5.4 and 5.5 have gone in some depth into the requirements for external data processing, the interlinkage of several machines, and production control methods. And although these factors were considered primarily from the point of view of the direct technical and physical problems with which the measuring systems must cope, their suitability for any particular purpose must, however, be judged on more extensive grounds than is the case for the internal data processing in machine tools. At the end of this chapter some general conclusions will be drawn concerning the first group (Chapters 3, 4, 5.1, 5.2, 5.3) thus preparing the way for the further discussion of internal data processing that is to take place in Chapters 6, 7, 8, and 9. The second group (Sections 5.4 and 5.5) cannot be considered usefully until after Chapters 11 and 12.

The first factor to be considered when comparing systems is the cost of the computer, as shown in Fig. 93. For computers, however, conditions are somewhat different from those for the measuring systems used in machine tools. In particular, the somewhat arduous conditions under which machine tool measuring systems have to operate result in some factors other than cost playing a major role; the most important of these are:

a) Reliability of operation, possibly in conjunction with the largely standardised control systems produced by various manufacturers:
b) Ease of fitting to various machine tools;
c) Simplicity of maintenance
d) The desire of users of machine tools and control systems to keep their stocks of spare parts to a low level and for as many standard items as possible to be included in these stocks.

Whereas a few years ago it was the principle of operation of measuring and control systems that was often of primary interest, nowadays it tends to be

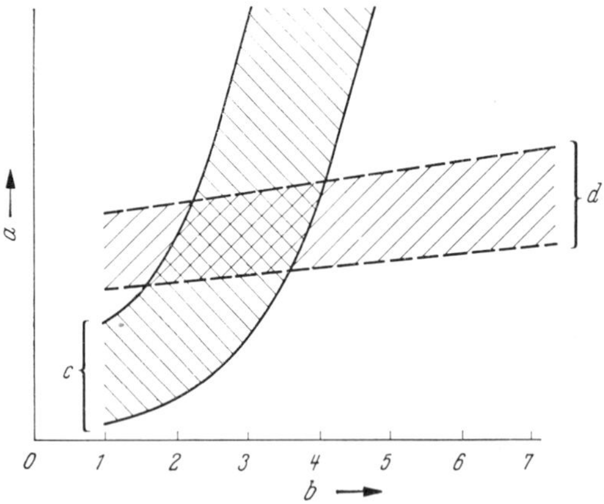

Fig. 93 Cost trends as a function of accuracy for analogue and digital computers (qualitative indication (68, 69))

a Costs (capital costs, rental). *b* No. of decimal digits. *c* Range of analogue computers available. *d* Range of digital computers available.

more the operational characteristics which decide the choice of system. For this reason some of the conclusions that have already been presented in earlier chapters will be repeated here to enable a complete survey to be made. The factors that are of major importance in making a comparison are indicated by vertical lines.

Digital-incremental process

1. The incremental process is a displacement measuring process; the instantaneous position of a machine slide can be determined only by the summation of all the increments. The system must be protected from spurious pulses when it is used for the control of machine tools. This requires careful design and manufacture of all the components together with the adoption of protective measures of greater or lesser degree of complexity. Whether the method then retains its advantage of simplicity compared with other possible systems will have to be checked against the points made in Chapters 4 and 5.
2. It is very simple to set the datum-point on the machine tool.
3. Because of the high rapid-traverse feeds that are encountered, displacement measuring systems should employ contactless scanning methods. Magnetic scanners offer certain advantages (high reliability, low maintenance, etc) for rotating systems; only photoelectric scanners are currently suitable for linear systems.
4. Careful consideration must be given to the need for providing a numerical display if a low-cost control system is desired. It seems probable that digital position read-outs will continue to be widely employed for certain applications, even without automatic comparison of the actual and desired values, since they simplify the task of making measurements from a central position whilst the machine is running (reduction in idle times required for making measurements).
5. On the question of the most suitable type of counter, and bearing in mind the need for reliability and low cost, a start will be made by considering the kinematics of the machine slide and dividing the motions and the associated control principles into three categories:

Group I: On-off circuits without reversal of counting direction.
Group II;   On-off circuits with reversal of counting direction.
Group III: Continuous-control circuits with constant change of direction and continuous monitoring of the slide motions.

6. Since continuous control circuits are required only in continuous-path control systems, Group III will in practice be encountered only in this application. The question of the suitability of digital-incremental control circuits in conjunction with internal interpolators and magnetic-tape control systems will be discussed again in Chapters 8 and 9.

7. Arrangement of the systems in Groups as proposed in Point 5 applies only if an individual optimum solution is being sought. If the matter is regarded more from the point of view of the control system manufacturer and his production problems, a standard range of components that can be used to build up a suitable system is preferable, even if in some cases this leads to a slightly more complex arrangement. With semiconductor devices, to reinforce the tendency for the prices of electronic equipment to drop, it is advisable to work towards the standardisation of components and groups of components, and to avoid excessive differentiation. This also simplifies the task of maintaining stocks of spare parts.

For the *digital-absolute process* it was found that:

8. It is possible to make coded linear scales and scanners (as a development of the uncoded linear measuring scales used in the incremental process) at an acceptable cost and technical outlay; at present these scales are, however, commercially available only for the binary system.
9. In practice, rotary position measuring systems employing photoelectric scanners are most frequently used at present. It is necessary for:
   a)   a suitable code to be chosen, which is suited to subsequent processing in the associated comparator, interpolator, etc.
   b)   the numerical pick-up system to be reliable and free from ambiguity. This is usually achieved by the adoption of V-scanning.
10. Virtually only semiconductors are used in the construction of numerical comparators, in some cases in the form of integrated circuits. It is necessary to distinguish between three types, primarily for economic reasons:
   a)   Simple numerical co-incidence checkers which are not direction-sensitive (for the construction of simple on-off circuits, or in some cases for the construction of stepped on-off circuits).
   b)   Numerical three-point controllers.
   c)   Numerical proportional controllers.
   The number of digitally-operating components and the cost of the circuits increases as one progresses from a) to c) above.
11. The greater number of components and coupling elements required as compared with the digital-incremental displacement measuring system makes it necessary to take into account the size of the machine (maximum displacements) and the minimum kinematic requirements of the machine slides (choice of displacement element Δs. Decision as to the need for employing on-off circuits with stepped operation,

continuous and intermittent control loop circuits, etc) requires careful consideration. To bring machine slides having appreciable masses to a stop it may be necessary to provide four or five preliminary stopping signals (cf. Chapter 7), so that the difference in the costs of S type and R type comparators tends to disappear.

12. The cost of the scanners and the conditions that the comparator is required to meet usually make it uneconomical to provide additional tracks for a redundant measuring code system.

13. If a datum-point shift is required this will prove relatively expensive compared with the incremental process; it can, however, simplify the work considerably when a numerically controlled machine is used (cf. Chapters 7, 8, and 9). The trend towards the use of standardised integrated circuits will probably cause this initial cost disadvantage of the absolute system to disappear.

Analogue process (taking Sections 4.2 and 4.3 into account)

14. Only contact-less measuring systems are suitable for machine tool applications; here inductive equipment predominates.

15. The working range of any single device (e.g. a synchro) for making absolute measurements is comparatively small; if angles or displacements exceeding this range are measured there will be cyclic repetition of the measurements.

16. Linear inductive displacement measuring systems can be regarded as multi-pole synchros and must be treated as such electrically.

17. As a result of the periodicity, there are two possible ways in which inductive analogue measuring systems can be used as displacement data transmitters:

   a) By close coupling of a system with a continuously variable command per axis by means of a carefully designed follow-up control loop circuit; this is referred to as 'synchronous operation'. The follow-on control loop can be either of analogue or digital type.

   b) Extension of the measuring range for analogue-absolute displacement measurement by using gearing between a number of individual systems.

18. Subdivision into three systems (coarse, medium, and fine) is predominant. In practice the resolution capacity of such an arrangement is about $1 : 10^5$.

19. Synchros form a very simple, relatively robust, and inexpensive form of displacement measuring system.

20. Higher frequencies (400 to 2000 Hz) result in small electrical components and in low moments of inertia for synchros.

21. The analogue comparator designs are very simple for all applications (on-off circuits, continuous and non-continuous control circuits); in

      addition they can readily be arranged to produce preliminary signals (e.g. for speed changes).

22. The basic task of coupling continuous displacement, angle, or position information with a numerical value is transferred away from the measuring system itself into a separate digital/analogue converter, in which respect this process differs from the digital-absolute measuring systems. In many instances this merely also transfers the cost and the technical problems. Digital calibration is nowadays often preferred (cf. Chapters 8 and 9).

23. Shifting and setting the zero-point can easily be carried out by means of differential synchros.

24. It is possible to display the actual values, but this will be more expensive than with a digital process.

At the beginning of this chapter three processes were described separately to supplement these twenty-four points. Of these, the first opened up a completely novel approach; the effective coupling of two independent numerical slide control systems by contact-less dog operation. The principal conclusion to be drawn was that such a process could be an attractive proposition in conjunction with high precision positioning control systems on large machine tools. The other two systems are basically modifications of the synchro principle; in both cases it is possible to achieve 'synchronism' with an elecronic computer (cf. Chapter 8) if a well-designed analogue or digital control circuit is used, so that these can be regarded as examples of 'synchronous operation'.

If the various conclusions and comments are to be summarised, it is clear that the desirability of each of the many alternatives will depend on whether the point of view of the designer (of machines or control systems) or that of the user of NC machines is decisive.

Cost comparisons serve no useful purpose, since any measuring system can be judged only as a part of a control system or of a complete NC machine; in any event an investigation of this type has already been carried out (10). Despite these reservations, the student should perhaps have some basis on which to form a judgment, and some of the advantages and disadvantages of the various systems are therefore listed in Table 10, together with the probable applications. One important factor for the selection of a displacement measuring system is the size of·the machine tool or the magnitude of the maximum slide movement that is to be measured. The complexity of the machine is also of importance, i.e. the number of axes that are to be controlled numerically and their functional relationships. Table 11 therefore lists three groups of slide movement ranges, and the most likely measuring systems, taking the required type of control system into account, are allotted to each. To do this it will be necessary to anticipate some of the matters discussed in Chapter 8 (internal interpolators), Chapter 9 (input equipment and data carriers), and Chapter 11 (external interpolators) since the choice of measuring system also depends on factors which are to be

	Digital Systems		Analogue Systems		Two-part Systems
	for incremental measurements	for absolute measurements	for absolute measurements	for synchronous operation	
Advantages	Simple measuring scales for linear and rotary scanning. Simple comparators for use in on-off circuits. (Pre-selector counter). Simple selection and shifting of datum. Simple numerical display of actual values. Measuring accuracy of 0.01 mm attainable even by comparatively simple means.	Position measuring process; cumulative measuring systems cannot be formed. Simple numerical display of actual values (position monitoring).	Simple measuring systems for rotary pick-up (synchros); relatively simple measuring scales for linear pick-up (e.g. Inductosyn, Accupin). Simple datum shift using analogue techniques; for operational reasons increasing use is being made of digital datum correction with aid of small computers (cf. Chapters 8 and 9). Very simple comparators for on-off and continuous control circuits.	Very simple measuring system (e.g. one synchro). Very simple datum shift in analogue range. Simple comparators using analogue techniques; more complex, but more sensitive comparators using digital techniques.	High-precision requirements limited to a few components (e.g. precision measuring scale with widely-spaced graduations) and small displacements (e.g. displacement of auxiliary scale). Free choice of remaining measuring systems, so that it forms relatively simple extension of existing standard types (mainly digital-absolute or analogue-absolute).
Disadvantages	Displacement-measuring, not position-measuring system. Summation errors may arise when cumulative measuring systems are formed. Precautions taken to avoid these errors may reduce the simplicity of the process. Continuous control circuits (digital-synchronous operation for continuous-path control systems) complicated relative to analogue systems.	Linear scales for binary system commercially available, but costly. Comparators are relatively costly, depending on the functions they are required to perform. For economic reasons the requirements must be classified into groups. Relatively complicated datum shift (computers).	Relatively complex digital/analogue converters (transfer of design and cost difficulties from measuring system and comparators to D/A converters). Stringent construction requirements of mechanical coupling elements (gearing). Display of actual values is possible but more complex than with digital systems	Machine stops or produces defective components if 'electrical shaft' breaks. A continuous command must be present, realisable in practice only by an electronic computer (internal interpolator) or a magnetic tape (external interpolator). Impulse generators will also suffice for continuous path control systems and digital control circuits.	Precision scale with widely-spaced graduations must be specially made. Not suitable for continuous-path control systems. Poorly suited to machines requiring a rapid succession of position changes.
Appears technically and economically suitable for:	Single-stroke motions. Continuous-path control systems with digital internal interpolators and magnetic tape controls. Coordinate table settings with small and medium displacements and also small to medium amount of positional data.	Positioning, straight-line, and continuous path control of heavy machine tools of all types in which increased cost of the control system is of secondary importance because of the undoubted advantages of this method.	Positioning, straight-line and continuous path control systems on small and medium-sized machine tools.	2-D to 5-D and multi-axis continuous-path control systems using internal and external interpolators (cf. Chapters 8 and 11) on large machine tools with long slide traverses or with a large number of simultaneous motions.	Two-and three-dimensional positioning and straight-line control systems on large machine tools where a high degree of accuracy is required (e.g. jig borers).

Table 10. Comparison of Characteristics of Measuring systems.

Range of slide displacements	Positioning and straight-line Controls		Continuous-Path Controls Two and Multi-Dimensional	
	One-Dimensional	Two and Multi-Dimensional		
Less than 400 mm	Digital-incremental		Digital-incremental digital-absolute	(e.g. for connection to internal interpolators cf. Chapter 8)
400 to 1000 mm	Digital-incremental	Digital-incremental	Synchro-synchronous operation	(e.g. with magnetic tape control, cf. Chapters 8, 9 and 11).
1000 to 2500 mm		Digital-incremental digital-absolute	Digital incremental digital-absolute	(e.g. for connection to internal interpolators cf. Chapter 8)
		analogue-absolute	synchro-synchronous operation	(e.g. for connection to internal interpolators and with magnetic-tape control
Above 2500 mm		Digital-absolute analogue-absolute	Synchro-synchronous operation	(cf. Chapters 8, 9, 10 and 11)

Table 11. Main areas for the application of displacement measuring systems.

discussed later; this is particularly the case with continuous-path control systems. This anticipation of later chapters is, therefore, necessary to enable the section on "Displacement Measuring Systems and Comparators" to be concluded.

Not only students, but also designers of NC machines, should find the approach presented in Tables 10 and 11 of interest. It is also possible to consider a question which is of considerable practical importance, namely which process could be used as a 'standard process' for the majority of machine tools. This question is closely connected with investment decisions that are made by the users, and has already been discussed in detail elsewhere (10). Today the installation of NC machines and of data processing equipment is becoming widespread as a means of modernising manufacturing plants. And during the next few years there will be many companies about to acquire their first or second NC machine and others who already have an appreciable complement of NC machines and the corresponding 'know-how'. The desire for standardised control systems - and hence for standardised measuring systems - will be much more marked with the second group of users than with the first, because they must be able to arrange their maintenance, instructional, and spare parts services as economically as possible. Tables 10 and 11 can therefore serve as rough guides only for systems analysis and synthesis.

In sections 5.4 and 5.5 some problems of external and integrated data processing have been discussed in some depth, although they will be dealt with in detail only in Chapter 12. They will therefore be summarised in Chapter 12.

6   Drive systems

Mention has already been made in the first chapter that every control process represents a specific association between signal and energy (28). Up to the present only the signal components (in the widest sense of the term) have been discussed; mention has been made of desired values, actual values, processing of measurements, etc., and the respective signals have played an insignificant part as the conveyors of energy. In this chapter, however, a closer look will be taken at the energy components of the control process.

Ultimately all control processes in machining and forming operation have to regulate the flow of mechanical energy. The transistion from (predominantly) electrical signals of low power to a mechanical movement normally occurs in three stages:

amplification - control of energy flow - energy transformation.

Equally it is often possible to divide the equipment into three distinct groups:

amplifier - regulator - actuator.

Electrical amplifiers need not be discussed in this book, as they are well covered elsewhere (70). It seems desirable, on the other hand, to discuss the available forms and development trends of regulators and actuators in some detail. One reservation must be made; the discussions will be concerned with the feed drive systems on machine tools rather than the main spindle drives, even though the latter transmit the major proportion of energy. The feed drive is an essential feature of numerical control because it is primarily concerned with the movements of the machine slides and their operation.

In view of the many requirements that feed systems have to meet, and the methods that are available for satisfying them, it is convenient to divide them into five groups, and to discuss them as follows:

6.1   Stepped electro-mechanical drives.
6.2   Infinitely-variable electric drives.
6.3   Stepped and infinitely-variable hydraulic drives.
6.4   Comparison of infinitely-variable electric and hydraulic drives.

6.5   Stepping motors for incorporation in digital-incremental control systems.

6.6   Pneumatic-hydraulic drives.

6.7   Summary.

6.1   Stepped electro-mechanical drives

Feed systems employing stepped drives, which may be of many types, are widely used for positioning and straight-line control systems in conjunction with on-off switching circuits. Their robustness and simplicity gives the electro-mechanical devices many advantages under the difficult conditions associated with workshop use. Here reference will be made to only one possible design feature, namely the use of remote-controlled clutches in stepped drives.

Remote-controlled clutches in stepped drives

Clutches can be remotely-controlled by means of oil under pressure, compressed air, or solenoids. Whichever of these three methods of actuation is used makes no difference to the basic character of the clutch as a means of controlling the flow of mechanical energy. Electro-magnetic actuation offers a number of advantages:

a) Electrical input signals can easily be passed through logical linkage circuits (interlocks) to prevent erroneous coupling combinations being established.

b) The switching times of mechanical clutches are about 10 to 100 ms under load, depending on the design and the load to be transmitted, and such times are well within the acceptable limits.

c) No auxiliary power supply (compressed air, hydraulic pressure) is required to operate the clutches.

Points a) and c) often result in a desirable minimum outlay (Chapter 1), as compared with other types of remote-control for clutches, so that there is every justification for limiting these considerations to electro-magnetic clutches. Figure 94 shows two of the basic forms of electro-magnetic clutch. In Type A) steel plates are generally used. The magnetic flux emitted by the solenoid *a* can be closed by way of the clutch plates and the armature disc *f*. Due to the remanent magnetism (electro-magnetic clutches are always excited by direct current) and the need for lubrication [1] the disengagement time

---

[1]   Steel plates of the same material tend to fret if they rub together when dry.

varies and is of relatively long duration; this should be borne in mind when designing the control circuit. The simple and robust construction, and the ease of cooling by the oil circulation, makes this type of clutch particularly suitable when high power values have to be transmitted and frequent switching is required. For this reason they are often used for remote-controlled main spindle drives. If they are carefully designed and due allowance is made for their particular characteristics, these clutches can also be used in feed drives. Their main advantage is that they are self-adjusting and require virtually no maintenance.

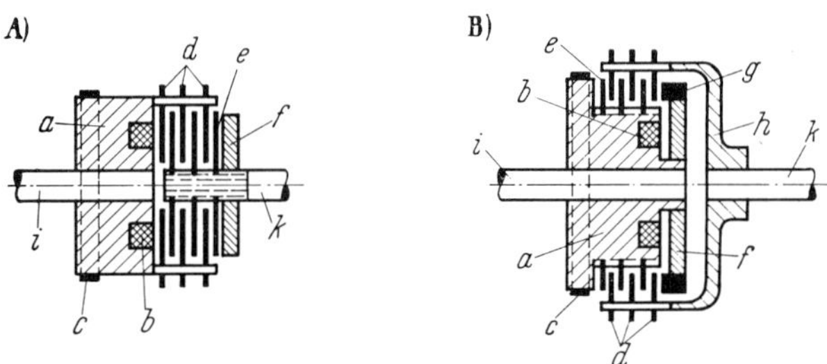

Fig. 94   Schematic diagram of operation of electro-magnetic clutches.

A) Electro-magnetic clutch in which the magnetic flux passes through the stack of plates. B) Electro-magnetic clutch in which magnetic flux does not pass through the stack of plates.

a Magnet body (soft iron, solid). *b* Winding. *c* Power supply (1 or 2 slip rings). *d* Outer plates. *e* Inner plates. *f* Armature disc (soft iron). *g* Adjustable pressure ring (usually made of non-magnetic material). *h* Carrier for outer plates. *i* Driving shaft. *k* Driven shaft (splined in clutches of Type A) to carry the inner plates).

Figure 94B) shows an electro-magnetic clutch in which the stack of plates is completely unaffected by the magnetic flux; as a result any materials can be used for the plates such as sintered bronze, copper-asbestos friction linings, etc., depending on the frictional effects required. These clutches can therefore be run dry while the absence of any remanence or oil drag ensures that accurate, short, and especially uniform clutch release times are attained. On the other hand, clutches of this type are larger and heavier than clutches of the type shown in Fig. 94A) for the same duty, and they require frequent adjustment. Because of their good switching characteristics when running dry, they are frequently used in conjunction with two-point controllers for large copy milling machines (5, 71, 72). In theory they could also be used for numerical continuous-path control and the associated follow-up control circuits to avoid the need for a continuously-running infinitely-variable drive. Because of the large number of alternative solutions, which will be described later, this arrangement is, however, not very suitable (irregular operation, maintenance of clutches, etc.). In certain instances the combination of a clutch-operating mechanism of this type with electric continuously-variable drives may result in design simplifications for very low speeds (cf. Section 6.2). With positioning and continuous-path control systems rapid traverses of 10 to 12 m/min are by no means uncommon to reduce the idle times to a minimum; while inching speeds of 3 to 4 mm/min, to avoid overshooting a desired position where on-off switching circuits are used, are also found ("Accuracy". Chapter 10). In such cases the speed of the leadscrew, and hence that of the feed gearbox shaft, must be adjustable over a ratio of 3000 : 1 and more. With positioning control systems (e.g. for the positioning of coordinate tables on boring machines) theoretically only one rapid traverse speed and one inching speed are needed for each feed drive to solve the movement problem. If on-off switching circuits are used it is essential that the moving part does not overshoot the desired position, since it is not possible to make corrections such as are possible with cam-operated control systems. On the other hand, to avoid wasting time through using inching speeds for large displacements, several low feeds can be engaged before the final inching operation begins, so as to absorb gradually the kinetic energy of the moving masses[2]; the number of low feeds to be provided in any particular case will depend on the size of the machine.

With these positioning controls, the control system should be designed to ensure that the gradual speed reduction down to the final inching speed is performed automatically (through suitable reduction gears and electro-magnetic clutches), so that apart from stating the dimensions and giving the switching command 'rapid traverse ON', no further information need be supplied by the programmer (cf. Chapter 10).

Matters are more complicated with straight-line control systems in which a tool is in contact with the workpiece along one axis of the machine at all times. It is now necessary to arrange the machining feeds between the rapid traverse and the inching speed in such a manner that they meet the required machining conditions (drilling, countersinking, milling, reaming, etc, of various materials). The number of feeds that is considered desirable on a machine tool and their range varies to an extraordinary extent (5, 71, 73). The requirements range between about four machining feeds on the one hand, and a wide range of infinitely-variable speeds on the other hand. Remote-controlled feed gearboxes using more than, say, 8 or 10 electro-magnetic clutches are no longer competitive with other possible methods which will be described later. From the control point of view, each speed of a gearbox is selected by means of a switching command; the electrical switching circuits connected in series with the electro-magnetic clutches can, of course, be very largely designed to suit the operating conditions (safety precautions, interlocking with other machine functions,

---

[2] In most instances the rotating masses present a particular problem.

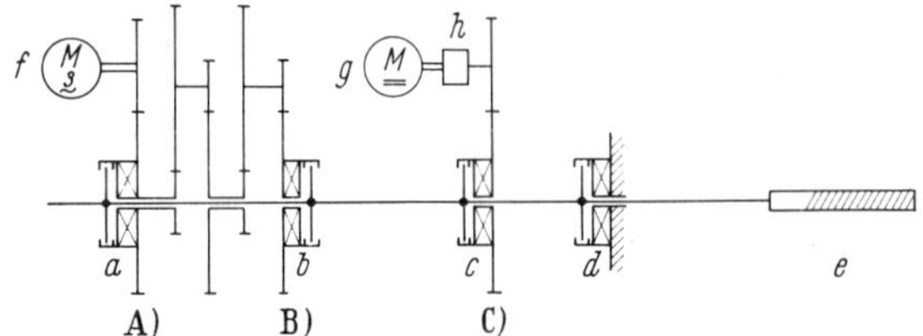

Fig. 95 Principle of operation of a feed and positioning gearbox for straight-line control (74)

A) Rapid traverse.   B) Inching speed.   C) Feed (machining) speeds (continuously-variable).

a Electro-magnetic clutch for rapid traverse. *b* Electro-magnetic clutch for inching speed. *c* Electro-magnetic clutch for machining feeds with straight-line control systems. *d* Electro-magnetic brake. *e* Leadscrew. *f* Three-phase motor (e.g. n_0 = 1500 rev/min) for rapid traverse and inching speeds. *g* D.c. motor for infinitely-variable speed adjustment (50 - 1500 rev/min) for machining feeds. *h* Intermediate gearing.

etc). Figure 95 shows the principle of operation of a feed and positioning gearbox for straight-line control which combines effectively electro-magnetic clutches and three-phase and d.c. motors each with their characteristic features, thereby arriving at a minimum cost solution (74).

6.2   Infinitely-variable electric drives

As has already been mentioned in Chapter 2, infinitely-variable drives can be used with positioning and straight-line control systems, although they are not essential; with continuous-path control systems, on the other hand, infinitely-variable drives and easy reversibility are an essential requirement. The main spindles of machine tools can be driven independently by means of stepped drives.

Three-phase power supplies, which are nowadays almost universally available, cannot be used directly for providing an infinitely-variable reversing drive; the normal three-phase commutator motors are unsuitable for this purpose. Electro-mechanical solutions (mechanical infinitely-variable gears driven by a simple three-phase squirrel-cage motor) are also not suitable since difficulties are encountered with rapid reversal and with remote control. The solution that is most commonly adopted at present for electric drives is, therefore, to use a d.c. motor, and to convert the available three-phase current supply into d.c. current. Even though the construction and method of operation of a d.c. motor will be generally familiar (75), it seems advisable to mention the main features briefly at this point to make it easier to follow the later arguments and conclusions when comparing this solution with other possibilities.

6.2.1 The d.c. motor

Construction and operational characteristics

Figure 96 shows schematic sections through a d.c. motor of conventional design; motors and generators are of similar construction. The most important winding arrangements, circuits, and standard symbols are shown in Fig. 97. For our purpose the following features are important:

a) The parts that require the greatest attention are the brushes g and the commutator b because of the friction and hence mechanical wear. Damage can also occur easily if static or dynamic overload occurs (e.g. arcing at the brushes if current is excessive). Regular maintenance is essential.

b) The principal characteristics determining the size of the machine from our point of view are solely
the armature diameter d_1 and
the armature length l_1.

Over the relatively small power range that is currently encountered in feed drives for numerically controlled machine tools, the permissible

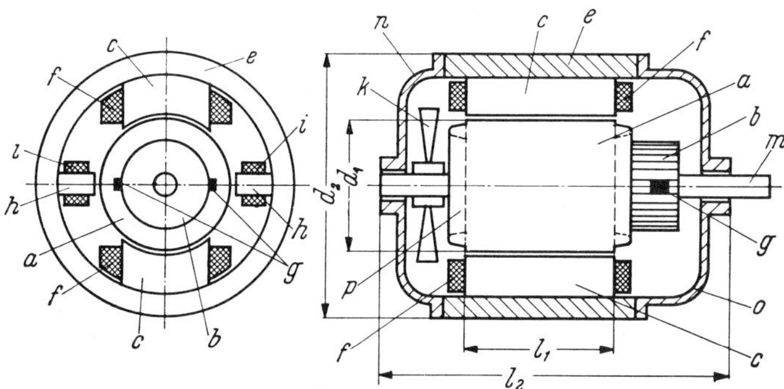

Fig. 96 The d.c. motor

a Armature (laminated). b Commutator. c Main poles. d_1 Armature diameter. d_2 External diameter of motor. e Yoke (magnetic return path and also motor casing). f Main pole winding. g Brushes (for supplying current to or from armature via commutator). h Reversing poles, i Reversing pole winding. k Cooling fan. l_1 Armature length (effective iron only). l_2 Overall length of motor. m Motor shaft.

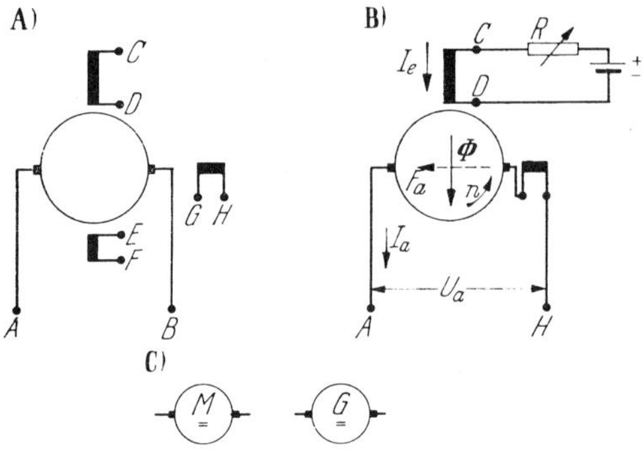

Fig. 97 Circuit diagrams showing operation of d.c. motors.

A) Designation of terminals and windings to VDE 0570. B) Circuit for a remote-excited shunt-wound motor. C) Circuit-diagram symbols for d.c. machine.

A-B Armature circuit (brush connections). *C-D* Excitation circuit, shunt winding (main poles). *E-F* Main current winding on main poles (compound windings). *G-H* Reversing pole winding. I_a Armature current. I_e Excitation current for main poles. F_a Armature field (proportional to the armature current, armature back e.m.f.) *n* Speed and direction of rotation (for given directions of F_a and ϕ) *R* Adjustable series resistance in excitation circuit. U_a Armature voltage. ϕ Magnetic flux of main poles (produced by excitation I_e).

utilisation of the material, as far as the current density and the magnetic induction are concerned, may be regarded as being approximately constant. For the torque, which is the characteristic of most interest for feed drives, the following simple relationship therefore holds for d.c. motors:

$$M \approx c_1 \cdot d_1{}^2 \cdot l_1 \quad (\text{if } \alpha \cdot B_L = \text{const.})\,{}^3$$

Since the armature of a d.c. machine can be regarded as a cylindrical drum, the radius of gyration of which is almost equal to the armature

---

[3] The output of a d.c. machine is

$$P \approx 4 \left(p \cdot \frac{n}{60} \right) \cdot (d_1 \cdot \pi \cdot \alpha) \cdot \left(\frac{\pi d_1}{2p} \cdot 0{,}7 \cdot l_1 \cdot B_L \right) \cdot 10^{-8} W \quad [99].$$

$$P \approx C_3 \cdot d^2 \cdot l_1 \cdot (a \cdot B_L) \cdot n$$

where B_L = magnetic air gap inductance under a pole
     α = ampere-turns per cm armature circumference
     p = Number of pole pairs
     n = Speed in rev/min

radius owing to the concentration of the copper at the periphery, the moment of inertia is:

$$I_\mathrm{p} = \int r^2 \ \mathrm{dm} \approx c_2 \cdot d_1{}^4 \cdot l_1$$

These simple relationships between the torque and the moment of inertia show clearly that for the case of high feed forces (high torque on feed drive) and high dynamic loadings (large accelerations and decelerations in all speed ranges) d.c. motors will present difficulties. These can only be overcome to some extent by measures taken at the design stage of the machine tool (low frictional losses in the leadscrews and guide ways, etc., keeping all moving masses to a minimum (Chapter 7). In general, it can be said that the use of d.c. motors becomes more difficult as the cutting conditions become more severe, but is more favourable where light alloys are being machined, e.g. in aircraft manufacture. In limiting cases use will have to be made of special d.c. motors having armatures of small diameter and long length, with the cooling air supplied by external fans (Fig. 98) (74, 94, 102).

c) With normal d.c. motors (with a relatively small speed range) it will usually suffice if a fan *k* (Fig. 96) is fitted to remove the heat produced by the losses in the motor. When continuous-path control is

Fig. 98
Example of the use of a
special d.c. motor on the slide
of a n.c. machine tool

a D.c. motor having a particularly small armature
diameter and long armature
length (low moment of
inertia). *b* Built-on three-phase
fan motor for external cooling
of d.c. motor. *c* Tacho-
generator for stabilising the
control circuit (speed feedback). *d* Recirculating-ball
screw to reduce frictional
losses (cf. chapter 7).

applied to a machine tool (where the motor may run for long periods at very slow speed in either direction of rotation) such a fan becomes ineffective; with heavily-loaded machines it is generally necessary to provide external cooling with the fan driven by a separate motor (Fig. 98).

d) The speed n of a d.c. motor is given to a first approximation by the relationship:

$$n = C_4 \; \frac{U_a}{\phi} \quad \text{(Fig. 97)}$$

and the torque M that is produced is calculated from the electrical and magnetic quantities I_a and Φ by the relationship

$$M = C_5 \; . \; I_a \; . \; \Phi \quad \text{(Fig. 97)}$$

C_4 and C_5 are machine constants; the losses are ignored for the present. The result of this consideration is shown in Fig. 99 for one direction of rotation. The nominal torque M_n (100%) is determined by the maximum possible magnetic flux Φ_n, which is limited by the saturation of the iron for any given dimensions, and by the maximum permissible continuous current in the armature $(I_a)_n$ (which for any given dimensions is limited by the temperature rise of the windings and the commutator loading). The speed can be varied by two methods: variation of armature voltage U_a and weakening of flux Φ_n. An increase is impossible because the iron is saturated).

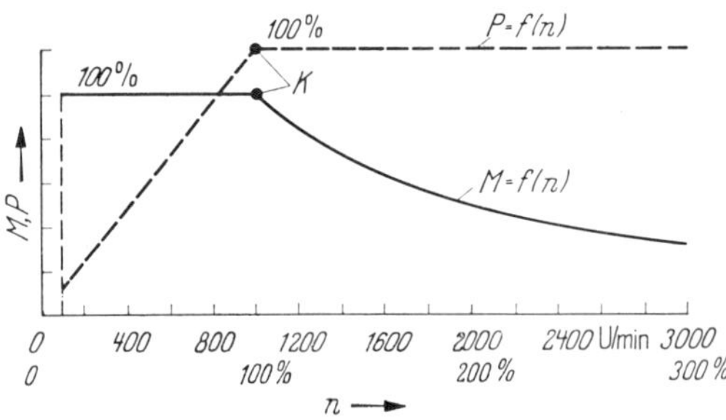

Fig. 99   Torque and power variation with change in speed of a d.c. motor over a range of about 1 : 30 (losses ignored).

K Point of inversion (rated values). *M* Torque. *P* Power. *n* Speed.

The magnitude of the armature voltage is, however, also limited electrically (insulation, brush arcing, etc), so that here again it is possible to speak of a rated voltage $(U_a)_n$ for any given machine dimensions. The rated power P_n is then given by

$$P_n = (U_a)_n \cdot (I_a)_n = C_6 \cdot M_n \cdot n_n.$$

In Fig. 99 P_n has also been made equal to 100%.

If the armature voltage is altered (flux Φ assumed to remain constant) the power will increase up to the point K (constant torque, increasing speed); from that point on it is not possible to increase the speed by varying the armature voltage since this has reached its upper limit.

From this point on it is only possible to reduce the flux Φ. The power will then remain constant, but not so the torque. The speed variation that can be achieved at acceptable cost by reducing the flux is about 1 : 3 (100% to 300% of n_n). Since similar conditions obtain with hydrostatic motors (cf. Fig. 114) the decision where the point K should be located is of great importance in the design of infinitely-variable drives.

e) Conditions are especially critical at speeds near zero. Figure 100 shows, in somewhat simplified form, the torque-speed characteristics for a d.c. motor, operating under reversing conditions and with the losses taken into account. If it is assumed that between idling and full load ($M \triangleq 100\%$) the speed drop due to the current-dependent voltage-drop in the armature, the armature back e.m.f., etc., amounts to about 8% of n_{max}, the drop will become much more marked at lower speeds (lines a_1 to a_5). Near the point of reversal (zero speed) no stable speed conditions will obtain unless special precautions are taken. Effective means of avoiding this unwelcome phenomenon include the provision of compensation windings to nullify the ill effects of the armature field (armature back e.m.f. zero) and the introduction of speed-stabilising control circuits. Lines b_1 to b_4, for

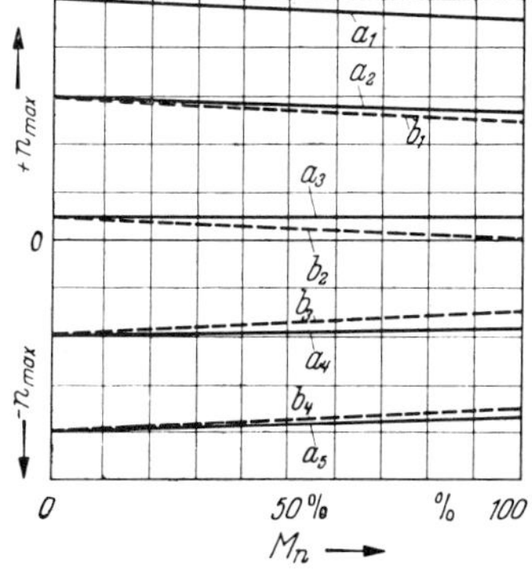

Fig. 100
Speed of a d.c. motor as a function of load for a very wide range of speed variation (taking losses into account) of $+n_{max}$ through 0 to $-n_{max}$.

a_1 to a_5 Speed variation when a speed-stabilising control circuit is employed to maintain a constant percentage speed drop. b_1 to b_4 Speed variation when a speed-stabilising control circuit is not used. n Speed. M Torque.

example, represent the stabilised load characteristics for a speed variation of about 2% of the actual initial value (not of n_{max}). The cost of this may, however, be high, especially if high torques have to be delivered at the lower speeds. But the design can be simplified by the incorporation of a reduction gear equipped with an accurately operated clutch of Type B (Fig. 94), where the clutch is controlled intermittently by a two-point controller (3, 12).

6.2.2. Control of d.c. motors

From what has been said, the problem can be described in general terms as follows: A variable armature voltage U_a and an armature current I_a (to transmit the energy) as well as a small excitation current I_e, variable approximately over a range of 1 : 3 (for building up the flux Φ in the motor) must be derived from an existing three-phase supply. Since a number of new components have been introduced during the past few years, three methods will be briefly described and compared in this chapter. Functionally all these three units may be regarded as regulators; the d.c. motor with which they are associated acts only as an energy converter.

Control by means of a Ward-Leonard set

The principle of this old-established method of control is illustrated in Fig. 101. The heart of the set is the Ward-Leonard generator d (regulator) which is

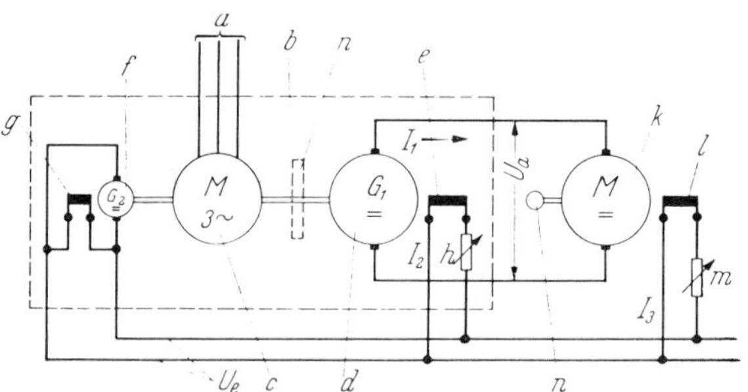

Fig. 101 Basic circuit of a Ward-Leonard control

> a Three-phase supply. b Ward-Leonard set. c Three-phase motor. d Armature of Ward-Leonard generator. e Field coil of Ward-Leonard generator. f Armature of exciter generator. g Field coil of exciter generator. h Variable series resistance. k Armature of controlled d.c. motor. m Variable series resistance for this n Tachogenerator for accurate speed measurement (provides actual value for use in speed-stabilising control circuit). I_1 Armature current. I_2 Excitation current of Ward-Leonard generator. I_3 Excitation current of controlled d.c. motor.

driven at approximately constant speed by the three-phase motor x. In many cases, especially in the larger models, a flywheel is mounted on the shaft joining the generator and the motor to prevent transient heavy shock loads (e.g. in planers, rolling mills, etc) from affecting the three-phase supply. The exciter machine f is used in this example only for generating a constant d.c. voltage for control purposes; it could equally well be replaced by a dry rectifier. Altering the resistance h causes the U_a to be varied (variation in armature voltage, constant-torque operation in Fig. 99). Controlling the exciter resistance m causes the excitation current I_3, and hence the flux in the output drive motor K, to be reduced (flux variation, constant-power operation in Fig. 99). The tachogenerator n can be connected to a

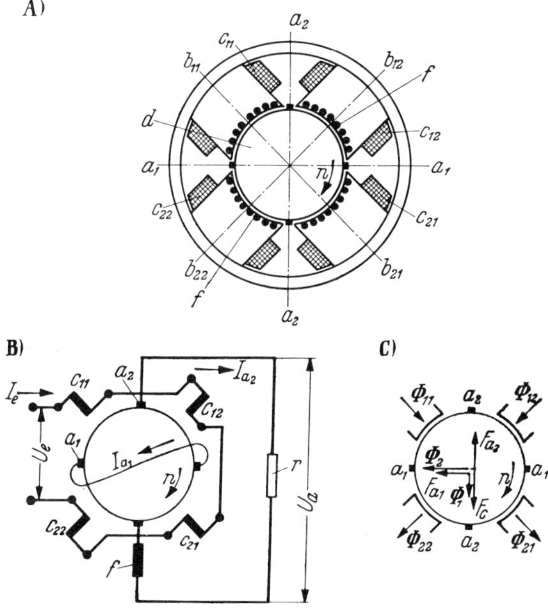

Fig. 102 The basic design of an Amplidyne (amplifier machine)

A) Mechanical construction of machine. B) Circuit. C) Field distribution.

a_1 Brush system 1. a_2 Brush system 2. b_{11} ,b_{12} , Corresponding pole halves (e.g. North poles). b_{21}, b_{22} Corresponding pole halves (e.g. South poles). c_{11}, $c_{12}$$c_{21}$, c_{22} Field coils. d Two-pole wound d.c. armature. f Compensation winding on pole shoes. n Speed and direction. r Load resistance (consumer). F_{a1} Armature current field 1 (armature back e.m.f. in brush axis a_1). F_{a2} Armature current field 2 (armature back e.m.f. in brush axis a_2). F_c Field of compensation winding f (to neutralise F_{a2}). I_{a1} Short-circuit current through brush system 1. I_{a2} External load current on brush system 2. I_e Excitation current. U_a Variable armature voltage. U_e Excitation voltage. ϕ_{11} , ϕ_{12}, ϕ_{21}, ϕ_{22} ,Magnetic fluxes in main poles (produced by coils c_{11}, c_{12}, c_{21}, c_{22} and the excitation current I_e).

speed-stabilising control circuit, to which it feeds back the instantaneous actual speed value, and which acts on the excitation current I_2 (3, 58).

Semiconductors (e.g. transistor amplifiers) are being used to an increasing extent to control the excitation current I_2. By suitably choosing the dimensions of the control amplifier (coupling the actual speed n, which forms the controlled variable X with the excitation current I_e which forms the regulating variable Y), it is often possible to match the time constant of the set as a whole to the requirements of a machine tool feed drive. In view of the basic remarks that have been made regarding the suitability of d.c. motors for machine tools and the competition that is offered by hydraulic drives (Section 6.4), it is always necessary to check whether in an 'electrical solution' of this type the cost would be justified.

One of the major advantages of the Ward-Leonard system is not only the ease with which the armature voltage can be reversed by suitable design of the control circuit for the excitation current I_2, and the consequent easy reversal of the direction of rotation; but also the very simple energy reversal. In the event of electrical (regenerative) braking of the motor k (rapid reversal, for example), this acts as a generator and supplies current by way of the Ward-Leonard set back into the three-phase supply. This braking effect cannot be achieved with the same simplicity by any other control method.

Control by means of an amplifier machine

In the course of efforts that were being made to keep the excitation (control) current of a d.c. generator as low as possible, and the armature current and voltage as high as possible, several types of amplifier machine were developed many years ago. These have become less important with the introduction of reliable pre-amplifiers using semiconductors and thyristor control systems, but since they are still occasionally installed they will be briefly described. Figure 102 shows diagrammatically one of the best known designs of this type of machine, the Amplidyne. Like the Ward-Leonard generator shown in Fig. 101, this is driven by a three-phase motor; the two units (driving motor and Amplidyne) are often located in the same casing. The armature d usually has two-pole windings and it rotates in an apparent four-pole field. The coils, c_{11} and c_{12}, are connected up in such a way that the corresponding pole systems are of the same polarity (e.g. *N-N*), whilst the wiring of coils c_{21} and c_{22} is such as to result in the opposite polarity (e.g. *S-S*). The pole sequence is then *N-N-S-S*, and not *N-S-N-S* as in a normal four-pole generator. The corresponding poles are, in effect, only split, and these generators are also known as split-pole converters. The two brush systems a_1 and a_2 are independent of each other. The method of operation is briefly as follows: the excitation current I_e produces four partial fluxes of equal magnitude, Φ_{11}, Φ_{12}, Φ_{21}, Φ_{22} in the windings c; these can be combined to form the primary flux Φ_1. With the direction of rotation n as shown the short-circuit current I_{a1} will then flow through the short-circuited brush system a_1. This produces a powerful field F_{a1} in the armature (armature back effect). Assisted by the pole arrangement with its good magnetic return path, this builds up the secondary flux Φ_2, which in turn acts like a field excitation on the brush system a_2. There a voltage U_a is produced; if a load r is applied, a current I_{a2} will flow. This builds up the secondary armature field F_{a2}, which opposes the primary flux Φ_1, and which would suppress it unless suitable precautions were taken. To counteract the effects of the secondary field F_{a2}, grooves for the compensation winding f are provided in the pole shoes of the excitation poles. This winding represents what is, in effect, a mirror image of the armature winding and is dimensioned and arranged so as to ensure that the primary flux Φ_1 cannot be affected by the secondary flux F_{a2} ($F_c = -F_{a2}$).

In practice, the effect of the arrangement is that quite small excitation powers ($U_e \cdot I_e$) produce high armature powers ($U_a \cdot I_a$) (hence the term amplifier machine). Whereas with a Ward-Leonard set as shown in Fig. 101, the amplification factor will be about 1 : 30 (or considerably higher if a transistor amplifier is used in the excitation circuit), with amplifier machines, the factors will have a value of about 1 : 2000. In addition, the time factor is much superior to that of a simple Ward-Leonard set since the decisive build-up of the flux Φ_2 takes place by way of the very low resistance short-circuited armature. The time constants that result from the inductances are very small. On the other hand the design of the machine and its electrical and magnetic calibration are a little complex. The two possible solutions that have been described suffer from the common disadvantage that they involve electrical machines which require a certain amount of maintenance. The trend for future developments is however, definitely in the direction of stationary control systems. In principle these can take two possible forms:

> Transductor control (magnetic amplifier) and
> Thyristor control.

A few years ago transductor motor control systems were serious competitors of thyristor controls (78, 79); they have now, however, been largely replaced by semiconductor control systems employing thyristors, at least for NC machines. There is, therefore, no need to consider transductor control systems any further at this point (58).

Control by means of silicon controlled rectifiers[4]

The use of gas-filled valves (Thyratrons) for the control of d.c. motors on machine tools has been known for about 20 years and has also been frequently adopted in practice (80). During the last five or six years, however, developments have taken place in the semiconductor field which also appear very promising for the control of d.c. motors. These will be discussed fairly

---

[4] These are nowadays generally termed thyristors.

fully at this point so as to make it easier to judge the effects of this development later when comparisons are being made of the various systems. However, it is not necessary to go into too much detail at this point (58, 81).

The substances known as elementary semiconductors (primarily germanium and silicon) possess the property of being very poor conductors of electricity in the pure state. If very small quantities of accurately determined 'impurities' are added (doping) it is possible to produce well-defined conductive properties. A distinction is thus made between p (positive conductive) silicon and n (negative conductive) silicon. If four very thin wafers of silicon crystals doped in these two ways are joined together as shown in Fig. 103A), three boundaries c_1, c_2, and c_3 will be formed. If a d.c. voltage (which may nowadays be up to about 500 volts) is applied to the combination no current will flow initially. But if a positive voltage is applied to one wafer (switch b closed) a control current i_1 will flow, and this will immediately cause the whole arrangement to become conductive. A high working current i_2, which is limited only by the external resistance r_2 will then flow through it. This current flow will continue even if current i_1 is shut off (switch b opened). A short control pulse will therefore suffice to cause the working current to flow. This current will become zero again only when the applied voltage is removed. When a positive voltage is again applied to the top end of the four-layer combination a fresh control pulse i_1 will be required to cause the working current to flow. If the polarity is reversed no current will flow, and the rectifier is blocked. If an alternating voltage U_1 is applied to a controllable cell of this type, and if a control impulse i_1 is fed into the cell at the same point in each positive half cycle by means of some suitable device, this will strike, and a current i_2, which depends virtually only on the alternating voltage U_1 and the load resistance r_2, will flow. This current will

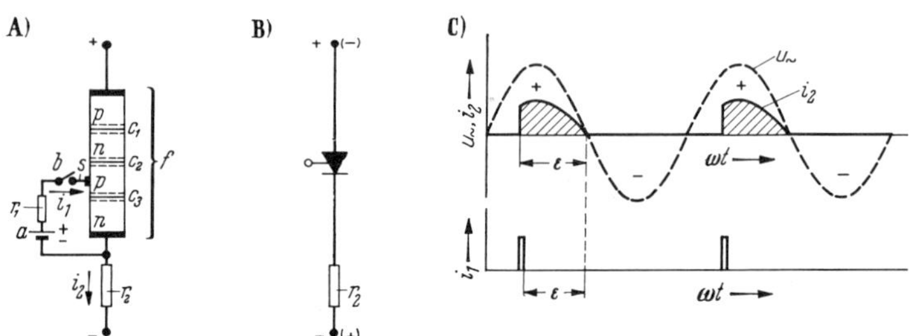

Fig. 103 Method of operation of a controlled rectifier (thyristor)

A) Diagrammatic construction. B) Circuit symbol. C) Variations of voltage and current with time.

a D.c. source for control. b Impulse switch, c_1, c_2, c_3 Boundary zones for electron exchange. f Thyristor. i_1 Control current (can be pulsed). i_2 Rectified a.c. (load current). n Negatively doped silicon wafer. p Positively doped silicon wafer. r_1 Series resistance in control circuit. r_2 Load resistance. $U\!\sim\!$A.c. voltage. t Time. ω Angular frequency $= 2\pi f$. a Current flow angle. s Control wire.

continue to flow until the alternating voltage next passes through zero, and it will remain off during the negative voltage half-cycle. The cell then acts as a blocking rectifier. Its behaviour is very similar to that of the thyratron tube or the mercury vapour rectifier (58). In these the gas charge is visibly 'struck' by applying a control impulse to the control grid, and as a result a larger or smaller section of the positive half wave is 'cut out'. Again the tube acts as a rectifier and blocks the negative half wave. With the thyratron and the mercury vapour rectifier the characteristic operation involving a gas discharge leads to the use of the term 'striking angle control', the term 'striking' can also be employed with reference to the silicon controlled rectifier, but it has also been proposed that the term 'current flow angle' control be used in this application.

Although a comparatively large number of cells is required to provide a direct current of reasonably acceptable wave form (Fig. 103) for the reversible operation of a d.c. motor, the small amount of space required and the high efficiency makes it nevertheless desirable to give serious thought to the possibility of using these controllable cells for reversing drives in continuous-path control systems for machine tools. Figure 104 shows the theoretical circuit diagrams for 6 and 12 cells (58).

Hydraulic drives are major competitors to the stepped and infinitely-variable feed drive systems that have been discussed up the the present. They will be dealt with in the next section, and will then be compared with the electrical systems.

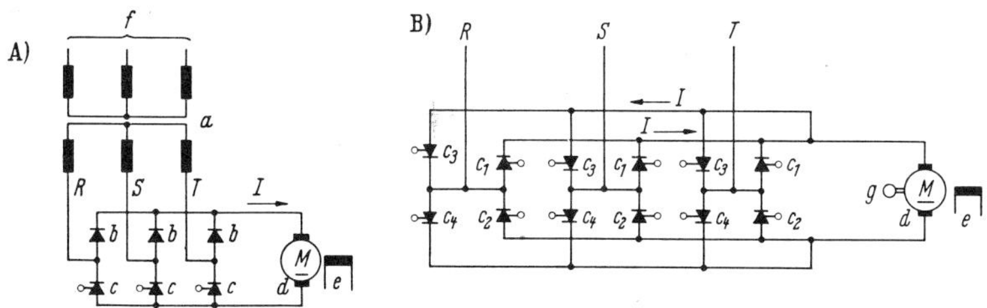

Fig. 104 Three-phase bridge circuits using silicon controlled rectifiers (theoretical circuit diagrams) (58, 81)

A) Bridge circuit for one direction of rotation of the motor where only half the rectifiers need be controlled. B) Bridge circuit for two directions of rotation of the motor (anti-parallel circuit) where all the rectifiers must be controllable.

a Three-phase transformer to match d.c. motor d/e to the three-phase supply *f*. *b* Non-controllable rectifiers (since these are in series with the controllable rectifiers, only one controllable rectifier per phase is needed). *c* Controllable rectifier. *d* Armature of d.c. motor. c_1, c_2 Controllable rectifiers for direction of rotation 1. c_3, c_4 Controllable rectifiers for direction of rotation 2. *e* Excitation winding of d.c. motor. *f* Three-phase supply . *R*, *S*, *T* Three-phase system. *g* Tachogenerator for speed stabilisation (3).

6.3   Stepped and infinitely-variable electro-hydraulic drives

In a brief survey of this type it is possible to deal with only a few of the
general principles and basic designs selected from the wide field covered by
this title. In many cases it will be essential to make a careful study of the
literature (82, 83, 84, 85).

Some years ago a popular catchphrase in a German firm was

"Electricity provides the nerves - hydraulic power the muscles".

There is, in fact, a great deal of truth in this brief phrase so far as the
numerical control of machine tools is concerned. As will be shown below, the
'hydraulic muscles', i.e. the power units, prove in many cases superior to all
other solutions for NC machines. Very simple and elegant solutions to the
data processing problems can be obtained electrically, and this has in fact
been demonstrated on several occasions in the earlier chapters of this book.
Special consideration should, therefore, be given to the interfaces between
the electrical and the hydraulic systems, i.e. the electro-hydraulic controllers.

One disadvantage of any hydraulic system is the need to introduce another
energy source in the form of oil under pressure. To keep the cost and
complication to a minimum (Chapter 1) it would be desirable to transform
the available electrical energy (three-phase power) into controlled mechanical
energy (slide movements) as directly as possible. Electro-mechanical systems,
with their direct drive by electric motor, are ideal in this respect, and the
introduction of a hydraulic system appears to be in the nature of a diversion
from these principles. If, however, as will be shown, this 'diversion' brings a
number of major advantages in its train, then it is necessary to counteract the
adverse effects of increasing the number of components by making absolutely
certain of the reliability of the hydraulic elements.

In this chapter only three major groups of elements will be considered;
these are the pumps used to pressurise the hydraulic fluid, the
electro-hydraulic controllers, and the hydrostatic drive units.

6.3.1   *Pumps and auxiliary equipment (pressure and flow controllers)*

Electrical energy is converted into hydraulic energy by means of pumps.
The power required to drive a pump is given by:

$$P = \frac{Q.\Delta p}{612.\eta} \text{ (kW)}$$

where P = power in kW
        Q = Flow in litres/minute
      Δp = Pressure in kp/cm^2, and
       η = pump efficiency (about 0.6 to 0.9) (85)

From this equation it is apparent that the energy in the hydraulic fluid is built up from two components:

Q, the pump delivery, (or the quantity taken by the motor); the mechanical analogue is the speed and the electric analogue the voltage.
p, the working pressure; the mechanical analogue is the force or the torque, and the electrical analogue the current.

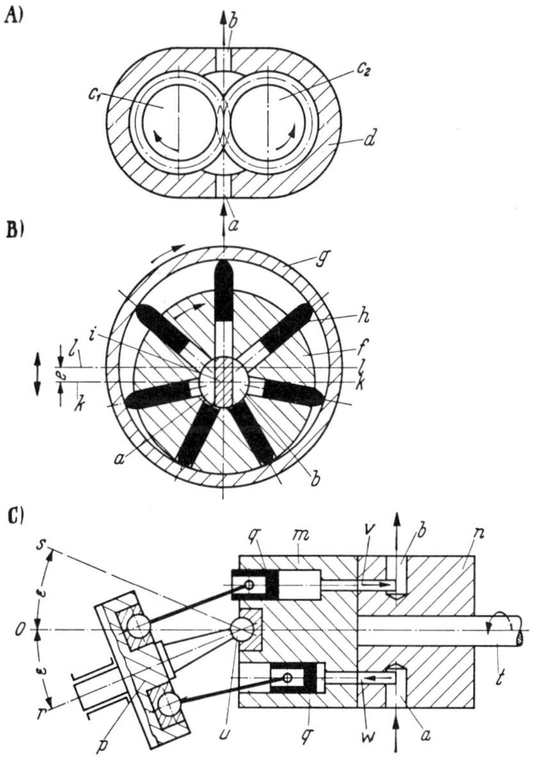

Fig. 105 Diagrammatic representations of three common types of hydraulic pump (82).

A) Gear-type for constant delivery. B) Radial piston pump for variable delivery. C) Axial piston pump for variable delivery.

a Inlet. b Delivery. c_1 and c_2 Gears. d Gear pump casing. e Eccentricity - for adjusting delivery in radial piston pump. f Rotating piston block. g Rotating ring, radially adjustable, to guide pistons h. h Pistons (odd number). i Fixed shaft with suction and delivery chambers (distributor). k Centre-line of shaft i and rotating piston block f. l Centre-line of ring g. m Rotating cylinder block. n Fixed distributor (with suction and delivery chambers v and w). o Axis of rotation and zero position of item p. p Swash plate. q piston. r, s End positions of travel ϵ. t Drive shaft. u Universal joint. v, w Distributor slots. ϵ Travel of pivoting disc p (for adjustment of delivery in the axial piston pump).

Several designs of pump are available, and these can be grouped from various points of view. From the special requirements for numerically controlled machine tools (minimum maintenance, high reliability, long life despite high utilisation of machine in automatic operation, suitability for remote control, etc) it seems advisable to form two large groups for present purposes. These are pumps with a *constant* delivery and pumps with a *variable* delivery. Three of the most widely used designs are shown diagrammatically in Fig. 105. Figure 105A) shows the well-known gear-type pump. The two gear pinions, which run in a casing, draw the oil in at *a* and discharge it at *b*. If the speed at which the pinions rotate is constant (e.g. drive by three-phase squirrel-cage motor), the delivery will be constant[5]. The pump is of simple robust construction, and if used in conjunction with a squirrel-cage motor comes very close to meeting the requirement for a simple and inexpensive unit. The disadvantage of this pump is that the units that it supplies (e.g. hydraulic rams or motors) often not only require a variable supply of hydraulic fluid (for speed variation), but also a pressure which can be varied as required. In most instances gear pumps therefore have to supply the units through pressure or flow control valves.

Pressure control valves may be:
Pressure-limiting (relief), check, or pressure-reducing. Components of these types are shown in the circuits illustrated in Figs. 113, 114, and 117; their operation will be clear from the illustrations and the specialised literature (e.g. (82)).

Flow control valves are usually not strictly controllers (i.e. elements equipped with feedback). The best known types are throttle valves with a throughput that is independent of the pressure (see Fig. 106, for example). But the throttling process is accompanied by losses which cause a rise in temperature of the hydraulic fluid. Because of the resultant changes in viscosity and flow behaviour it is often necessary to make provision for temperature stabilisation.

There is, therefore, an obvious inducement to design the pumps so that they will have a variable delivery. Two examples of such pumps are shown diagrammatically in Fig. 105. Figure 105B) shows the basic elements of a *radial piston pump* with internal oil supply. The piston block *f* rotates about the fixed shaft *i* with the two chambers *a* and *b* (grooves with connections for pipes). In its radial bores (odd number) it carries pistons *h* which can move freely. These bear against a ring *g* which can be moved off-centre by distance $\pm e$[6]. As the ring and block rotate the piston will slide if the ring is

---

[5] Gear pumps are currently made for pressures up to about 160 kp/cm² and deliveries of up to 160 litres/minute. The speeds are usually about 1500 rev/min.

[6] The way in which the piston is attached to the ring *g* varies from one design of pump to another. Details such as this are not shown in the figure which is intended only to show the method of operation.

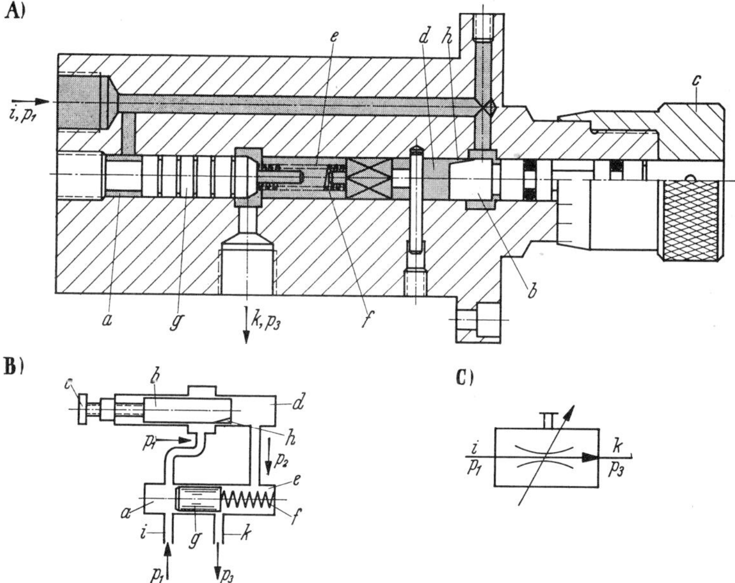

Fig. 106 Example of a flow controller which is largely independent of pressure and temperature (Illustration courtesy of *Heller/Hydraulikring*).

A) Section through controller. B) Schematic diagram of operation. C) Symbol (for use in hydraulic circuit diagrams).

a Pressure chamber for initial pressure p_1. *b* Throttle slide with axial slot *h*. *c* Handwheel for adjusting throttle. *d*, *e* Pressure chambers for intermediate pressure p_2. *f* Compression spring. *g* Differential plunger and balancing valve. *h* Tapered axial slot. *i* Inlet. *k* Outlet. p_1 Initial pressure. p_2 Intermediate pressure. p_3 Final pressure.

off-centre. For the direction of rotation shown, the oil is drawn in by way of the chamber *a* as the space between the plungers increases, and is then forced out through chamber *b* to the delivery pipe as the space decreases. If the eccentricity of the ring *g* is changed, the quantity of oil drawn into the pump and delivered will vary. If the eccentricity of the ring *g* is zero, no oil will be delivered; while if it is moved in the opposite direction (beyond *e* = 0) the inlet and delivery chambers will be reversed, so that the direction of flow through the pump reverses. The flow rate Q can thus be varied in this simple manner. Pumps of this type operate at pressures of up to 150 kp/cm² and deliver up to 500 litres/minute (85).

The method of operation of the *axial piston pump* shown in Fig. 105C) is similar to that of the radial piston pump. The rotating cylinder block contains the cylinder bores for the pistons, of which there are usually an odd number. The distribution functions of the chambers *a* and *b* in Fig. 105B) are now undertaken by the circumferential slots *v* and *w* in the fixed block *n* to which the suction and delivery pipes are connected. If the swash plate *p* is moved through the angle $\pm \epsilon$ the effect is the same as the eccentric displacement of the ring *f* in Fig. 105B). The flow rate is thus infinitely variable. These axial piston pumps operate at pressures of up to about 150 kp/cm^2 and deliver up to about 300 litres/minute.

In all pumps with variable flow rates certain difficulties are encountered in arranging for the remote control of the flow rate. If electrical control currents are to be used it is necessary to fit a small electric motor to make the necessary adjustments (eccentric displacement or swash plate of disc). Whether it is worthwhile to employ such an arrangement on a numerically-controlled machine would have to be determined in each particular case. Although the use of these pumps eliminates the need for flow control valves, pressure control valves (e.g. pressure relief valves) will, of course, have to be fitted.

6.3.2   Electro-hydraulic controllers (energy switching devices)

The term 'controller', which is taken from the terminology employed in control technology, is deliberately used here, since it describes the function that these components perform within the installation. In practical engineering these components are, however, also referred to as 'multi-way valves', this latter expression being more descriptive of the actual design of the elements. Depending on the details of the design it is also possible to refer to slide valves, rotary valves, etc.(85).

It is a requirement for every electro-hydraulic controller design that the hydraulic fluid shall exert no appreciable forces on the control element itself; in other words the forces exerted by the hydraulic fluid within the valve shall not tend to disturb the positions of the control element which are selected by the electrical control signal. The size of the solenoid that actuates the valve must be determined solely by the inevitable frictional forces, and not by the working pressure of the hydraulic fluid.

Since it is desirable for the controller to be capable of being linked directly to the electric circuits of the internal data processing equipment, no consideration will be given to controllers equipped with mechanical or hydraulic actuators. It is more important to distinguish between non-continuous (on-off) controllers and continuous (infinitely-variable) controllers.

Non-continuous controllers (on-off valves, multi-way valves with fixed positions)

There are vitually only two types of electro-hydraulic controller that fall into this category; these are *two-position* and *three-position valves*. They possess

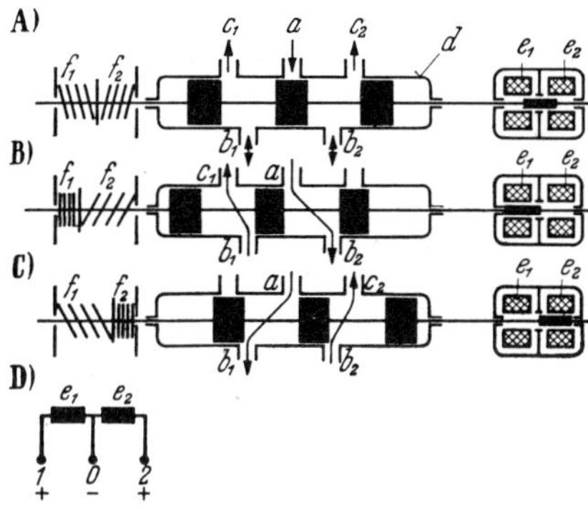

Fig. 107
Principle of operation of an electro-hydraulic controller having a switching characteristic (four-way three-position system, 4/3-way valve)

A) Slide position for zero control current. B) Slide position for full excitation of coil e_1. C) Slide position for full excitation of coil e_2. D) Coil circuit.

a Oil inlet. b_1, b_2 Connections for hydraulic power unit (ram, hydraulic motor). c_1 c_2 Oil return lines. d Valve casing. e_1, e_2 Solenoid coils. f_1, f_2 Return springs.

the common feature that they can occupy only the pre-determined positions, and no others, including intermediate positions. In view of this binary or ternary nature, these valves are especially suited for connection with electric circuits in positioning or straight-line control systems. Two-position valves are used as on-off valves or as changeover valves (e.g. from rapid traverse to working feed speed). Three-way valves can be used for reversing the direction of the flow of the hydraulic fluid, e.g. where there are reciprocating motions. Hydraulic switches can be connected in just the same way as electrical switches (see the extensive literature). Here reference will be made only to Fig. 107, which illustrates the most important points, taking a three-way valve as an example.

No unduly high standards of accuracy are required in the manufacture of pistons and ports of hydraulic valves that are used for switching purposes, since there is always sufficient margin in this type of equipment to ensure that the valve is firmly closed or fully open (cf. the conditions in the continuous controllers described below). As a result of this relaxation in the design conditions, switching valves are appreciably cheaper than continuous controllers; on-off circuits are therefore also of advantage in the majority of instances where hydraulic drives are used in conjunction with positioning and continuous-path control systems. The electrical input of signals (Fig. 107D) must also have an on-off characteristic matching that of the hydraulic component; in the example shown there are only three possible states for the coils:

> No voltage applied to either terminals 1 or 2.
> Positive nominal voltage on terminal 1.
> Positive nominal voltage on terminal 2.

It is not permissible for any intermediate conditions between the full voltage (rated current) and zero to arise in the electrical circuit. It is therefore convenient to regard the whole arrangement as a switching or on-off circuit, and to use the same terminology for both the electrical and the hydraulic components.

In this connection it is also very important to arrange for any storage capacity for the switching information that may be needed.

Either the valve can be provided with suitable pawls which hold it in the appropriate positions (when an electrical impulse will suffice for control purposes), or the storage effect is incorporated in the electrical part. In the latter case the valve solenoid must be supplied with current by way of a self-energising relay or similar device for the time the information is to be retained. For the practical design of an electro-hydraulic control system, this ability to provide the storage capacity in either the hydraulic or the electrical component is of considerable importance.

Very different conditions to those obtaining with non-continuous controllers are encountered in the operation of continuous controllers.

Continuous controllers (continuously-variable valves, electro-hydraulic servo valves). (87, 167, 237)

The principal feature of these components is that the oil flow can be adjusted to any value from a maximum in one direction, through zero, to a maximum in the other direction, even though the pump delivery and the inlet pressure to the controller remain constant. The signals used to regulate the controllers can also vary continuously; in the ideal case there will be linear proportionality between the control current and the oil flow.

To illustrate the operation of these devices some examples are illustrated in Figs. 108 to 111. Figure 108 shows in diagrammatic form the working of a single-stage servo valve of rotary-slide rotary-solenoid type, which is illustrated in Fig. 109. This valve, in which the edges of the rotary slides and the ports are arranged so that continuous control is possible, is of compact construction.

Figure 110 shows a simplified section through a continuously-variable proportional electro-hydraulic controller having a longitudinal slide. This is a two-stage servo valve with an auxiliary valve (82). The most interesting feature of this design is that it requires very low electrical inputs (about ± 300 mA through 40 ohms), and the hydraulic amplification provided by the auxiliary valve system enables working pressures of up to 100 kp/cm^2 and flow rates of up to 80 litres/minute to be handled. This device is made by AEG who term it a 'Tauchspulen-Regler' ('Dipping-coil Controller') (87). Its method of operation is briefly as follows:

Four coils of extremely light construction dip into the powerful field of an electro-magnet or permanent magnet. The coil circuit is shown in Fig. 110C). The two outer coils (f and i) are excited by means of an adjustable bridge circuit, and act as calibration coils for the entire electro-hydraulic system (in place of mechanical springs). If coil f is excited more than coil i the whole coil assembly will dip further into the magnetic field; if coil i is excited more than coil f the assembly moves up closer to its undeflected position. The d.c. supply l for the calibration circuit should preferably have a slight a.c. superimposed on it to cause the auxiliary valve p to vibrate slightly for overcoming any frictional effects. The two control coils g are located between the calibration coils. They are connected to the electrical control system (e.g. phase-sensitive bridges as shown in Fig. 82).

The whole of the dipping coil assembly is connected to the auxiliary valve p by means of the rod e, which should also be made as light as possible. In the lower part of the illustration note the distribution of the oil under pressure and the unpressurised oil in the various passages and chambers. The auxiliary valve p is located inside the main control valve o. The latter can move up and down in the casing n and 'floats' between chambers w and x, which contain oil under pressure. The auxiliary valve is designed so that it closes the two inlet ports for the oil under pressure only partially by means of its inner edges. Some oil can therefore always pass to the unpressurised section at this point. The main valve (slave valve) therefore always 'floats' in its central position if the pressure in chambers w and x is the same. If the auxiliary valve

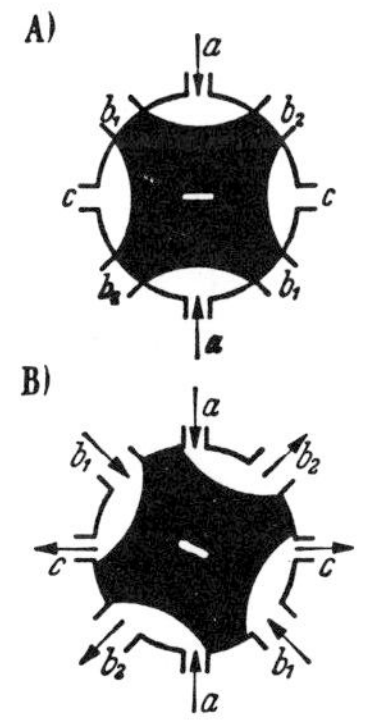

Fig. 108
Principle of operation of an electro-hydraulic controller of rotary-valve type providing a continuously variable characteristic (single-stage servo valve, hydraulic section only)

A) Valve closed (torque motor not energised). B) Valve fully open (torque motor fully energised in one direction; any intermediate positions can be obtained).

a Supply of hydraulic fluid under pressure. b_1, b_2 connections for hydraulic power unit (e.g. ram as shown in Fig. 113). c Oil return path.

Fig. 109
Example of a continuously variable electro-hydraulic controller employing a rotary valve. (Overall height about 130 mm. Photo courtesy of *Sperry*).

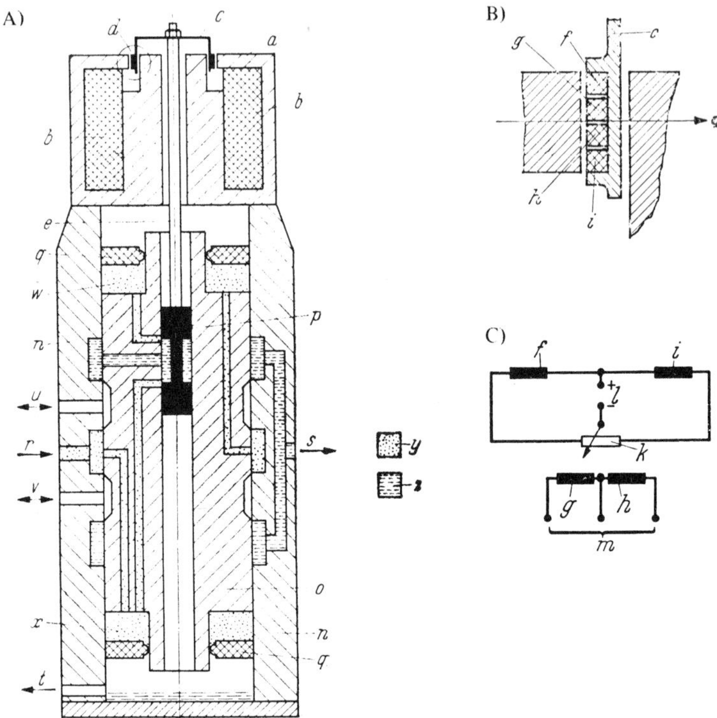

Fig. 110 Principle of operation of an electro-hydraulic controller of the longitudinal
slide type with a continuously-variable characteristic (AEG Dipping Coil
Controller)

A) Section through controller. B) Arrangement of dipping coils in magnetic air
gap. C) Coil circuits.

a Magnet casing. *b* Energising coil to produce magnetic flux ϕ (can also be
replaced by a permanent magnet). *c* Holder for dipping coils *f, g, h, i. d* Part
section shown to larger scale in drawing B). *e* Piston rod for auxiliary valve *p. f*
and *i* Calibration coils. *g* and *h* Coils carrying signal current. *k* Calibration
resistance. *l* Source of calibration voltage (d.c. with a.c. superimposed to
overcome friction of valve *p*). *m* Input for signal voltage (d.c., taken, for
example, from a phase-sensitive bridge as shown in Fig. 82). *n* Valve casing.
o Main valve slide. *p* Auxiliary valve slide. *q* Seal. *r* Inlet for oil under pressure.
s Oil return path. *t* Drain for leakage oil. *u* and *v* Connections for hydraulic
power unit (e.g. hydraulic motor, ram). *w* Upper pressure chamber. *x* Lower
pressure chamber. *y* Oil under pressure (up to 100 atm). *z* Unpressurised oil.

p is moved upwards by a displacement of the dipping coil, more oil can
escape from the upper chamber *w* than from the lower chamber *x*; this
disturbs the equilibrium. The pressure in chamber *x* will then be higher than
that in chamber *w* so that the main valve also moves in the upward direction,
and it follows the auxiliary valve until the pressures in the two chambers are

again the same. The same occurs in the reverse direction if the auxiliary valve is moved downwards. The main valve will thus rapidly and accurately follow every movement of the auxiliary valve. This hydraulic amplification enables the oil under pressure which is supplied to the inlet opening *r* to be controlled very accurately by means of a very small force (control current). The remainder of the principle of operation can be followed by tracing the various passages, the edges of which are designed so as to ensure that the direction of the oil flow from passage *r* to either passage *u* or passage *v* is continuously variable and proportional to the valve displacement. Using a device of this type as an example it is readily possible to build up electro-hydraulic control circuits of any type. Figure 111 shows a practical example. Owing to its robust construction the time constant of this device may prove critical in extreme cases, (natural frequency only 30 Hz), but its ruggedness is itself an advantage as is the possibility of using a relatively high working pressure (100 kp/cm^2). It is, incidentally, advisable to investigate the behaviour of the control device when it is operating in conjunction with the drive unit; this point has also formed the subject of published work and need not be further considered here (88, 89). Mention may also be made of other designs that serve a similar purpose (115, 167, 237, 302).

Fig. 111
Example of an electro-hydraulic controller using a longitudinal slide valve and having a continuously-variable charateristic. (Photo courtesy of *AEG*).

6.3.3   *Hydraulic drive units (actuators)*

The two possible basic types are illustrated diagrammatically in Fig. 112. They are:

Feed drive unit using hydraulic cylinder and piston and
Feed drive unit using hydraulic motor.

Feed drive unit using hydraulic cylinder and piston

There is not a great deal that needs to be said about this simple type of power unit (82). This design offers the great advantage that the fluid flow is transformed directly into a longitudinal motion, without the intervention of rotating motors and leadscrews. Basically this allows much faster rapid traverse motions to be achieved than is possible when using rotating leadscrews. This proves especially convenient when positioning the work table or slide. Special attention must, however, be given to certain factors when using a hydraulic piston arrangement for the feed drive unit:

Seals: At high working pressures, which may be necessary to produce the feed forces required for a continuous-path control system in a machining operation, very robust seals will be needed. The resultant friction can be appreciable. It also causes stick-slip which can prove particularly troublesome at low feeds (inching speeds). For positioning control systems (e.g. positioning of coordinate tables) there are no machining forces to overcome, and a lower oil pressure can be employed which reduces the difficulties encountered with the seals.

Oil compressibility. The length of a column of oil is reduced considerably if the pressure is increased. The compressibility factor is approximately:

$$\beta \approx 7.10^{-5} \ \frac{m}{m \, . \, \text{kp/cm}^2} \qquad (85)$$

If, for example, a column of oil of length 2 metres is subjected to a rise in pressure of 50 kp/cm^2 its length will be reduced by 7 mm. The elastic expansion of the piping is also subjected to the pressure rise. This is not

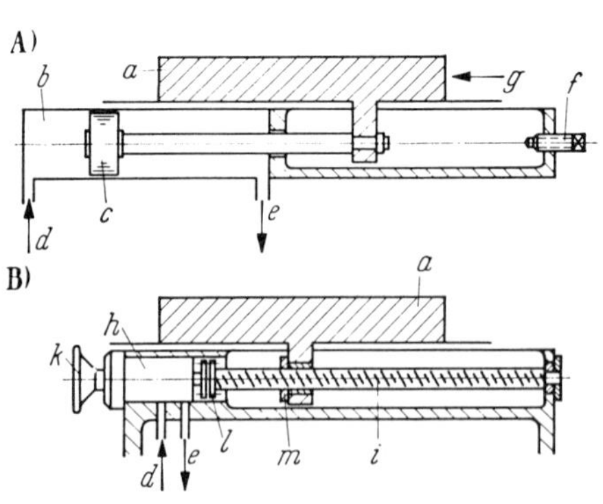

A) Hydraulic cylinder and piston. B) Hydrostatic hydraulic motor, hydromechanical feed drive. *a* Machine carriage. *b* Hydraulic cylinder. *c* Piston. *d* Oil supply. *e* Oil return path. (*d* and *e* can be interchanged at will). *f* Stop screw. *g* Resistance (e.g. machining force). *h* Axial piston motor. *i* Leadscrew (self-blocking or with recirculating ball system as shown in Fig. 128). *k* Handwheel for manual setting of carriage. *l* Clutch. *m* Feed nut.

Fig. 112 Schematic diagrams of two types of hydraulic feed drive unit (85)

a serious factor for small uniform pressures and short displacements. If, however, the working load at a machine table varies considerably (e.g. intermittent cuts in the case of continuous-path controls), compressibility of the oil will certainly be a disadvantage if the displacements are large. On the other hand, compressibility does not play a major role with positioning controls and machining operations (slide moved into contact with a stop). Figure 113 shows the hydraulic circuit diagram for a large numerically-controlled jig borer (cf. Fig. 218) with a hydraulic piston drive. The sliding parts *a* and *b* are clamped hydraulically when in the correct positions.

Attachment of displacement measuring system: It will be clear that where a hydraulic ram feed drive is fitted to a slide it is not possible to use a rotating displacement measuring system directly. If such a system is nevertheless to be

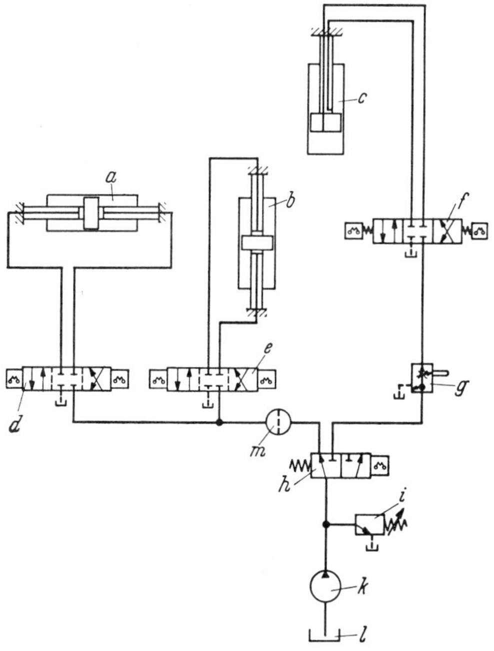

Fig. 113 **Hydraulic circuit diagram for a drive system for a cross slide (using hydraulic pistons). The symbols used conform to DIN 24300. The drive system is for the machine shown in Fig. 301.**

a Hydraulic cylinder with fixed piston (*x*-axis). *b* Hydraulic cylinder with fixed piston (*y*-axis). *c* Hydraulic cylinder with fixed piston *(z*-axis). *d, e, f* Electro-hydraulic controllers (Figs. 108 and 109). *g* Throttle valve. *h* Electro-hydraulic changeover valve (2-position 3-way valve). *i* Pressure regulating valve. *k* gear-type hydraulic pump. *l* Oil reservoir. *m* Filter.

used it is possible to fit a rack and pinion or some similar device (cf. Fig. 11C). The machine shown in Fig. 218, for example, is fitted with a linear Inductosyn measuring scale as illustrated in Fig. 65, with, in addition, two synchros for the medium and coarse measuring ranges, these in turn being connected to the machine table by means of racks and pinions (Chapter 4).

Feed drive unit using hydraulic motor

A characteristic feature of this type of drive is the presence of a leadscrew i and nut m (Fig. 112). The leadscrews and nuts can be irreversible (normal acme or square threads) or can be of the low-friction recirculating ball type. Both versions have advantages and disadvantages (Chapter 7). Here, however, it is the hydraulic motors that are of primary interest. As with pumps, there are various different designs of motors that can be used:

> Gear-type
> Vane-type
> Radial-piston
> Axial-piston
> Roll-vane
> etc. (82, 83, 84, 85, 132).

It is beyond the scope of this book to discuss the differences and details of these various types. Only the axial piston motors, which are frequently employed, will be described as an example.
> Their main advantages are:
>> High efficiency (up to 95%).
>> Compact design.
>> Low moment of inertia. (90)
>> (as a result of the axial form of construction the polar moment of inertia is only about 1/10 that of a comparable d.c. motor, cf. Section 6.4).
>> High torque even at very low speeds (despite the small size of the motor)[7])

---

[7]   The speed is calculated from: (82)

$$n = \frac{500}{F \cdot z \cdot r \cdot \sin\alpha} \cdot Q \text{ (rev/min)}$$

The theoretical torque (without losses) is given by

$$M = \frac{F \cdot z \cdot r \cdot \sin\alpha}{\pi} \cdot p \text{ (cmkp)}$$

Where F is the piston area in cm^2, z is the number of pistons, r is the radius producing the plunger movement, α is the inclination of the swash plate, Q is the supply of hydraulic fluid in litres/min, and p is the pressure in kp/cm^2.

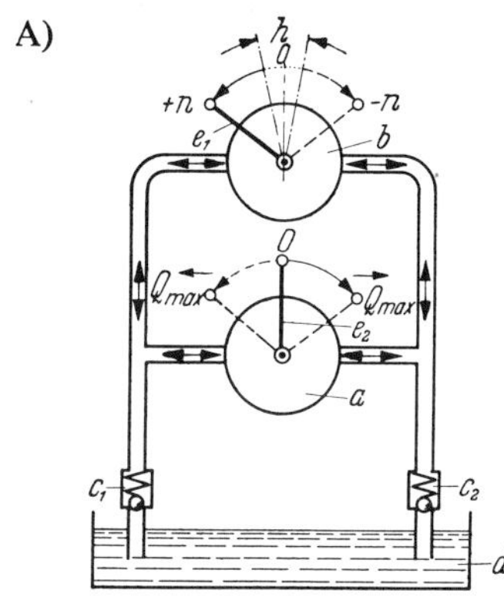

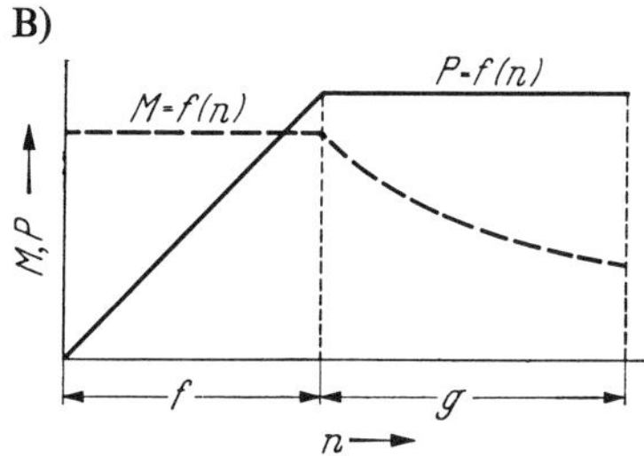

Fig. 114
Closed hydraulic circuit with pump and motor

A) Schematic layout of hydraulic circuit. B) Torque and power variations with change of motor speed in one direction of rotation.

a Pump with variable delivery. b Motor with variable throughput. c_1 and c_2 Non-return valves. d Oil reservoir. e_1 Lever to control motor throughput. e_2 Lever to control pump delivery. f Speed variation at constant torque (delivery Q is varied, motor throughput/rev remains constant). g Speed variation with constant power (pump delivery Q remains constant, motor throughput/rev is varied). h Unstable region. n Speed. M Motor torque. P Power of hydraulic motor (cf. electrical analogue in Fig. 99). Q Pump delivery.

In principle, the designs of the axial piston motors can be precisely the same as that of the pumps shown in Fig. 105C); the pistons can also work against a swash plate which may be set at a variable angle (however, the motor shown in Figure 115, has a fixed swash plate). Used in conjunction with a variable-delivery pump each of these models will exhibit a working characteristic similar to Fig. 114. As in the Ward-Leonard control of a d.c. motor (cf. Figs. 99 and 101) there are again two ways in which the speed-torque relationship of the motor can vary:

1. With the motor throughput/rev maintained constant (swash plate angle fixed) and variable pump delivery Q, the motor torque remains constant over the speed range $+n_{max}$ to $-n_{max}$, and the speed will vary in proportion to the flow rate at which the hydraulic fluid is supplied. This arrangement is thus a constant-torque drive (range f in Fig. 114B) similar to that produced by a d.c. motor in which the armature voltage is varied while the field is maintained constant (Fig. 99).

2. If the pump delivery remains constant and the motor throughput/rev is varied (variable-angle swash plate) the torque will reduce as the speed increases; the lowest speed will occur when the angle of inclination of the swash plate is a maximum. Over a certain range this arrangement provides a constant-power drive (region g in Fig. 114B).

The conditions are similar to those obtaining with a d.c. motor with a weakened field; the stable region extends over a speed ratio of about 1 : 3 to 1 : 4. It is easy to change the direction of rotation by switching the fluid flow, but it is not possible to provide infinitely-variable control of slow positive and negative speeds not greatly different from zero. Here again there is a similarity with a d.c. motor with a weakened field.

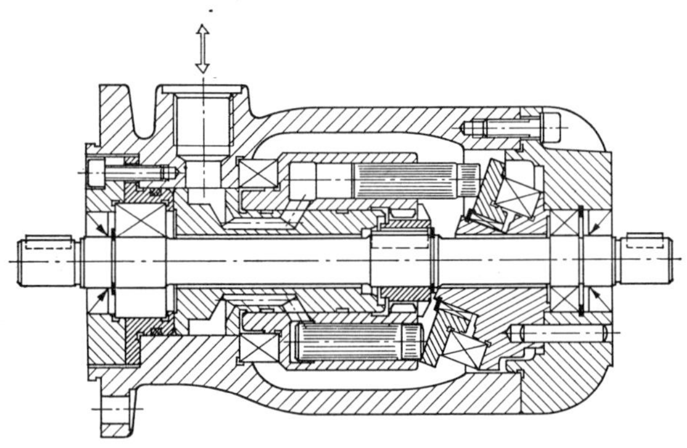

Fig. 115 Axial piston motor with inclined swash plate. Manufactured by *Heller* (82).

For feed drive mechanisms on machine tools equipped with continuous-path control, hydraulic motors with adjustable swash plates represent an unnecessary complication, since all the necessary control can be exercised as well or even better by means of a pump or valve control unit. Figure 115 shows an axial-piston motor with a fixed swash plate angle (constant throughput/rev) of the type which is very frequently employed in hydraulic feed drive units. These motors are made with torque outputs of about 1.5 mkp to 12 mkp. The design shown in Fig. 116 is even more suitable for higher torques (82); advantages are in particular obtained during slow running. Roll-vane motors are especially suitable for use with machine tools. They are being made in the USA for torques of up to 27 mkp at pressures of up to 140 kp/cm^2 and speeds of up to 2000 rev/min. One of their major advantages is their low moment of inertia (82, 132, 302).

Finally Fig. 117 is a diagrammatic representation of a feed drive system of the type which has been adopted for the great majority of numerical continuous-path controls as a result of the various points that have been mentioned. No reference is made in this illustration to the particular nature of the pump or pressure producer g. To enable the maximum speed to be set to any desired value it is possible to use:

a) An adjustable pump (Fig. 105B) or C).

b) A constant-delivery pump, e.g. a gear-type pump, in series with a flow control valve (Fig. 105A).

In both instances the question remains as to whether and under what circumstances the feed speed must be remotely controllable. A satisfactory answer to this question cannot be given until Chapter 7 (Design of machine tools) and Chapter 10 (Programming) have been studied. If, for example, the whole hydraulic circuit in conjunction with the electro-hydraulic controller e is designed so that a constant delivery at the controller e will suffice, then any additional expenditure on pumps, etc., is unnecessary, and a simple gear-type pump will suffice. In the example a simple pressure relief valve h is shown as the means of adjusting the pressure; it would, of course, be possible to use some other design. The most important and, in the limiting case, the only remote-control element in the circuit is the electro-hydraulic controller e. No matter how this is designed in detail, (see Figs. 108, 109, 110, 111 for examples), this always contains the input for the electrical signal voltages f, and hence the connection to the 'nervous system'. The tachogenerator l is also shown; it serves for measuring the actual speed, if required, for use in a

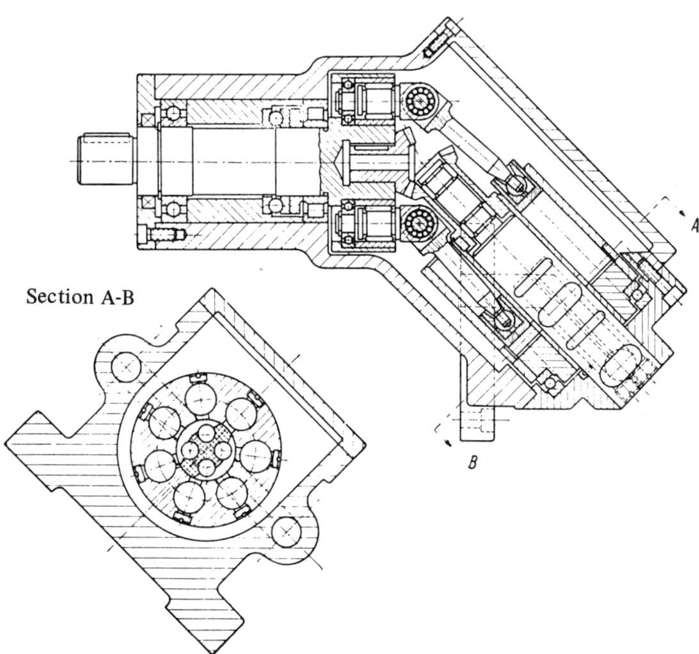

Fig. 116 *Heller* **axial-plunger motor, inclined type (62).**

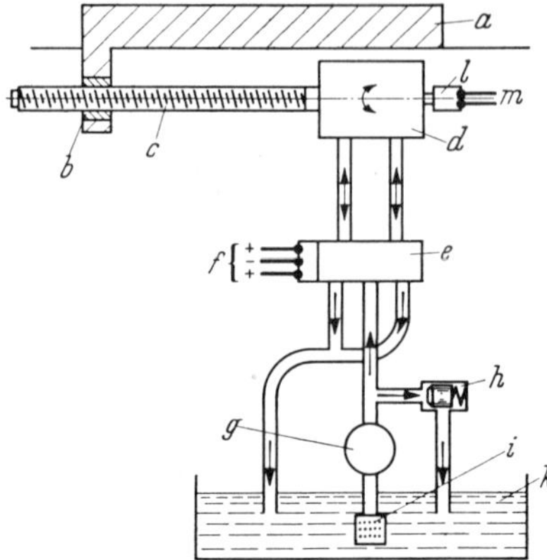

Fig. 117
Principle of operation of a hydromechanical feed drive system with
the controller arranged before the
power unit (primary system) (85).

a Machine slide. *b* Nut. *c* Leadscrew. *d* Hydraulic motor.
e Electro-hydraulic controller.
f Electrical signal input. *g* Adjustable pump or fixed-delivery pump
with flow control valve.
h Pressure relief valve. *i* Filter.
k Oil reservoir. *l* Tachogenerator
m Speed sensor.

control circuit for speed stabilisation. More details regarding the design of
parts *a, b,* and *c* will be found in Chapter 7.

To conclude this discussion it is necessary to make a few remarks regarding
the means of energy transmission, namely the hydraulic fluid.

6.3.4   *Hydraulic fluids* (82)

Invariably, high-quality mineral oils with certain specific chemical additives
are used. The specific gravity is about 0.9 kg/dm^3. The flash-point and pour
point are not of particular interest in this application. The most important
properties are as follows:

1. The viscosity (kinematic viscosity) governs the leakage and slip losses
   that occur. An average value is about 360 stokes at 50° for the fluid
   heated up to its normal operating temperature.
2. Temperature variations must have only a slight effect on the viscosity.
   It has already been mentioned that throttling and compression
   processes in pumps and valves, etc., produce heat which causes the
   temperature of the hydraulic fluid in the circuit to rise rapidly. If the
   viscosity varies too much with changes in temperature, the operating
   characteristics of the various components may change to an
   unacceptable extent. It may therefore be necessary to maintain the
   whole of the hydraulic system at as uniform a temperature as possible
   by means of thermostats. With some items of equipment it is also
   possible to provide temperature compensation by means of relatively
   simple design measures.
3. Low foaming (mixing with air). The adverse effects of the
   compressibility of the oil, which have already been mentioned,

become more severe if even finely distributed air inclusions are present in the oil.

Other desirable properties in the hydraulic fluid include resistance to ageing, good lubrication properties, low sludge formation, possibility to remove air, gas, or water inclusions. Furthermore, the fluid and its additives should not attack the seals in the system.

6.4 Comparison of infinitely-variable electric and hydraulic drive systems

The competing electric and hydraulic solutions to a given problem were discussed in Sections 6.2 and 6.3, many references being given to the appropriate specialist literature; At this point students, who have specialised in various subjects, will be given some basic information to enable them to make a satisfactory choice from among differing designs and operating systems. In practice, it appears that about one-half of the numerically controlled machines that have been installed throughout the world to date are equipped with electro-hydraulic drive units. International exhibitions which have been held in recent years (92, 93) have also shown that both these possible types of solution enjoy roughly equal popularity, with no clearly defined trend to one or other. This provides a further reason for seeking a theoretical basis for the current 'bipolarity'.

There are basically three criteria that can be adopted for comparing various types of drive system:

1. Reliability (including nearest possible approach to the requirement that the complexity be kept to a minimum).
2. Economy (i.e. low initial costs together with low maintenance and repair costs). With positioning or straight-line control systems these two criteria can be covered fairly simply by comparing stepped and continuously-variable drives, and suitable solutions can be found. A third criterion is, however, required for continuous-path control systems, which are becoming increasingly popular.
3. Time response. In this connection it is always necessary to consider the controllers and the drive units together. Further discussion will therefore be based on this requirement.

Electrical or electronic controllers having a short time response are in general simpler and cheaper than hydraulic controllers. When controllers and drive units are combined to form complete drive systems the time response of the drive unit is always appreciably larger than that of the controllers, since they are powerful energy transformers and so inevitably contain appreciable masses. In an earlier study a theoretical investigation was therefore made into the fundamental behaviour of such systems based on the physical features of normal d.c. motors and axial-piston motors (91). The results, which at first sight are somewhat surprising, are as follows:

A coefficient is introduced to represent the time response as governed by the machine and the task that is being performed (the main factors are the machine tool, its size, and the machining operation). This coefficient is the ratio:

$$\epsilon = \frac{\text{Moment of inertia of motor}}{\text{Rated torque of motor}} = \frac{\Theta}{M}$$

Even a rough calculation will then provide a satisfactory basis for comparison, if ϵ_e and ϵ_c for the two alternatives are calculated. Since there are comparatively close limits to the extent to which the dimensions of electrical machines can be reduced, owing to the saturation of the iron and the temperature rise of the copper under continuous-load conditions, the axial-piston motor is better placed, since there is much more scope for varying the working pressure p than, for example, the saturation of the iron. If typical values for the materials employed and for the designs normally encountered in both electric motors and hydraulic motors are inserted in the somewhat complex calculations one arrives at the following approximate relationship:

$$\epsilon_h \approx \frac{13}{p} \cdot \epsilon_e$$

where the numeral 13 represents an average between 8 and 18, which in turn is the range of variation that can occur with various types of hydraulic motor. Comparison of the two types of drive using this relationship indicates:

1. That hydraulic systems offer no advantages at low pressures.

2. That with a suitable choice of high pressure, a hydraulic motor of the axial piston type can have a moment of inertia of only 1/5 to 1/10 that of a normal d.c. motor that produces the same torque.

3. The more difficult the machining conditions, i.e. the higher the continuous rated torque M becomes, the more desirable it is to use hydraulic motors if a good time constant for the control circuit is also required. There is some limit to this for large machines because long oil pipes present certain difficulties at high pressures.

The investigation referred to was carried out in 1963/64 to form a basis for interdisciplinary discussions in the field of adult education. It does not take into account the technical improvements that have been made during the last six years in the hydraulic sector [8] , in thyristor control systems (81), or in electrical machines[9] . Taking an overall view it appears at present that the possibilities opened up by thyristor control systems (ability to handle transient current peaks for rapid acceleration of the motor) in conjunction with special low-inertia d.c. motors will shift the advantage in the direction of electrical solutions, especially as the use of the high pressures of more than 120 kp/cm^2 that are required for the hydraulic systems also have their disadvantages (sealing problems, noise, considerable experience required, etc.).

In conclusion it is, however, necessary to point out that the choice between the two alternative systems often depends solely on operational considerations, with the attitudes and experience of the people involved playing a large part.

6.5   Stepping Motors

Several references have already been made to this basic design (Fig. 1 and Chapter 5). The method has long been known (95), and it is occasionally referred to as delta-modulation (3). The new feature is that in recent years it has been found possible to produce these motors for torques which are high enough to open up the possibilty of using stepping motors in machine tools. It may be pointed out straight away, however, that it is unlikely that a purely electrical solution will be satisfactory in the majority of cases, since, as was shown in the case of the d.c. motor (Section 6.4), electrical machines become much too heavy owing to the limits imposed by magnetism. If higher forces (torques) are to be employed the dynamic performance becomes poor. The solution will clearly again be some form of electro-hydraulic device (82, 96).

To explain the operation of this type of motor, the principles of an electric stepping motor are illustrated in Fig. 118 (95, 97); several different

---

[8]   The Roll-Vane motors made in the USA have in recent years also become known in Europe as Hartmann motors, from the name of their manufacturer. They are made for torques of 3.6 to 27.7 mkp at a pressure of 140 kp/cm^2 and are notable for their good slow-running characteristics and high stalling torque (82, 132). The speed can range up to 2000 rev/min. One of their main advantages is their low moment of inertia; while their disadvantage is the high degree of precision involved in their manufacture, which is reflected in the price.

[9]   These include the Japanese Minertia motors (BBC, 102), which contain no or very little iron and the disc rotor motors made by the French firm of CEM-Servalco (103). The normal torques that can be attained are at present only about 160 to 1000 cmkp, however. Using special d.c. motors, which contain iron, but in which the ratio of armature length/armature diameter is about 2 it is at present possible to obtain a torque of about 4.0 mkp (74) (Fig. 98). Low-inertia d.c. motors known as Ro-tran or Inland motors are also made in the USA.

designs are, however, employed. The number of slots in the stator and the rotor is the same in Fig. 118, 24 slots and 24 teeth being provided (Fig. 118C). Measured at the air gap δ (rotor diameter = internal diameter of stator, since the air gap is only about 0.15 mm) in effect:

Width of stator slot = width of stator tooth = width of rotor tooth.

Teeth having the same dimensions will thus always be opposite each other. Three sets of stator laminations are arranged side by side axially, care being taken to ensure that the teeth are accurately in line axially. There are also three sets of rotor laminations arranged side by side axially on the shaft (Fig. 118B). Unlike the stator sets, however, the rotor sets are each displaced 2/3 of the tooth width relative to each other (Fig. 118D). If the individual stator teeth are wound as shown in Fig. 118 A), so that the teeth are alternately of opposite polarity, the individual stator teeth become unit poles, and the pitch of the slots (distance $(f + g)$ in Fig. 118 D) becomes the pole pitch τ_p; 24 poles will then be produced. The rotor acts as the magnetic return path and need carry no windings. The overlapping of the staggered rotor teeth with the stator teeth is shown in Fig. 118 D). If the three stator windings c are connected together as shown in Fig. 118 E), and if the common point 0 is connected, for example, to the negative pole of a d.c. supply, the following will occur; If terminal l is connected to the positive pole of the supply, magnetic fields are built up in the stator laminations k_1 which complete their circuit by way of the rotor teeth. The rotor will then rotate into a position such that the magnetic resistance becomes a minimum, i.e. such that the teeth of sets of laminations k_1 and d_1 are precisely opposite each other. If the d.c. supply is disconnected from terminal l and connected to terminal m, rotor d_2 will rotate by the distance $\tau_p/3$ anti-clockwise, since its teeth h_2 already overlap one of the stator teeth by 1/3 the tooth width. In our example where there are 24 teeth, the motor rotor will thus turn through the angle

$$\Delta\alpha = \frac{1}{3} \cdot \frac{360}{24} = 5°$$

anti-clockwise, and since it is then in a position where the magnetic flux path is an optimum it will remain there. If the voltage is then applied to point n the armature will again turn $5°$ anti-clockwise. The cycle then starts again with an impulse applied to l. If pulses are applied to the motor in the sequence l-m-n-l and so on it will therefore rotate anti-clockwise in $5°$ steps. If, on the other hand, a current pulse applied to terminal l is followed by one on terminal n, and then by one on terminal m, the rotor will rotate clockwise in steps of $5°$. The angular step of $5°$ per impulse depends only on the pole pitch (number of teeth) of the motor, and so is a machine constant. If different angular displacements per impulse are required the motor must be provided with a different number of teeth. The total number of impulses to be applied will always determine the overall angle of rotation, while the speed at which the pulses are applied controls the speed of the motor. The similarity between the principles of operation of this system and that used in the Bendix control system has already been mentioned in Section 5.3.

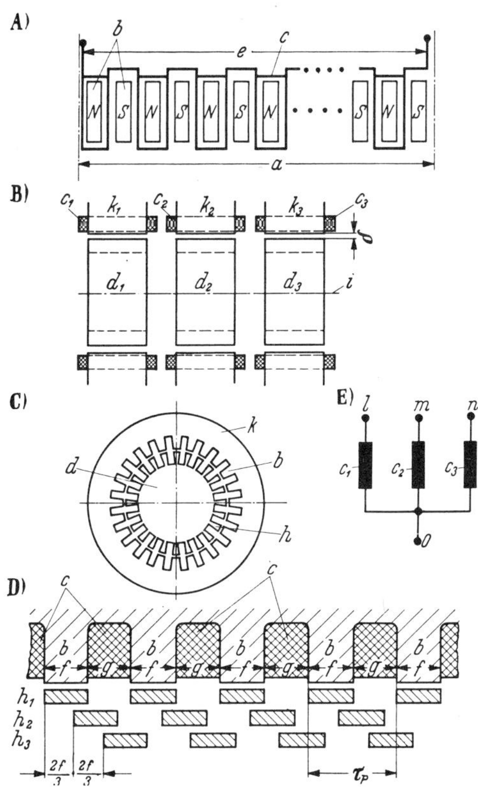

Fig. 118 Diagrams showing the principle of operation of an electric stepping motor (example of one possible form (95).

A) Arrangement of stator poles and windings. B) Arrangement of the three axially-displaced parts of the motor. C) Stator and rotor grooves in one part of the motor. D) Angular displacement of the three rotor sections relative to the unstaggered stator sections. E) Circuit for the stator windings (the rotor need not carry windings).

a Developed view of stator bore. b Stator teeth - stator poles. c_1, c_2, c_3 Stator windings. d_1, d_2, d_3 Rotors on common shaft, axially shifted, rotationally displaced by $2/3\,f$ relative to each other. e Stator d.c. supply (pulsed). f Stator tooth width=rotor tooth width (measured in air gap δ). g Stator groove width (in the air gap δ, $g = f$). h_1, h_2, h_3 Rotor teeth (symbolised) of the three rotors. i Motor shaft. k Set of stator laminations. l, m, n Starting points of stator windings. τp Stator pole pitch ($\tau p = f + g$). δ Air gap (in practice about 0.15 mm in stepping motors).

It will be obvious that this principle provides an ideal means whereby required digital values can be transformed directly into mechanical movements. The practical difficulties encountered in realising this principle fall into three main groups:

a) The magnetic tangential force at the circumference of the rotor is very low. If it is borne in mind that the magnetic force amounts to a total of only about 1 kp/cm^2 for an air gap induction of 5000 gauss (99), and that the greater part of this force appears as a radial component with only a small proportion being available as a tangential component, it will be clear that large forces (torques) require very large dimensions, since the upper limit of the air gap induction is probably about 7000 to 8000 gauss (saturation of iron teeth). The forces admittedly increases with the square of the induction, but even so it is not possible to obtain any significant torque with this arrangement if the moment of inertia (diameter of the rotor) is to be kept small.

b) The magnetic lines of force that appear as a circumferential force act like rubber springs in conjunction with a mass. If the mass of the rotor is large, a slow sequence of pulses can lead to undesirable oscillations about the correct tooth position; while if the impulses are too short, the armature can overshoot the required positions to such an extent that the direction of rotation may not be clearly defined.

c) The most critical conditions occur when the rotor is being accelerated or decelerated; it is essential that it is not allowed to get out of step if it is to perform its function of a quantity summator, e.g. in conjunction with a digital computer (100).

Electric stepping motors having a wide range of technical characteristics and of many different designs have been available in the United States for some years (97). They are, however, nearly all designed for very low torques, and so are not suitable for use in machine tools. The more powerful of these motors have very high moments of inertia and can be used only in conjunction with low pulse frequencies. Intensive work is, however, also being carried out in this field in the USSR and in Japan (96, 98).

If an electric stepping motor is used to operate a suitable hydraulic actuator, this can be connected to one of the hydraulic motors that has been described; the quantity of oil delivered is controlled by the stepping-motor/oil-valve combination and the hydraulic motor will follow the electrical signals precisely (Fig. 119). This is, therefore, a direct practical application of the suggestions made in Section 6.4.

For positioning and straight-line control systems, the electrical section of the electro-hydraulic unit is controlled by a small electronic control device (101); while for continuous path control systems, control is effected directly by impulses produced by the internal interpolator. In the hydraulic slave circuit *d* the required angle of rotation is provided by the electric stepping motor which rotates and simultaneously opens the hydraulic valve *c*. This valve allows the oil to flow to the hydraulic motor in such a way as to produce the correct angle of rotation. As it rotates the axial-piston motor closes the valve. When the angle through which the hydraulic motor has rotated corresponds to the angle that the stepping motor has produced as the

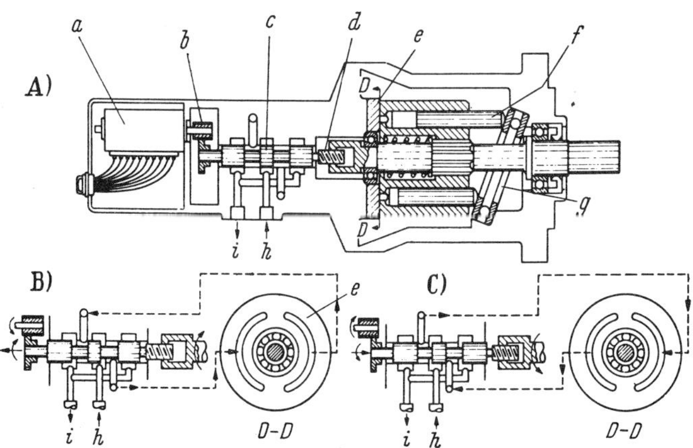

Fig. 119 Principle of an electro-hydraulic stepping motor (74, 82)

A) Cross-section. B) Valve position for clockwise rotation. C) Valve position for anti-clockwise rotation.

a Electric stepping motor. *b* Reduction gearing. *c* Hydraulic valve. *d* Threaded pin. *e* Valve disc. *f* Plunger of axial-piston motor. *g* Fixed swash plate (cf. Fig. 115). *h* Oil inlet connection. *i* Oil outlet connection.

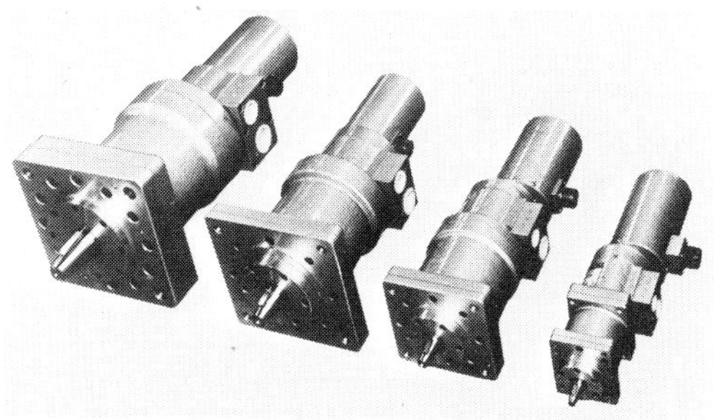

Fig. 120 Range of commercially available stepping motors (series of 8 types, photo courtesy *Siemens/Fujitsu*) (101)

Max. torques 0.35 to 6.4 mkp. Working torques 0.15 to 2.5 mkp (for continuous-path control, 2000 pulses/s and oil pressure of 70 kp/cm²). Stepping frequency 4000 to 16000 pulses/s. Speeds (max.) 1000 to 3200 rev/min.

required value, the valve will have been closed again by the action of the threaded pin d, and the hydraulic motor has closed its own valve. All the while no impulses are applied to the electric motor the hydraulic motor will remain stationary. The maximum valve opening that is possible forms a mechanical store that can accommodate at least ± 300 steps. This integrating effect results in the rotation of the hydraulic motor being continuous in nature down to very low speeds; only at the lowest speeds does the hydraulic motor carry out a stepping movement. The action of the threaded pin d causes a lagging error to arise in the follow-on operation, and this increases approximately proportionally with the speed; at a rate of 2000 electrical pulses per second this lagging error amounts to about 15 steps. This lagging error produces no dimensional deviations for straight-line motions in one axis; it is also negligibly small for large radii. For smaller radii a lower pulse frequency will have to be employed to keep the lagging error small. Figure 120 shows some models from the range of motors made in Germany by Siemens under licence from Fujitsu Limited, of Tokyo.

When selecting the type of motor to use it should be borne in mind that when hydraulic motors are operated in this way they are unable to produce even short-term torques that are higher than their rated torque. The static loading due to friction and the cutting forces should therefore not exceed 40% of the rated torque to ensure that adequate reserves are available for acceleration, etc. Careful attention should also be paid to the information given on the data sheets produced by the manufacturers or licensees.

6.6   Pneumo-hydraulic positioning systems with open loops

For various reasons pneumatic alternatives to electrical and electronic elements have been adopted for practical applications in recent years. The previous sections have discussed the possibilities of combining electrical and hydraulic systems, and this section deals with the possibility of combining hydraulic systems with compressed air as a means of transmitting signals and energy. Owing to its compressibility, air, in general, plays only a comparatively secondary role in machine tools (clamping operations, clutch operation, feeding devices for tools and workpieces, etc), but in many instances low-pressure systems have become almost indispensable for processing signals. Owing to its simplicity and reliability and the ability to use it in explosive atmospheres, pneumatic control equipment has long been widely used in the mining and chemical industries (116). In the last 5 or 6 years it has also been found possible to build up logic elements on a fluid flow basis (fluidics (117)) and to produce these at low cost thanks to the progress that has been made in the plastics industry; and it has been apparent that in many instances these would compete with electronic elements. Fluidics offer certain major advantages over electronic components under adverse ambient conditions; among these is their complete insensitivity to radiations of all types and the fact that the ambient temperature has very little effect on them. The physical laws that govern their operation are such that fluidics will inevitably be slower in operation than electronic components, but in the micro-second range they are very well suited to the requirements of machine tools.

Where power is to be produced it seems convenient to establish an analogue to the electro-hydraulic systems that have already been mentioned; hence the derivation of the term pneumo-hydraulic. The problems that arise in the use of this system for this purpose clearly fall within the scope of this chapter. Where the primary interest is in the pneumatic processing of signals, on the other hand, it is more convenient to discuss the problems and their solutions in Chapter 9 (117, 118).

In view of the desirability of keeping cost and complexity to a minimum, it seems advisable to use pistons similar to those illustrated in Fig. 112 A) for the hydraulic power units of positioning and straight-line control systems and where the slide displacements are not large. Under these conditions it is possible to take advantage of the fact that with a piston very rapid positioning movements can be attained so that idle times of the machine are reduced. The main field of application for the pneumo-hydraulic control systems that are currently available is therefore likely to continue to be for small to medium-sized drilling machines (113).

In the two systems that are to be described a further step has been taken; there is no feed-back of the actual value and no automatic comparator, and the system is thus of the open-loop type as illustrated in Fig. 1 A). Instead of an incremental process as used with the stepping motors described above, absolute dimensions are fed into the system, and the design of the position-measuring power units ensures that they are maintained when the table is positioned.

6.6.1   *The Moog hydraulic positioner system* (114,115)

Only the positioner drive system will be discussed here; it is already of comparatively long standing (developed in the USA in about 1950). It will also suffice if the application of the drive to one axis only is considered. The maximum stroke for the X-axis is 20 in (510 mm). Accordingly 200 holes are drilled along the length of the cylinder spaced exactly 0.10 in (2.5 mm) apart. A slide valve in the upper part of the cylinder uncovers only one hole at a time (Fig. 121 A). The servo-piston b in the cylinder a, which moves the machine slide, is provided at the end with a helical collar e having a pitch of 0.10 in (2.5 mm), but which extends only 288° round the circumference. The remaining gap permits the oil delivered into the cylinder by the hydraulic pump f to flow back into the reservoir (from right to left in the illustration), during the course of which a pressure of 105 kp/cm^2 builds up in the system. This pressure acts on the collar e and in the first instance attempts to force the servo-piston b to the left. This cannot move, however, unless the oil has first flowed out of chamber d which is formed by the coaxial secondary piston c, which is fixed axially, and the bore in the servo-piston.

If the servo-piston is in its end position to the left, and one of the holes in the cylinder wall is uncovered (e.g. that shown in Fig. 121), the cylinder space is connected to the servo-piston chamber, which is now also subjected to the pressure in the system. The area of the end face of the chamber is, however, made twice as large as the effective area of the collar and the shoulder formed on the servo-piston shaft. The resultant differential force will therefore cause the servo-piston to move to the right, towards the open hole. Shortly before the helical collar reaches the hole, the shoulder reduces the area through which oil can flow into chamber d so that the reduced oil flow causes the speed at which the servo-piston moves to drop. The system thus incorporates a built-in changeover to an inching speed. When the helical collar is at the centre of the hole there will be openings (segments of the hole) left on either side if the dimensions are correctly chosen; these act as throttles and permit the pressure in the chamber to drop to half that of the system. As a result, the forces acting on the servo-piston are equalised, so that it is in hydraulic equilibrium and remains stationary. The collar centralises itself in this way precisely over the open hole; any deviation would upset the hydraulic equilibrium and cause the collar to move back towards the centre.

In practice there is not just the one hole in the cylinder wall, as shown in the simplified sketch, but three, disposed helically round the circumference to avoid the need to make the collar so narrow that it is smaller than the hole

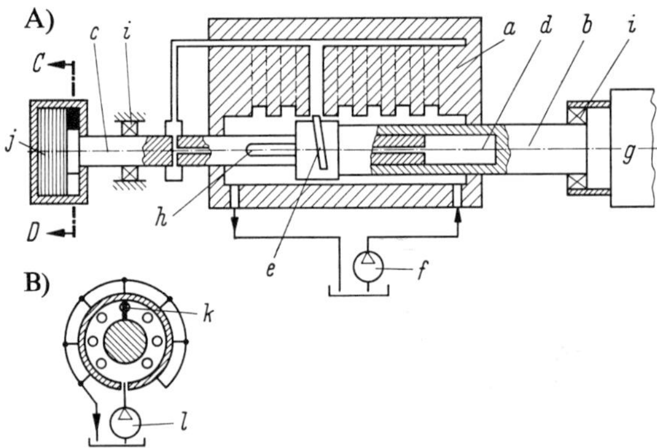

Fig. 121 Principle of the *Moog* hydraulic positioner (114, 115)

A) Coarse positioning by hydraulic cylinder to 0.1 in (2.5 mm). B) Precision positioning by rotation of piston, Section C-D.

a Cylinder casing with slide valve. *b* Movable servo-piston for moving table. *c* Fixed secondary piston. *d* Servo-piston chamber. *e* Helical collar extending over an angle of 288°. *f* Hydraulic pump. *g* Machine slide. *h* Key to prevent relative rotation of *b* and *c*. *i* Ball bearing (also capable of absorbing thrust). *j* Rotary disc valve as shown in B). *k* Rotary vane. *l* Hydraulic pump.

diameter, and in order to obtain an adequate cross section for the flow.

The hole that is exposed by the slide valve thus provides for a coarse positioning of the servo-piston, and hence the machine slide, which is then located centrally with respect to the hole, or to within 1/10 inch.

Fine positioning is effected by rotating the secondary piston c, the rotation being transmitted to the servo-piston b by means of the key h. Because of its helical form this causes the edges of the collar to be displaced from the centre of the hole, and this upsets the hydraulic equilibrium. As a result, a longitudinal movement of the servo-piston takes place to counteract this displacement of the edge of the collar. The secondary piston is rotated by a rotary vane system, as illustrated in Fig.121 B), which is a section through C-D. A hydraulic pump l applies pressure uniformly to both sides of the rotary vane k by way of the casing. The wall of the casing is provided with 100 holes, so that the effective pitch of 1/10 inch of the collar e is divided into 100 equal parts. This enables fine-positioning to be carried out in steps of 1/1000 inch (0.0254 mm).

If only one of the holes is uncovered, the hydraulic fluid can escape through it into the reservoir. The drop in pressure on this side of the chamber then upsets the hydraulic equilibrium on the vane, and this moves towards the open hole. There it aligns itself accurately with the centre of the hole, since its width is less than the diameter of the hole, and the equilibrium condition is restored with the oil flowing out uniformly through the two uncovered portions of the hole (similar to the coarse positioning illustrated in Section A) of the illustration). The holes are opened by means of a rotary disc valve, only one hole being exposed at a time. The principles on which the

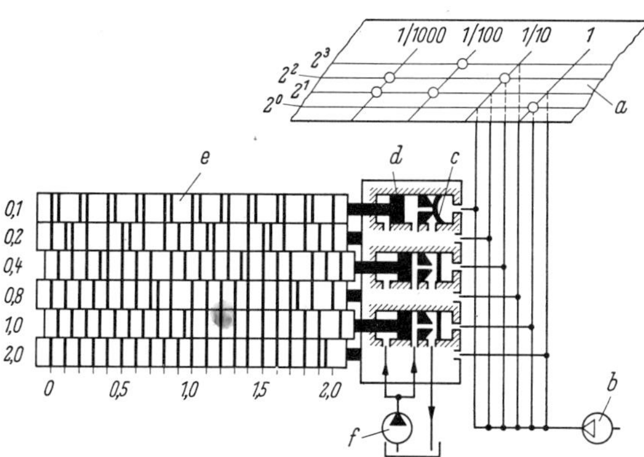

Fig. 122 Schematic diagram of a pneumo-hydraulic signal transformation for the control of a slide valve as shown in Fig. 121.

a Punched tape. b Compressed-air supply. c Diaphragm. d Piston. e Valve slide. f Hydraulic pump.

Note: To simplify the drawing, the three pistons for the punched-tape tracks 1×2^1, 0.1×2^3, 0.1×2^1 are not shown.

rotary disc valve operates are similar to those governing the slide valve as described below.

The slide valve represents the interface between the pneumatic signals produced by the punched-tape reader (Chapter 9) and the hydraulic servo-system; strictly speaking this valve is the regulator. In Fig. 122 this valve is shown in simplified form for a servo-piston stroke of only 2 in (50 mm). It consists of a number of superimposed plates with vertical holes. Each plate can be displaced by 0.05 in (1.25 mm) from its rest position. The 21 holes in the cylinder spaced at intervals of 0.1 in (2.5 mm) which represent the decimal numbers for the positions 0, 0.1, . . . 2 inches are located below the lowest plate. The plates with their holes, on the other hand, each represent 1 bit of a binary-coded decimal number system, the rest position of the plate (to the left) indicating that this bit is not present (O), while the displaced position of the plate (to the right) represents that it is present (L). The 21 decimal numbers from 0.1 to 2.0 can be produced by six bits or plates, there being 2 plates for the unit decimal places and 4 plates for the one-tenths decimal places $(1 \times 2^0, 1 \times 2^1, 0.1 \times 2^0, 0.1 \times 2^1, 0.1 \times 2^2, 0.1 \times 2^3)$. The zero position is obtained when none of these bits is present, i.e. when all the plates are in the rest position.

For 200 decimal numbers, corresponding to the 200 holes spaced along the length of the cylinder, 10 plates will therefore be required. Each plate is drilled in such a manner that when it is in its rest position there is a hole vertically above the decimal numbers which do not require this particular bit. If the bit is required the hole is located 0.05 in (1.25 mm) in front of the decimal number so that the path to the cylinder is opened only when the plate is displaced. The only vertical path through the stack of plates that is open is therefore that which is formed by the sum of all the bits that are present, i.e. all the plates that are displaced.

The holes in the plates are represented in Fig. 122 by heavy black lines, it being borne in mind that in practice three holes will be required to correspond with the holes in the cylinder.

The plates are displaced by means of small pneumatic-hydraulic cylinders, each piston d of which is connected to a plate. A pressure of about 5 kp/cm^2 is applied uniformly to both sides of these pistons by a hydraulic pump f. A diaphragm c, which is acted on by compressed air controlled by the punched-tape reader, seals off the opening to the discharge passage on the right-hand side of the piston so that oil cannot escape when the air pressure is applied. The presence of the piston rod on the left-hand side of the piston reduces the effective piston area, so that the force exerted on the right is larger, and the piston moves over to the left, as shown for the uppermost piston in Fig. 122.

If the air pressure on the diaphragm is relieved (due to the presence of a hole in the punched tape a), the discharge passage is opened, and the pressure on the right-hand side of the piston drops. The piston d thereupon moves over to the right-hand side of the cylinder thereby displacing the plate by 0.05 in (1.25 mm) from its rest position.

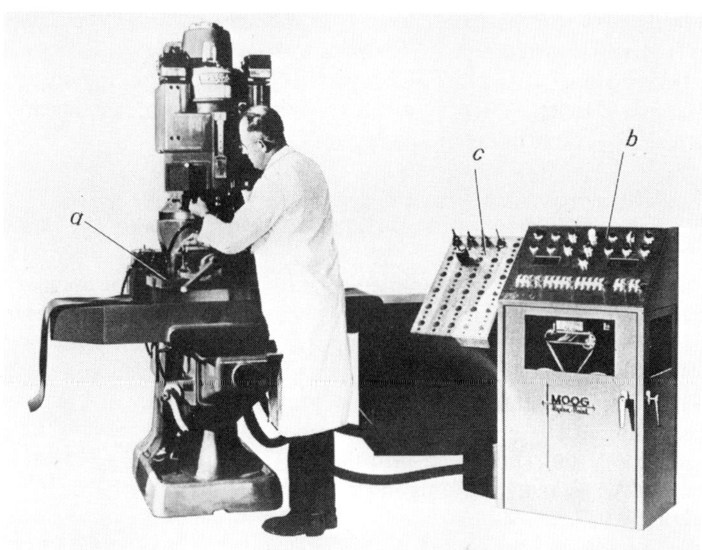

Fig. 123 Example of a pneumo-hydraulically controlled milling machine (photo courtesy of *Moog, Hydra-Point/Bridgeport*).

a 3-D milling machine (positioning and straight-line control). *b* Control console (Chapter 9). *c* Tool magazine for manual tool changes (Chapter 7).

Whether an air pressure is applied to the diaphragm *c* or not is therefore determined by the punched tape *a*. As shown in the illustration, the compressed-air lines run from the supply connection *b* both to the diaphragm chambers and to the punched tape. If there is no hole in the punched tape for one of the lines a pressure will be built up, which is then also present in the diaphragm chamber. If a hole is present, however, the air can flow out freely, the pressure drops, and the diaphragm opens the discharge passage.

The punched tape, which is coded in the BCD system, is read in blocks (cf. Chapter 9). Each block contains columns for the decimal places 10^0, 10^{-1}, 10^{-2}, and 10^{-3}. For the actual stroke of 20 in (510 mm) it will also, of course, be necessary to provide columns for the decimal places 10^1 and 10^2, but these are not shown in this drawing. The coarse information in the columns down to 10^{-1} is 'read' by the compressed-air lines for the coarse setting, and is applied to the slide valve; while the 10^{-2} and 10^{-3} columns are required for the precision settings and are used to control the rotary disc valve.

In Fig. 122 a coarse adjustment to 1.4 (35 mm) is to be made. For this, holes are required in the punched tape in positions 1×2^0 and 0.1×2^2. These cause the plates 1.0 and 0.4 to be displaced and as a result only the hole corresponding to 1.4 in is open right through the stack.

The slide valve therefore also acts as a code converter which converts the displacement information that is present in BCD code on the punched tape into the decimal code required for positioning the piston. Figure 123 shows a 3-D milling machine which employs these principles; the machine is also available for use with the metric system of dimensions.

6.6.2   *Plessey hydraulic positioning principle (from about 1967) (119)*

Figure 124 represents in simplified form the positioning of the machine slide along one axis. Slide a is displaced by means of the hydraulic cylinder system b. The slide also carries the indexing rod h for accurate positioning and the slide c with the shoe d by means of which signals are produced.

The displacement information carried by the punched tape, which is punched in BCD code, is read off pneumatically by the punched tape reader l (Chapter 9), and is split up into the 10^2, 10^1, and 10^0 decades, which are used for the coarse measuring device, and into the 10^1, 10^0, 10^{-1}, and 10^{-2} decades, which are supplied to the precision measuring device (this is for the metric system of dimensions).

Both groups of compressed-air signal are amplified in pneumatic storage amplifiers, and, as a result of the storage, are transformed into a continuous air flow. The information required for producing the signals is decoded in the decoder, i.e. the binary-coded decimal numbers are converted into the decimal system and are applied to the signal rod r. This acts as the signal emitter and carries 25 pairs of holes for the positions 0 - 500 mm, i.e. the hole spacing is 20 mm. The shoe d, which has a round pocket and a slotted pocket, moves over this measuring rod. The shoe is attached to the slide c which moves in a slot in the measuring rod and which can block off the air inlets in the measuring rod (which come from n) from the outlet pipe e. If the shoe d is in the position illustrated and a position signal is produced at, for example, hole 4, the blocking effect of slide c prevents it being passed to the outlet pipe e, and no signal is produced at this outlet. The absence of a signal is interpreted by the logic element o to which the outlet is connected as the command 'move machine slide to the right'. The fluidic logic system (Chapter 9) acts on the hydraulic control unit q which controls the oil flow in such a way that the piston in the hydraulic cylinder b moves the slide to the right. If, on the other hand, a signal is produced at hole 1 this can be transmitted on to the logic unit since it is not blocked by the slide c. The presence of a signal at e is interpreted as 'move machine slide to the left', and pressure is applied to the opposite side of the hydraulic cylinder to achieve this. Line e can thus be regarded as a signal link for the direction of motion of the machine slide; the slide will always move towards a signal hole that is carrying a signal.

The signal air flow can always escape through the pair of holes corresponding to an inlet. As the shoe approaches a signal will, therefore, first be generated in the slotted pocket of the shoe; this signal passes along line g to the logic unit o where it is interpreted as 'reduce speed'. This results in the

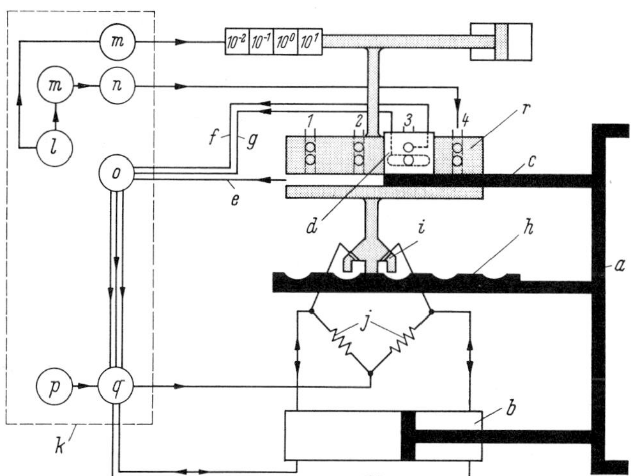

Fig. 124 Plessey pneumo-hydraulic positioner, coarse position adjustment in one axis (118, 119).

a Machine slide. *b* Hydraulic cylinder and piston. *c* Slide. *d* Shoe. *e* Signal line 'direction of motion'. *f* Signal line 'speed of motion'. *g* Signal line 'Stop'. *h* Precision graduated rod. *i* Oil jet unit. *j* Hydraulic resistances in bridge circuit. *k* Control cabinet. *l* Pneumatic punched-tape reader (cf. Chapter 9). *m* Pneumatic signal amplifier with storage capability. *n* Decoder. *o* Pneumatic logic unit (fluidic), (cf. Chapter 9). *p* Hydraulic pump. *q* Hydraulic control unit with pneumatic triggering.

Note: For clarity many of the details have been greatly simplified. It is not possible, therefore, to draw any conclusions regarding the actual design details from this diagram.

flow of oil to the hydraulic cylinder being reduced by *q*. When the round pocket of the shoe is over the hole through which air is emerging, the resultant signal is led to the logic unit by means of line *f* and is there interpreted as 'stop machine slide'. This causes the supply of oil to the hydraulic cylinder to be shut off at *q* and also for the air bearings to be evacuated and the hydraulic table clamps to be applied. Rigidly coupled to this signal emitter is the oil jet unit *i* which is provided with two oil outlets opposite the graduated rod, which moves beneath it and is equipped with transverse grooves. If both these outlets are equidistant from two grooves, the oil at both outlets encounters the same flow resistance and therefore emerges at the same pressure. The outflow resistances form a bridge circuit in conjunction with two hydraulic resistances, and are arranged so that in this case the bridge resultant is zero. If the oil jet unit changes its position, one of the outflow resistances will increase and the other will be reduced. This upsets the bridge balance, and the resultant oil flow acts on the piston of the hydraulic cylinder in such a manner as to cause the machine slide to move and to bring the graduated rod *h* back into the equilibrium position. The 'oil jet unit/graduated rod' system can therefore be regarded as a form of hydraulic pressure equalising valve. This concludes the coarse positioning to within 20 mm of the desired value.

The fine adjustment of the oil jet unit and the signal emitter within the range of 20 mm is effected by the pneumo-mechanical fine measuring system, which is shown schematically in Fig. 123, and which is also controlled by the tape reader acting through storage amplifiers. The stroke of a piston in the pneumatic cylinder j is limited by a number of gauge blocks, which all have the same length, and which are placed end-to-end in the casing d. In addition to this main set of gauge blocks there is also the row b made up of gauge blocks of differing lengths. Each two adjacent gauge blocks of these rows form a pair which are supported in the half-round grooves of the common trolley e. The differences in the lengths of the gauge blocks forming each pair is such that, measured from the uppermost in the drawing to the lowermost, it covers in practice only three decades of a binary-coded decimal system. Since each decade requires four units, the total requirement is only twelve pairs of gauge blocks and twelve trolleys. The smallest difference in the lengths of a pair of rollers is then 0.02 mm and the largest 10 mm. If in the absence of pressure at j one pair of gauge-blocks moves to the right, the subsidiary block takes the place of the main block (Fig. 125 B). The length of the row of main gauge blocks is shortened accordingly when they are again

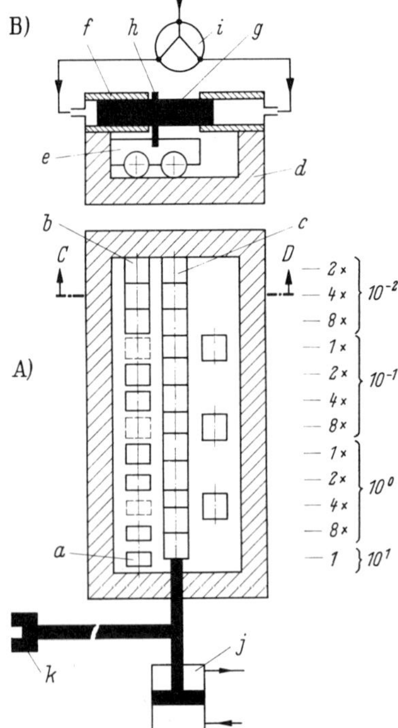

Fig. 125
Fine positioning in one axis - supplement to Fig. 124

A) Mechanical gauge block system for the range 0.02 to 19.98 mm. B) Gauge block changing mechanism, Section C - D.

a Gauge block roller. b Subsidiary row of gauge blocks. c Main row of gauge blocks. d Casing. e Trolley. f Pneumatic cylinder. g Piston. h Drive pin. i Signal amplifier. j Pneumatic cylinder with piston for applying pressure to gauge blocks. k Oil jet unit (corresponding to part i in Fig. 124).

pressed together by means of the piston rod of the pneumatic cylinder *j*. In the illustration the length of the row of blocks has been reduced by 4.90 mm; to achieve this the trolleys carrying the pairs of blocks with length differences of 4.0, 0.8, and 0.1 mm have been moved to the right. The fine measuring device thus also acts as a decoder.

The displacement of the trolleys is by means of a small pneumatic cylinder *f* and piston *g*, which is connected to the trolley by means of a pin *h*. When the signals for 4 mm, 0.8 mm, and 0.1 mm are obtained from the punched tape, the signal amplifier *i* directs the compressed air to the left-hand sides of the appropriate pistons, which then move to the right taking their trolleys with them. If there are no signals for the other binary numbers the compressed air will remain applied to the right-hand sides of the pistons and the trolleys will not move.

When the appropriate rollers have been moved into position, the main row of gauge blocks is pressed together by means of the piston rod *j*; as a result, the change in length is transmitted to the oil jet unit *k* and the signal emitter. The readings of the fine measuring device are governed by a pneumatic cycle timer. This is located in the control cabinet and is not shown in Fig. 124 to avoid undue complication of the diagram. It alternates with the logic unit and

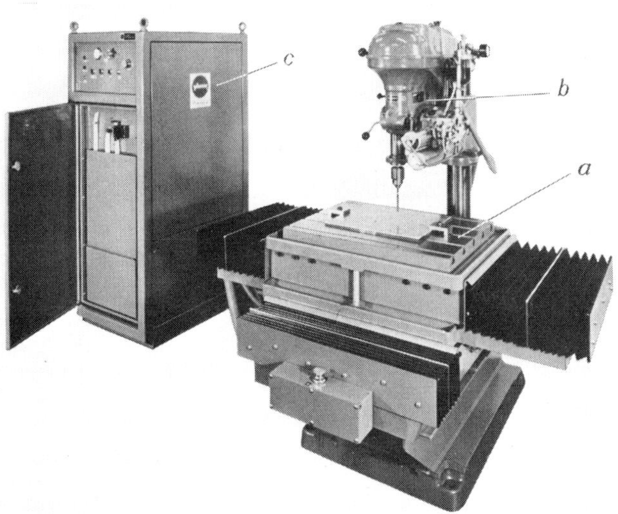

Fig. 126 Example of a pneumo-hydraulically controlled coordinate table used in conjunction with a conventional drilling machine (photo courtesy *Plessey* '*Fluidic 220*').

a Coordinate table 635 x 470 mm (25 x 18½") travel 500 x 360 mm (20 x 14"). *b* Conventional column drilling machine. *c* Control cabinet with pneumatic punched tape reader and low-pressure logic system employing Maxalog elements (Chapter 9).

also controls the coarse positioning unit as well as switching off the air supply to the air bearings of the machine table and the clamping of the table when it is correctly positioned. In principle this is a two-part measuring system of the type already described in Chapter 5 with reference to electro-mechanical components. As was mentioned then, these systems are especially suitable for positioning controls. Figure 126 therefore shows an example comprising a coordinate table which has been installed under a conventional drilling machine. It should be noted that, as in the case of the Moog system (Fig. 121, 122, 123) there is no feed-back in the normal control technology sense of the term, and the coding has been transferred into the drive system. This enables very robust and inexpensive machines, that possess sufficient accuracy for many applications, to be obtained. It is therefore probable that in the future further machines employing pneumo-hydraulic principles will be introduced and that these will appeal to particular classes of user (10, 113). These coordinate tables are also available for the inch system of measurement.

6.7   Summary of drive systems

This chapter can be summarised as follows:

1. Electro-mechanical feed mechanisms all have the advantage that the control acts directly on the mechanical power system, and so the energy switching and transformation points required are reduced to a minimum.
2. Electro-magnetic clutches can conveniently be used with stepped main spindle drives on all types and sizes of machine tool. The individual speeds to be engaged are selected by the switching information. Drive systems with reduction gears and electro-magnetic clutches can also be used successfully in the feed drive systems of machines with positioning control and straight-line control; the designs for the gearing have to meet different conditions for these two types of control. It may be advantageous to select a suitable combination of three-phase a.c. and d.c. motors.
3. Unless there are some special production requirements, it will be necessary to provide infinitely-variable drives for the machine slides only where continuous-path control systems are in use.
4. If d.c. motors are used in infinitely-variable drives, their relatively high moment of inertia must be taken into account. There are also certain limits to time constants when there are large moments of inertia and also with electronic control equipment unless special motors are to be used (74, 94, 102).
5. The advent of new semiconductors (e.g. silicon controlled rectifiers for heavy-current use appears to provide solid-state control systems

for d.c. motors in feed drive systems with technical and economic advantages over the competing solutions that employ electrical machines.

6. Owing to their compact design hydraulic axial-piston motors that operate with high oil pressures have a moment of inertia only about one-tenth that of a comparable d.c. motor; conditions are even more advantageous with rotary-vane motors, especially where high nominal torques are concerned.

7. The introduction of thyristor drives enables speed control circuits with good time constants to be designed in conjunction with special d.c. motors.

8. With electro-hydraulic feed drives it is also necessary to make a sharp distinction between on-off operation (positioning and straight-line control) and continuous operation (continuous-path control).

9. If the slide movements are not too large it is possible to produce very satisfactory positioning and straight-line control systems using piston and cylinder drives, since the longitudinal displacements are produced directly. The working range (slide displacement) that can be handled is larger for positioning control than for straight-line control systems (absence of forces opposing the feed motion).

10. The main line of development in the production of electro-hydraulic continuous-path control systems is likely to be concentrated on electro-hydraulic controllers with good time constants to enable full advantage to be taken of the low moment of inertia combined with a high torque, and the development of a high torque even at speeds close to stalling, that are features of hydraulic motors.

11. A disadvantage of all hydraulic control systems is the variation in the viscosity of the oil with temperature. For larger installations it may prove necessary to adopt the necessary measures to keep the temperature stable. One result is that it may take some time for the equipment to warm up when it is first switched on. Long pipes also prove a disadvantage in high-pressure hydraulic systems.

12. In view of the comparatively low continuous torque ratings of d.c. motors (if they are to have a suitable time constant) on the one hand, and of the effects of the compressibility of the oil and the expansion of the pipes of extensive high-pressure hydraulic systems on the other hand, it may well prove the most convenient solution to adopt electrical equipment on large continuous-path controlled machines where the cutting conditions are relatively light (e.g. the machining of light alloys in the aircraft industry) and hydraulic equipment of compact design for medium-sized machines that have to cope with more arduous cutting conditions.

13. Further development of high-performance electro-hydraulic stepping motors could change the whole trend of the design of numerically-controlled machine tools (e.g. elimination of displacement measuring systems).

14. It will be interesting to follow the future development of pneumo-hydraulic feed drives, since these are likely to provide simple, reliable, and inexpensive solutions. In the first instance these will, however, be limited to positioning and straight-line control systems, and they may not necessarily be suitable for incorporation in complete numerically-controlled production lines (Chapter 7). They will presumably appeal to a special class of NC-machine user, who requires only a few simple high-performance NC machines and who, either because of the size of the firm or because of the nature of the product, is not so dependent on computer techniques (see Appendix).

This discussion of the drive systems to some extent concludes the first part of the book. The theoretical basis was developed in Chapters 1 and 2, while Chapters 3 to 6 have described the results of 20 years of technical development work.

The *displacement measuring systems* can be regarded as being fully developed, and future work seems likely to be in the fields of stepping motors and fluidic control systems which require no measuring system.

So far as *drive systems* are concerned the relatively small field of applications for remote-controlled clutches used in conjunction with reduction gears is completely defined, and in the case of continuous-path control systems, which will probably become of increasing importance, the only factor of importance is the competition between hydraulic and d.c. systems, both of which can be controlled remotely without difficulty. At the present moment these seem to share the market more or less equally, but the introduction of thyristor control systems seems likely to lead to a come-back of the d.c. motor in a different form.

By and large, however, it seems unlikely that there will be any revolutionary changes introduced in the foreseeable future in either displacement-measurement systems or drive systems. The minor improvements and changes that will undoubtedly occur do not appear likely to be of major importance although they will no doubt have an effect on the fields of application for NC machines.

The situation is very different so far as Chapters 7 to 12, which follow, are concerned. Here the effect of computer techniques is unquestionably the dominant factor, both in the speed of technological development and the range of applications. The subject first started to become clearly defined about 3 to 4 years ago, and can be described as 'integrated data processing in the production field'. It affects the design of machine tools, the nature of their controls, the way in which they are programmed, and the control of the entire flow of materials and information through the plant. The need to view the developments as a whole is, therefore, even more marked than was the case in the earlier chapters of this book, and a limit is soon reached as to the amount of material that can be included in a basic text book, especially as this soon becomes dated owing to the rapid rate at which developments are taking place. Ideally, there should be annual supplementary volumes, so that the text book indicates only the main lines along which development is proceeding.

7   The machine tool and numerical control

There can hardly have been any previous instance where a new development in a 'foreign' field has had such a stimulating and far-reaching influence on the further development and the future utilisation of machine tools as the introduction of the idea of numerical control did in the early 1950's. Development work in this field has by no means come to an end; it will continue for a long time to come, as it is based in the main on the constantly increasing penetration of electronic data processing (EDP) into all aspects of the production process (10). Apart from a general survey of the current position, it is, therefore, not possible to do more in this chapter than to make some suggestions for further work and to indicate the general trend of developments. The machines chosen for the examples have been selected solely for the factors they illustrate, and there is no suggestion that they are superior to others of their type.

The problems of the interaction between the machine tool on the one hand and the numerical control system as a part of the EDP system on the other can, in the main, be regarded from two points of view:

a) The selection of the machine elements and their adaptation to the changed conditions, and

b) the development of new types of machine tool influenced by the ability of these control systems both to extend the machine kinematics and to integrate the individual machines more effectively into the flexible flow of information and material in a production plant engaged on one-off and small batch production.

In view of the trend that the development is taking, it is also advisable to take a fresh look at many facts and definitions associated with the classical design and construction of machine tools, and to formulate these anew to take account of the incorporation of the machine within a larger system structure. In the limiting case, it is necessary to look on machine tools as special types of data processing equipment, and they can, indeed, in a sense be regarded as digital/analogue converters.

7.1   Special design features required to match machine tools to numerical control systems

If it is assumed that the required 'numerate' machine is available and fully up to requirements of reliability and of incorporation into the remainder of the production process (i.e. no waiting times due to interruptions in the flow of materials), the following situation could already obtain:

All the machining information required to produce a workpiece is stored in the data carrier (e.g. a punched tape); the machine can turn out a constant stream of workpieces of invariable quality at optimum speed. 'Man', who is the principal fatigue factor, has virtually been eliminated. If the machine is changed over to produce a different workpiece (different program), the setting times[1] are in many cases much lower than those for conventional machine tools; the machining cycles follow in quick succession and idle times are short in relation to the system as a whole. If the required number of workpieces increases the machine can supply them, possibly without the need

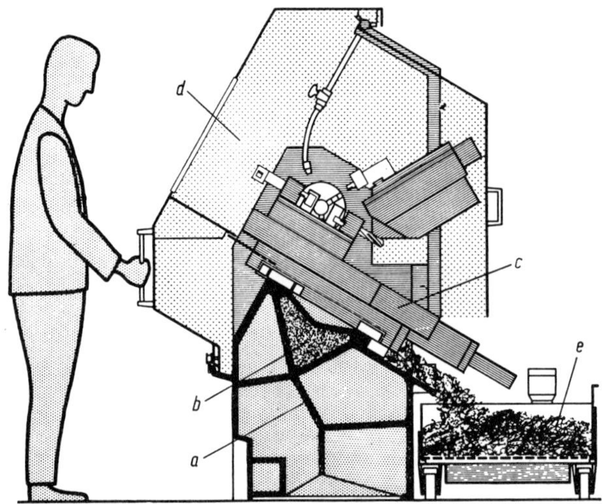

Fig. 127 Schematic diagrams of a high-capacity numerically-controlled lathe

a Heavily-ribbed base. *b* Hollow space filled with moulding sand to damp down vibration. *c* Inclined bed to ensure rapid disposal of hot swarf and warm coolant. *d* Sliding guard made of acrylic material to protect operator from flying hot chips and swarf. *e* Large mobile trolley for swarf and coolant collection, not attached to machine, to simplify swarf removal.

---

[1] Even with numerically-controlled machines the setting times will vary from case to case according to the shape of the workpiece and the fixtures (clamps, alignment devices) that are available. (See Section 7.3 and Chapter 9).

for additional skilled labour so that a changeover to multi-shift working would be easier.

This brief review is sufficient to show that a numerically-controlled machine of this type will have much lower idle times than a manually-controlled or less highly automated machine of the conventional type; the machine utilisation and productivity will be greatly increased.

Every care must be taken in the design of these machines to make them very robust and to reduce the likelihood of wear. The design must also make provision for the removal of what may be very much larger quantities of swarf. This is illustrated by Fig. 127, which shows the basic features of a lathe that has been designed in accordance with the latest ideas. In addition many of the movements, even of larger masses, have to be carried out much more rapidly than for a manually-controlled machine, (e.g. the positioning of large coordinate tables); as a result the loading conditions on the feed mechanisms and on the guides are much more severe. Since sensitive measuring systems are fitted, and these are required to operate satisfactorily even when the machine is taking heavy cuts (e.g. photo-electric linear scales, Chapter 3), the machine has to meet very stringent demands concerning stiffness and freedom from vibration. Another problem is that the principle of measuring the slide movements can result in the acceptable workpiece tolerances being exceeded after a short time owing to elastic deformations and wear of the guides unless the design of the machine and the selection of the materials employed take account *ab initio* of the higher stresses to which the machine is subjected.

These preliminary considerations lead to two conclusions:

1. If numerically-controlled machine tools are to have the highest possible degree of utilisation and to achieve high accuracy they must be of very robust and solid construction (5, 7, 130, 362).
2. Fitting a numerical control system to a conventional machine tool does not usually produce the full performance of which the control system is capable. In many cases the cost is then no longer justifiable by the results obtained.

Use of special machine elements

Mention was made in Chapter 6 of the great importance that the frictional resistances in the machine have on the dimensions of the drive systems. If anything, it is even more important for the coefficient of friction to remain constant than for it to be as low as possible. Both continuous-control circuits (continuous and intermittent) and on-off circuits will be reliable in operation only if the frictional resistance remains constant.

In particular, it is essential that stick-slip (5, 7, 20) is avoided at all costs. This effect arises at very low sliding speeds (both translatory and rotary) and is due to the constant alternation of static friction and sliding friction. At low sliding speeds, i.e. at values between, say, 5 and 20 mm/min, the thin lubricant film can break down, and static friction will come into play for a short time. During this period the power transmission elements (e.g. leadscrews, hydraulic fluid columns, etc.) will deform elastically until the static friction is overcome. The elasticity of the deformed element and the restored sliding friction then cause the moving machine part to be accelerated in the direction of motion. Immediately after this it may stick again, and the whole process is repeated. Since the stick-slip phenomenon occurs only at very low speeds, it is a particular nuisance in positioning and straight-line control systems as well as near points of reversal in continuous-path systems. Some improvement can be obtained by using suitable materials (including plastics) for the surfaces of the guides and by suitable additives to the lubricants (e.g. molybdenum disulphide).

In many instances, it is, however, also desirable to reduce the coefficient of friction to a very low value, particularly to keep the temperature rise down at high speeds of displacement where leadscrews are being used, and to avoid certain difficulties associated with the dimensioning of the drive systems (Chapter 6). In principle, the adoption of elements that produce rolling friction enables the power requirements to be reduced, the temperature rise to be kept low, and stick-slip phenomena to be avoided.

Figure 128, for example, shows a recirculating ball screw with a pre-stressed double nut. An arrangement of this type has two effects:

a) The friction between the threaded screw and the nut is much reduced; the efficiency rises to over 90%, and the temperature rise will remain low even under heavy loads;

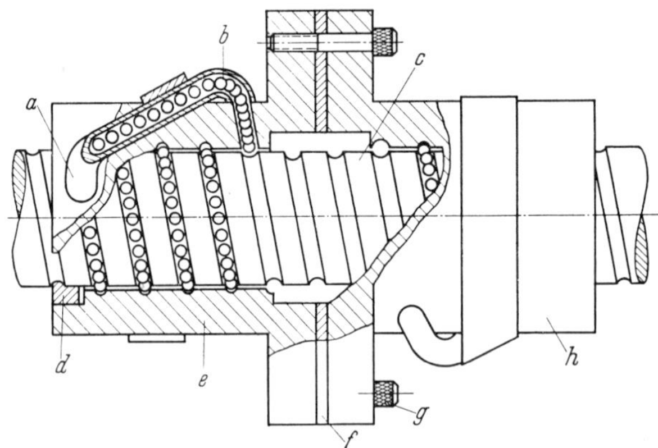

Fig. 128 Example of a recirculating ball screw with shim to adjust pre-load and hence eliminate backlash. (Courtesy of *Bristol Siddeley Engines*).

a Recirculation tube for balls. *b* Circulating balls. *c* Ground precision screw. *d* Seal and scraper. *e* Recirculating ball nut 1. *f* Pre-loading shim. *g* Connecting screw. *h* Recirculating ball nut 2.

b) use of an accurately ground shim, *f,* to pre-load the nuts eliminates backlash without causing any appreciable increase in the friction and hence in the temperature rise.

Recirculating ball systems of this type are nowadays available for spindle diameters from 6 to 150 mm, and for spindle lengths from 50 to 8000 mm (121). If the individual elements are sufficiently accurate, the excellent service characteristics of this type of system often make it possible to fit a rotary displacement measuring system directly to a recirculating ball screw used as a leadscrew (Chapters 3 and 4). In this way it is often possible to produce solutions which are not only economically attractive, but which are also sufficiently accurate for many purposes even under the adverse conditions to which machines are subjected in practical service in the workshop (c.f. the machines shown in Figs. 171, 179). This argument is probably more relevant to positioning control systems (where cutting forces do not affect the control system) than for continuous-path control systems. In the latter, the cutting forces, especially if they are not of constant magnitude, can produce elastic deformations and vibrations in the recirculating ball system especially if this is not suitably dimensioned. These would then have an unacceptable effect, not only on the displacement measuring system, but on the machining process itself.

A)

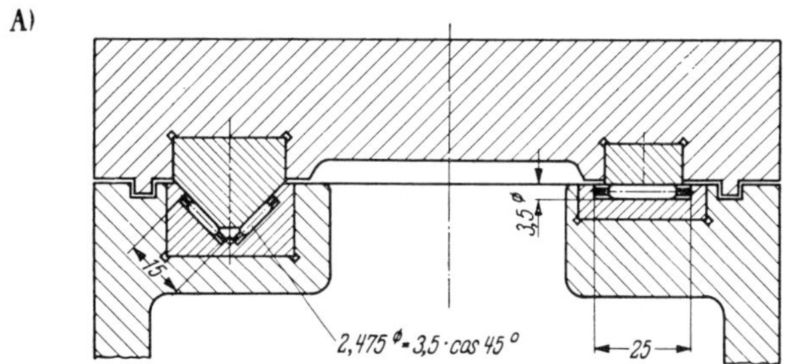

B)

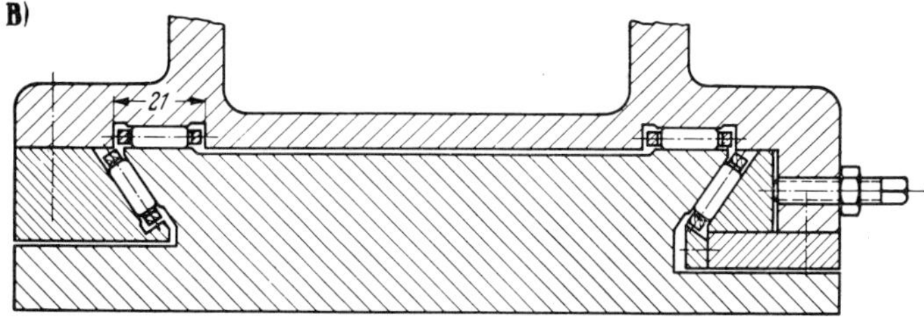

Fig. 129 **Machine slides using anti-friction bearings** (7, 128)

A) Open slideways, slide supported on needles of different diameters.
B) Dovetail slides with four separate sets of rollers.

The high efficiency of recirculating ball screws enables the longitudinal movements to be converted back into rotary movements with comparative ease. This may prove very useful for certain purposes (e.g. for fitting a rotary displacement measuring system to a measuring spindle that does not itself transmit power) (122); in other cases this absence of any irreversible effect can be a disadvantage, and some arrangement must be made on the spindle for slowing down and stopping effectively the masses that are moving and rotating at high speeds (e.g. by a suitable choice of drive elements or by fitting brakes). The choice of the device to be employed will depend on whether it is applied to a positioning, straight-line, or continuous-path control system.

There is a much wider choice of methods for producing low-friction slides than there is for lead screws. The arrangements shown in Fig. 129 are two examples of some of the many designs that are possible using slide mechanisms with anti-friction bearings. With long tables, difficulties are encountered in constraining the thin rollers (needles). Even though this type of design enables the frictional forces to be reduced, and eliminates stick-slip, slides incorporating anti-friction bearings offer disadvantages for many applications. These difficulties are least with straightforward table positioning systems. The more severe the cutting conditions are in straight-line control

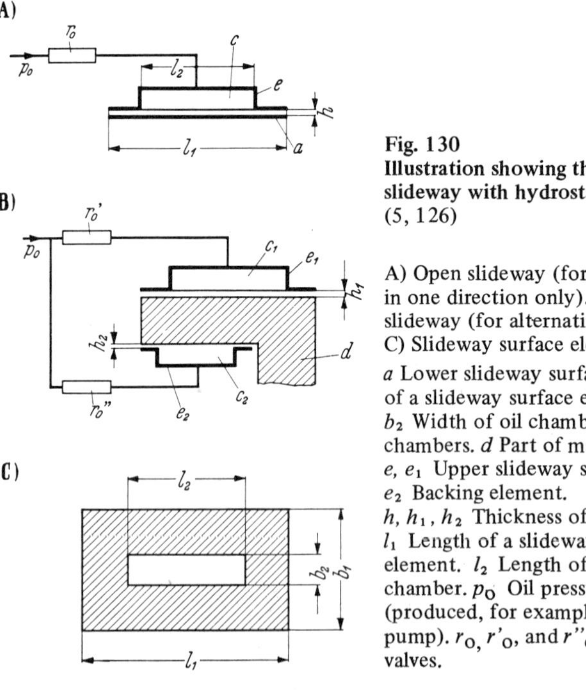

Fig. 130
Illustration showing the principle of a slideway with hydrostatic lubrication (5, 126)

A) Open slideway (for forces acting in one direction only). B) Enclosed slideway (for alternating forces). C) Slideway surface element.

a Lower slideway surface. b_1 Width of a slideway surface element. b_2 Width of oil chamber, c, c_1, c_2 Oil chambers. d Part of machine bed. e, e_1 Upper slideway surface. e_2 Backing element. h, h_1, h_2 Thickness of lubricant film. l_1 Length of a slideway surface element. l_2 Length of an oil chamber. p_0 Oil pressure available (produced, for example, by gear-type pump). r_0, r'_0, and r''_0 Reducing valves.

systems, however, the more disturbing the low vibration damping properties of the balls or rollers prove to be. Similar difficulties can also arise with continuous-path control systems on precision machines. Careful consideration must be given to the conditions in each particular case before adopting elements (spindles, nuts, slideways) with rolling friction in place of those with sliding friction.

There have been many attempts to develop slide-ways that would exhibit similar frictional characteristics to those obtained with slides incorporating rolling bearings. One such possibility is the use of hydrostatic slideways. The principle is illustrated in Fig. 130 (5). In the simplest examples, the slideway surface consists of a single rectangular plate with internal pockets (oil chambers) through which oil under pressure can emerge (Fig. 130 C). If the design condition

$$l_1 - l_2 = b_1 - b_2$$

is met, the oil will flow uniformly through the gap h (Fig. 130 A); during this passage the pressure of the oil will drop to atmospheric pressure. The two surfaces of the slideway are thus separated by a film of oil under pressure of thickness h. The pressure is adjusted to the correct value by means of the reducing valve r_0. The requirements for the correct operation of an oil lubrication system of this type are that

a) oil can be supplied at the rate at which it flows out through the gap;

b) no dirt can penetrate into the guides;

c) the slide cannot move until the necessary oil pressure has built up between the slideways.

The last condition can easily be met by incorporating a logical AND condition in the normal control system.

In difficult conditions it will be necessary to fit a backing element as shown in Fig. 130 B). The machine shown in Fig. 301 (Appendix) is equipped with similar hydrostatic slideways. With positioning control systems it will be necessary to clamp the slide in the required position; in the machine in Fig. 301 this clamping is done electro-hydraulically. In addition to the use of suitable machine elements, it is likely to be of advantage to adopt a modular construction system when designing these machines to make it possible to meet different customers' numerical control requirements as economically as possible; this latter point is too well covered to be considered further at this stage (5, 130, 220).

7.2   Tools as information stores

7.2.1   Tools as Form-Stores

In many cases tools that are used in metal cutting or metal forming
operations are such that they store the details of some accurately defined
geometric form.

Among such tools are:

Profiling tools (form milling cutters, hobs for a fixed tooth section, form
turning tools, complete set-ups of turning tools for conventional automatic
lathes (133, 134), etc.); twist drills (which store, for example, the form
specification 'hole diameter 9.8 mm'); reamers (which store, for example, the
form specification '10 mm dia. H7'); countersinking tools and counterbores,
taps, and similar items.

Additional members of this group include all types of drill jigs, templates,
masters, cams, etc. (6). It is necessary to understand clearly the 'information
storage' nature of these well-known tools if the machine tools, which are the
'users' or carriers of these tools, are to be regarded as data processing systems
and are to be incorporated ultimately into a larger information flow
structure. All press tools, drawing dies, and the like that are used in metal
forming processes and for processing plastics have a similar storage function.

Since it is normally not possible, or at least very difficult, to alter the
information content of such tools they are sometimes referred to as 'rigid
(fixed) form stores'. Later in this chapter it will be shown that these
fixed-form stores themselves can be produced more effectively on
numerically-controlled machine tools than on conventional machines. In this
way, numerically-controlled machine tools help indirectly to increase the
productivity and improve the flexibility of use both of automatic lathes and
of metal-forming and plastics processing machinery.

In contrast to these 'form-storing' tools, there are tools which have little or
no special profile, and which for the most part provide point contact, so that
the whole operation of shaping depends on the relative motion between the
tool and the workpiece. This group includes many lathe tools and end mills,
as well as the oxygen jet of a flame-cutting machine (127, 158). This
difference between the two types of tools is illustrated by the extreme
examples shown in Fig. 131. There are, of course, also intermediate types,
especially among lathe tools, in which the shank of the tool is shaped in such
a manner as to reduce the danger of touching the workpiece (Fig. 131 a-c).

From the information processing standpoint, this variety in the types of
tool leads to a variety of types of machine tool and associated control
systems, together with differences in the trends of development.

In the first case, using form cutters, the tools that are needed for a
particular task must be kept 'on call' in a specially designed store (capstan
head, magazine, etc). This introduces a number of subsidiary problems when
numerically-controlled machine tools are used (automatic tool changing,

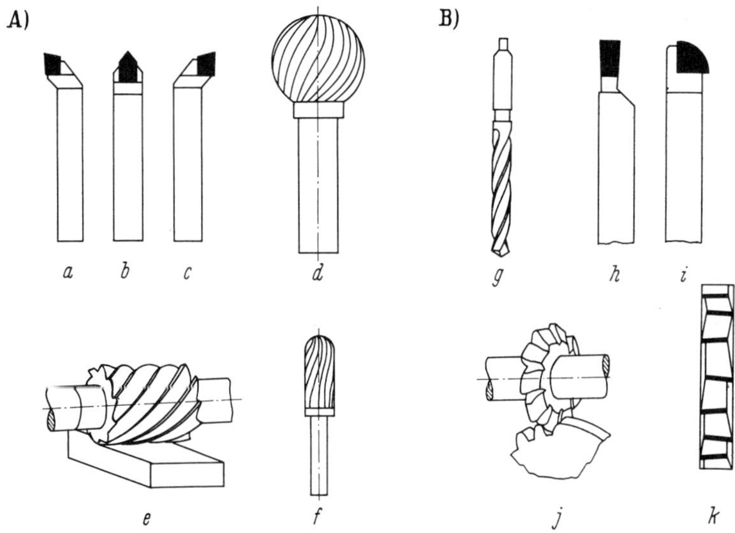

Fig. 131 Examples to illustrate the various types of tool (6, 127)

A) Examples of tools having little or no form-storage capacity. B) Examples of tools having a high, rigid, form-storage capacity.

a, b, c Lathe tools. *d* Spherical end mill (for performing milling operations on machines similar to that shown in Fig. 173). *e* Plain milling cutter, could be replaced by end milling cutter of type shown in Fig. 135. *f* Finger mill with hemispherical end (e.g. for form milling on copying and 3-D-NC machines similar to Fig. 175). *g* Twist drill (e.g. for hole 10 mm diameter). *h, i* Lathe tools with marked form storage capacity. *j* Form cutter for involute gears (e.g. Module 3). *k* Side and face cutter for a given slot width.

presetting of tools, etc), and these will be discussed in greater detail below.

In the second case, where the tools used are not profiled to suit the work in hand, the whole forming of the workpiece has to be performed by a relatively complex data processing operation; in the majority of instances this involves a contour control system with internal and external interpolators (Chapters 8 and 11). A practical intermediate solution is produced by the adoption of sub-programming techniques in which copying techniques are suitably combined with the numerical process (cf. Figure 141).

If these basic considerations are to be summarised it seems advisable to adopt in the main the existing data-processing terminology (6, 33), to establish a link with the data-processing equipment, and hence to simplify the continuing integration of machine tools and computers (Chapter 8) (10). In the following sections the tool stores that have so far been developed will be regarded in general as mechanical information stores, and will be classified according to their capacity and their access time. There are, however, some specific questions that arise with the mechanical information stores, which are not encountered in electronic information stores (magnetic tapes, magnetic discs):

a) What are the sizes and weights of the tools and their holders?
b) With what accuracy are the tools required to operate?
c) To what extent is tool changing on numerically-controlled machines to be automated?
d) What preliminary organisational work is needed to establish such a system of operation and to run it economically?

The next sections can only touch on the basic problems, indicate the possible solutions that are practicable in the current state of technology, and illustrate the trend of developments to encourage interdisciplinary discussions on a common basis in working groups established for the purpose.

7.2.2. Tools in turrets

The oldest and best-known method of arranging tools with form-storage capacity in groups and calling them up on demand is to use a turret (5, 6, 133, 134). The turret is also probably one of the most accurate types of tool store, since the tools remain rigidly clamped in it and, in the majority of instances, are set up and adjusted while clamped in place. Setting-up can be done in the conventional way by hand (possibly using a small hammer or mallet), using a dial gauge or making trial cuts, or numerically by decade switches using computers (Chapter 8 and Fig. 168). The accuracy then depends largely only on the indexing accuracy of the mechanical turret positions. Since turrets can be rotated comparatively quickly and easily, the 'access time' for a new tool is short; for this reason all attempts to provide a solution to the problem encountered in changing tools invariably rely in some shape or form on the old turret principle (Figs. 154, 155, 157).

The great disadvantage of a turret is, however, the limited number of tools that can be stored on it without interfering with each other. The first consideration should, therefore, be to reduce the number of tools to the minimum actually required for machining the workpiece. In the majority of cases this is more a question of organisation and internal standardisation than of meeting the design requirements for the workpiece. Figure 132 shows a statistical analysis of the frequencies of the groups of tools that are actually required; these numbers could no doubt be reduced in many plants if the necessary investigations were carried out in the design and planning offices (129).

The turret head type most commonly employed is the star type (Fig. 133); the turret positions are controlled by the switching information. The maximum number of tools that can be fitted to a turret without the danger of interference is about 8; in most instances there is a limit of 6 tools. Figures 134 and 135 show examples of turrets on drilling and milling machines, there

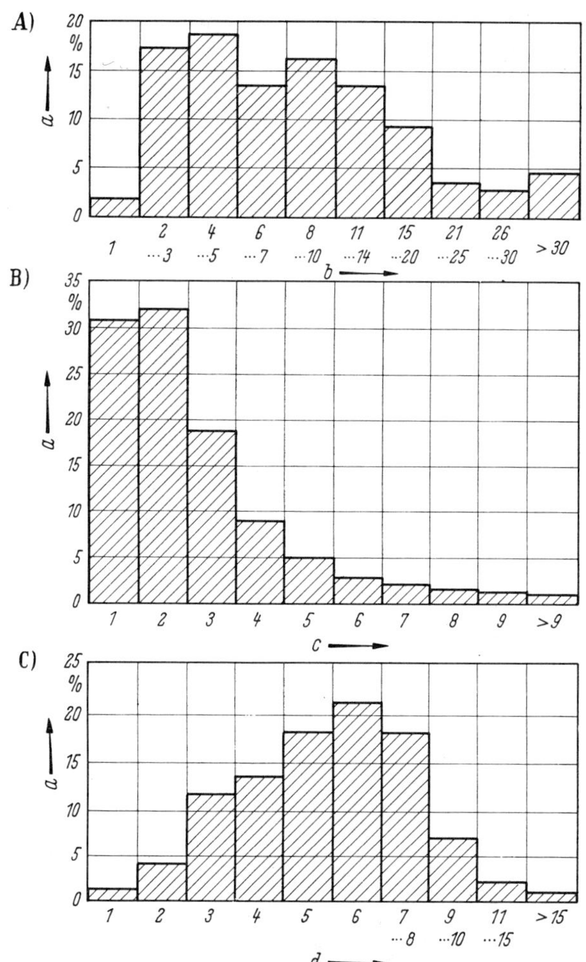

Fig 132 Statistical distribution of the requirement for various types of tool (129)

A) Histogram of the requirement of drills per workpiece. B) Histogram of the requirement of milling cutters per workpiece. C) Histogram of the requirement of lathe tools per workpiece.

a Frequency in %. *b* Different drills per workpiece. *c* Different milling cutters per workpiece. *d* Different lathe tools per workpiece.

is a similarity in the design in Fig. 135 with that shown in Fig. 144, due to the size of the tools. From the statistical analysis in Fig. 132 it will be obvious that only a fraction of the drilling and milling operations required can be covered with the use of a 6-way turret, so that it may well be economically justifiable to provide larger stores in certain cases (cf. for example Figs. 142, 148, 154, etc).

With lathes, 4-way and 6-way turrets are at present most frequently encountered. For three reasons the conditions are somewhat more complicated with these machines than with drilling or milling machines:

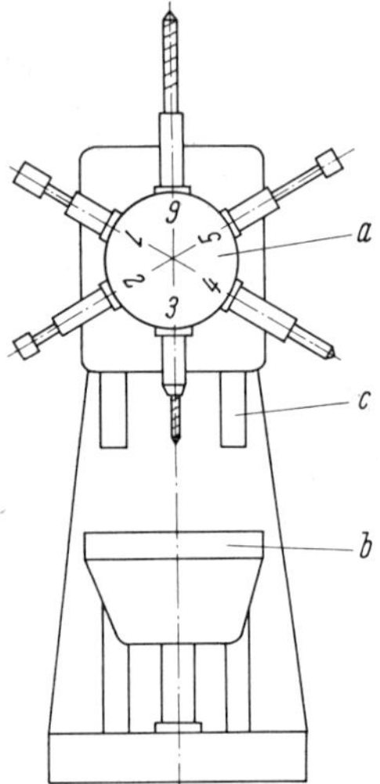

Fig 133
Schematic diagram of a turret boring and milling machine

a Swivelling turret with 6 to 8 firmly fixed pre-set tools; the head can be moved vertically along the slideway *c* (Z-axis) *b* Numerically-controlled co-ordinate table (*X-Y*-plane). *c* Slideway for turret head.

Fig. 134
Special turret boring machine with 3 spindles for each turret position (Photo courtesy of *Burgmaster*).

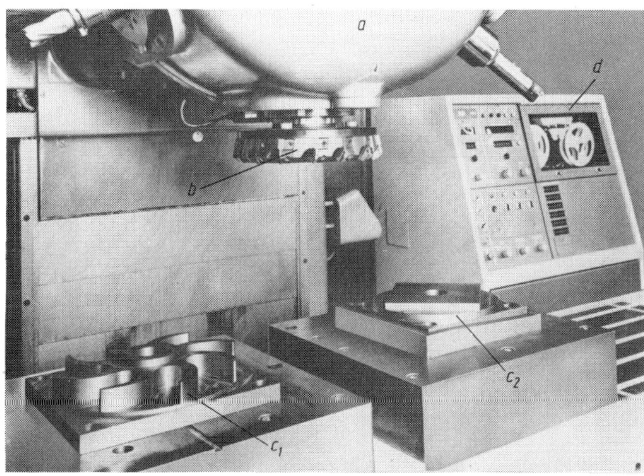

Fig. 135 Heavy turret boring and milling machine with hydraulic feeds. (Photo courtesy of *Heller*) (136)

> *a* Turret head with six tools which are connected by means of a Hirth tooth clutch (5, 130) to the robust vertical drive spindle (which does not move round with the turret). *b* Large end milling cutter coupled to the spindle. c_1, c_2 Different workpieces which can be fitted to the long work table (2500 mm) and are machined alternately. *d* NC control unit; in this case a 3-D continuous-path control system (Chapter 8).

a) In general the working areas are more cramped, so that more attention must be paid to the danger of collision than is the case with a drilling machine for example;

b) to ensure that numerically controlled machines are of universal

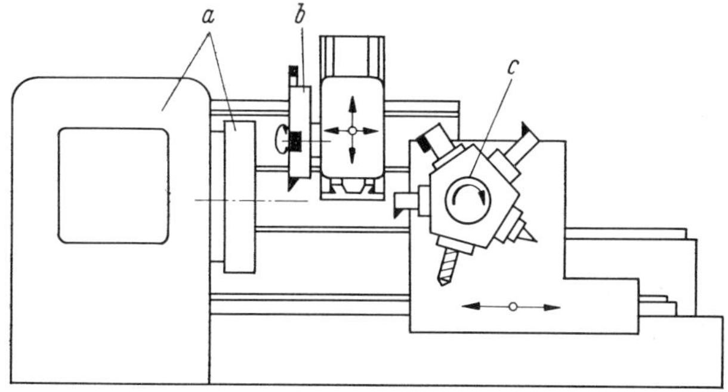

Fig. 136 Schematic diagram of a turret lathe with two slides

> *a* Headstock with workpiece holder (collect chuck). *b* Turret 1 carrying, for example, 4 tools and mounted on a cross slide 1. *c* Turret 2 carrying, for example, 5 tools and mounted on a slide 2.

A)

B)

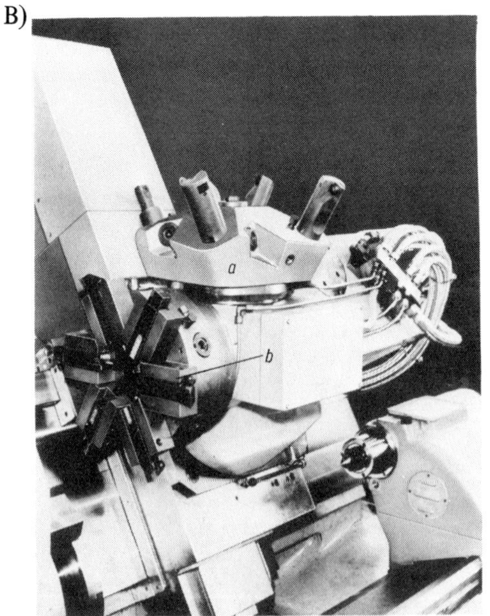

Fig. 137
Universal tool head mounted on a single cross slide for machining both chucked components and bar stock (Photo courtesy of *Heyligenstaedt*).

A) Position of universal tool head for internal machining. B) Position of universal tool head for external machining.

a Four-way turret for internal machining (boring bars). *b* Six-way turret for external machining.

application, provision must be made for tools for internal and external machining;

c) in view of the small working area, the nature of chip formation and the swarf disposal arrangements are of major importance.

In addition, to save machining time it is desirable to perform several operations simultaneously. This, however, leads to complications both with regard to interference and in the programming (Chapters 10 and 11).

Figure 136 illustrates diagrammatically the technological ideal in simplified form. Here two turrets are fitted, one for tools for external machining and the other for tools for internal machining. The two turrets are mounted on slides which can move independently. Turret 1 is mounted on a cross slide and can be moved in two coordinate directions; in this case turret 2 can be moved in one coordinate direction only (primarily for boring tools) while it can be indexed about a horizontal axis. The turret axis could equally

A)

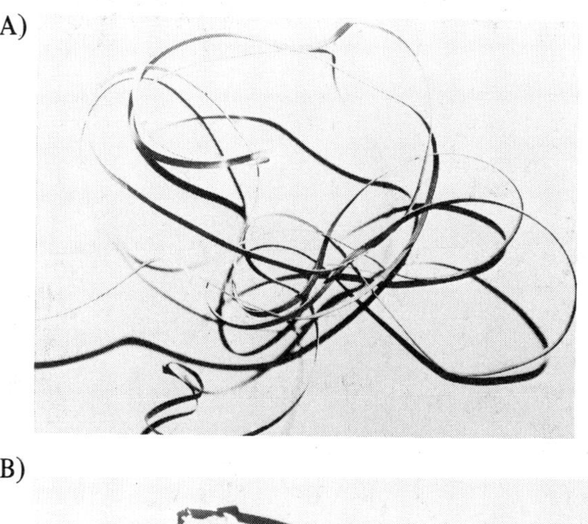

B)

Fig. 138 Effect of the machine kinematics (e.g. by producing suitable impulses by means of stepping motors) on chip formation with tough materials normally producing very long chips (Photo courtesy of *Siemens*).

A) Chip formation with continuous feed. B) Chip formation with intermittent feed (pulse gaps).

A)

B)

Fig. 139 Flat table turret developed from the drum-type turret under the influence of numerical-control techniques (Photo courtesy of *Pittler*) (135)

A) Flat table turret with four double-acting tools. B) Original drum turret with 16 positions.

a Standardised tool holders which can be mounted in accurately defined and precisely reproducible positions relative to the mounting table *b* by the use of locating pins. *b* Mounting table which can be coupled to a numerical straight-line or continuous-path control system by means of two linear inductosyn scales (Chapter 4). *c* Special lathe tools with a right-hand and a left-hand principal cutting edge so that both sides of the pre-set tool can be used to perform different functions. *d* Workpiece.

A)

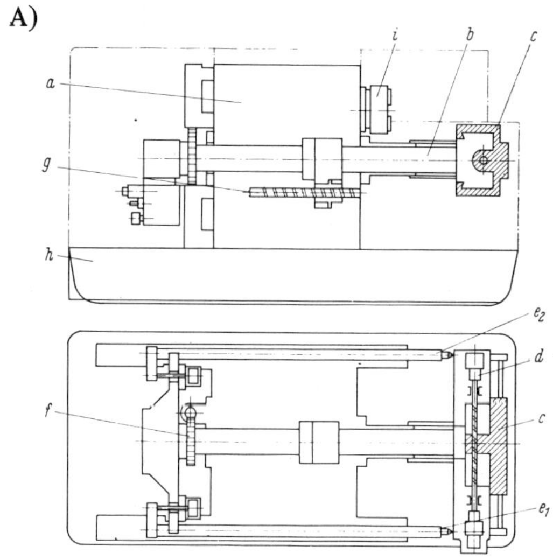

B)

Fig. 140 High-precision chucking lathe adapted for numerical control (Photo courtesy of *Monforts*)(138)

A) Schematic diagram of machine. B) Turret head.

a Stiff, vibration-free body. *b* Turret axis, about which the turret unit can be swivelled through 180° intervals. *c* Turret unit with slides for surfacing work and carrying the two turret heads *k*. *d* Surfacing drive (Y-axis). e_1 and e_2 Indexing pins, widely spaced, for positive location of turret unit *c*. *f* Swivelling drive. *g* Longitudinal drive (X-axis or Z-axis, Chapters 10 and 11). *h* Coolant tray. *i* Chuck. k_1 and k_2 Two five-position turrets to carry up to 10 different tools.

Note: All available types of control and displacement measuring systems can be fitted.

well be arranged vertically and also be mounted on a cross slide to enable any internal diameters to be machined using boring bars. In the example illustrated in Fig. 297 (Appendix) both turrets are located on the same inclined plane, and turret 2 is, in fact, capable of being moved only along the longitudinal axis. It is interesting to note that in this example the machine is programmed by means of a plug board which enables any danger of interference to be recognised at the time the machine is being set up.

In the example shown in Fig. 137 no attempt has been made to arrange for the simultaneous movement of two turrets working indendently; the tools for internal and external machining are carried by separate turrets mounted on a common cross slide, and can be brought into use as required by the adoption of a common carefully selected indexing axis. This both reduces the danger of interference and simplifies programming. At the same time the universal tool head can be used both for machining bar stock and for chucked components.

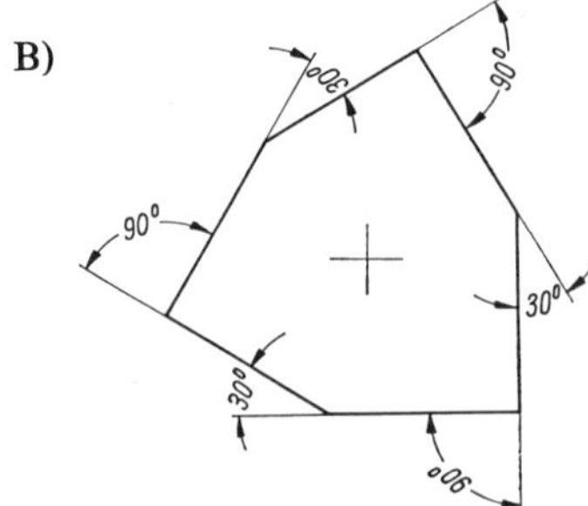

A) Work area of the machine. B) Basic geometry of the irregular turret for 3 to 6 different tools.

a Turret equipped with 3 tools for longitudinal turning, mounted on a straight-line controlled cross slide. *b* Hydraulic copying attachment for forming a profile to match a template which is scanned by a probe. *c* Headstock. *d* Drive coupling.

Fig. 141 Combination of numerical-control technology and copying methods (Section 7.4) (Photo courtesy of *VEB Grossmaschinenbau '8. Mai'*) (139)

It is highly desirable for the swarf produced on lathes to be broken up into short lengths, to assist in its disposal and because of the cramped conditions in the working area. This can usually be achieved by providing chip breaker grooves in the lathe tools. Difficulties may nevertheless occur with very tough materials. For this reason Fig. 138 draws attention to the possibility of achieving a chip breaker effect by modifying the feed kinematics. If the feed drive is operated by electro-hydraulic stepping motors (Chapter 6) it will be possible to break the swarf into short lengths by arranging electrically-controlled pulse gaps.

The design of the well-known drum-type turret (Fig. 139 B) enables 10 or more tools to be mounted on a single turret. If numerical control techniques were adopted there would, however, probably be difficulties when changing the tools (precise, easily reproducible tool positioning in the turret) and when carrying out profiling operations under continuous-path control. To suit numerical control techniques the drum turret was therefore developed into the flat-table turret (Fig. 139A) in which it is not only possible to fit four separate tools in precisely defined positions on a coordinate table that can be moved accurately in any direction; but also each tool has two cutting edges, so that there are effectively eight tools available for machining operations.

The further development of a well-known automatic lathe for producing short components and embodying numerical-control techniques is shown in Fig. 140. A feature of this machine is that the advantages of the rigid 5-position turret can be extended to a set-up of ten tools by precision rotation through 180°.

These three examples will suffice for the present to show the efforts and ingenuity that have been put into extending the storage capacity of turrets, while retaining their advantages of holding the tools rigidly and accurately, with a view to increasing the accuracy and versatility of the machines to the maximum (172). Finally, there is the approach, discussed again separately in Section 7.4 of mixing of numerical-control processes with other methods (also discussed further in Section 7.4). Considerable use is made of this approach to reduce costs of simple machines. In the example under consideration a copying attachment is fitted to a lathe to replace a continuous-path control system. Where simple profiling operations recur at frequent intervals (e.g. rounded edges of constant radius, tapers of constant pitch, etc), it may well prove worthwhile to store these simple forms on a metal template; to scan them by means of a probe; and then to transfer them to the workpiece using a hydraulic or electrical copying device (follow-on circuit). The remaining numerical control need then merely be a straight-line control system. Figure 141 A) shows a combination of this type which even nowadays proves very effective.

The turret on the cross slide, which moves under the action of the straight-line control system, need then not carry any tools for the profiling operation (e.g. Lathe tools as shown in Fig. 131 b), and can be matched more readily to the other operations (Fig. 141 B). Whether a combination of this type will still prove an economical proposition in five or six years time is another question in view of the downward trend for the costs of continuous-path control systems (Chapter 8 and (10)).

7.2.3   *Tools in magazines*

The term 'magazine' is not defined exactly in this context; it is desirable to distinguish between at least two basic designs which differ both in the number of tools that they can accommodate and the tolerances that can be maintained. Figure 142 is a diagrammatic sketch of one design in which a number of complete, coded, spindle heads are mounted on a semi-circular or straight rail in such a manner that they can be called up by the switching information and moved to the engaging position *d*. There the selected spindle head is coupled both to the spindle drive and to the feed drive, so that a complete spindle unit carrying a specified boring or milling tool is produced. The individual spindle heads can be changed manually, so that the tools can be fitted to the spindles and adjusted outside the machine. Depending on the precision of the jigs that are employed a very high degree of accuracy can be obtained and this is maintained when the spindle head is fitted to the machine. Figure 143 shows an example of an automatic boring and milling machine which works on this principle and which is equipped with 18 spindle heads. The number of tools that can be changed automatically is now much larger than is possible using a star-type turret, while the machine is capable of achieving similar accuracy. The access time for a spindle head that is located remotely from the coupling point may, however, be longer than for a turret. Movement in the Z-axis can be controlled by cams, or if a higher degree of automation is required, it can also be numerically controlled; all movements

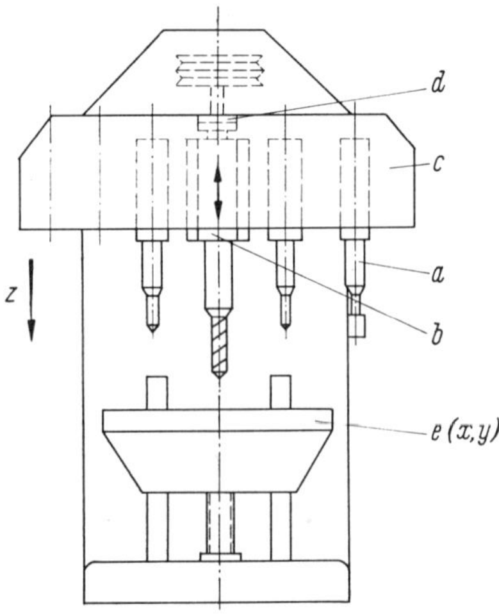

Fig. 142
Schematic diagram of machine equipped with spindle-head magazines.

a Spindle heads with drilling spindle mountings, with fixed, pre-set, tools and robust longitudinal slides. *b* Feed slide with feed and spindle drives to which the spindle head in the operational position is coupled. *c* Holder for several spindle heads (magazine). *d* Clutch for the drive system. *e* Numerically-controlled cross slide, (straight-line control in X-Y plane). Z Feed device (straight-line control, can either be numerical or by means of cams, cf. Section 7.4).

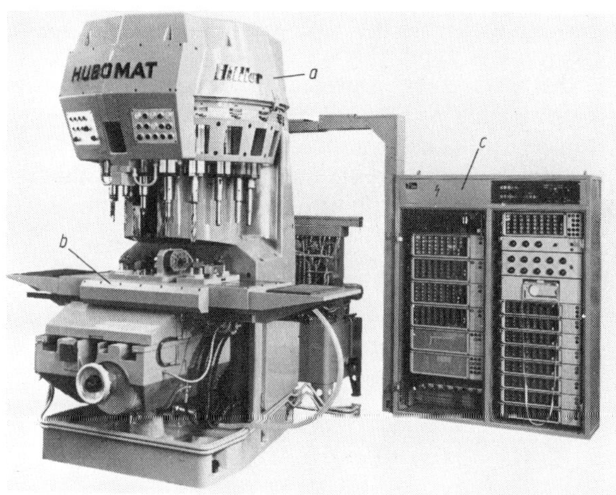

Fig. 143 Automatic boring machine with a linear spindle-head magazine (Photo courtesy of *Hüller*) (140)

> *a* Spindle-head magazine with 18 interchangeable spindle heads (power for driving main spindle 5.5 kW, max. spindle stroke 315 mm, *Z*-axis). *b* Coordinate table (mounting area 1540 x 420 mm, max. table displacement *X* = 800 mm, *Y* = 450 mm). *c* Numerical control unit for 3 axes with free choice of system.

in the *X-Y* plane are numerically controlled. In this sense this machine should also be included in Section 7.4.

The machine shown in Fig. 144 operates on similar principles. In this instance, there is only one fixed spindle head equipped with a tool-holder sleeve. In this up to 12 tools, which are housed in the magazine, can be clamped in turn in precision guides. The tool magazine itself takes the form of a shallow truncated cone, and because of the relatively low weights of the individual tools and their holders, it can be rotated rapidly. The access time is probably somewhat lower than in the machine shown in Fig. 143, but on the other hand the tool holders rather than the spindle heads are gripped. The whole tool changing mechanism is housed within the magazine casing and is well protected. The machine shown in Fig. 135 employs similar principles for enabling heavy milling cuts to be undertaken using a fixed spindle head. Both these machines (Figs. 135 and 144) can be equipped with continuous-path control systems.

If the number of tools that is to be stored increases still further, it is necessary for mechanical reasons to develop different types of magazine in which pre-set tools and their holders are kept 'ready to hand', either coded or uncoded. From this stage of automation onwards it is essential that a logically considered tool numbering system be worked out in good time; it is still very important to make every effort to keep the number of tools circulating within a plant system as low as possible to save tooling and storage costs.

The capacity that should be provided in tool magazines attached to machine tools is not merely a technical question but one that also requires economic and management considerations. During the last three years there has been a very marked increase in the number of solutions that have been put forward; there seems hardly any limit to the design ingenuity that is being

A)

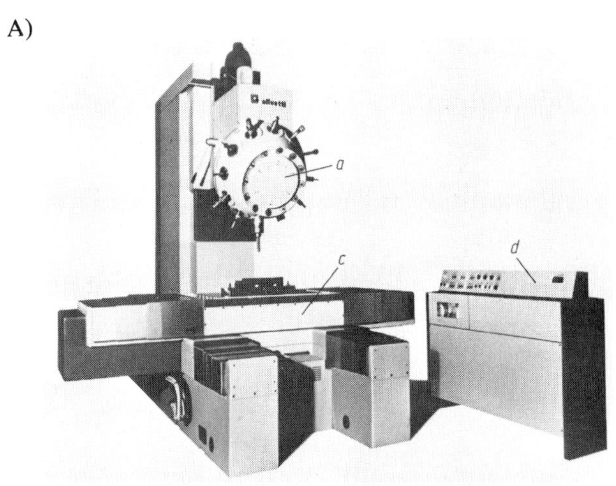

B)

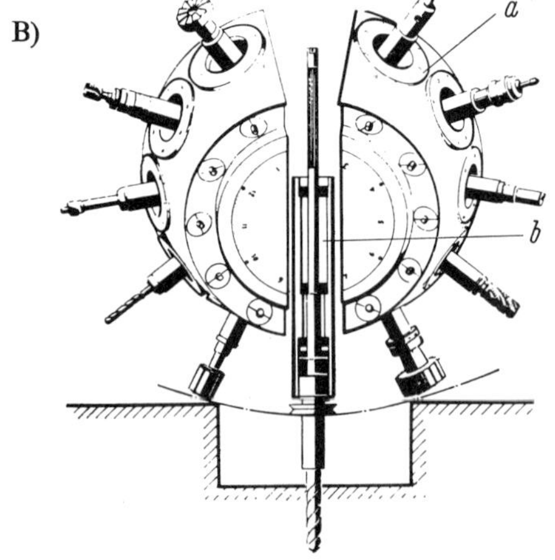

Fig. 144
Automatic boring and milling machine with a fixed spindle head and a cone-shaped magazine for 12 tools (Photo courtesy of *Olivetti*) (141)

A) General view of machine. B) Illustration showing principles of operation of spindle head and magazine.

a Rotary cone-shaped magazine with 12 preset tools in standard holders. *b* Fixed spindle head with sliding sleeve and clamp for the tool holders (*Z*-axis. Displacement measured by rotary pick-up). *c* Coordinate table. (Max. displacements X = 800 mm, Y = 500 mm. Displacement measured by linear inductosyn, Chapter 4). *d* Olivetti CNZ control system (Chapter 8).

applied or to the methods that are being adopted to circumvent competing designs or even patents.

Under these circumstances it seems undesirable to introduce a rigid classification scheme at the present time, or to describe details which may soon become dated. It is important, however, to illustrate some of the basic approaches that are being followed by quoting examples to enable the student to develop his own critical faculties in this field, so that he can make a positive contribution at a later date when investment policies are being discussed.

Machine tool magazines that are currently available have capacities ranging from about 9 to 100 tools. It may be that during the course of the next few years a preference for an average of about 30 tools will develop, the way in which the tools are changed remaining very varied. This figure should therefore be taken as a guideline for the subsequent discussion.

7.2.4   Mechanical and manual tool changing systems

Pick-up system

Although this principle has many disadvantages it probably provides the simplest means of changing the tools automatically and it is illustrated diagrammatically in Fig. 145, which represents a radial boring machine. The tools required for a particular production process are suspended in numbered positions in a turntable magazine from which they are picked up by the boring head and to which they are returned after use. With simple boring

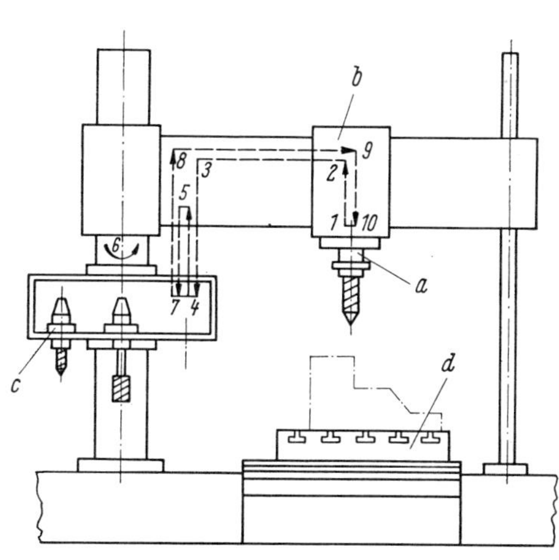

Fig. 145
Schematic diagram of a radial boring machine with a tool magazine and tool changing by means of the pick-up system.

a Machine tool spindle (with a quill to permit movement in the Z-axis).
b Transverse slide (in its movement is coincident with the Y-axis).
c Rotary tool magazine with standard tool holders (the selection of the tools is part of the positioning process).
d Work table (movable in X-axis, i.e. perpendicularly to plane of illustration).

**Fig. 146
Machining centre in
which the tools are changed
by the pick-up process. Detail
from the machine illustrated
in Fig. 308 and 176. (Photo
courtesy of *Marwin*) (143)**

a Box-shaped tool magazine
containing 2 x 20 tools of very
different sizes in standard
holders. Twenty tools are
stowed on each of two
opposing sides, and the
magazine is mounted on a
turntable which can be rotated
through 180°. Tools are
selected by positioning the
spindle head. *b* Spindle head
with quill that can move in
Z-direction. *c* Tool being used.
d Rotary mounting table
(B-axis, cf. Table 12). *e* Workpiece being machined.

**Fig. 147
Lathe with a tool magazine
and tool changing using the
pick-up method. (Photo courtesy of *VDF/Gebr. Boeh-
ringer*) (142)**

a Rotary magazine for 9 preset
tools in standard holders (the
tool holders are positively
coded by means of dowel pins,
Fig. 166). *b* Machine slide
which is also used to pick up
required tools.

machines this form of pick-up would probably result in comparatively long
access times, and there would be the danger of inaccuracies due to dirt
between the mating surfaces.

Figure 146, on the other hand, shows a practical example in which the pick-up principle is used on a large machining centre (Fig. 308). Twenty preset tooling units are held ready for use on each of two sides of the box-shaped tool magazine. A programme similar to a positioning control system causes the spindle head to move to the desired tool location, where the spindle nose with the mounting taper and the taper bore in the tool holder are cleaned by compressed air; the tool is then automatically positioned on the spindle by a drawbar. When the machining operation has been completed the tool in last use is returned to its correct location and the spindle moves to the location for the next tool. In this example the pick-up system is justified since it enables even large and bulky tools to be loaded into the machine without the use of complicated gripping systems. Under workshop conditions the blast of compressed air is essential if the desired accuracy is to be maintained.

Figure 147 shows a similar arrangement on a lathe. In this case 9 tools are stored in coded standard holders in which the tools are preset to form part of a standardised tooling system and then clamped (cf. Figs. 166A, B), C)). The magazine is rotated into the required position by means of a programme, and the slide then picks up the tool, which is cleaned by a blast of compressed air; the holder is clamped into position hydraulically and the tool then moves to its programmed working position (cf. Fig. 306).

Manual tool changing

It is a question of economics and labour availability whether mechanical tool changing will, in fact, always be necessary. If the machine is equipped with quick-chucking devices (hydraulic or electrical) it may be possible to change the tools just as quickly manually as mechanically, it may also be cheaper, and also more reliable. The number of methods that can be adopted and the variations in the ease with which they can be used are currently also large. Figure 148 A), for example, shows a numerically-controlled boring mill with a loading and working station and with clearly arranged tool stands with numbered tool places. The numbers of the tool locations are programmed on the punched tape and are called up by numbers on indicators on the control console (Chapter 9). Figure 148 B) shows a commercially available tool stand for this type of application.

It is, of course, possible to make the indication of the numbers more convenient and fool-proof. Figure 149 A) shows a simple arrangement with illuminated numbers directly on the machine control panel, while Figure 149 B) shows a tool storage system that reduces the likelihood of mistakes to a minimum. In this system every tool location is provided not only with a number, but also with an illuminated signal controlled by a punched tape. If the operator nevertheless draws a tool from the incorrect location the machine will not start up again. If, on the other hand, either through

A)

B)

Fig. 148
Machining centre with manual tool changing.

A) General view of machine (Photo courtesy of *Wadkin)* (144). B) Commercially-available tool stand with 40 numbered tool locations (Photo courtesy of *Kelch)*.

a Spindle head capable of movement in $Y-Z$ plane (1020 x 700 mm) and with quill to provide movement in W-axis (cf. Table 12).
b Machining station with indexing turntable (B-axis. Max. turntable load 2250 kg).
c Loading station with indexing turntable (the two stations are interchangeable and machining takes place at each alternately, since the machine column *h* can be displaced through 2540 mm). *d* Tool stands with numbered tool locations for manual tool changing (the tool numbers required are entered on the punched control tape and are indicated on the control cabinet *e* (Chapter 9). *e* Control cabinet with punched-tape reader and tool number indication (primarily straight-line control).
f Conveyor for discharge of swarf. *g* Transportable swarf container. *h* Machine column (can be moved max 2540 mm in X-direction).

negligence or on purpose, the incorrect tool is inserted in the correct magazine location it is not possible for the error to be detected unless the operator notices it either from the drawing of the required workpiece or because of his familiarity with the work.

A)

Fig. 149
Machining units with manual tool changing.

A) Number of required tool is indicated on column of machine. (Photo courtesy of *Lindner*).
B) Punched-tape controlled tool location indicator (Photo courtesy of *Herbert-De Vlieg*).

a Boring and milling machine with 3 controlled axes (straight-line control). *b* Workpiece. *c* Tool magazine with 52 (or more) numbered tool locations equipped with punched-tape controlled illuminated signals and punched-tape controlled indication of tool number required at magazine. If necessary an electrical interlock (logical AND circuit) can be combined with the next starting signal for the machine.

B)

Coding the shanks of the tools

It is thus clearly desirable to adopt some means of identifying the tools positively in a way which can also be 'recognised' by the machine control system. Since all the interchangeable tools are preset in standard holders it is advisable to have the coding carried out by a responsible expert as the tools

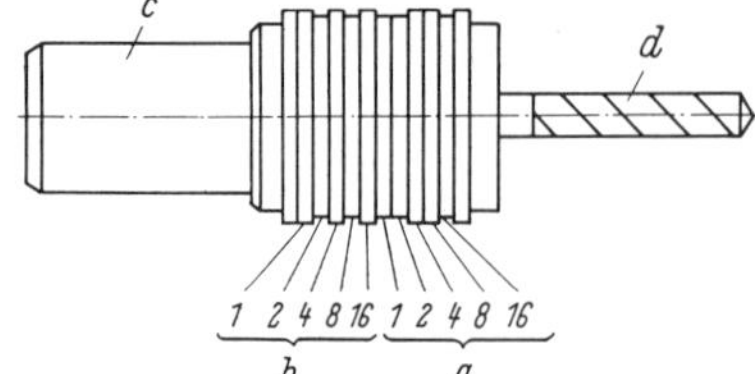

a Five dual positions to indicate the particular tool number (31 numbers possible). *b* Five dual positions to indicate the tool group number (also 31 group numbers, so that with this coding arrangement a total of 31 x 31 = 961 tools can be automatically distinguished). *c* Standard tool holder. *d* Preset tool (Figs. 161 to 167).

Fig. 150 Schematic diagram of a coded tool. A binary digit can be coded either by the convention 'Large ring diameter = L, small ring diameter = 0' or if rings of equal diameters are used and a suitable material is employed by the conditions 'magnetic = L, non-magnetic = 0'.

are set up. Figure 150, for example, shows the coding system applicable to a selection of 961 tools in which the number of coding rings governs the extent of the tool system that can be handled automatically. The coding can be mechanical or magnetic. Figure 151 shows a work station at which the tools are preset and coded and which includes the tool store (see also Fig. 160).

The types and numbers of tools that are manually set in a machine tool magazine and which are then changed by the machine will depend on the workpieces that are to be machined or on the progress of the work during lengthy machining processes.

Fig. 151 Work station used for coding and presetting tools and placing them in magazines (Fig. 160. Photo courtesy of *Siemens*).

Double-gripping system

The principles of this system are clearly illustrated in Fig. 152. It is very simple and is designed to reproduce closely the operations of the manual methods. Since it is to a large extent protected by patents despite its simplicity, many of the alternative designs that have been proposed are clearly intended only to evade the patent restrictions. These legal difficulties will continue for some years, and it is therefore at present not possible to reach a verdict regarding the technically 'ideal solution'. Figures 153 A) and B) are of historic interest in that they show the first tool magazine with coded tools and a double-grip system, which was introduced in about 1959, and so started the wave of 'machining centres' which dominates present-day technical exhibitions. This was the first time there was a departure from the

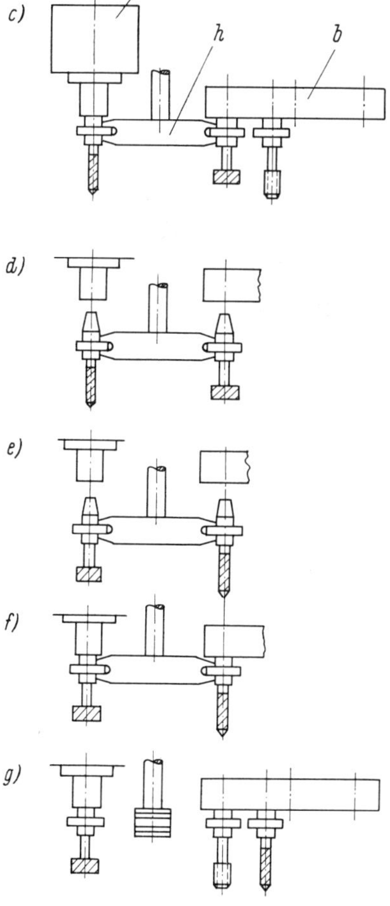

Fig. 152.
Schematic diagrams showing the construction and operation of the double-gripping system of tool changing.

a Main spindle of machine. *b* Tool magazine with coded (cf Fig. 150) or uncoded (but fixed-location, cf Figs. 148 and 154) tools. *c* Grip swings through 90° from rest position and grips both tools. *d* Grip withdraws tools from mountings. *e* Grip rotates through 180°. *f* Carrier inserts tools. *g* Grip swings through 90° to rest position. The tool magazine then indexes until the next location is reached or until the automatic selector has detected the next coded tool. Next time a tool changing order is given the cycle will be repeated.
h Double-grip system with clamps.

Fig. 153
Earliest example (1959) of a tool magazine for coded tools (Photo courtesy of *Kearney & Trecker*).

A) Double grip system in the rest position.
B) Double grip system in operation.

a Rotating drum with up to 30 tools with shanks coded as shown in Fig. 150 to avoid confusion. *b* Tool in the work spindle.
c Automatic tool-changing device (double-grip system).

rigid 'transfer line principle' involving the movement of the workpieces to the flexible 'machining centre principle' with the workpieces accurately rotated and turned into position and the tools moving. As will be discussed in greater detail in Section 7.3.4, the current trend is towards establishing a synthesis of these two principles and determining their limiting conditions.

Disadvantages of the double-grip system are the relatively long access times which may amount to 3 to 5 seconds, and the limitations on tool size. In the practical examples shown in Figs. 154, 155, 156, and 157 attempts have been

Fig. 154 Location-coded tool magazine for 30 tools used in conjunction with a double turret for high-speed tool changing (Photo courtesy of *Burkhardt & Weber*) (145)

a Double turret with two positions: one tool is in the machining position, and one tool is in the changing position. The time-consuming task of returning tool *c* to the magazine *h* and replacing it by another tool takes place during the machining cycle of tool *b*; the access time for a new tool is thereby reduced to about 2 seconds. *b* Tool in use. *c* Tool in changing position. *d* Hydraulic tool changing attachment. *e* indexing turntable (*B*-axis), 500 mm dia. with 24 indexing positions. *g* Workpiece. *h* Rotary tool magazine to accommodate 30 preset tools in standard holders. The 30 magazine locations, and hence the tools, are dual coded and are called up by the switching information (location coding). *i* Motor for spindle carrying tool *b*.

made not only to evade the relevant patents, but also to find solutions which are not subject to these disadvantages. In all cases an attempt has been made to combine tool turrets with tool magazines. Figure 154, for example, shows a location-coded tool magazine for 30 tools in conjunction with a quick-indexing double turret. The comparatively slow operation of changing the tools (which is done hydraulically) takes place during the machining cycle, and so is of no practical importance. The access time is only that required to index the turret, which amounts to about 2 seconds. This idea has been taken a stage further in the machine shown in Fig. 155. Here the magazine consists of a circulating chain which can hold up to 45 tools. Since experience shows that there is also a requirement for large and heavy tools which will not fit readily in a magazine, use is made in this case of a 4-way turret. Two diametrically-opposed light-weight tools are changed automatically and brought into service by indexing the turret through 180° while the heavy tools located at 90° to these two can be changed manually during the machining cycle and can be brought into use by indexing the

Fig. 155
Chain-type tool magazine
used in conjunction with
4-way turret for rapid tool
changing (Photo courtesy of
Fritz Werner) (173)

a Section of a location-coded
tool magazine of the chain
type to accommodate up to
45 tools. *b* Four-way turret
with two tools (*c* and *d*) that
are changed automatically
and two tools (*e* and *f*) that
are changed manually.
c Tool in changing position.
d Tool in machining
position. *e* Large tool 1 to be
changed manually. *f* Large
tool 2 to be changed
manually. *g* Hydraulically
operated automatic tool
changer.
Note: The tools that are
handled automatically are
changed rapidly by indexing
the turret through 180°, as
in the machine shown in Fig.
154.

turret through 90°. The design principle of the machine shown in Fig. 156, which has a location-coded chain magazine for 32 tools, is similar. In this case, however, location coding is by means of magnetic codes on a magnetic drum that turns in synchronism with the chain, and this is so flexible and capable of being extended to such an extent, that the effectiveness approaches that of tool coding. The example illustrated in Fig. 157 has even a 6-way turret which carries two small tools in diametrically-opposed locations which can be drawn automatically from the disc magazine, which has 24 coded tool locations, and also 4 heavy tools (e.g. boring bars, large cutter heads, numerically-controlled facing tools as shown in Fig. 174, and similar types), which are changed by hand and which can be adjusted if necessary. These four examples show clearly how a skilful mixture of manual and mechanical tool changing methods makes it possible to bring out the specific advantages of the tool changing system and ensure that the work is performed with the required accuracy.

Figure 158 shows part of the operation area of a large machining centre, a general view of which is shown in Fig. 309, where it is also described. Here the chain-type tool magazine has a capacity of 63 tools for drilling, countersinking, milling, tapping, reaming, etc., so that, used in conjunction with a continuous-path control system a universal machine suitable for

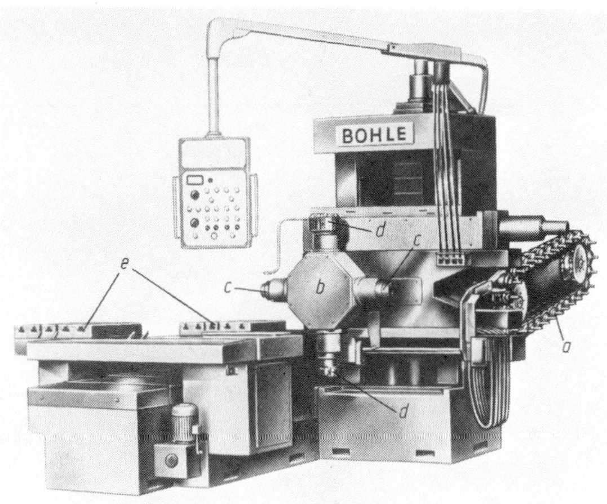

Fig. 156 **Machining centre with a chain-type tool magazine used in conjunction with a 4-way turret for changing 36 tools automatically (Photo courtesy of** *Messrs Bohle* **) (146)**

a Tool magazine. A drum capable of being magnetised moves in synchronism with the chain; the code marks for the tools can be entered on this drum and be detected by means of Hall generators, so that the tools are location coded. *b* Four-way turret. *c* Light-weight tools intended for automatic changing. *d* Large heavy tools intended for manual changing. *e* Standard pallets 500 x 500 mm for transporting workpieces held in chucks or fixtures to subsequent machining operations. (For combination into complete machine lines see Fig. 176).

Note: The feed mechanisms are driven by stepping motors of the type shown in Fig. 119.

complicated machining operations has been produced. Furthermore, special arrangements enable the spindle heads to be changed, so that if necessary multi-spindle heads can be used. The large tool store with a capacity of 100 tools employed on the example shown in Fig. 159 probably represents the upper economical limit for storage capacity on a single machine.

The arrangements shown in Figs. 160, 176, and 177 therefore follow an entirely different concept which in some cases can provide an alternative solution. To supplement Fig. 151, Fig. 160 shows an installation in practice. In the background is the tool preparation platform with the clearly arranged large tool store with its many preset and coded tools, another view of which was shown in Fig. 151. In the foreground can be seen the upper part of a machining centre with a store for only 20 tools. Depending on the operation plan for a particular workpiece, up to 20 tools can conveniently be loaded by hand into the magazine from the platform and as the machine is running, these are automatically changed by a grip system. With many workpieces the

Fig. 157 Machining centre with a location-coded disc magazine used in conjunction with a 6-way turret for changing 25 tools automatically and 4 bulky tools manually. (Photo courtesy of *Heller*)

a Disc-type tool magazine to accommodate 24 location-coded tools. *b* Six-way turret for rapid tool changing. *c* Tools for automatic tool changing. *d* Large, heavy tools for high-precision manual changing (also for numerically-controlled boring and facing heads similar to Fig. 174). *e* Workpiece on turntable (*B*-axis).

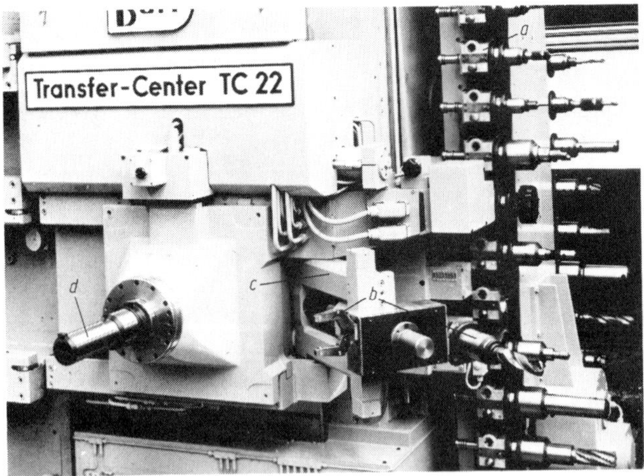

Fig. 158 A portion of the large machining centre shown in Fig. 309 (Photo courtesy of *Ludwigsburger Maschinenbau/Burr*) (147)

a Chain-type tool magazine carrying up to 63 coded tools. *b* Double-grip system of type shown in Fig. 152. *c* 90° pivot mechanism for transfer from the plane of the chain to the plane of the main spindle *d*. *d* Tool in main spindle ready for use.

**Fig. 159Large machining centre with a storage capacity of 100 tools (Photo courtesy of
Hüller) (148)**

> *a* Large tool store for 100 tools comprising 4 storage trays with coded positions.
> *b* Vertically sliding tool selector located between the tool trays and the feeder *c*
> and forming an intermediate transport device for the interchangeable tools.
> *c* Horizontally sliding feeder for the double-spindle head *d* which engages with
> the tool in the change position. *d* Double-spindle head similar in function to the
> double turret of Fig. 154. *e* Tool in working position. To save weight the main
> spindle is driven by an infinitely-variable hydraulic axial-piston motor, since the
> whole drive unit has to conform to motions in the X-Y plane. *f* Fixed work
> table; if required it can be fitted with turntable equipped with clamping and
> machining stations for alternative machining. X, Y, Z Axes of motion,
> principally with numerical continuous-path control and analogue displacement
> measuring systems.

number of tools required is certain to be less than twenty; the magazine will
therefore prove fully adequate for such cases. If, on the other hand, a
workpiece requires a total of 30 or 40 tools to machine it fully, the operator
can remove the first 10 or 15 tools manually from the magazine when these
have been used and can replace them with others. Since the operator has to
be present in any event, changing the tools in the magazine in this way forms
no significant addition to his tasks; and the comparatively small machine is
effectively equipped with a very large store which can deal with the majority
of machining problems. A convenient combination of manual and automatic
tool changing thus enables a very economical solution to be obtained if the
work is appropriately organised and the workshop layout is suitable.
'Full-scale automation' is thus of limited practical importance.

Fig. 160 Work station where a mixture of manual and automatic tool changing takes place with a large number of coded tools. (Photo courtesy of *Siemens*).

a Store of preset and dual-coded tools arranged by decade-numbered plates (see also Fig. 151). *b* Drum type tool magazine on a numerically-controlled machining centre; several machining centres share the one tool store and presetting bench (Fig. 151). *c* Tool changing device of the double-grip type (cf. Fig. 152). *d* Frame of machining centre. *e* Steps leading to tool platform (about level with tool magazine) and presetting position from workshop floor (height difference about 2 metres).

7.2.5 *Tool presetting and tool adjustment*

Several of the principles for presetting tools are illustrated diagrammatically in Fig. 161, and the various accuracy ranges for these are indicated. Boring and milling tools need often be set to an accuracy of only a few tenths of a millimetre within a given length. Figures 162 and 163 show examples of adjustable holders, while Fig. 164 is a simple jig for making mechanical length adjustments against adjustable stops. The presetting device shown in Fig. 165A) with its one or two dial gauges is capable of more accurate work. Figure 165 B) shows examples of preset tools with their holders as used, for example, on the machine shown in Fig. 158. Figure 166 shows as an example a complete and largely standardised tool system for lathes; the tool holders are coded by means of pins and fit into the magazine shown in Fig. 147. Even more convenient is the microscope shown in Fig. 167 with its viewing screen.

Despite the comparatively complex arrangements that have been described it is impossible to eliminate entirely all errors in the indexing of the tool holder. As mentioned in Chapter 1, it is the major difficulty with all slider displacement measuring systems that it is necessary to be able to reproduce the position of the cutting edge relative to the tool holder accurately for

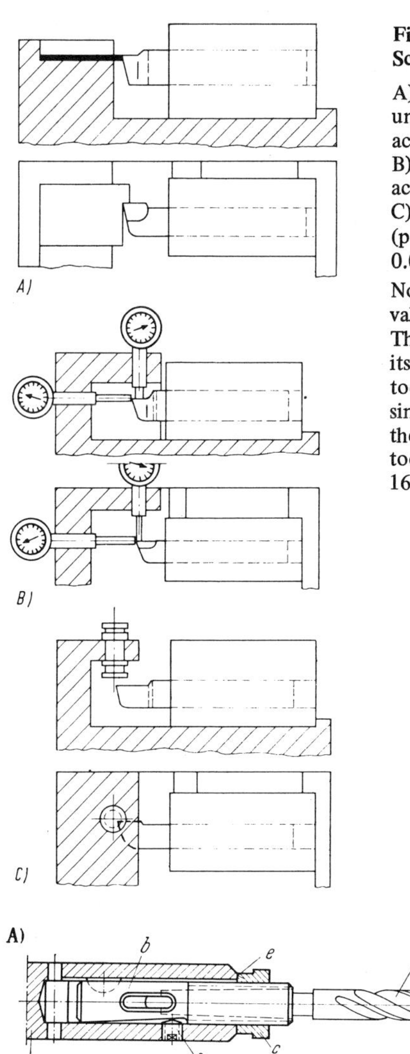

Fig. 161
Schematic diagrams of tool presetting

A) Use of templates viewed by the unaided eye (possible setting accuracy of about 0.2 to 0.1 mm).
B) Use of dial gauges (possible setting accuracy of about 0.03 to 0.02 mm).
C) Use of setting microscopes (possible setting accuracy of about 0.02 to 0.01 mm or even better).

Note: The figures quoted are average values based on practical experience. They relate to the presetting process itself, not to the accuracy of the tools when in place on the machine, since then the accuracy with which the tool holder is positioned on the tool slide has some effect (cf. Fig. 168 and Chapter 9).

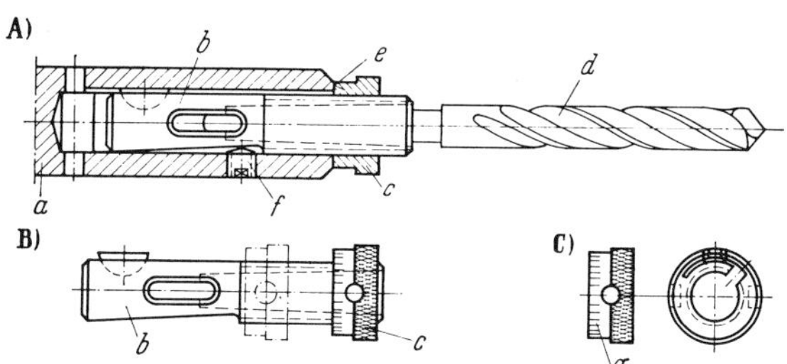

Fig. 162 **Holder with drill length adjustment to VDI 3249**

A) Tool, manually inserted in machine spindle. B) Tool holder with adjusting nut. C) Adjusting nut.

a Machine spindle. *b* Tool holder with keyway. *c* Adjusting nut. *d* Twist drill (to illustrate operation). *e* Presetting reference edge. *f* Clamp screw (for securing holder manually in drilling spindle). *g* Fine adjustment scale.

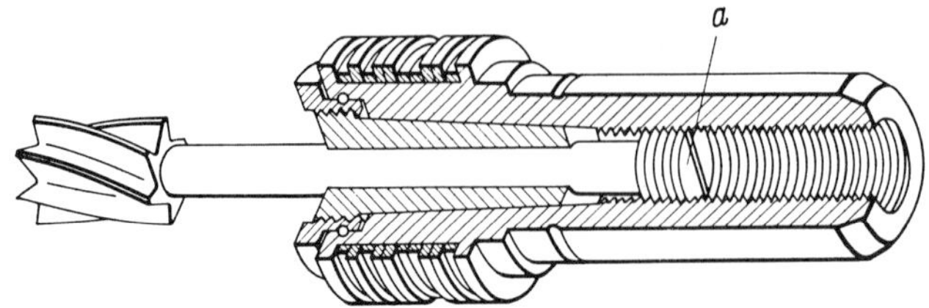

Fig. 163 Tool holder collet with code rings and length adjustment screw for magazines of type shown in Fig. 153.

a Setting screw and thread for drawbar for automatically securing tool holder in machine spindle.

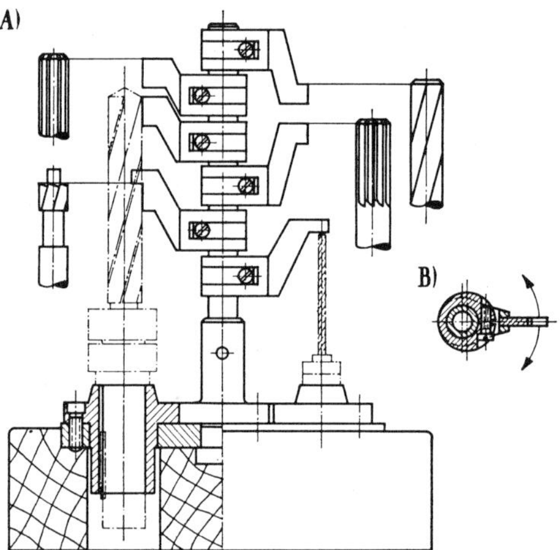

Fig. 164
Schematic diagram of a simple length setting jig using adjustable stops
(131)

accurate results to be achieved. Without numerical-control techniques it would be necessary to adopt the time-honoured method of adjusting the machine mechanically after a trial run. The problem can be solved more elegantly using small computers. These enable the position of the tool holder to be corrected from the control panel in the event of any deviations arising between the actual and the desired positions of the cutting edge of the tool. Figure 168, therefore, refers to the possibility of using data-processing methods for adjusting the tools; in this case three tools on a lathe can be corrected in the X and Z directions after they have been fitted to the machine, by means of coordinate transformation. Details of this technique are described in Chapters 8 and 9. It will be obvious that numerical correction of the tool position in the machine in this way becomes more difficult the greater the number of tools that are stored.

A)

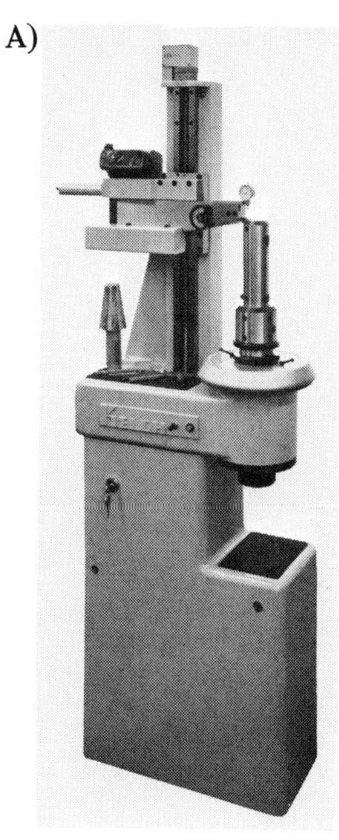

A) Setting unit with same mounting arrangements for tool holders as the numerically-controlled machines that are served.
B) Examples of preset tools with their holders for the machine illustrated in Fig. 158.

B)

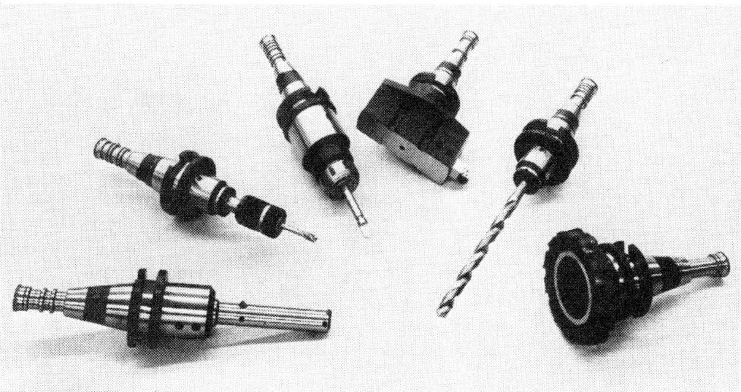

Fig. 165 Tool presetting using dial gauges (as shown in Fig. 161B) specifically for setting boring bars (Photo courtesy of *Kelch*) (149)

It can be undertaken comparatively easily for machines equipped with turret heads, but becomes virtually impossible where there are large stores. Mention may finally be made therefore of an attempt to synthesise tool

A)

B)

A) Tool holder for shank-type lathe tools with shank cross sections of up to 32 x 25 mm. B) Tool holder for boring bars up to 60 mm diameter. (Both tool holders are coded by means of dowel pins so that they will fit into one unique position of a magazine of the type shown in Fig. 146). C) Setting microscope, magnification 20X. If required this instrument can be fitted with a projection screen as shown in Fig. 167.)

C)

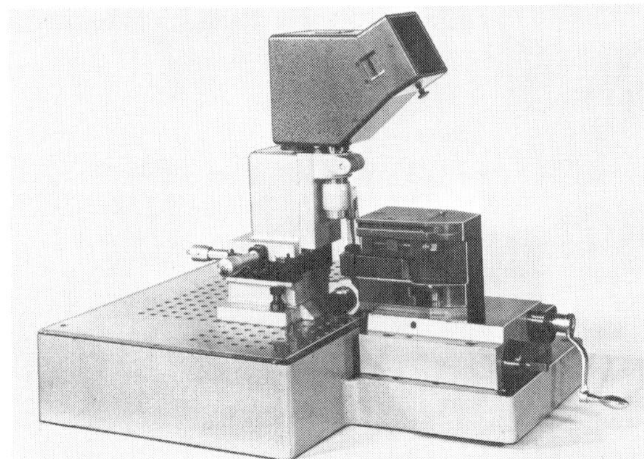

Fig. 166 Tool presetting device with a microscope reader (as shown in Fig. 161C) and two standardised lathe tool holders to fit the magazine shown in Fig. 146 (Photo courtesy of *VDF, Gebr. Boehringer*) (142, 160)

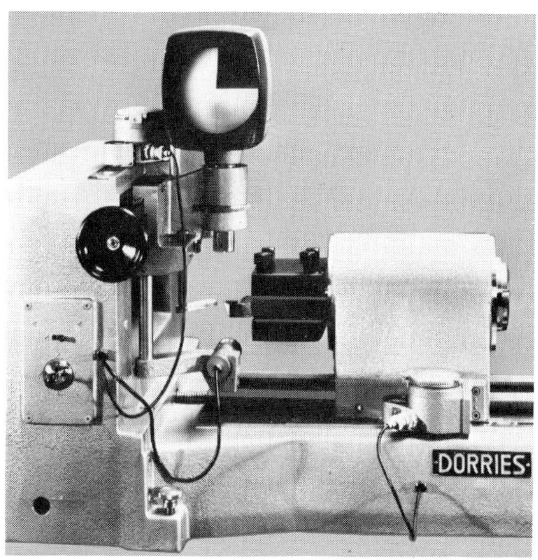

Fig. 167 Setting microscope with a projection screen (Photo courtesy of *Dorries*)

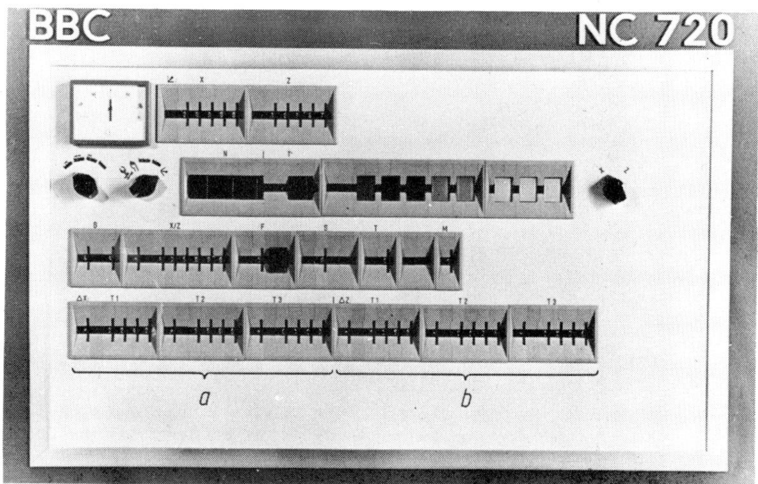

Fig. 168 Adjustment panel on the control panel of a lathe (Chapter 9) to enable presetting and positioning errors of the tool holder relative to the slide to be corrected. (Photo courtesy of *Brown, Boveri*)

a Input for correction (or of absolute value relative to machine reference point) for the *X*-axis and three tools (e.g. in a turret). *b* Similar input for *Z*-axis, (for *Y*-axis, cf. Chapters 10 and 11).

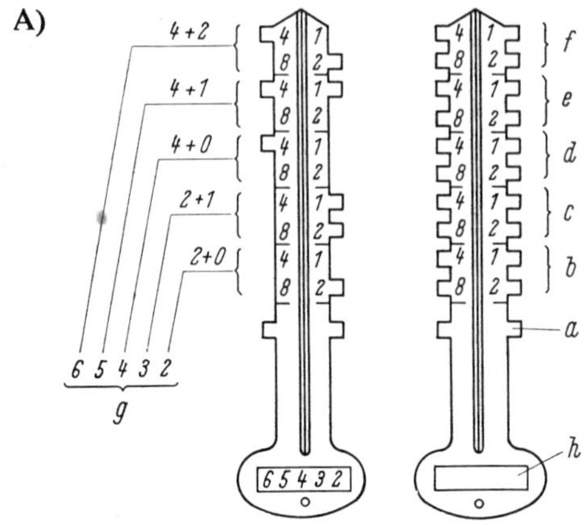

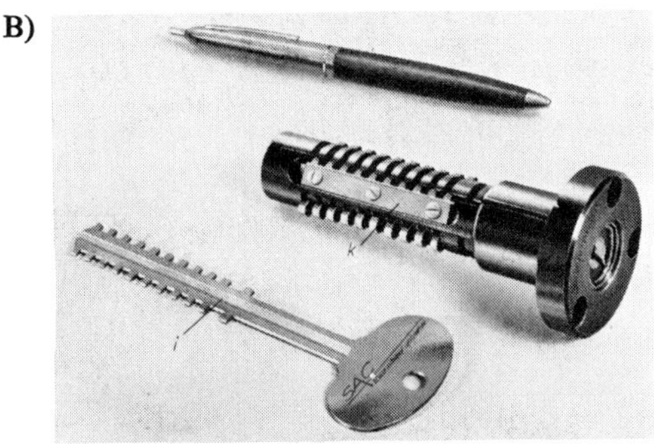

Fig. 169 **Tool coding key for identifying a very large number of tools located in large tool stores (Figs. 159 or 312).**

A) Uncoded key. By breaking off teeth following a BCD system (Chapter 1), it is possible to identify without ambiguity up to 99 999 tools (166). B) Key with contact bush (Photo courtesy of *Scully-Jones*).

a Teeth for operating an electrical contact block. *b* Teeth for the decimal place 10^0. *c* Teeth for the decimal digit 10^1. *d, e, f* Teeth for the decimal digits $10^2, 10^3. 10^4$. *g* Example showing coding of tool No. 65432. *h* Label area showing tool number in easily-read decimal form. *i* Key with binary coding. *k* Contact bush.

Note: Every key relates to a specific tool. When the latter is preset it is also mounted in its holder, so that no mistakes can occur. Each magazine location, especially on disc type magazines, is provided with an electrical contact bush, the contacts being splayed by the remaining teeth on the key. When the key is inserted and turned after the tool has been placed in the required position in the magazine, the correct contacts will be splayed, thus enabling the required magazine location to be identified automatically.

coding and location coding and at the same time to achieve a major increase in the number of tools that can be handled. Figure 169 A) shows a key that can be coded and on which up to 99 999 numbers can be represented in binary-decimal form. The coded key must permanently accompany the tool to which it relates. It is attached to the tool by means of a wire to reduce the possibility of it being lost. Figure 169 B) shows the corresponding contact bush of the type fitted to all magazine positions, e.g. on a disc magazine (see machine illustrated in Fig. 312). When a preset tool in its holder arrives, the tool is inserted in the magazine and the key is pushed into the contact bush and turned. As a result the contact springs corresponding to the coded number are splayed out, and the location of the corresponding tool can be found automatically without difficulty by the machine (166).

The possibilities and limits of the devices described in Figs. 159, 160, and 169 show the need for the efforts that are being made to keep the number of tools that circulate in a factory and that are actually required as small as possible. Otherwise automatic tool changing becomes very expensive and will prove economical only in special cases.

7.3   Extension of the kinematic possibilities

The remote-control of the machine slide, in conjunction with numerical monitoring of its position, opens up a wide range of possibilities for considerably extending the kinematics of many types of machine; this does not apply only to metal cutting machines, as is shown in more detail in Section 7.5.

7.3.1   *ISO Axis Definitions – System of Symbols for Machines and Control Systems*

It was of primary importance to reach international agreement as early as possible on standard definitions and symbols to be employed for defining the directions of the axes of the machines and the movements of their slides. This was all the more urgent because the programming of NC machines, as shown in more detail in Chapters 10 and 11, requires a good deal of abstract thought and demands the use of symbols to enable verbal descriptions to be given of the kinematic processes in the machine. International discussions, based on the American EIA draft standards issued in 1961 to 1964, were held in the ISO/TC97/SC8 committee, and these led to the issue of ISO Recommendations ISO/R 1058-1069, on which, in turn, the four draft standard sheets DIN 66025 and the VDI Regulations 3255 were based (168). At this point the requirements of Sheet 4 of the draft DIN standard and the VDI Regulations 3255 are of most interest (169).

The first important matter is the symbols for the directions of the axes. These are identical with the address letters used in programming, which will be discussed later (Chapter 10). The latter are shown in Table 12 — while some of the axis directions are marked in the illustrations that follow to show the way in which these symbols are used.

To enable the machine kinematics to be described with sufficient accuracy, some further definitions have been laid down in DIN 66025 Sheet 4. To some extent these cover definitions relevant to machines and control systems, and to some extent they deal with programming techniques; they should therefore be studied in the context of both this chapter and Chapter 10. Because of the large number of examples illustrated in this chapter it is convenient to discuss these matters at this point.

According to the International Rules adopted so far, the symbol defining a numerically-controlled machine tool consists of at most four letters and three decimal numbers. The letter at the beginning of the alphabetical group (P,L,C, or D) indicates the type of control system:

> P = Positioning
> L = Straight-line
> C = Continuous path
> D = Continuous-path with facilities for positioning.

The second letter of the group (A,T, or S) indicates the method of writing the words in programming (Chapters 9 and 10 and DIN 66025, Sheet 1):

> A = Address
> T = Tabulator
> S = Tabulator-Address

The third letter (M or I) indicates the system employed for giving linear dimensions:

> M = Metric units, millimetres and fractions thereof
> I = Imperial units, inches and fractions thereof.

The fourth letter in the group (D or R) indicates the system employed for giving rotational dimensions;

> D = Degrees and decimal fractions thereof
> R = Revolutions and decimal fractions thereof.

If any of these letters are not required for any individual machine, the symbol '–' is to be given at the appropriate place in the group.

Symbol	Meaning
A**	Rotation about X-axis
B**	Rotation about Y-axis
C**	Rotation about Z-axis
D	Rotation about a further axis or third feed
E	Rotation about a further axis or second feed
F*	Feed
G*	Displacement condition
H	(May be used as required)
I*	Interpolation parameter or thread pitch parallel to X-axis
J*	Interpolation parameter or thread pitch parallel to Y-axis
K*	Interpolation parameter or thread pitch parallel to Z-axis
L	(May be used as required)
M*	Supplementary function
N*	Sentence number
O	(not to be used)
P	Third movement parallel to X-axis or parameter for tool correction
Q	Third movement parallel to Y-axis or parameter for tool correction
R	Third movement parallel to Z-axis or quick-feed motion in the direction of the Z-axis or parameter for tool correction
S*	Spindle speed of rotation
T*	Tool
U**	Second movement parallel to X-axis
V**	Second movement parallel to Y-axis
W**	Second movement parallel to Z-axis
X**	Movement in the direction of the X-axis
Y**	Movement in the direction of the Y-axis
Z**	Movement in the direction of the Z-axis

*Important programming address letters
**The most important axis definitions and their address letters

**Table 12. Address letters as given in DIN 66025 Sheet 4
(March 1970 position)**

The group of three figures denotes the movements of the machine tool that can be carried out under remote control.

The first figure gives the total number of all the positioning movements of the machine tool that can be remote-controlled or called up, i.e. all the movements that can be called up by means of displacement and switching information. This includes all axes with cam control systems, sub-programmes for copying motions, etc. (Section 7.4).

The second figure gives the number of positioning movements of the machine tool that can be programmed by means of words for the coordinates, i.e. this relates solely to the numerically-controlled axes controlled by means of displacement information as described in Chapter 2.

The third figure gives the number of positioning movements of the machine tool that can be carried out simultaneously and that are programmed by displacement information; it relates in the main, therefore, to simplified subsets of positioning and straight-line control systems in which, as an economy measure the computing and comparison system is to be used once only, after which a transfer is made to the individual axes with their measuring systems.

A drilling machine with numerical positioning control in the X-Y plane similar to that shown in Figs. 179 to 181 is therefore described by the following symbols:

PTM-32 or PAM-321,

depending on whether it is the address or the tabulator method (Chapter 10) used during programming, or whether the control system employs one or two comparators so that the movements in the X-Y plane can be performed simultaneously or in sequence.[2]

From these remarks it will be apparent that it is possible to use the symbols to describe the kinematics of the machine with adequate accuracy. The symbols clearly indicate the fundamental unity of the machine and its control system, and they should never be regarded separately. The subdivision into chapters employed in this book has been undertaken solely for the sake of clarity.

7.3.2   Multi-axis machines

To illustrate the remarks made above, the following figures not only have the letters defining the directions of the axes marked on them, but also give the possible symbols for the machine and for the control systems with which it can be fitted. In many instances alternatives for the latter are possible, since the majority of modern machines can be 'married' to various control systems produced by different manufacturers.

To demonstrate the basic extension of the use of the term 'automatic machine' beyond the case of the conventional automatic lathe, Fig. 170 shows the principles of operation of a machine of the latter type, large numbers of which are still required nowadays for performing certain operations (10, 133, 134). In these machines the automatic mechanism takes the form of rigid-form storage devices such as cams. Their flexibility can be increased only indirectly by the use of NC machines similar to that shown in Fig. 172.

---

[2] The symbols that have been described are at present only recommendations, and they may be subject to modification.

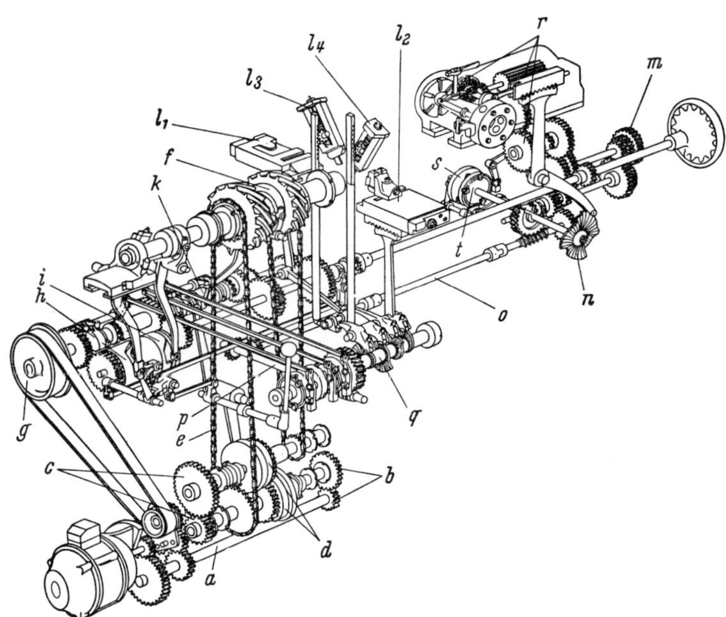

Fig. 170 **Diagrammatic representation of a mechanical automatic lathe (e.g. INDEX B 42) (133)**

> *a* Main drive shaft. *b* Pick-off gears for high-speed rotation. *c* Pick-off gears for low-speed rotation. *d* Mechanically controlled clutches for high-speed and low-speed. *e* Driving chain for main spindle. *f* Clutches for clockwise and anti-clockwise rotation of the main spindle. *g* Control clutch shaft. *h* Cam for tool feed. *i* Cam for workpiece chucking. *k* Cam for reversing direction of rotation. l_1 to l_4 Cross slides. *m* Pick-off gears for driving control shaft. *n* Turret slide control cam. *o* Transfer shaft. *p* Front worm shaft. *q* Cross-slide cam shaft. *r* Turret indexing drive. *s* Quick-feed cam. *z* Turret cam.

Figure 171 shows a boring and milling machine with straight-line control in which as usual it is possible not only to move the coordinate table with the workpiece in the *X*-direction towards the tool, but also to move the boring quill in the *W*-direction towards the workpiece by remote control. In addition, the rotary table can either take the form of an indexing table with indexing pins or can rotate continuously about the B-axis and have its own measuring system (cf. Chapter 4). Depending on the solution adopted, the fourth letter will be a D or R, and the numerical control with displacement measurement changes the figure in the 2nd decimal digit from 3 to 4, it being assumed that the quill movement is controlled by a cam.

Figure 172 shows a special machine for milling cams using a polar coordinate system, since the distance between the centre of rotation and the periphery depends on the angle. This machine can, for example, use a 2-D

Fig. 171 **Five-axis boring and milling machine with straight-line control (Photo courtesy of** *CNMP***). Axes defined in accordance with Table 12, i.e. relative motion of the workpiece and tool in the z-direction can be achieved in two ways: by movement of the table against the tool (Z-axis) or by movement of the boring quill against the workpiece (W-axis). Symbol LAMD 532**

Fig. 172 **Cam milling machine with 2-D continuous-path control. This machine is used for rapid production of control cams for mechanically-controlled automatics (Fig. 170) (Photo courtesy of** *Philips/SIG***) Symbol CAMD 322**

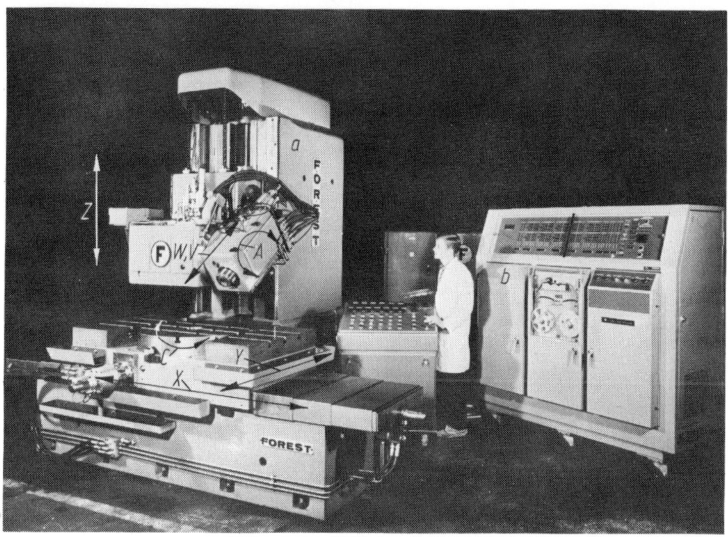

Fig. 173 Universal profile milling machine with numerical 6-axis control. (Photo courtesy of *Forest*) (161)

a Milling machine Type U.N. 800 BRC (Table size 1600 x 800 mm, spindle power 11kW). *b* Numerical control system with integrated circuits Type Bunker-Ramo 3000, numerically-controlled axes *X, Y, Z, W, A, C* (Table 12). Symbol CAMD 663.

continuous path control system in the *Y-C* axis, while it is manually adjusted in the *Z*-axis to suit the thickness of the cam. Figure 173, on the other hand, shows a machine capable of performing a very versatile range of movements with a 6-axis control system. This could be allotted the symbol CAMD 663 if the control system employed provides for simultaneous 3-D continuous-path control. Such machines can be used very effectively in all cases where complicated workpieces are to be machined out of the solid — eliminating the need to employ any joints (e.g. by adhesives or welding), typical applications are airframe construction (especially of prototypes) and press dies (cf. Figures 312, 314).

7.3.3   NC Boring heads

A recent application of remote-control and NC techniques is to the boring and facing head shown in Fig. 174 (171). Fine numerical adjustment of the cutter can easily be performed by means of a stepping motor; because of the low torques that are required, an electrical stepping motor (Chapter 6) will prove quite adequate. Some difficulties in definition arise when it comes to

Fig. 174 **Numerically-controlled boring and facing head (Photo courtesy of *Scharmann*)** (150, 171)

a Slide, positioned by stepping motor which advances 2.5 μm per step. *b* Boring bar or facing tool (e.g. *H*-axis). *c* Electro-magnetic stepping motor, Type Slo-Syn (cf. Fig. 118). *d* Boring head unit complete inserted in the spindle nose of the boring machine.

the choice of the address letter, since the movement of the cutter is always perpendicular to the *Z*-axis, i.e. in the *X-Y* plane. There is, however, no separate address for this; Table 12 shows that letters H and L can be used as required and these can be used for such purposes, the details being agreed between the makers of the control gear and those of the machine tool. The unit can be used for such tasks as milling slots or making taper bores.

The machine and the sample of the work shown in Fig. 175 illustrate that the extended machine kinematics in conjunction with 3-D continuous-path control systems can result in the reduction in tool variety recommended in Section 7.2 and at the same time can even lead to technological improvements in the process. It is, for example, possible to produce large holes of various diameters not by boring, but by milling, so that several separate tools or an NC boring head of the type shown in Fig. 174 can be replaced by a single milling cutter. This brings with it the further advantage of using multiple cutting teeth (162). An essential requirement is, of course, that the interpolation accuracy is sufficient to ensure that a true circle is cut and not a polygon (Chapter 8). Nearly all modern machining centres (e.g. as shown in Figs. 154 to 159) are arranged so that they can also be operated under continuous-path control; this can, in many instances, save valuable and expensive tool storage capacity. The decisive factor will, however, always be the customer's requirements regarding the shape of the components that are to be produced.

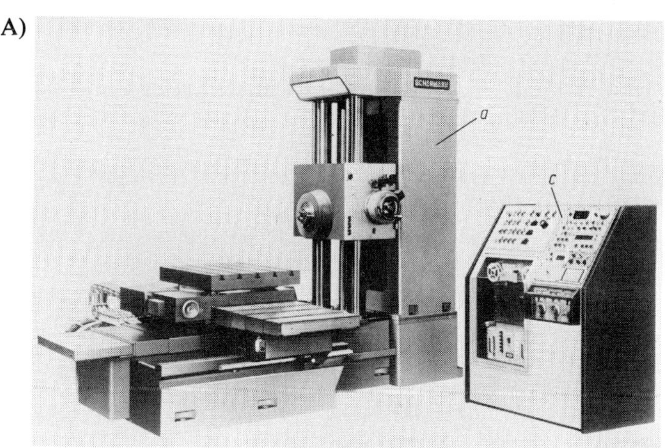

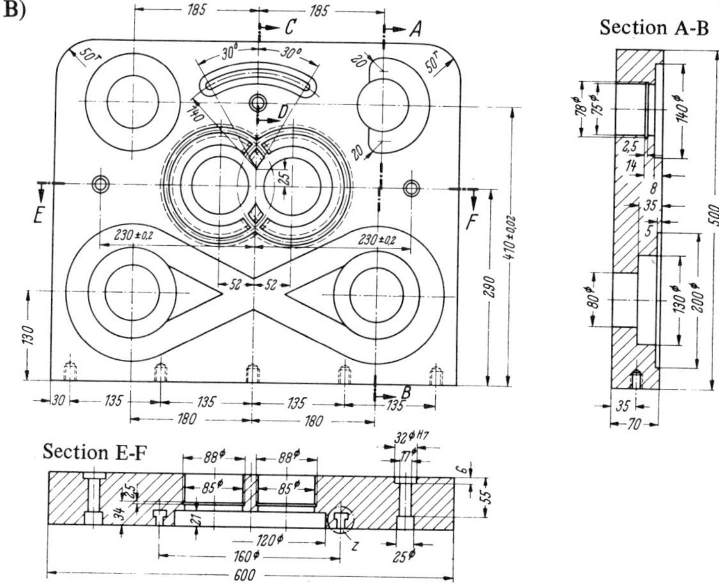

Fig. 175 Continuous-path controlled boring and milling machine with example of work-piece (Photo courtesy of *Scharmann*) (162, 163, 164)

A) Machine with control cabinet, symbol CAMD 43. B) Example of workpiece. The large holes are milled, using the continuous-path control system (multiple cutting blades and reduction in variety of tools required).

a Boring and milling machine with four degrees-of-freedom of movement (X, Y, Z, B). *b* Cincinnati-Acramatic IV-3−D continuous-path numerical control system capable of being extended to 12 axes (in part with straight-line controls).

7.3.4   NC Machining centres and machining lines

The increase in productivity that can be achieved using simple NC machines depends, in the main, on three factors:

a) The absence of form-storing components (drilling jigs, templates, patterns, etc).

b) Reduction in idle time due to the elimination of many measuring operations and to more rapid positioning and also, if the clamping devices are correctly standardised, the reduction of setting-up times. Both these savings in time result in the machines being more flexible than conventional machines and suitable for rapid changes of programme.

c) Improvements in quality, since operator fatigue is no longer a factor; while at the same time there is a wider choice of those who can be used as operators.

If the use of tool magazines, together with full utilisation of the new kinematic possibilities (continuous-path control and multi-axle drive), is employed to extend the field of operations of the individual machine tool, this transition to a machining centre introduces two further factors:

d) Elimination of the need to chuck and release workpieces, to transport them to other machines (principally for boring and milling, together with other associated operations such as countersinking, reaming, tapping, etc) or to reposition them to enable several sides to be machined.

e) Tools are to a greater or lesser extent changed automatically and as a result there is a reduction in the possible sources of error.

This development probably represents the highest productivity that can be achieved using a single machine. The economic and technical limits on the sizes of machining centres and tool stores on a single machine have also been discussed and alternative solutions have been described.

Increasing the productivity of a single machine is however not in itself a virtue; unless it is combined with a free flow of materials and a good management organisation, expensive idle and waiting times will ensue. (10) In the transfer line technique, the flow of materials is rigidly specified for a large number of workpieces, such as is encountered in the motor car industry, for example. The questions now are: Would it not be possible to find groups of machines which complement each other and which are connected together by a computer-controlled flexible flow of materials? Would this not result in the individual machines becoming simpler? Would it not also be possible to arrange for the necessary information to be passed on automatically, so overcoming the difficulties that result from the increasing stream of paperwork that accompanies each job through the plant? Questions such as this are very relevant in this age of computer-based production control systems (10, 124). The problem could also be formulated differently: The critical order of size of machining centres is in many instances already being attained and to increase productivity still further it would be better to join together a relatively small group of simpler, but complementary, NC machines by a computer-controlled flow of materials, and possibly also of tools, to form a self-contained unit. An NC machining line of this type will have to be carefully thought out in each particular case so that the combination of machines is such as to avoid, despite the increased flexibility of the installation, any economic drawbacks. An installation of this type would probably not be very easy to programme, and very high costs would be incurred if the plant were to be stopped for any length of time due to inadequate programming or insufficient orders (Chapters 11 and 12). Installations of this nature must be investigated simultaneously from the technical, operational, and economic points of view if realistic judgment criteria are to be obtained.

Figures 176 to 178 illustrate the first trends towards this development in the form of published models and sketches.

In all cases it is a requirement that the workpieces are mounted on standard pallets, these being coded so that not only does the pallet move to the correct machine, but that once there the correct machining programme is selected. Figure 176 shows a model of an NC machining line for large

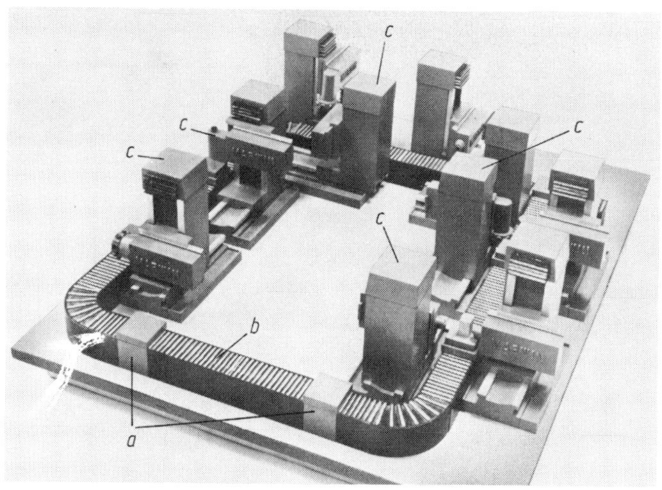

Fig. 176 Model of a NC machining line for large workpieces (cf. also Fig. 147). Photo courtesy of *Marwin*)

a Loading and discharging station for the coded pallets to which the workpieces are clamped. *b* Roller conveyors for the computer-controlled transport of the pallets and workpieces. *c* Linkable machine units (e.g. as shown in Figs. 156 and 308).

workpieces: the loading and discharging stations for the pallets are shown in the foreground. Machines of the types illustrated in Figs. 146 and 308 could, for example, be incorporated in this line, it being an open question at present whether a central tool magazine should be employed for all the machines or whether individual special machines with fixed tooling should be employed. Figure 177 shows clearly the dual role played by the central computer for controlling the material flow and for the direct control of the machines. The advantages and disadvantages of direct on-line numerical control are discussed in greater detail in Chapter 9; with this system all data carriers (punched tape or magnetic tape) are dispensed with to increase reliability. Suggestions regarding the number of machine tools that should be linked to the computer and their nature are deliberately not put forward here so as not to inhibit any experimental installations; it is probable that in the early stages not more than 8 or 10 machines will be linked together in this way to avoid difficulties in data transmission over long distances. Figure 178A) shows a possible arrangement of NC machines, pallet conveyor and computer centre which has been under consideration since 1967 (152). The special feature of this is that the machining units themselves are controlled by magnetic tape (Chapters 9 and 11), while the material flow (movement of pallets and workpieces) is controlled directly by the computer, the whole system being conceived

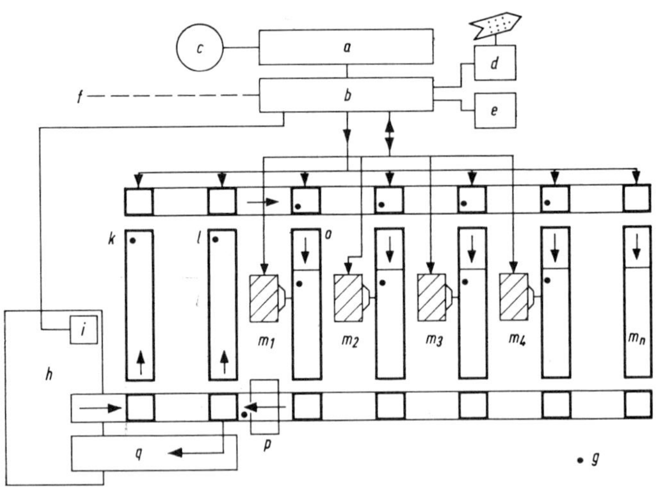

Fig. 177 Schematic diagram of an NC machining line (Courtesy of *Siemens*)

a Process computer. *b* Process element for direct control of a number of NC machines and of the flow of materials. *c* Magnetic disc store. *d* Punched tape input. *e* Operator's chart recorder. *f* Store. *g* Pallet loading stations. *h* Pallet clamping station and workpiece loading. *i* Inspection unit. *k* Preliminary buffer store. *l* Semi-finished components (intermediate buffer store). $m_1 \ldots m_n$ Linked NC machines. *o* Machine store. *p* Washing station. *q* Finished components.

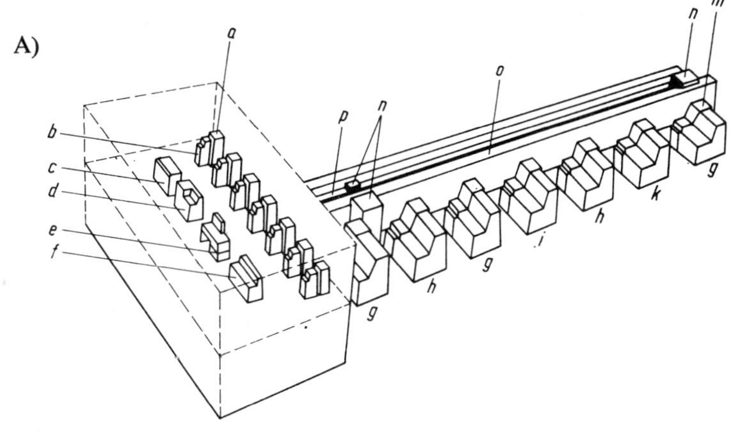

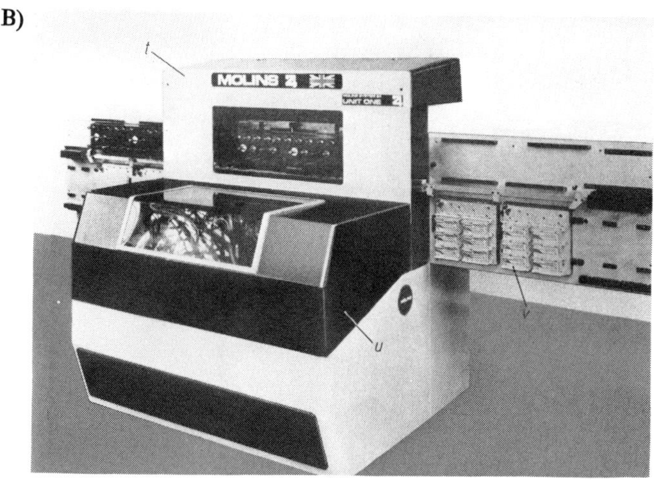

Fig. 178 NC machining line for small light-weight workpieces (Photo courtesy of *Molins*)

A) Schematic diagram of the Molins System 24. B) Example of a magnetic-tape controlled boring and milling unit (Unit One) with 3-D continuous-path control for machining light-alloy workpieces which are mounted on pallets. Maximum workpiece size 300 x 300 x 150 mm. Main spindle driven by oil-hydraulic Pelton wheel turbine at speeds of up to 24 000 rev/min (151, 152).

a Numerical control unit. *b* Magnetic tape store. *c* Printer *d* Card punch-reader. *e* Computer. *f* Operating console. *g* Milling units. *h* boring units (Unit One). *i* Measuring machine. *k* 6-axis unit. *m* Workpiece loading device. *n* Pallet transport trolley. *o* Pallet support rack. *p* Conveyor belt. *t* Machine tool with casing. *u* Casing to provide protection against swarf and oil splashes. *v* Coded pallets with light-alloy workpieces.

primarily for small light-alloy workpieces (max. 300 x 300 x 150 mm). Figure 178 shows a fully-enclosed machine tool (Unit One for boring and milling operations with continuous-path control for 3-D machinings of light-alloy components) with the pallet input.

It is probable that this trend towards the development of NC machining lines will continue during the course of the next few years. Difficult as it is to predict the economic factors for NC machining centres, this difficulty increases still further where NC machining lines are concerned (Chapter 12). In the arrangement of systems proposed in (10) such devices are grouped in Stage 8, admittedly with the proviso that adaptive controls, which will not be discussed until Chapter 8 of this book, are also admissible.

7.4   Combination of NC techniques with other methods

7.4.1   *Combination with cam controls and copying systems*

Various references have already been made to this basic technique, which is frequently employed:

Combination with copying process, Fig. 141 A)
Combination with cams, for reference see DIN 66024 Sheet 4 and
Machine classification in accordance with Section 7.3.1.

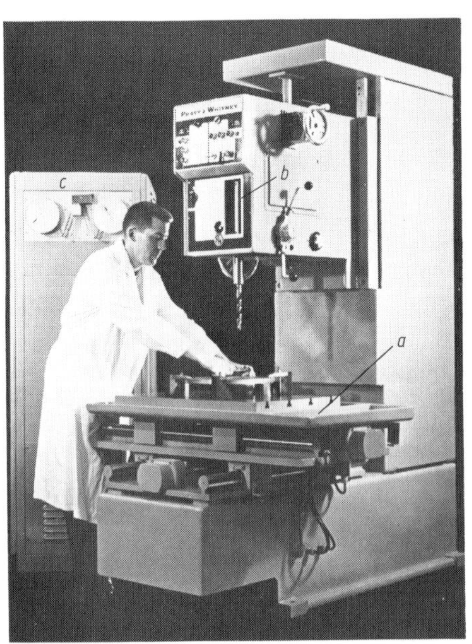

Fig. 179
Simple single-spindle drilling machine (Photo courtesy of *Pratt & Whitney*)

a Coordinate table, numerically controlled by punched tape. *b* Cam-controlled drilling spindle. *c* Control unit (digital-incremental) with punched tape reader.

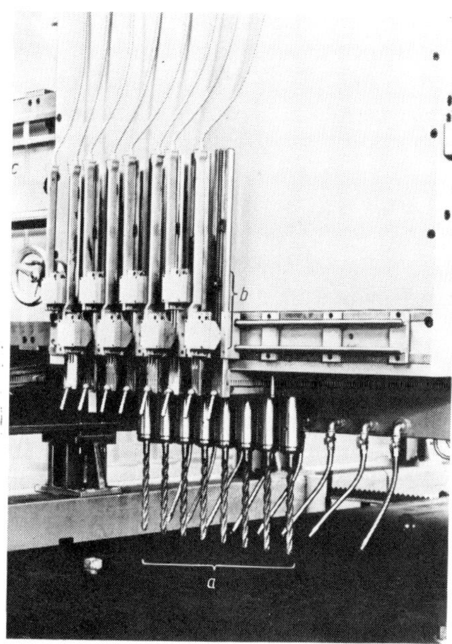

Fig. 180
View of drilling head of an
8-spindle numerically-
controlled plate-drilling
machine (for a general view
of the machine see Fig. 302).
(Photo courtesy of *Kolb*)
(153, 154, 155)

a Eight drilling spindles with
hydraulic feeds, indepen-
dently operated. The
spindles are arranged in line,
and the spacing between
them is manually adjustable.
b Cams to control drilling
depth, freely adjustable, but
constant once set. *c* Transverse member carry slide-
ways for the drilling head
(Fig. 302).

Fig. 181 View of a multi-spindle plate-drilling machine (8 to 18 spindles), arranged in
groups in a frame and selected by switching information) for drilling printed
circuit boards for the computer industry (Photo courtesy of *Burkhardt*).

a Groups of 4 spindles. *b* Combined pressure pads and guide bushes for the
slender carbide-tipped drills which have a long tool life. *c* Two workpieces for
simultaneous machining.

Figures 179 to 181 illustrate further applications of this combination
technique. In the machine shown in Fig. 179 the drilling spindle can be
operated manually, even though the table is positioned by a punched tape; if

the depth of hole remains constant it is also possible to use a manually set cam. With the multi-spindle machines shown in Figs. 180 and 181 cams are used for limiting the movement in the Z-axis, although the spindle spacings are manually adjustable to produce various hole patterns in plates (154).

7.4.2   2½-D continuous-path control systems

Since there is a major cost increase in the transition from 2-D to 3-D continuous-path control systems (Chapter 8), and since the programming also becomes more complicated (Chapters 10 and 11), it is likely that there will continue to be a trend towards achieving simplifications by performing three-dimensional (3-D) machining tasks by means of a two-dimensional (2-D) machining system. To do this, the three-dimensional body is divided into planes located between 0.5 and 2.0 mm apart, and each plane in turn is machined using a 2-D control system. After each plane has been machined the necessary constant feed movement in the Z-axis to bring the cutter to the level of the next plane is produced by the switching information. The workpiece machined in this way approaches the required shape in a series of steps, which can then be smoothed off manually, for example using a rotary file. In many ways this process is similar to that employed with many copy-milling machines.

7.5   NC Techniques for Non-Cutting Machinery

Although numerical control systems were originally developed for use with metal-cutting machine tools, it was clearly only a matter of time before their use would also be extended to other fields. Of the many possibilities only a few examples will be described here, partly to show even more clearly the way in which the numerical control systems are merely components of the general data-processing system, and partly in order to arouse the reader's imagination and to encourage him to exercise his ingenuity.

7.5.1   NC Machines for metal forming

Even though there are at present already a large number of machines of this category in use (flanging machines, pipe-bending machines, etc), because of the limited amount of space available only two press working examples will be described. With NC presses *fewer* complicated tools are required than with conventional presses. Figure 182 shows a multiple-die punch; the turret can accommodate up to 18 tools, since there can be no danger of interference between the tools as with metal cutting machine tools. The coordinate table

A)

B)

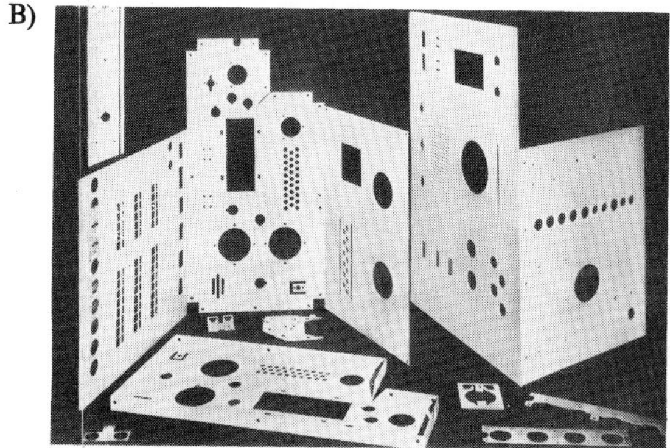

Fig. 182 NC multiple-die punch having a tool store (turret) with a capacity for 18 punch tools together with examples of the work produced (Photo courtesy *Behrens*).

A) Multiple-die punch set-up
B) Examples of work produced
a 2-D coordinate table *b* Numerical positioning control system for table.

carrying the sheet to be punched is positioned by means of a 2-D positioning control system; and when both the workpiece and the turret are in their correct positions, a punching stroke is initiated by the switching information. For the most part these machines are used for making chassis for electronic equipment; some typical examples of workpieces of this type are shown in Fig. 182 B).

An important adjunct of a press of this type is the complex punch tool which by its very nature is a form-storing element. Such tools can, however, be produced very rapidly on numerically-controlled grinding machines. Just as with the machine shown in Fig. 172, here again a metal cutting machine

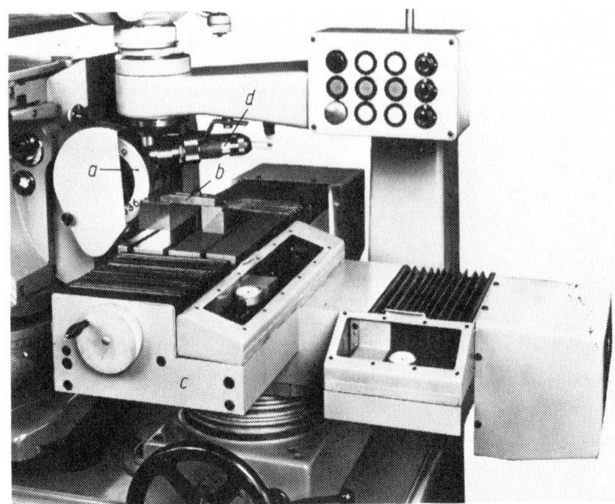

Fig. 183 View of work area of a profile grinding machine with numerical continuous-path control and visual inspection by means of a projection screen (Fig. 300 shows a view of the complete machine), used for the production of accurately profiled punches (Photo courtesy of *VEB Mikromat*).

a Grinding wheel. *b* Workpiece (punch). *c* Continuous-path controlled coordinate table. *d* Grinding wheel dressing attachment to maintain grinding wheel profile constant.

Fig. 184 Example of a continuous-path controlled punch and nibbling machine forming a sheet-metal working centre with manual tool changing (Photo courtesy of *Trumpf*)

a Workpiece (sheet-metal) clamped in coordinate guides (max. working range 1300 x 2000 mm). *b* Pole-changing main motor for 200 and 400 strokes/minute. *c* Support table for large workpieces (sheet-metal panels). *d* Control cabinet for the numerical continuous-path control system (in this case a circle straight line internal interpolator that controls the electro-hydraulic stepping motors (Chapters 6 and 8).

tool, similar to those shown in Fig. 183, can be used to increase the flexibility of other items of production equipment.

In the example of the machine shown in Fig. 184, the use of numerical-control systems has been extended still further. This punch and nibbling machine is not only equipped with a 2-D continuous-path control system, but it also has a tool store (Fig. 185) for manual tool changing; the tools can be changed within 10 to 15 seconds. The combination of form-storing tools and a continuous-path control system enables a very wide range of sheet-metal shapes to be produced very rapidly. Two characteristic hole patterns in sheet metal are shown in Fig. 186.

Fig. 185 Tool magazine for manual tool changing on the machine shown in Fig. 184. Time required to change tools about 10 to 15 seconds (Photo courtesy of *Trumpf*).

7.5.2   NC Flame-cutting machines

Numerically-controlled flame-cutting machines have been in use for some time, particularly in shipyards; their predecessors and current competitors are copy-tracing flame-cutting machines in which a self-acting optical follower traces a precision drawing and controls one or more flame-cutting nozzles by means of follow-up circuits. (3) Figure 187A shows the general arrangement of a copy-tracing flame-cutting machine with an optical follower; while Fig. 187 B) is an example of a large flame-cutting machine (158) equipped with numerical controls and several cutting heads. Both types of equipment are offered as alternatives by almost all the major firms engaged in this field; a critical discussion regarding the best type of data carrier to employ was published in (159). The importance of all these processes must be judged in the light of the constantly increasing replacement of castings by welded structures. The drawings required as templates are usually produced on NC drawing machines (Fig. 189).

7.5.3 Other applications of NC

A new application of NC techniques is in conjunction with spark-erosion machines as shown in Fig. 188 A). It is in the nature of this process (170) that the operator cannot observe the erosion as it takes place; the machine must be reliable in operation without being monitored visually. With this method it is possible to produce complicated precision profiles (Fig. 188 B) even in hardened steel or hard alloys, so that there is no fear of distortion due to subsequent heat treatment and the grinding operation can be eliminated (Fig. 183). This, again, is a very important machine for producing tools for the metal forming industry.

Finally mention may be made of the possibilities of graphical data processing using equipment of the type shown in Figs. 189 and 190; further details are given in Chapters 11 and 12. With the precision drafting machine shown in Fig. 189, drawings and data encoded on punched tapes are drawn on paper, or, if even higher precision is required, are engraved by means of a small scriber on distortion-free plastic film.

A)

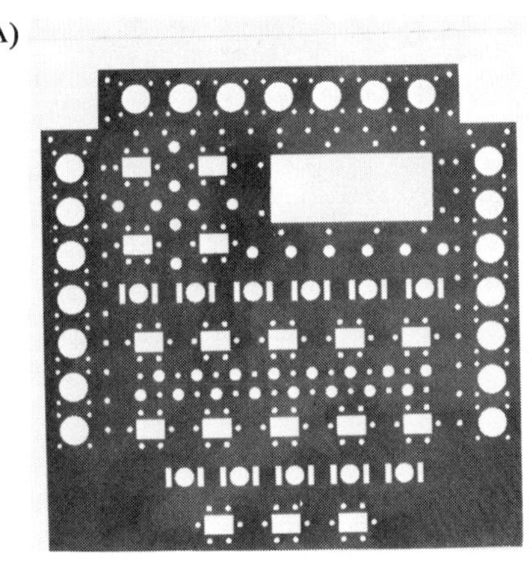

A) Workpiece size 500 x 500 mm, sheet thickness 1.25 mm. Number of punched holes 354, number of nibbled holes 1. Number of tools required 6. Unit time 3.9 minutes.
B) Workpiece size 500 x 1000 mm, sheet thickness 2.5 mm. Number of punched holes 333, number of nibbled holes 9. Number of tools required 5. Unit time 7.5 minutes.

B)

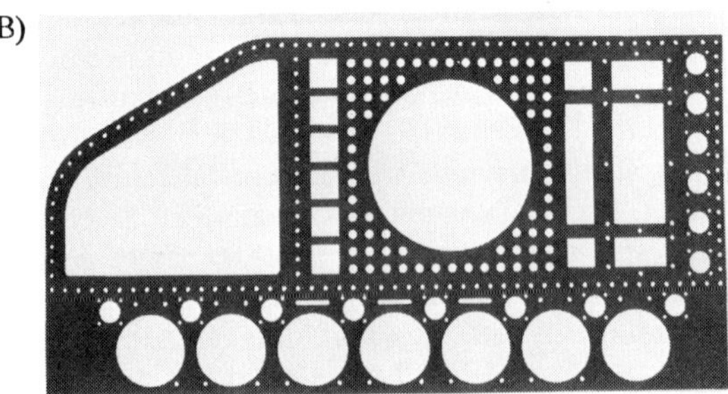

Fig. 186 Examples of workpieces that are produced on the machine shown in Fig. 184 (Photo courtesy of *Trumpf*).

The machine shown in Fig. 190 operates in the reverse manner. Here a given drawing or curve is followed, converted into numerical values, and these are then indicated and recorded. The corresponding extension to the three-dimensional field is shown in Fig. 316 in the Appendix.

A)

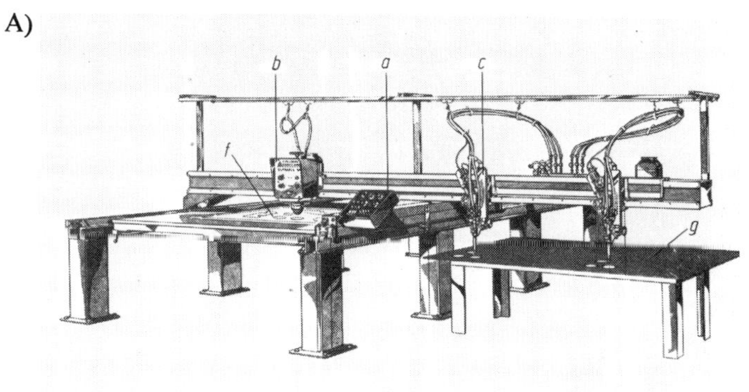

B)

Fig 187 Flame-cutting machines for gas and plasma cutting with two possible control methods.

A) General arrangement of photo-electric tracing of a precision drawing with the cutting nozzles guided by means of follow-on circuits (illustration courtesy of *Kjellberg-Eberle*) (157). B) Example of a large-scale numerically-controlled flame-cutting machine (continuous-path control of several cutting nozzles that work simultaneously). (Photo courtesy of *Messer-Griesheim*) (158, 151).

a Control console. *b* Photo-electric follower. *c* Cutting nozzle holder. *d* Cutting nozzle. *f* Precision drawing (as made on a machine of the type illustrated in Fig. 189). *g* Workpiece. *h* Main switchboard.

A)

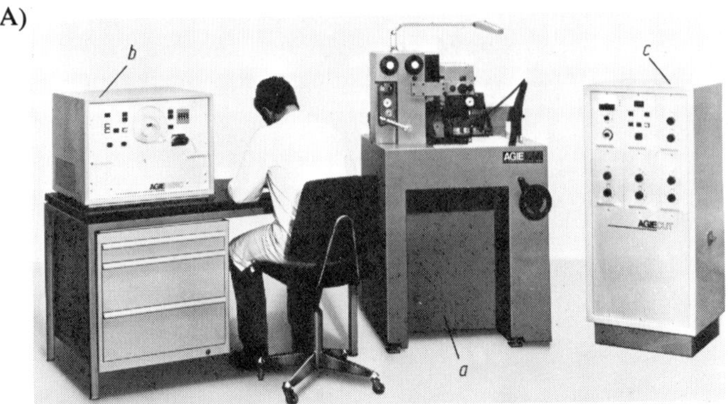

B)

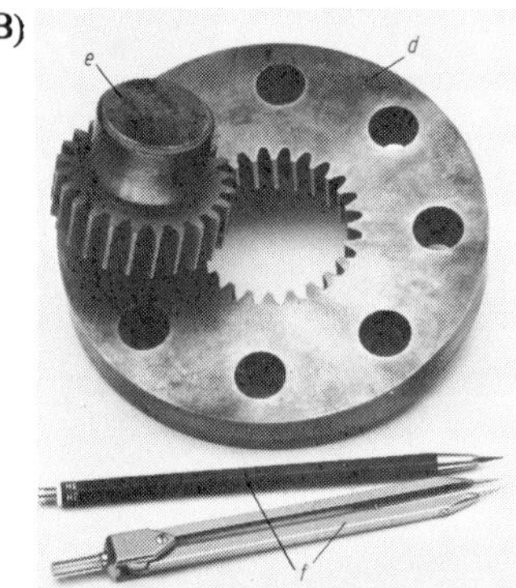

Fig. 188
Spark-erosion machine with numerical continuous-path control (Photo courtesy of *AGIE*) (174)

A) Erosion machine with accessories.
B) Examples of work pieces

a AGIECUT spark erosion màchine.
b Work bench with AGIEMERIC numerical continuous-path control. *c* Erosion generator. *d* Punching die. *e* Punch.
f Articles to convey impression of size.

Fig. 189 Numerically-controlled precision draughting table AEG-GEAGRAPH (Photo courtesy of *AEG Aristo*)

a Control console with indication of actual coordinate values and 2-D continuous-path control system. *b* Numerically controlled drawing head. *c* Drawing (normally drawn with ink on non-shrink paper, but if increased precision is required can be engraved by scriber on coated plastic film).

Fig. 190 Numerical 2-D coordinate follower unit for transforming analogue drawings or curves into digital values. (Photo courtesy of *Ferranti*).

a Follower templates. *b* Digital read-out. *c* Recording chart recorder.

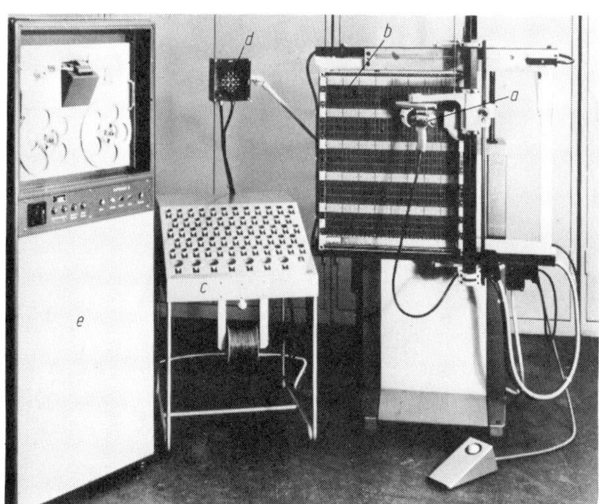

Fig. 191 Semi-automatic NC wiring machine (Photo courtesy of *Siemens*).

a Hand-operated wire-wrap pistol. *b* Reverse side of an assembly frame with pins for mounting wiring harness. *c* Magazine to hold various coloured wires. *d* Punched-tape controlled indication for next direction of motion. *e* Control cabinet with punched-tape reader for wiring programme.

Figure 191 shows a semi-automatic machine for wiring up electronic units, e.g. plug boards with printed or integral circuits. The wire-wrap pistol for winding the bare end of the wires round the contact pins is manually operated. It can only be switched on, however, if the pistol is in the position specified by the punched tape, so that it is subject to a numerical positioning control. The indicator merely shows the direction in which the wire-wrap pistol has to be moved to bring it to the next joint.

7.6   Summary

To summarise the information contained in this extensive chapter it is convenient to sub-divide it into a number of subjects:

> 1 Design features
> 2 Tools and tool stores
> 3 Machine kinematics
> 4 Other NC machines
> 5 Relationships with plant organisation.

The following points then emerge:

Subject 1 – Design features

1. The increased productive capacity of continuously-running numerically-controlled metal cutting machine tools makes it almost essential to strengthen the normal conventional machine tools to ensure that they are capable of meeting the more severe requirements regarding stiffness, resistance to vibration, thermal stability, nature and power of drives, swarf removal, etc. Only in exceptional cases will it prove economically sound to fit numerical control systems to existing conventional machines (e.g. NC coordinate tables fitted to conventional drilling machines).
2. To improve dynamic behaviour, low-friction slideways and transmission elements (recirculating ball bearings, piston and cylinder systems, hydrostatic guides, etc) are recommended, unless there are technological objections to their use (176).

Subject 2 – Tools and tool stores

3. It has been found advantageous to make a systematic analysis of the tools in use in a plant with the object of reducing the different shapes to the minimum number essential to meet design and production

requirements. It may be convenient to divide the tools into those with form-storage capacity and those without.

4. To achieve high machining accuracy and short access time to a tool having a greater or lesser degree of form-storage capacity, attempts should be made to make use of turrets. With drilling and milling machines and lathes, these are almost invariably of the star type; their main disadvantage is the limited number of tools (usually not more than 8) that can be accommodated owing to the danger of mutual interference. There are special design features associated with stationary tools (lathes) on the one hand and rotating tools (drilling and milling machines) on the other hand. The limited number of tools enables comparatively simple means to be adopted for adjusting the tools individually, thereby reducing the likelihood of errors.

5. Where the use of a large variety of tools on a machine cannot be avoided, or where the tools are required with very high cutting capacities it is advisable to adopt tool magazines for which tool changing is automated to a greater or lesser degree. The double possibility of error (tool presetting of limited accuracy and the danger of interference due to dirt at the interface between the tool holder and its mounting) together with the difficulty of subsequent adjustment by numerical means leads to the danger that the accuracy of working may be less than on machines equipped with turrets.

6. The variety of design methods by which these major difficulties can be overcome is very large; they range from changing the complete spindle head, to automatic cleaning processes using compressed air, the adoption of enclosed magazines and clamping devices, up to the time-consuming and expensive adjustment of the tools once they are clamped in place on the machine.

7. For economic reasons it is always advisable to bear in mind the possibility of adopting manual tool changing, and so deliberately to limit the extent to which the machine is automated (10).

8. Both for technical and for organisational reasons, the decision as to whether coded tools (or tool shanks) are to be adopted is of fundamental importance. It depends to a large extent on the number of tools that are actually in use.

9. To ensure that short access times are achieved where the tools are stored in magazines, it is often very useful to combine the magazines with turrets. This also enables large and heavy tools to be changed by hand without any loss in productivity.

10. The capacities of the tool magazines fitted to individual machines is currently between about 9 and 100. It may be that an average figure of 20 to 30 tools will prove satisfactory (see Point 16).

11. Depending on the accuracy required, presetting can be done by the naked eye or by using optical aids; this is not a specific requirement for NC machines alone, but can also prove useful with copying machines or mechanical automatic lathes (10) ,

12. It is possible to adjust tools that are clamped in place on the machine by means of numerical control. The number of tools that can be adjusted and the nature of the adjusting device are economic factors (10) and will be considered further in Chapters 8 and 9.

Subject 3 – Extended machine kinematics

13. In order to encourage international trade in machine tools and to enable them to be coupled with the widest possible choice of numerical control systems of various makes, National and International Standards authorities have adopted a standard scheme for the designation of axis directions and for symbols for the machine kinematics.

14. The extension of the kinematic possibilities of machine tools of all types as a result of applying NC techniques is based in the main on two factors:
    a)   The possibility of reducing the number of tools in use in the plant;
    b)   simplification and speeding up of the production of jigs and tools, which results in an indirect improvement to be attained in the flexibility of metal and plastics forming machines and of cam-controlled mechanical automatic machines;
    c)   manufacturing very complex three-dimensional bodies from the solid without introducing any jointing techniques (welding, adhesives, etc). This proves of particular advantage where the workpieces concerned will be subjected to very high stresses.

16. The increasing complexity of machining centres, the increase in size of their stores, and the requirement for a carefully planned flow of materials lead to the suggestion that thought be given to the development of NC manufacturing lines as a further extension of the NC techniques (220). These would offer a number of advantages, for example the possibility of making use of simpler specialised machines, of employing a central large tool store to serve several machines, and to arranging the flow of materials so that it can readily be controlled by a data-processing device. The technical economic, and programming problems involved in this type of plant are very large and, at present in many cases are at the experimental stage only or are being kept confidential.

17. Another possibility of simplifying NC machine tools is to adopt subprogramming techniques in which the slides are moved along the various axes without numerical monitoring, the motions being started by the switching information system. The subsequent machining cycle can then be governed by conventional and well-known automation techniques (cam control, copying control, etc).

Subject 4 – Other NC machines

18. During the last 6 to 8 years numerical control techniques have been adopted for a number of other applications. To some extent these can be described as 'integrated data processing in production technology.' Among such applications are:
Pipe bending machines
Multiple-die punches
Flame-cutting machines
Nibbling machines (for sheet-metal cutting)
Spark-erosion machines
Automatic wiring machines
Testing machines, etc.

19. A special place in this review must be allotted to the NC draughting machines; in a sense these form the link between comprehensive plant-wide information processing and graphical data processing which is discussed in greater detail in Chapters 11 and 12.

Subject 5 – Relationships with plant organisation

20. The close links that exist between NC machines and the organisation of the plant have been mentioned on six occasions in this chapter:

a) The necessity of keeping down the number of tools in use in the plant combined with the trend towards taking steps to ensure that the number of tools required to produce a workpiece is kept small at the design stage;

b) the possibilities of machining a workpiece extensively without damping and having to introduce a second operation; this saves time, machines, and transport media;

c) the introduction of tool catalogues and tool numbering systems;

d) possibilities in the rationalisation and increase in productivity in jig and tool making;

e) the possibility of coupling the flow of information that passes through the production system with data-processing installations;

f) the transition to 'integrated data-processing' using graphical data output units (e.g. NC draughting machines), NC monitoring devices, computers, and NC machines of various types (Chapters 11 and 12) (10, 86).

21. From the discussions in this chapter it will be clear that it is quite justifiable to regard the machine tool itself, together with its tools, its kinematics, etc., as a closed loop system (175).

Mention was made in the first two chapters of the fact that the question whether it is necessary to maintain a functional relationship between the movements in the individual coordinate directions or not is of major importance in determining the construction of a numerically-controlled machine tool. The replacement of the mechanical scanning of rigid form stores (templates, patterns, etc.) and of the associated scanning control systems by the numerical processing of curve data can be achieved only by the use of electronic computers. When NC machines were first introduced this problem aroused the main interest (200). As this new type of machine became more widely used, however, more and more tasks arose which encouraged the adoption of small or larger computers to enable full advantage to be taken of the possibilities offered by the concept of numerical control. The importance of the fixing and shifting of the zero point was referred to in Chapters 3 and 4; while in Chapter 7 the need arose for adjusting tools clamped in position on the machine so as to provide a simple means of correcting any inaccuracies that might arise in presetting and to increase the accuracy of the work performed by the machine.

As will be discussed in greater detail in Chapter 10, it is particularly advisable when performing profile milling operations to establish the programme in accordance with a particular cutter radius and then to make as little modification as possible to the punched tape. In this case some means must be available at the machine for correcting for any variations in the cutter diameter, due to sharpening for example, to prevent any errors arising in the workpieces.

Basically, it is possible to perform all these calculations either with analogue or with digital circuits. The decision as to which is preferable will depend on economic considerations and on the accuracies that can be achieved. There is, however, a definite trend towards the adoption of digital computers. There are two main reasons for this:

1. Where high accuracy is required the cost of a digital computer is considerably less than that of an analogue computer; Fig. 93 in Chapter 5 gave an indication of the relevant relationships.
2. With the introduction of integrated circuits (Chapter 3) there has been a shift in the cost/performance ratio in favour of digital computers.

In this Chapter with one exception in Section 8.4, only digital computers and computing methods will therefore be discussed[1].

8.1 Calculation of corrections by means of digital computers

Before the details of the circuits required for performing the calculations are discussed in Section 8.4, the nature and extent of the calculations will be described.

Figure 192 shows the relationships between the required value, the zero-point shift, and the tool radius correction for straight-line control systems.

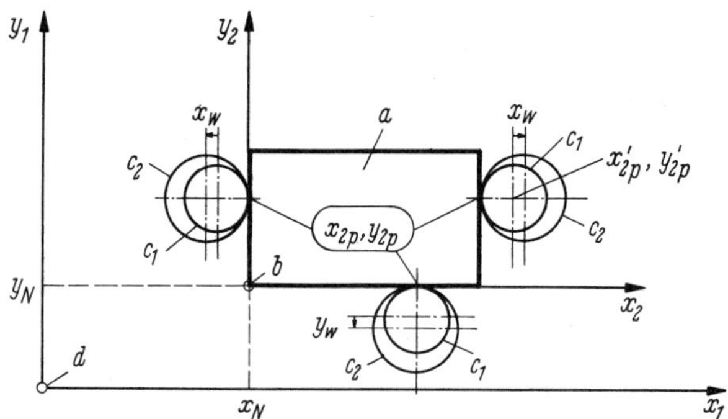

Fig. 192 Zero-point shift and tool diameter correction for straight-line control systems (Example: Peripheral milling)

a Workpiece. *b* Workpiece zero point. c_1 Tool 1. c_2 Tool 2. *d* Machine zero point. x_1, y_1 Coordinates of machine measuring system. x_2, y_2 Coordinates of workpiece measured from workpiece zero point (drawing data serves as programming basis). $x_N y_N$ Zero-point correction values (zero-point shift). x_W, y_W radius correction if there is a deviation from programmed cutter diameter. x_{2p}, y_{2p} Required values for workpiece contour. x'_{2p}, y'_{2p} Required values for tool centre.

Zero-point correction

It is first necessary to distinguish between the machine zero point and the programme zero point. This also involves the differences between incremental

---

[1] In the 1st Edition of this book it was necessary to devote a complete section to the principles of analogue internal interpolators. The developments that have taken place during the last six years have shown that analogue techniques are now of virtually no importance in this field, and that this type of solution is now only of historical interest (181, 182, 183).

and absolute displacement measuring systems (Chapter 3). With an absolute system the machine zero point is fixed by the construction of the machine and by the manner in which the displacement measuring systems are fitted. The programme zero point relates by definition to the workpiece that is to be machined and that could, in theory, be clamped at any point on the machine table. In Chapter 3 one of the advantages of the incremental measuring system has already been mentioned, namely the fact that the starting point can be moved to any desired position on the table surface. As a result the two zero points can easily be made to coincide. In this instance, therefore, no calculations of the zero-point correction are needed.

Circumstances are different with the absolute measuring system; for simplicity Fig. 192 shows the conditions for machining surfaces which are parallel to the axes in the two-dimensional region. The workpiece a, for which the zero point b is fixed by the machining programme (Chapter 10), is secured anywhere in the $X_1 Y_1$ plane on the machine table; the zero point for the machine coordinate system $X_1 Y_1$ is c. The control system can work only with machine dimensions; for all programmed workpiece dimensions the coordinates must therefore be transformed in accordance with the magnitude of the zero-point shift from b to c. This requirement could be met, for example, by providing an adding mechanism for each axis — if care is taken to ensure that point b is clearly defined on each workpiece. Moreover the point must be clearly recognisable so that the operator can align the workpiece zero-point with the cutter axis with sufficient accuracy. This basic disadvantage of the measuring control system based on slide movements was mentioned in Chapter 1. It can be overcome by the use of suitable clamping arrangements (e.g. precision chucks on lathes) and by providing the workpieces with smooth and clearly defined edges. In principle it is, however, necessary always to provide the means for numerical coordinate transformation, the only question being how large the range of transformation should be; whether it should extend over the whole area of the table or need cover only portions of it. To ensure that the numerical control system should be of universal application, it is usual to design the computer so that the zero point can be shifted to anywhere within the range of adjustment of the machine, thus providing the user with the same freedom of zero-point selection as he has with the incremental method.

As the cost of digital computer elements tends to fall, and to achieve greater standardisation of the control cabinets, it is usual to arrange for the zero-point shift to be carried out digitally even where an analogue displacement measuring system is in use. This means that a digital zero-point shift is located before the D/A converter, although the problem could also be solved by analogue methods (Chapter 4).

Tool dimension corrections

This heading covers several problems and possible solutions which can best be understood with reference to Chapter 7.

a) *Tool length correction* makes allowance for variations in the dimensions along the axis of the tool, and so compensates with drills, for example, for wear or regrinding. By definition (Section 7.3) these corrections affect therefore the coordinate dimensions in the Z-axis. They can be fed into the computer manually using decade switches (Chapter 9) or automatically. In the latter instance, the tool-carrier for the new tool of unknown length is automatically moved to contact a reference surface located at a known distance from the zero point along the Z-axis. The length of the tool as measured in this manner is stored in a core or drum store (Fig. 218) and is automatically taken into account when the corrections in the Z-coordinate are being calculated. The machine shown in Fig. 218 is equipped with an automatic device of this type which varies for correcting for tool length. This will, of course, simplify the operation of such a machine; on the other hand a new idle time 'tool measurement' is introduced for each machining cycle in which a new tool is used, and the costs of the control system are increased.

b) *Tool position correction* affects two axes and is mainly employed on lathes. Despite all the efforts that are made when presetting and grinding lathe tools, it is impossible to avoid minor errors in the tool radius after changing a tool; these errors would make precision turning impossible.

In many instances where numerical control is applied to lathes a tool correction in the X-and the Z-axis is therefore provided for each turning tool. In theory, it would again be possible to move the tool into contact with a reference point, and to adjust each tool by means of two small computers, and this solution is in fact sometimes adopted. To increase the accuracy it is, however, frequent practice to use the new tool to turn an external or internal surface, to measure this by means of a mechanical instrument (e.g. a micrometer screw), and to set a decade correction switch on the basis of this measurement to obtain an accurately turned dimension. Depending on the number of tools accommodated on a turret it is necessary to provide 2 x 5 or 2 x 6 tool correction switches and the corresponding computer circuits (Chapter 9). This is disliked by many engineers, and strengthens the determination of some users to avoid the use of punched-tape control systems. They prefer to equip turret lathes, for example, with manual means of inputting switching and numerical displacement information, as shown on Fig. 297 in the Appendix. There are numerous references to this in the literature (e.g. 124, 201); in many cases identical lathes are available with several input methods.

c) *The tool diameter (or radius) correction* applies in the main to milling machines; the problems involved in milling round a periphery using a straight-line control system are shown schematically in Fig. 192. Basically two methods can be applied:

c 1) The *path* to be followed by the centre of the cutter is programmed for a particular cutter diameter, and a small computer is incorporated in the control system which automatically calculates the corrections needed for a different cutter diameter. These differences can occur because the tool has been reground or because a different tool is employed. If there is a tool of the same diameter used, no correction will be needed.

c 2) The final contour required for the workpiece is entered in the programme, and by means of decade switches the diameter of the cutter that is being used is manually introduced into the control system and its computers.   The latter case is the more general in its applications and seems to be becoming the standard solution. The amount of computing required is not much greater than for case c1), and programming is simplified if it is based on the final contour of the workpiece. With straight-line control systems it should be noted that the sign of the correction will depend on the side on which machining takes place. It can be determined on the basis of the displacement conditions (Chapter 10).

In the case of continuous-path controlled 3-D milling machines, which for the most part use hemi-spherical or spherical cutters as shown in Fig. 131 *f* and *d*, the amount of computing required will in any event be larger; the determination of the path to be followed by the centre of the cutter is then made part of the external data processing, and only small ranges of correction are provided at the control consoles (Chapter 11).

As soon as the corrections described in a) to c) above have been calculated the required coordinate values for the positioning device are known. These must be compared with the actual coordinate values. This task is carried out by means of the comparators described in Chapters 3 and 4. All corrections mentioned up to here have been applied to the programmed values as actually required.

Another feasible solution would be to apply the tool dimension correction to the programmed required value and the zero point correction to the instantaneous actual value. The advantages of this solution become apparent if the control system is to be provided with an indicator for showing the actual values; when the tool has been brought into a working position, the corrected actual value will then coincide with the programmed required value, so that it is possible to make a visual check on the correct operation of the numerical control system. Some notes on the principles underlying the design of computer circuits are given in Section 8.5.

8.2   Internal interpolators

In Fig. 8 (Chapter 2), two model system structures were illustrated which contain the limiting cases for the numerical curve generation and interpolation. In the model shown in Fig. 8 B) the entire computation was carried out in the internal data processing system, whereas in that shown in Fig. C) it was transferred to the external data processor.

In practice the trend has been towards the use of control systems with internal interpolators[2]. These special computers are inevitably much more extensive and complex than the correction computers for positioning and straight-line control systems that were referred to in the previous section.

Some years ago the costs of internal interpolators were very high compared with the costs of the remaining parts of the system, including the machine tools. The rapid adoption of micro-electronic techniques for numerical control systems has not only resulted in a reduction in the volume and weight of these systems, but has also increased their reliability and reduced the costs of the individual units. As a result the differences in the price of continuous-path and straight-line control systems for lathes, for example, has become so low that nowadays control systems available for lathes are mainly of the continuous-path type. No matter how the internal interpolators are built up in detail they all have one common factor: There is a close relationship between the time occupied by the calculations and the issuing of the interpolated data on the one hand and the duration of the various technological functions of the machine tool. These on-line systems thus form part of the group of real-time computers.

The principal requirements that must be met by an internal interpolator can be summarised as follows (179):

1. At the end points of the interpolation sections (reference points) continuous interpolation should, in general, not produce linear steps tangential to the continuous curve that is being generated. (However, see the exception in Fig. 102).
2. The amount of data required for the interpolation of a curve (starting and end points, and interpolation parameters such as centres of circles) should be as small as possible.[3]
3. It is necessary for the velocity along the path to be constant irrespective of direction.
4. The curve must pass accurately through the numerically specified reference points to prevent any cumulative errors.

---

[2] Only very recently have there been signs of a trend in the opposite direction. These are discussed briefly at the end of this chapter and in greater detail in Chapter 9.

[3] This condition is of particular importance with reference to the possibilities of manual programming (Chapter 10).

5. The curves produced by the interpolator must approximate as closely as possible to the required workpiece contours.
   Since in practice the outlines of many workpieces consist of a combination of straight lines and arcs of circles it must be possible to interpolate curves of this type accurately and simply.
6. With milling machines it must be possible to take into account changes in the tool diameter, the correction value for which (Chapter 9) is introduced manually into the control system.

The various processes and types of interpolator can now be considered. It is basically possible to design interpolators for any three-dimensional curves that can be calculated. In only a very few instances, however, would the complex control and programming arrangements involved be economically justified. It is, therefore, usual to restrict the interpolation to curves that lie on the principal planes as defined by the coordinate axes. This is referred to as a control system with plane interpolation or as a 2-D continuous-path control system, with no indication as to whether the *XY, XZ* or *YZ*-plane is involved. As has already been shown in Chapter 7, many three-dimensional tasks can be performed by two-dimensional machining processes. Such an arrangement is referred to as a 2½-D continuous-path control system. This is considerably less complicated than the three-dimensional or 3-D continuous-path control system. For simplicity, the remainder of this section will be concerned solely with two-dimensional matters.

Mathematically, interpolation is the process of drawing as smooth a curve as possible through a number of fixed points. For a machine-tool control system, the requirement is to calculate intermediate points between the starting and end points of an interpolation section such that these points lie on a curve of a given function (180). The functions most frequently encountered on engineering drawings are straight lines and circles. One of the major features of these curves is that, if a change of path is required to accommodate a tool radius correction, they are transformed to curves of the *same* type; straight lines are displaced parallel to themselves, while circles only change their radius. There are also interpolators which interpolate parabolic curves. The variations in curvature of these enable smooth curves to be drawn through three fixed points. Interpolators for functions of higher orders are at present of no practical importance, although parabolas of the 3rd order give the most satisfactory equivalent polynomials (including points of inflection).

As has already been mentioned, interpolation can be performed both by digital and by analogue methods; here only digital interpolation will be considered. No matter what the length of the curve section that is to be interpolated, the digital interpolator will, as a matter of principle, have a constant accuracy. This depends primarily on the smallest unit of displacement on which the calculations are based. The advantages of digital techniques to which reference was made in Chapter 3 also apply here. An interpolator is a self-contained functional unit which can take two forms of circuitry. Either it is a permanently-wired circuit which performs the calculations in accordance with an algorithm that remains fixed, or it is a freely-programmable small computer which is then completely integrated into the numerical control system. The advantage of the former is its reduced complexity, and for this reason this type has been almost universally employed in continuous-path control systems up to the present.

The main advantage of a freely-programmable data-processing installation is its great flexibility. In theory it enables the functions which are used for the interpolation to be altered at will. Whether in view of the current trend towards machines which do not employ punched tape (Chapter 9) this alternative offers any major practical advantages still needs to be established.

It is possible to distinguish between two basic methods of operation of digital interpolators:

1. The DDA (digital differential analyser) process, which works from the differential equation of the contour curve.
2. The direct function calculation which depends on the analytical representation of a curve $y = f(x)$.

In the discussion of these processes, attention will, for simplicity, be confined to machining processes that can be solved by means of two-dimensional linear and circular interpolation.

8.2.1 The DDA Process (9, 184)

The analytical representation of a curve $y = f(x)$ does not of itself offer any possibilities for computing in an internal interpolator. It is not possible to make any positive determination of movements parallel to the y axis ($x = $ const.) nor is there any indication of the speed (in the example under discussion the feed velocity) at which the curve is to be generated. It is therefore necessary to introduce the time t as a parameter and to represent the above function as $x = p_1(t)$ and $y = p_2(t)$.

Only in this way is it possible to apply the necessary time and displacement controls to the machine-tool slide, i.e. to apply the necessary control signals to the follow-up control circuits for the various axes.

The internal interpolator thus has to fulfil two functions: In the first place it must interpolate with sufficient accuracy between the prescribed reference points, using a computing procedure that has been prescribed, and secondly it must apply these values of the required positions to the individual machine axes at the correct times.

Assume in the example considered that the displacement measuring systems of a machine tool supply the current positions of the slide as actual values in binary form (for example by means of code discs as illustrated in Fig. 48, Chapter 3); the required values must then be available in the same form, and the R type comparators must continuously determine the numerical differences which are required as regulating variables in the follow-up control loops of the machine. The binary system of numbers offers the advantage of minimum complexity for the general processing of the actual and required values (Chapter 1). At this point only the solution of the interpolation problem is of interest together with the proper distribution at the correct times of the numerical commands to the digital follow-up loops for the two machine axes.

Linear interpolation

In Fig. 193 A) the function $y = f(x)$ is drawn as a straight line of length R between the points P_1 and P_2. If a tool is to be moved along this straight line at constant velocity the components $(x_2 - x_1)$ and $(y_2 - y_1)$ must be traversed during the same time T with a linear time relationship. The movement with respect to time, resolved along the two coordinates, is shown graphically in Figs. 193B) and 193C).

For the solution of the interpolation problem, the intermediate values of x and y as a function of the time t are of importance, where

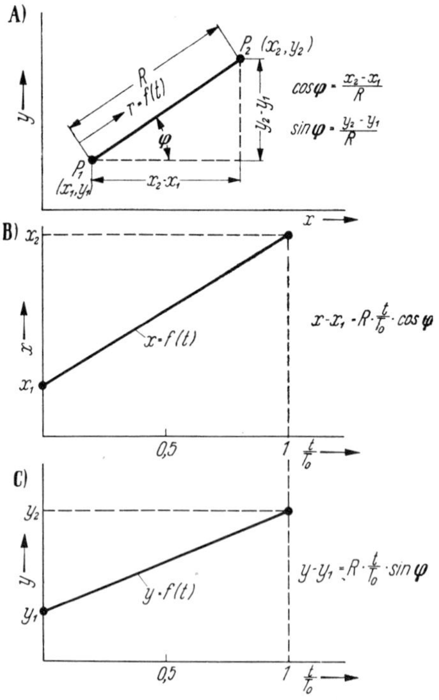

Fig. 193
Diagrammatic representation of the movement occurring with a straight-line tool path.

A) Spatial representation $y = f(x)$. B) and C) Variations with time.

P_1 Starting point of movement having coordinates x_1 and y_1. P_2 End point of movement having coordinates x_2 and y_2. R Total path between P_1 and P_2. $r = f(t)$ Part of path between P_1 and P_2. t Time. T_0 Time taken to traverse the straight section of length R.

$$x_1 \leqslant x \leqslant x_2$$

and $\quad y_1 \leqslant y \leqslant y_2$

The expressions for this can be derived from Fig. 193; they are:

$$x = x_1 + R \; \frac{t}{T_0} \; \cos \varphi = x_1 + \frac{t}{T_0} \; (x_2 - x_1),$$

$$y = y_1 + R \; \frac{t}{T_0} \; \sin \varphi = y_1 + \frac{t}{T_0} \; (y_2 - y_1).$$

If the total time T_0 is expressed as an integral multiple n of the time element t, then

$$T_0 = n \, . \, \Delta t \, .$$

If the orders of the successive quanta are designated by ν, then the following summation formulae are obtained:

$$x = x_1 + \sum_{\nu = 1}^{\nu = n} \left[\frac{x_2 - x_1}{n} \right]_\nu$$

$$y = y_1 + \sum_{\nu = 1}^{\nu = n} \left[\frac{y_2 - y_1}{n} \right]_\nu$$

These summation formulae can be used for digital interpolation. The number of steps between points P_1 and P_2 will be n, and $(n - 1)$ is the number of interpolated points. The relationships are shown graphically in Fig. 194. The summation equations for x and y can be solved fairly easily by digital computers. Since the basic operation consists always in forming the sums of quantified differences, devices of this type are also known as digital difference summers.

The difference $x_2 - x_1$ and $y_2 - y_1$ are obtained by subtraction. The sign of the differences causes the computer to add or subtract as appropriate. The quotients $(x_2 - x_1)/n$ and $(y_2 - y_1)/n$ are obtained with the aid of division commands. It should be noted that this type of calculation becomes extremely simple if n is a multiple of 2, since the division of a binary number by a power of 2 merely involves shifting the decimal point to the left by the appropriate number of places[4]. If a binary number is, for example, divided by $2^{10} = 1024$, the decimal point must be moved ten places to the left. For exact interpolation the computing circuit must be arranged so that account is taken of all the places that follow the decimal point. It is not possible at this point to go further into the construction of computers; information on this can be found in the literature (4, 9, 57, 184, 185, 186).

---

[4] The relationship is similar to that obtaining with a decimal number, where division by a power of 10 merely results in transposition of the decimal point.

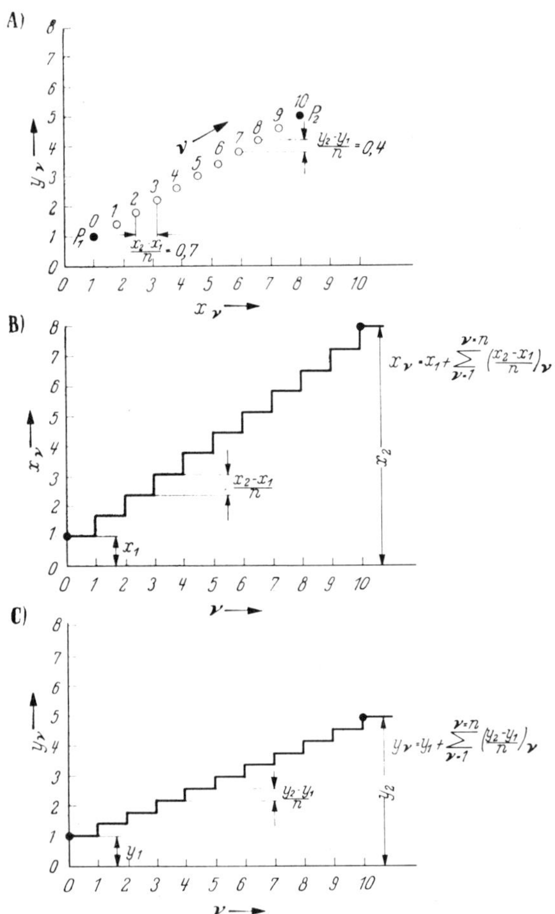

Fig. 194 Example of a linear numerical interpolation (illustration of principles) (184)

A) Spatial representation. B) and C) Variations with time.
P_1 Starting point ($x_1 = 1, y_1 = 1$). P_2 · End point ($x_2 = 8, y_2 = 5$). n Number of finite steps from P_1 to P_2 (in the example $n = 10$). v Order of individual steps.

So far no mention has been made of the technical possibilities of introducing the time function to digital internal interpolators. For this an electronic timing pulse generator is required, the frequency of which can be accurately adjusted and which is used to synchronise the adding stages, i.e. at each timing pulse an increment $(x_2 - x_1)/n$ is added to the existing value for the x axis, and an increment $(y_2 - y_1)/n$ to the existing value of the y axis. The value of n gives the number of steps between the starting point P_1 and the end point P_2. When the n individual steps have been added for each axis the point P_2 will therefore be reached. The interpolation will be mathematically exact if all the places following the decimal (binary) point are taken into account. The timing pulse frequency determines the rate at which the steps $\nu = 1, 2, 3, \ldots 10$ are covered (in the example shown in Fig. 194A). In this particular case this means that the resultant feed velocity of the machine tool is determined by the timing pulse frequency f_t. The following point must, however, also be borne in mind: According to paragraph 2 of the summary of design requirements outlined earlier, there should be a free choice of reference points, so that the distance between them is not constant. The resultant feed velocity is:

$$V = V_x{}^2 + V_y{}^2, \quad \text{where}$$

$$V_x = \frac{x_2 - x_1}{n} \cdot f_t \quad \text{and}$$

$$V_y = \frac{y_2 - y_1}{n} \cdot f_t$$

If the resultant feed speed V is kept constant for technological reasons, and the denominator n is kept constant for ease of computation, the above relationship means that the pulse frequency f_t must vary with the distance between the reference points[5].

$$f_t = \frac{n \cdot V}{(x_2 - x_1)^2 + (y_2 - y_1)^2}$$

With this process it is, therefore, necessary to provide an additional computing circuit for adjusting the timing pulse frequency to correspond with the distance between the reference points, assuming that the programme condition 'free choice of reference points' is to be met.

If the coordinates of the reference points and the computed steps are available in numerical form (e.g. as binary numbers), all the intermediate values will also be supplied by the interpolator in numerical form as commands to the follow-up control loop circuits in each axis of the machine. As has already been mentioned, the comparators must be of the numerical proportional type described in Chapter 3 (Fig. 51).

The digital-absolute method is accurate and reliable, but relatively complicated. In many instances the linear interpolator is, therefore, designed so that it produces the commands, but in the form of pulses which correspond to the displacement elements of the incremental displacement measuring systems (187, 188). The magnitude of the displacement elements in both axes is then constant and the pulse frequencies must be different for

---

[5] This statement is true for 'synchronous interpolation'. With 'asynchronous interpolation', the next computing cycle is called up from the follow-on control circuits when a given synchronising error is exceeded.

each axis to suit the values of the velocity components V_x and V_y; or alternatively the method described in Section 8.2.2 is adopted. If the pulse technique is employed the comparators should take the form of counters as shown in Fig. 43. This will reduce the cost, but against this the incremental process with its disadvantages (Chapter 3) has to be accepted. In all cases the output values of the digital interpolator are converted into analogue values in D/A converters (189, 190); the follow-up control loops can then be of the analogue type.

Circular interpolation

If the general case of a circle in a given coordinate system as shown in Fig. 195 is considered it is possible to write:

$$x = x_M + R \cos \varphi$$

$$y = y_M + R \sin \varphi$$

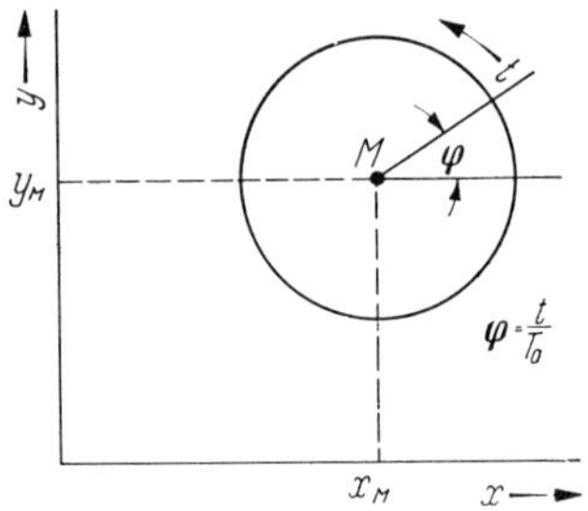

Fig. 195
Diagram illustrating some of the terms employed in circular interpolation.

M Centre of circle. T_0 Time to traverse unit arc ($\varphi = 1$ radian $= 180°/\pi$). $t =$ time (independent variable). x_M and y_M Co-ordinates of centre of circle for general position of circle. φ Angle at centre (in radians).

If φ is expressed in radians and T_0 is the time required to traverse the unit arc, then

$$\varphi = 1 \ \widehat{=}\ 180°/\pi = \text{approx. } 57.3°,$$

and it is possible to write

$$x = x_M + R \cos \frac{t}{T_0}$$

$$y = y_M + R \sin \frac{t}{T_0}$$

where t is the time and is an independently variable parameter. If the two equations are differentiated with respect to time, then

$$\frac{dx}{dt} = -\frac{R}{T_0}\sin\frac{t}{T_0} = -\frac{y - y_M}{T_0};$$

$$\frac{dy}{dt} = +\frac{R}{T_0}\cos\frac{t}{T_0} = +\frac{x - x_M}{T_0}$$

If an interpolation formula having general validity for time as an independent parameter is to be derived from this, then for infinitely small increments of time:

$$dx = -\frac{y - y_M}{T_0}\,dt$$

$$dy = +\frac{x - x_M}{T_0}\,dt.$$

For numerical processing, the differential dt must be replaced by the finite difference Δt. If $n = \dfrac{T_0}{\Delta t}$ is the number of steps required to traverse the unit arc $\varphi = 1\ (=57.3°)$, then

$$\Delta x \approx -\frac{y - y_M}{T_0}\,\Delta t$$

$$\Delta y \approx +\frac{x - x_M}{T_0}\,\Delta t$$

$$t = v\,\Delta t$$

$$v = 1, 2, 3 \ \dots\ \mu\ \dots\ v_2$$

where v is the order number of the time step (progressively changing variable)
μ is the instantaneous value of the stepwise change in v
v_2 is the last step in time.

Then the values of x and y at any time, starting from the initial values x_1 and y_1, are given by

$$x = x(t) = x(\mu) = x_1 + \sum_{v=1}^{v=\mu}\Delta x$$

$$y = y(t) = y(\mu) = y_1 + \sum_{v=1}^{v=\mu}\Delta y$$

$$x = x(\mu) \approx x_1 - \sum_{\nu=1}^{\nu=\mu} \frac{y\,(\nu-1) - y_M}{n}$$

$$y = y(\mu) \approx y_1 + \sum_{\nu=1}^{\nu=\mu} \frac{x\,(\nu-1) - x_M}{n}$$

Here ν_2 is both the last step in time and also the order number of the final step required to move along the arc to reach the point P_2 from P_1. In the case of linear interpolation n is the number of steps required to reach the end point, whereas in the case of circular interpolation as described here n is the divisor for unit arc and has no connection with the point P_2 as the end point of an arbitrary circular arc.

The special features of the interpolation process are clearly shown in Fig. 196; the geometrical relationships also explain the equations given above. In Part A) of the figure a single step $\Delta\varphi = 1/n$ is first considered in more detail. Suppose the point B of order number ν lies precisely on the circle of radius R_ν. The first geometrical position for the point D, which is of order number $\nu + 1$, is on the radius $R_{\nu+1}$, which is displaced by the angular element $\Delta\varphi$ from R_ν. The second geometrical position for D is on the tangent to the circle at point B. The arc BE is thus replaced by the tangent BD. The point D no longer lies on the circle. The error ED becomes less the smaller $\Delta\varphi$ is, i.e. the larger the value of the divisor n. In the limit $n \to \infty$ the summation formulae given above become integrals and only then is the process mathematically exact. Since digital summers are only capable of dealing with finite numbers, further efforts must be devoted to keeping low the errors which arise as a result of the approximations, without involving excessive technical complexity (n must not be too large).

Before discussing an improved interpolation formula it seems advisable, however, to take a closer look at the existing summation formulae with reference to an example. In Figure 196A), triangle DCB is similar to triangle MAB; the linear reduction is $1/n$. The base and perpendicular height of the right-angled triangle DCB thus represent the increments along each axis for the νth position, corresponding to the above interpolation formula[6].

Figure 196B) shows the geometrical construction for the case $n = 2^3 = 8$ ($\Delta\varphi = \frac{1}{8}$ = approx. 7,2°) and $\nu = 10$. The total angle subtended at the centre by the line $P_1 - P_2$ is then $\varphi = 10 . 7{,}2° = 72°$.

---

[6] Since the construction of the tangent must always be carried out from the preceding point to arrive at the next point, the index for x and y following the summation sign must be $\nu - 1$.

A)

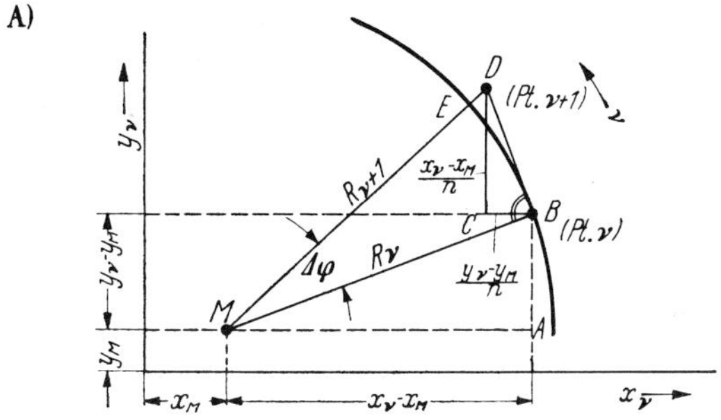

B)

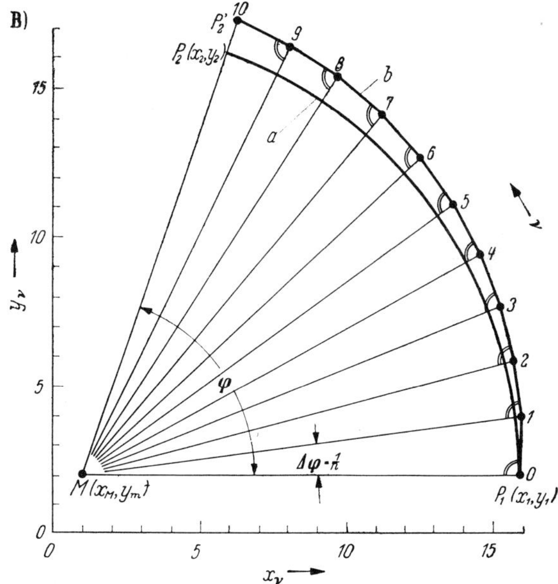

Fig. 196 Example showing the principles of circular numerical interpolation (184)

A) Geometrical representation of a simple numerical interpolation formula.
B) Example of the use of the simple interpolation process.

a Required curve (circular arc). *b* Substitute curve. *n* Divisor for the unit arc; *n* = Number of steps to produce unit arc (in the example $n = 8$). *M* Centre of circle with coordinates $x_M = 1$ and $y_M = 2$. P_1 Desired starting point of circular arc with coordinates $x_1 = 16$ and $y_1 = 2$. P_2 Desired end point of circular arc with coordinates $x_2 = 6$ and $y_2 = 16$. φ Angle at centre of circle between P_1 and P_2 (in the example $\varphi = 72$) $\Delta\varphi$ constant angular increment (angular element) *v* Order number for individual interpolation steps (in example $v = 10$).

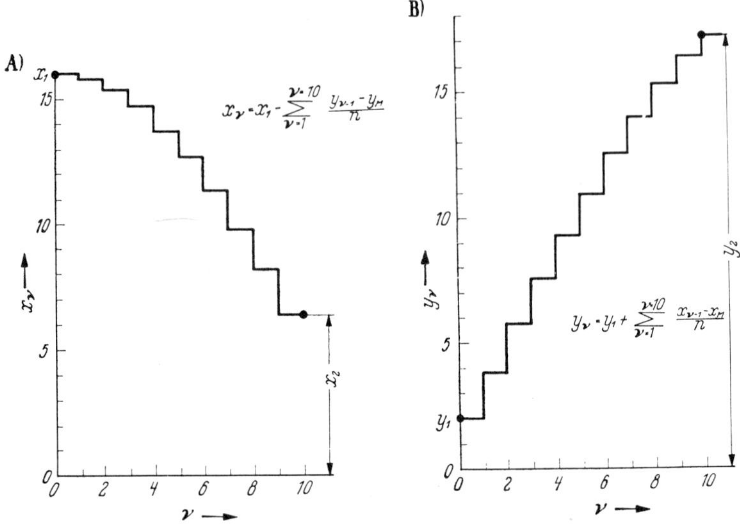

Fig. 197 Diagram showing the interpolation process in quantified time sequence (corresponding to Fig. 196)

A) Numerical changes in x. B) Numerical changes in y.

Figure 197 shows the stepwise changes of x_ν and y_ν as calculated by the computer; the adders are again controlled in synchronism by a timing pulse generator having a frequency of f_t, as with linear interpolation. As is also shown in Fig. 196 B), the end point P'_2 of the summation calculation is not on the desired circular arc; the error is represented by the distance $\overline{P_2 P'_2}$.

As has already been mentioned, the error can be kept down to any desired value by making the divisor n large, the complexity increasing considerably because of the large number of places after the decimal point that have to be taken into account. If the desired radius is R_0, and the 'incorrect' radius at the νth position is R_ν, it is possible to write for small values of $\Delta\varphi$ ($\Delta\varphi < 4°$):

$$\frac{R_\nu}{R_0} = 1 + \frac{\nu}{2}\left(\frac{1}{n}\right)^2. \quad [7]$$

Since $\nu_2 = \varphi\, n$ steps are required for a total angle at the centre of φ (in radians), the error after ν_2 steps amounts to

---

[7] From Fig. 198 it follows that $R_0/R_1 = \cos \Delta\varphi = \sqrt{1 - \sin^2 \Delta\varphi}$. Hence the above formula for small values of $\Delta\varphi$

$$\frac{R_{\nu 2}}{R_0} = 1 + \frac{\varphi}{2n}$$

The radial error therefore decreases only as the reciprocal of n, so that the divisor n for the unit arc must have a large numerical value if the desired accuracy is to be achieved by means of the procedure so far described. In practice, therefore, a different approach is adopted, the geometrical principles being illustrated in Fig. 198. Let the desired circular arc a have the radius R_0. If one proceeds from A (having the order number ν) by an angular step $\Delta\varphi$ using the previous interpolation formula (Fig. 196), one arrives at point B, and by halving the distance $\overline{AB}$ at point C. If the previous interpolation formula and the corresponding geometrical construction is again employed, starting at point C, one arrives at point D. If a line parallel to $\overline{CD}$ is marked off along it from A, one obtains point E which is the corrected interpolation point of order number $\nu + 1$. Expressed algebraically, the relationship for a circular arc which is to be interpolated with ν_2 steps and which has the initial coordinates $x_1,\, y_1$ (184) is:

$$x = x_1 - \sum_{\nu = 1}^{\nu = \nu_1} \left[\frac{y_{\nu-1} - y_M}{n} \right] + \left[\frac{x_{\nu-1} - x_M}{2n^2} \right]$$

$$y = y_1 + \sum_{\nu = 1}^{\nu = \nu_2} \left[\frac{x_{\nu-1} - x_M}{n} \right] - \left[\frac{y_{\nu-1} - y_M}{2n^2} \right]$$

The additional terms can be determined fairly easily, so that sufficiently accurate results can be attained using this improved interpolation formula

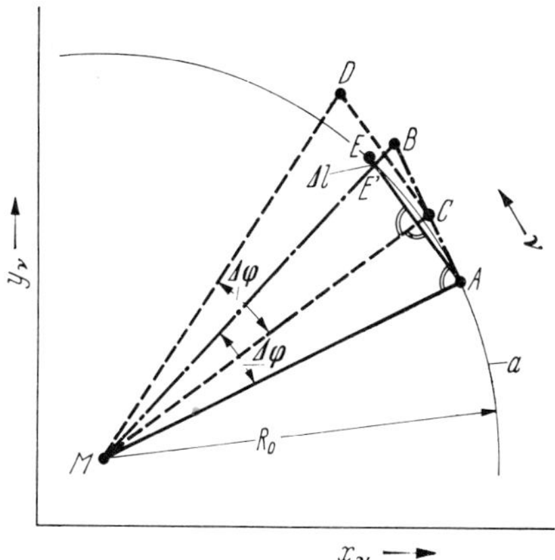

Fig. 198
Example of an improved circular interpolation (with correction term) (184)

a Desired curve (circular arc) of radius R_0. A Point having order number ν. E Point having order number ($\nu + 1$). Length $E'B$ Deviation from desired radius R_0 using the interpolation shown in Fig. 196. Length $\overline{MB} = R_1$. Lengths $\overline{AC} = CB = \tfrac{1}{2}\,\overline{AB}$. Length CD = Length $\overline{AE}$ and CD is parallel to $\overline{AE}$
$\Delta\varphi$ = angular step = $1/n$, where n is the divisor for the unit arc.

without the need for excessive values of n. The ratio of the radii is now

$$\frac{R_v}{R_0} = 1 + \frac{v}{8} \ (\frac{1}{n})^4.$$

i.e. the error is now only $\frac{1}{4} \cdot (\frac{1}{n})^2$ - times that which is given by the simple formula.

As these discussions have shown, the amount of computation required is larger for circular interpolation than for linear interpolation. The examples have, however, also shown that digital methods offer a number of possibilities for increasing the accuracy of computation. If for any reason analogue displacement measuring systems and comparators are used in the follow-up control circuits of the machine it will, of course, also be necessary with circular interpolation to insert D/A converters between the output of the digital interpolator and the input to the analogue comparator (189, 190). Figure 199 (192) shows the block diagram of a computer circuit which will perform both linear and circular interpolation by the DDA process. The circuit contains three multipliers (a) of which one merely reverses the sign, two adding circuits (b), and two registers for storing x and y. The paths followed by the signals can be followed with ease from the descriptions that have been given for the methods of operation of linear and circular interpolators. The transition from one type of interpolation to the other is effected simply by operating a changeover switch; this can of course also be controlled by the punched tape in accordance with the programme (Chapter 10).

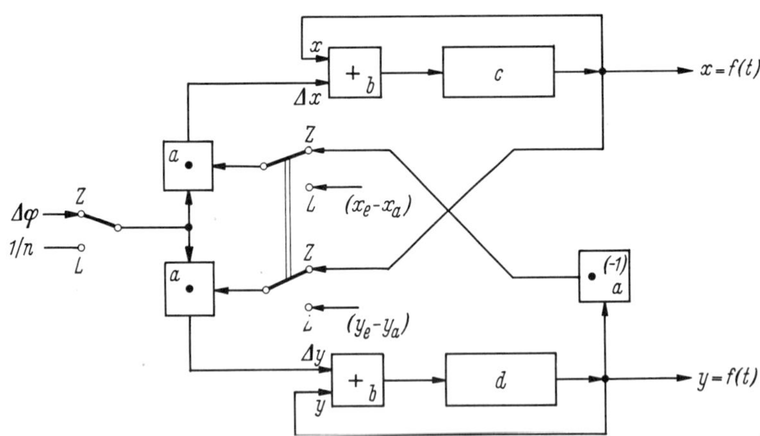

Fig. 199 Block circuit diagram of an interpolator for either linear or circular interpolation using the DDA process

a Digital multiplier. b Digital adder. c Register for x values. d Register for y values. Z Switch position for circular interpolation. L Switch position for linear interpolation.

8.2.2 *Direct calculation of function* (191)

The interpolation process, in which the interpolation function is calculated directly, uses constant step lengths in place of the variable increments employed in the DDA process. Here again the discussion will be limited to linear and circular interpolation in one plane.

If one starts with a curve which is capable of being described by a mathematical function, then for a curve in the x-y plane the analytical expression is $y = f(x)$, or in different terms $F(x,y) = y - f(x) = 0$ (Fig. 200).

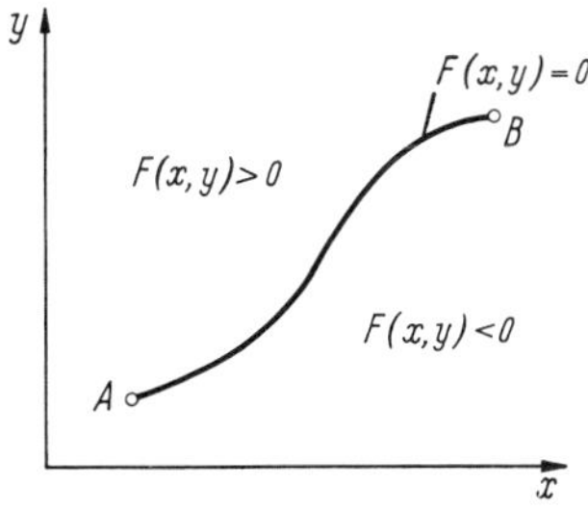

Fig. 200
Definition of the function $F(x, y)$

A Starting point, B End point of interpolation section.

The condition $F = 0$ is satisfied only by points that lie on the curve. If the point is located above the curve, F will be greater than zero; if the point lies below the curve, F will be less than zero. The sign of F thus gives an indication of the location of the points. If the function that describes a curve is known (e.g. straight line, circle, etc), the curve can, within limits, be generated by iteration using this principle; this accounts for the term 'curve generator' being used occasionally in connection with this method. The criterion can be used for the step-by-step production of curve sections provided that the curve has no maxima or minima other than at the starting or end points. The principles will be described in greater detail using the simple case of a straight line as an example.

Figure 201 shows a straight line running from point A to point B. Let point A be the starting point of the interpolation section. Let the length of a step be 1, the actual units of length being of no importance for this discussion of the fundamental principles. The step lengths that are used in practice range from 0.01 mm down to 0.001 mm. Since point A lies on the curve F will have the value zero. The first step can therefore be made in any direction. Since the interpolator requires an unambiguous decision even for $F = 0$, let the first step from the value zero in general have a positive sign. If the first step is in the $+x$ direction, the end point of the step will be below the curve; as a result F will be less than zero. The negative sign of F requires a movement of one unit in the positive y direction in order to bring the function $F(x,y)$ as rapidly as possible back to zero (step 2). Since the sign of F is now positive again, the third step will again be in the positive x direction. As a result the sign of F

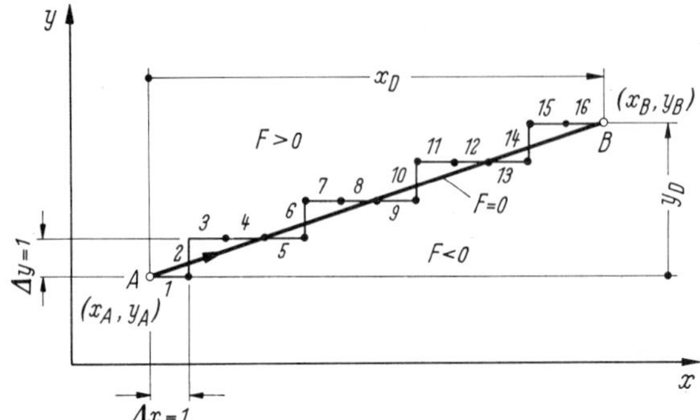

Fig. 201 Iteration procedure for a curve during interpolation using constant displacement increments (direct calculation of function) (191) cf. also Table 13.

$A(x_A, Y_A)$ Starting point. $B(x_B, Y_B)$ End point of interpolation section. F Function (in this example a straight line L). x_D Length difference in x coordinate direction. y_D Length difference in y coordinate direction. $\Delta x = \Delta y = 1$ Unit step (0.01 to 0.001 mm).

remains positive, so that the fourth step will also be in the x direction. These steps continue until the end point B is reached. The result is a stepped curve, the lengths of the individual portions of which are integrals of the step length. The maximum deviation of this curve in the normal direction from the straight line that is to be interpolated will always be less than one unit step. In practice, the steps are not noticeable on the finish-machined workpiece, since the mechanical features of the drive, the machine tool, and the geometry of the cutter all have an integrating effect as a result of which the minor discontinuities in the digitally generated curve are completely masked.

The mathematical description of the operation that has been described yields the following equations:

The equation of a straight line through the point A is $y = mx$ if y and x are referred to the point A and not to the origin of the coordinates. The expression for the slope m is

$$m = \frac{y_B - y_A}{x_B - x_A} = \frac{y_D}{x_D}$$

The expression that was defined above, namely $F(x,y) = y\text{-}mx = 0$ can therefore also be expressed in the following way as the equation for a straight line L

$$L(x,y) = x_D.y - y_D.x = 0.$$

As described in the discussion of Fig. 200, the sign of this difference will determine the further iteration of the straight line as progress is made one unit step at a time in the x or the y direction.

With movement equal to 1 unit parallel to the axis of length the corresponding differences ΔL_x and ΔL_y are given by the above equation. The value for a unit step in the x direction is

$$\Delta L_x = L(x + 1,y) - L(x,y) = -y_D.$$

and the value for a unit step in the y direction is

$$\Delta L_y = L(x,y \pm 1) - L(x,y) = +x_D.$$

No matter what the direction of the step, the values

$$\Delta L_x = -y_D \text{ or } \Delta L_y = +x_D \text{ are added to } L(x,y) \text{ at each step.}$$

The direction of the next step is determined solely by the sign of the sum.

The number of steps needed to reach the end point of the interpolation section, which is the same as the number of additions required, is given by the equation

$$n = \frac{x_D + y_D}{\text{unit step}}.$$

The values for the simple example shown in Fig. 201 are tabulated in Table 13.

The speed with which the curve is generated is determined by the number of additions in unit time. One of the major requirements for the computer of the appropriate interpolator is, therefore, that its cycle frequency should be externally controllable within wide limits; there was a similar requirement for synchronous interpolation in the case of the DDA process.

Figure 202 shows the block diagram of a computing circuit that will perform the operations which have been described (191). It includes a pulse generator i, two registers a and b for the lengths x_D and y_D, an adding element c for positive and negative values; a sum register d for the values of the function $L(x,y)$, a sign evaluator e, and a logic circuit for distributing the pulses produced by the generator. The action of the various elements can be followed by studying the description that has been given for the function of the equipment. If required the sums of the partial steps Δy and Δx can also be formed in separate registers g and h for checking purposes.

The interpolation of circular arcs can also be performed in accordance with the principle of direct calculation of the function. Before the circle interpolator is described brief reference must, however, be made to a compromise solution for curve interpolation. In general it is true that a linear interpolation is simpler and requires less equipment than any other form of interpolation. For this reason interpolators have been developed in which continuous curves are replaced by a series of straight lines, and linear

Step No.	$\Delta L_x = -y_D$	$\Delta L_y = +x_D$	Register content $L(x,y)$ in part d in Fig. 201
1	−4		$0 - 4 = -4$
2		+ 12	$-4 + 12 = +8$
3	−4		$+8 - 4 = +4$
4	−4		$+4 - 4 = 0$
5	−4		$0 - 4 = -4$
6		+ 12	$-4 + 12 = +8$
7	−4		$+8 - 4 = +4$
8	−4		$+4 - 4 = 0$
9	−4		$0 - 4 = -4$
10		+ 12	$-4 + 12 = +8$
11	−4		$+8 - 4 = +4$
12	−4		$+4 - 4 = 0$
13	−4		$0 - 4 = -4$
14		+ 12	$-4 + 12 = +8$
15	−4		$+8 - 4 = +4$
16	−4		$+4 - 4 = 0$
Number of interpolation steps	$n_x = 12$	$n_y = 4$	

Table 13. Interpolation values for the example shown in Fig. 201
($x_D = 12$, $y_D = 4$)

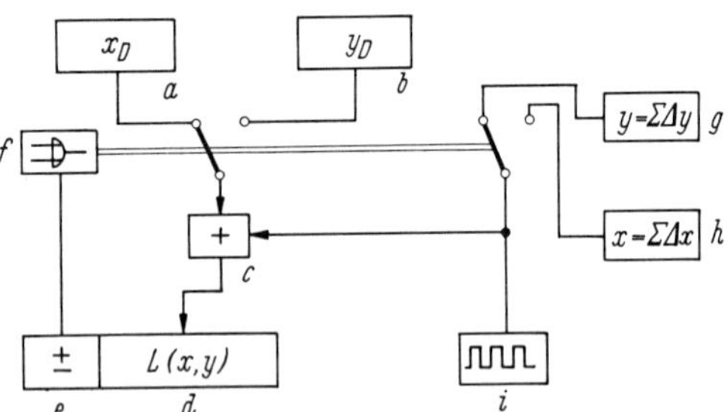

Fig. 202 Block circuit diagram of a linear interpolator for direct calculation of a function (191).

a Register for setting x_D. *b* Register for setting y_D. *c* Adding element for positive and negative values. *d* Sum register for the value of the function $L(x,y)$ *e* Sign register. *f* Logic circuit for evaluating signs in accordance with function. *g* Sum register for the *y* coordinate. *h* Sum register for the *x* coordinate. *i* Pulse generator (see also Fig. 205)

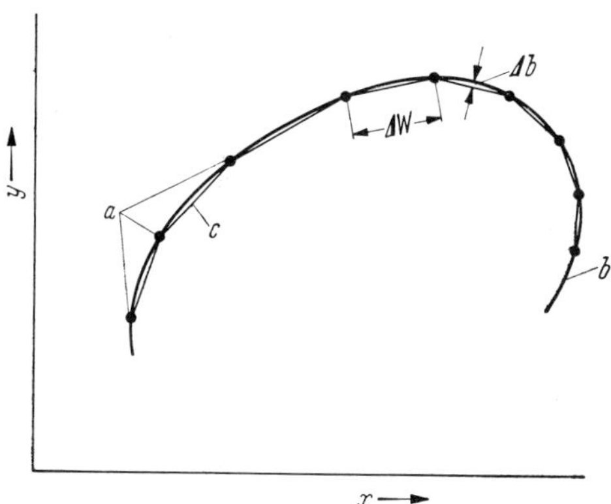

Fig. 203 Replacement of a curve of any shape by a series of straight-lines with linear interpolation.

a Reference points. *b* Desired curve. *c* Interpolation polygon (series of straight-lines approximating desired curve). Δ*W* length of line between two reference points. *x* and *y* Coordinate axes for two-dimensional representation. Δ*b* Permissible deviation between desired curve and straight-lines (tolerance field).

interpolation is performed between the points of intersection of these lines. The lengths Δ_W of the straight lines between the reference points (Fig. 203) will then vary, according to the curvature of the arc, so that the permissible deviation Δ_b from the required curve is not exceeded. If the steps are small the fact that a series of straight lines is used in place of a continuous curve is not apparent on the workpiece, since the digital control of the machine is in any event performed by finite steps, and with linear interpolation the curve is smoothed out by the integrating properties of the tool and of the machine and its drive. It is of course equally possible to use tangents or secants as the linear elements of the curve, and not merely chords as illustrated for simplicity in Fig. 203. Circular interpolation using constant advance values proceeds basically in the same manner as linear interpolation with the difference that the direction of motion is governed by the circle equation instead of the straight-line equation. If the centre of the circle coincides with the origin of the coordinates the implicit form of the circle equation for a circle in the *x-y* plane will be:

$$C(x,y) = x^2 + y^2 - r^2 = 0$$

Here again the changes in the function ΔC that result from movement by one unit in the *x* or the *y* direction are to be calculated (Fig. 204).

For a step in the *x* direction:

$$\Delta C_x = C(x_A + 1, y_A) - C(x_A, y_A) = +2x_A + 1 :$$

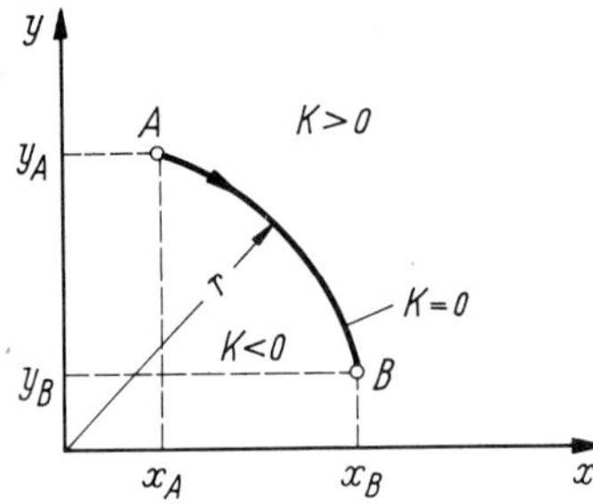

Fig. 204
Definition of the parameters for circular interpolation with direct calculation of a function.

Note: The arrow on the curve shows the direction of interpolation; in this case clockwise (Chapter 10).

For a step in the y direction:

$$\Delta C_y = C(x_A, y_A + 1) - C(x_A, y_A) = + 2y_A + 1.$$

These equations can be evaluated with ease by digital equipment if the values of x_A and y_A are given in binary form (192), since multiplication by two then corresponds to moving the point in the binary number one place to the left. One point that must be borne in mind regarding the results is that, as in linear interpolation, the sign of $C(x_A, y_A)$ is of greater importance than its precise value. It is therefore sufficient to calculate $\Delta C_x = + x_A + \frac{1}{2}$ for the change in function. If the equation is simplified still further to the form

$$\Delta C_x = + x_A \ .$$

this is equivalent to displacing the circle, and an error of magnitude equal to half a step unit is introduced. In general this can be neglected in view of the small individual step sizes which range from 0.01 to 0.001 mm. This simplified calculation can, however, be performed very easily by the computing circuit of the linear interpolator. It is only necessary to interchange x and y and to reverse the sign of x. (191). The restriction mentioned above, namely that within the interpolation region, apart from the starting and finishing points, there must be no maxima or minima of the curve, also applies. For this reason circular interpolation must always be restricted to a quadrant. Or in other words, if a circular arc extends beyond a quadrant, a separate interpolation will have to be performed for each quadrant.

The maximum deviation of the step curve that is generated from the desired circular arc will always be less than one displacement unit; the end point of an interpolation section is reached exactly.

8.2.3   *Comparison of interpolation processes* (191, 192)

Any comparison of the two interpolation processes must be based on the requirements for internal interpolators, that were given at the beginning of this section. The accuracy of interpolation is approximately the same for the two processes. When the direct calculation of the function is used the prescribed end point of an interpolation section is always reached exactly; whereas in the case of circular interpolation in a plane using the DDA process a spiral is in fact produced. The deviation from a circle can, however, be maintained as small as required.

Changes in the dimensions of the tool are allowed for by calculating the new end points and interpolation parameters of a curve section before the actual interpolation is performed. With both processes the method of compensating for changes in the tool diameter has thus no effect on the interpolation. Details regarding the method of correcting for changes in the tool diameter for continuous-path control systems will be found in the literature (190, 194).

The only major changes in the two types of interpolation are in the speed of movement along the curve. It is a basic feature of the DDA process that this is maintained constant over the whole of the curve section for both linear and circular interpolation. With interpolation using equal advance values the speed of movement along the curve depends on the slope of the curve. In unit time the interpolator feeds a given number of displacement increments into the follow-up control loops. The programmed speed along the curve v_0 is

$$v_0 = \frac{n \cdot a}{T}$$

where n = Number of steps
  a = Length of steps
  T = Time unit.

If the curve is parallel to one of the coordinate axes, then all the displacement increments are in this direction, and the actual speed along the curve, v, is equal to the programmed value

$$v = v_0$$

If, on the other hand, the curve is a straight line inclined at 45° to the axes, the displacement elements are applied alternately to the two follow-on control loops. The actual speed along the curve is then given by the equation:

$$v = \sqrt{\left(\frac{n \cdot a}{2T}\right)^2 + \left(\frac{n \cdot a}{2T}\right)^2} = \frac{v_0}{\sqrt{2}}$$

As far as circular interpolation is concerned this means that the speed along the curve varies between v_0 and $v_0/\sqrt{2}$. In the case of a three-dimensional curve the speed would even vary between v_0 and $v_0/\sqrt{3}$. These variations in the speed can be compensated for at the expense of additional interpolator complexity.

To keep the circuits in the interpolator as simple as possible, thereby keeping its costs low, it can interpolate a curve only over one quadrant at a time. Interpolators for both processes show no major differences in complexity. In practice the DDA process is, however, more widely adopted than the process that directly calculates functions and uses constant displacement increments.

8.3   Thread cutting

Apart from the special case of the turret lathe, the new NC techniques have been adopted only comparatively lately for lathes. It was not until about 1963 that the preliminary work, which had not previously been published, had progressed to the stage where it seemed likely to prove successful, and was first made public in the USA (166). The reasons neither the machine-tool manufacturers nor the production engineers have previously pressed for the adoption of NC techniques on lathes are interesting; basically there had been such a widespread use of other more conventional devices to enable lathes to operate automatically that these were far in advance of all other types of machine tool in this respect. Any new methods would therefore have to offer major advantages if they were to be widely adopted. The well-known 'auto', which is a type of lathe that operates without the direct intervention of an operator, had been developed more than 60 years earlier; the advantages of turrets had long been put to practical use, and since the early 1930's copying lathes had been developed to the stage where they proved economical in use for batches of 5 or more workpieces. It appeared that no further advantages were to be gained by the direct use of NC techniques; mention has already been made in Chapter 7 of the indirect use of these methods in tool-making.

It was not until special aspects of numerical control had been developed and perfected, together with an analysis of the weak points in the use of lathes, that a break-through in this field appeared likely. With hindsight it is possible to say that there were two basic reasons why the NC lathe was able to attain the important position which it holds at present (93, 166), and which will in all probability ensure that it becomes even more widely used in the future. These are:

1. The increasing simplification and reduction in cost of continuous-path control systems, and
2. A satisfactory solution to the problem of thread-cutting with the use of electronic control equipment.

It was obvious that in view of the many alternative solutions NC techniques would find successful application on lathes only if the universal lathes generally available in 1962/63 (166) could be made even more universal and flexible in operation. The reduction in the cost of 2-D continuous-path control systems, which is still continuing, enabled the problem of the simple turning of internal and external contours to be solved; the dispute as to whether NC lathes should be equipped with straight-line or continuous-path control systems has now faded into history. The production engineers also know fully well that thread cutting on lathes using single-point lathe tools is expensive, and that it can be considered only for one-off and small-batch production (7, 212, 215); there are too many fully-developed alternative solutions in daily use (thread chasers, dies, self-opening die heads, taps, etc.) (212, 214). Right from the start one of the major objectives was therefore the simplification of thread cutting with the use of electronic control equipment and also provision for this operation being more flexible than in the past, to enable it to cope with the wide variety of threads that will inevitably be encountered.

Basically the problem is very simple and indeed can be solved quite easily if a numerical continuous-path control system is available; it is only necessary to arrange for the control system being readily adaptable to suit the particular features of the many different designs of lathe. It is of course understood that the reliability of the control systems must be high, as indeed it must be for all numerically controlled machines; with one-off and small batch production scrap is not acceptable, nor are idle times due to defects in the machine tool or its control system. Without going into unnecessary details, an attempt will be made, using Figs. 205 and 206, to explain the basic problem and the methods of solution that are possible in terms which will be understood by readers of varying background and experience. Figure 205A) first shows the characteristic features of the mechanical solution. The positive connection between the rotation of the workpiece, containing item b, and the feed motion of the lathe tool in the tool holder k is of fundamental importance. Since the pitch of the leadscrew f is one of the machine constants, steps must be taken to ensure that the feed gearbox d can be operated rapidly and easily to meet all possible variations in the thread specification (pitch, multi-start, etc) so far as this is practicable without in any way diminishing the positive nature of the transmission from workpiece to leadscrew. There are many possible solutions (pick-off gears, Norton-box, sliding gears, etc) and these need not be described here, since they are covered extensively in the literature (e.g. 7, 8, 212); mechanical designs of this type are either time-consuming to operate or expensive.

Figure 205B) illustrates the electronic solution. As has already been mentioned, this is very simple in principle if an internal interpolator is available. It has already been shown in Section 8.2 that for all the digital solutions described, a pulse generator will in any event be required. If this is located on the main spindle b or l, the positive relationship between the feed rate and rotational speed of the workpiece will be assured, and the problem is solved in principle. The previous free movement of the machine slide in mm/min or mm/s becomes a constrained movement mm/rev or inch/rev.

Photo-electric or magnetic pulse generators of the rotary type have already been described in Chapter 3; the same designs are basically suitable for use in this application. In principle it makes no difference whether a mechanical reduction gear of ratio $1 : n$ is interposed between the machine element l and the pulse generator m or not; the only important requirement is that the pulse frequency at the output of the pulse-processing chain is suitable for the internal interpolator, and that this can be adapted to the requirements of the programme by means of the device t which is controlled by the input signal u. The device t is thus in a sense the electronic equivalent of the mechanical gearbox d. In the device t the thread pitch can be programmed in steps of

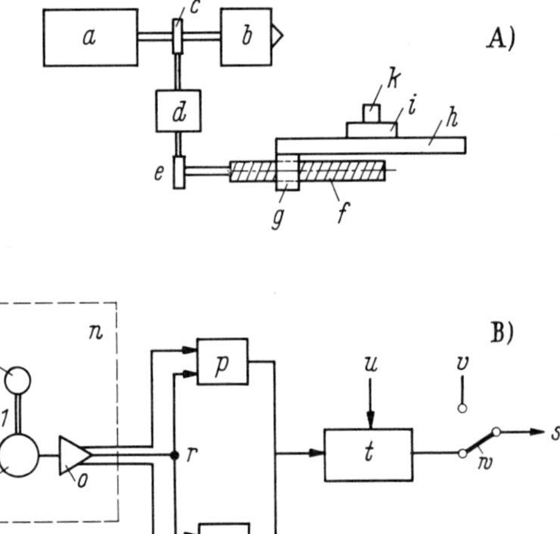

Fig. 205 **The principles of thread-cutting on a lathe**

A) Mechanical solution (7). B) Electronic solution (211).

a Drive. *b* Gearbox for main spindle. *c* Reduction gear to feed drive (subsidiary drive). *d* Gearbox with a wide range of gear ratios which can be selected manually (pick-off gears, sliding gears, Norton gearbox, etc). (212) *e* Gear drive to leadscrew *f*. *f* Leadscrew having fixed pitch (usually Acme thread with pitch between 3 and 24 mm or ½ inch). *g* Leadscrew nut (usually hinged). *h* Longitudinal (sliding) carriage on machine. *i* Cross (surfacing) slide on machine (can be moved perpendicularly to the plane of the drawing). *k* Tool holder with tool. *l* Main spindle. *m* Digital pulse generator. *n* Lathe housing. *o* Amplifier and impulse emitter stage. *p*-Unit for processing rapid pulse trains. *q* Unit for processing slow pulse trains. *r* Indexing pulses (1 pulse per revolution of spindle *l*). *s* To interpolator, pulse frequency corresponding to Fig. 202. *t* Function generator for frequency selection. *u* Adjustment of desired thread pitch (frequency). *v* Pulse frequency for normal interpolation (e.g. from unit *i* in Fig. 202). *w* Changeover switch operated by G functions (Chapter 10).

0.01 mm and less. As a result the electronic equipment has a much wider range of adjustment than a mechanical gearbox and is easier to control; it may also be cheaper, at any event in the future. The ease with which it is possible to control the electronic system enables it, if designed accordingly, to meet such requirements as suitability for inch, metric, or modular threads, multi-start threads, continuously-variable pitch, taper threads, transition from longitudinal turning to facing, e.g. changeover from thread cutting to the production of a spiral on a plane surface, etc. If it is borne in mind that the pulse generator *m* can be switched off merely by operating the switch *w* in order to use the machine for the production of any desired forms by means of the normal interpolator control system it will readily be understood that these facilities of the numerical-control system make it possible to produce universal lathes which are suitable for very flexible service conditions and can be used side-by-side with the well-known automatics (Chapter 12).

The principal operations are illustrated in Fig. 206 in such a way that they can be followed without any detailed programming knowledge; more information will be given in Chapter 10. The characteristic tool motion cycle illustrated in portion B) of the figure, which covers the operation "First cut of thread – withdraw tool – fast traverse return – feed-in tool for second cut of thread – withdraw tool – fast traverse return, etc." can either have the individual operations programmed separately, or can be called up as a complete cycle, depending on the way in which the machine is equipped; this will be discussed further in Chapter 10.

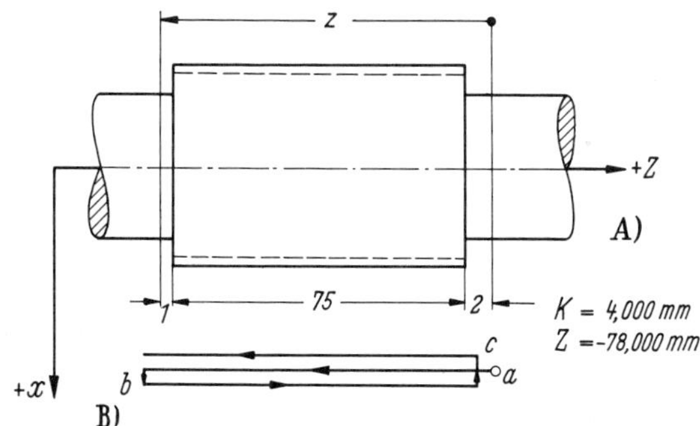

Fig. 206 Example of thread cutting on cylindrical workpiece using an NC machine (213, 214)

A) Workpiece. B) Tool motions.

X,Z Coordinate axes. *K* Address for thread pitch in Z axis (Table 12 section 7.3), address letters for programming (Chapter 10).

a Start point of cycle. *b* Tool withdrawal after first cut of thread. *c* Start of second cut of thread.

8.4   Optimisation of technological working conditions within machines
(Adaptive control, AC)

The effects of the developments in computers that have been described are leading to development trends in the numerical control systems for machine tools which, in conjunction with the more widespread adoption of automation, will enable the cycle times to be reduced and programming will be simplified. In the past one of the main aims of numerical control systems has been to reduce the setting and idle times of production plant. The stage is now rapidly approaching, however, where the additional savings likely to be achieved in this direction do not warrant the further expenditure that would be needed (Chapter 12).

In the numerical control systems that have been considered up to the present the cutting speeds and feeds required for machining a component are selected from tables in the planning department and are included in the switching information. Inevitably it is necessary to select a safe value that will be satisfactory under the worst conditions likely to arise during the course of a cut. This means that all possible adverse factors, such as variations in the properties of the material being machined, and tool wear, have to be allowed for. As a result a rigidly programmed machining process often operates at low efficiency.

This can be countered by adopting a further development of the numerical control process which one may describe as 'self-adjusting' or adaptive control (abbreviated to AC).

In the simplest case the function of such a control system may be to maintain certain of the parameters of the machine tool, e.g. the power applied to the main spindle, constant at a preset value despite external disturbing influences; this value may conveniently be the maximum value permissible for the machine or the tool. In a complicated case the adaptive control system can use analogue or digital means for calculating the parameters of the machine tool to which it is fitted in accordance with a predetermined optimisation law; in other words, an NC machine with adaptive control can then itself optimise its method of working. Examples of the optimisation targets include minimum manufacturing costs, maximum output, high surface finish, maintenance of extremely close dimensional tolerances, etc (175). Since these targets conflict to some extent it is necessary to compromise and to determine their relative priorities, and to select one as the basis for the optimisation law. To enable this to be done the complexity of the control system must be increased by offering optional equipment in the form of additional computing circuits and, in particular, of measuring devices for determining the machining characteristics, temperatures, vibrations, etc. At present it is, for example, already possible to measure the cutting forces and their variations and the spindle torque with comparative ease. Considerable instrumentational difficulties are still encountered, however, in determining by contactless methods the precise temperature at the point at which cutting is taking place and in measuring the tool wear during the actual machining process.

Further extensive theoretical discussion of the problems of applying adaptive control to machine tools would be beyond the scope of this book. Reference may therefore be made to the technical literature in this field (175, 195, 196, 197). The international literature so far available on adaptive control indicates that much development work is still being undertaken, and that the nomenclature is far from being standardised. There has been a considerable amount of work done towards stating and studying the problems, but little of this has as yet been applied in practice. The situation could, however, change rapidly, and because of the fundamental importance of this development its progress should be followed in the technical press. All that can be done here is to provide two examples to illustrate the range of this trend in so far as the systems that form the subject of this book are concerned. In both cases adaptive control units that employ analogue methods are used to supplement existing numerical control systems.

8.4.1 *Automatic depth-of-cut adjustment for individual cuts on a lathe* (198)

The object underlying the development of this simple adaptive control system was to increase the efficiency with which parts could be turned on NC lathes. Among the methods available for this are a reduction in the programming costs (Chapters 10 and 11) and of the actual machining times during which the tool is cutting. The part programming is to be independent of variations in the mechanical properties of the material being machined and of tolerances in the dimensions of the blank. Since the control programme contains the contour of the finished part, the adaptive portion of the control system must automatically select the depths of cut for the individual cuts or passes in such a way that the capacity of the machine is utilised to the full. The adaptive control system to be described is an extension of a two-dimensional continuous-path system. In addition to the normal positioning devices commonly used in numerical control systems, it includes a means of measuring the spindle speed; the simplest method of doing this is as an analogue quantity using a tachometer. The torque is measured by strain gauges located on a boss on the main spindle; which means that this measurement is also made by analogue means. The readings are transmitted from the rotating to the stationary parts of the machine by inductive methods. The advantage of this method of transmission is that it is free from wear and requires no maintenance. The product of the values of the spindle speed and spindle torque represents the power absorbed by the machine tool in machining the metal. It is possible to multiply these two voltages with sufficient accuracy by means of analogue methods. The value of the machining power and its changes are the factors that govern the adjustment of the instantaneous depth-of-cut and feed rate. The extreme values of these characteristics are fixed by means of electronic limit switches.

The machining cycle starts with the tool being fed towards the workpiece at the rapid traverse speed. The path followed by the tool is that which would be required if the finished workpiece contour were to be produced in a single pass. As soon as the tool contacts the blank the machining power absorbed rises almost instantaneously and exceeds the preset limiting value. To prevent the tool being broken the feed rate must now be reduced with the minimum possible delay. To enable this to be done the drives and actuators employed must have very low time constants (3). When the feed rate has dropped rapidly to a preset minimum limiting value, the depth-of-cut in the X direction is reduced at constant power, without any further feed in the Z direction taking place, until the machining power at the surface of the cylinder has dropped below a preset maximum; the feed rate in the Z direction then increases again. As the cut proceeds the adaptive control regulates the speed with which the contour is followed in such a way that the permissible machining power is fully utilised. Details of this automatic

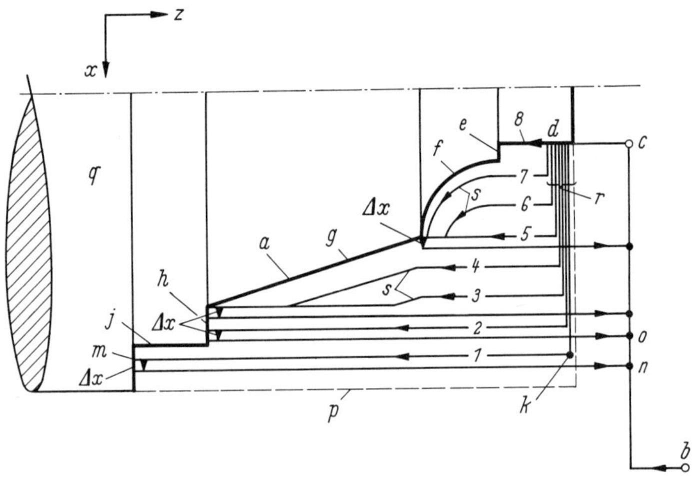

Fig. 207 Example of the use of a fixed-value adaptive control system on a lathe (198)

> *a* Programmed contour of finish-machined workpiece. *b* Machine zero point. *c* Point at which machining cycle commences. *d* to *j* Contour sections, described in six program sentences (Chapter 10). *k* Commencement of first cut in Z-direction. *m* End of first cut and withdrawal of tool by preset amount Δ*x*. *n* Point to which tool is returned after first cut. *o* Point to which tool is returned after second cut. *p* Contour of blank. *q* Workpiece (bar stock). *r* Seven starts of cut (in this example) producing overload and automatic withdrawal of tool by adaptive control system. *s* Cuts controlled by numerical control program (effect of unit *E* in Fig. 208). *X,Z* Directions of coordinate axes on a lathe (Table 12). Δ*x* Programmed amounts by which tool is withdrawn to clear workpiece on return to start line.

machining operation are shown on Fig. 207; some reference to the work covered in Chapter 10 (programming) is unavoidable.

The programme covers only the finish-machined contour at the end of the workpiece. The contour of the blank can be of any form; in the case illustrated a cylindrical shaft is to be provided at one end with a contour that is to be formed making the maximum possible use of the power available at the machine. The adaptive control system which employs analogue means of keeping the machining power at its maximum value acts on the numerical (digital) continuous-path control system which has already been discussed (Section 8.2.2). A block circuit diagram showing the arrangement of the various units is given in Fig. 208 and will be discussed later. First, however, the operation of the adaptive control system will be considered; this is superimposed on the normal numerical control system and is arranged to match the technological process. Assume that the tool is first moved at speed from its rest position b to c, the starting point of the machining cycle; here the programme for the finish-machined contour a commences. From the point c until the tool contacts the workpiece the machining power is equal to zero; the speed of calculation of the interpolator is therefore high and the tool moves at a rapid traverse rate from c towards the workpiece. As soon as the tool contacts the workpiece the machining power increases very rapidly; as soon as it attains a limiting value, the adaptive control system comes into action and causes a rapid reduction in the contour-following speed (pulse rate in interpolator). When the lower limiting value of the pulse rate is reached it is a sign that the depth-of-cut is too large. A second limit switch is actuated

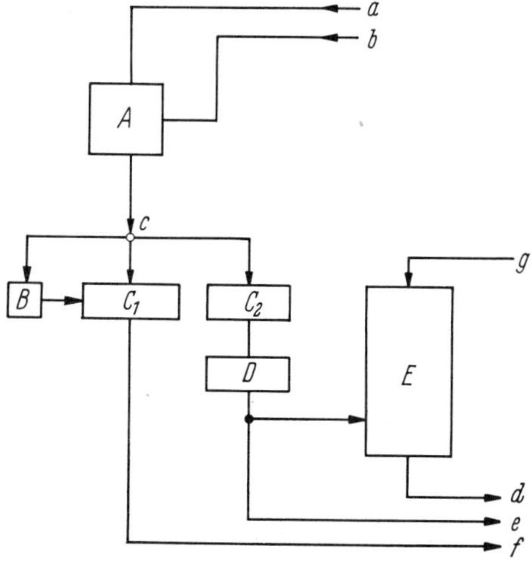

A Multiplier employing analogue values. B Limiting-value stage (A/D unit). C_1 Regulator for contouring speed. C_2 Regulator for changing depth of cut in the X-axis. D additional pulse generator. E Control logic system. Forms link between NC and AC sections. a Spindle torque. b Spindle speed. c Power determination (analogue multiplier). d Auxiliary functions for lathe. e Depth-of-cut variation. f Contouring speed (interpolator control system). g Control commands from NC section.

Fig. 208 Block circuit diagram of a SINUMERIK with adaptive control (198)

and the tool is withdrawn in the X direction, say to point k. It is assumed that at this point the correct depth-of-cut l is attained for the power developed by the machine, so that the cutting feed in the Z-direction can be employed; from this point on the contouring speed is controlled solely by the interpolator. For a numerical control system that is equipped with an adaptive control of this type, the actual path followed by the tool can thus initially be completely different from the programmed path. This deviation is stored in a special digital register (Item E in Fig. 208), and is allowed for during the next cut.

At the point m the programmed path comes to an end, and the tool is withdrawn from the workpiece by the amount Δx, and it then returns via point n to the starting point c. The value of Δx is predetermined by means of decade switches. In the meanwhile, the punched tape has been reversed to the start of the sentence that indicates the start of the adaptive control program. The program can then once more be run through. The tool once again advances at high speed against the face of the blank, withdraws in the X-direction, and then follows path 2. At the end of this cut the tool for the first time meets the programmed contour for the finish-machined workpiece, and performs the remainder of the cut under the control of a normal NC program. When the last programmed flank is reached the tool is again returned to the starting point c and the punched tape is once more re-spooled. The program can start once more. The machine then performs cut 3. During this cut a start is made with machining the programmed taper g, the maximum-value control again withdrawing the tool in the X-direction in good time to avoid over-loading the machine. When the flank h on the finish-machined contour is reached the tool is again returned via point o to the starting point c, and the program for the finish-machined workpiece is again read in. In the example chosen this cycle is repeated up to cut 8, on the completion of which the whole contour will have been machined out of solid material, the capacity of the machine having been utilised to the full. An important feature of the procedure is that those parts of the contour (program sections) that have been finish-machined are held in the digital stores and so are ignored when the program is run through again.

This short description of the process is sufficient to show the major advantages offered. By a suitable selection of the sign for the direction of feed in the X-axis it is possible to machine both internal and external contours in this manner. In drilling operations, the adaptive control system acts only on the feed rate in the Z direction. It will be clear that when the full performance capacity of the machine is utilised it is only possible to perform rough-machining operations. The final finish-machining must be performed with a different tool and a separate program.

As has already been mentioned it is, of course, possible to change the terms of reference for optimisation. The NC and AC installations that have been adopted to date can, for example, optimise on:

Minimum machining cost,
Constant (maximum) metal-removing capacity with the cuts divided
up as shown in the above example,
Constant deflection of the machine frame,
Constant temperature at the cutting edge of the tool,
Constant spatial position of the tool cutting edge,
Constant spatial position of the main spindle, etc (206).
It will be obvious that the additional cost of an adaptive control system will
increase as the number of measuring positions is increased, and with the
width of choice of the automatically controlled optimisation conditions. The
use of AC techniques is thus, in itself, an optimisation problem. If a large
number of measuring positions and parameters is employed, the point will
very rapidly be reached at which it will prove necessary to abandon analogue
techniques and to employ digital process computers instead (Fig. 93 for the
cost trends of analogue and digital equipment).

The block circuit diagram (Fig. 208) shows what units are required to
extend an NC system to an AC system, and how these units interact. The
multiplier A (3), which processes analogue values (voltages), determines the
machining power c as a product of the spindle torque a and the spindle speed
b. To obtain a faster control response it would be preferable to measure the
cutting force directly at the tool instead of making use of the spindle torque.
This would, however, complicate the measuring arrangements which must
also be capable of working reliably and without interfering with the operation
of the machine under the adverse conditions prevailing in a machine shop
(rough-machining lathes). As soon as the machining power exceeds a preset
limiting value the limiting stage B reduces the rapid traverse to a low cutting

Fig. 209 NC lathe with adaptive control (Photo courtesy of *Gildemeister/Siemens*).
a Workpiece with required contour (rough-machined) in chuck. *b* Tapered blank
(e.g. forging). *c* Cylindrical blank. *d* Control cabinet. *e* Lathe.

feed. This unit also performs the transition from the analogue to the digital regime; the electronic limit switch controls directly the regulator C_1, which is the unit that controls the computing frequency for the interpolator (Fig. 202). The second regulator C_2 for adjusting the depth-of-cut is also acted on by the limiting value stage B. An additional pulse generator D produces impulses which operate the stepping motor in the X-axis (depth of cut alteration) and at the same time cause the displacement increments that depart from the programmed contour to be stored in the registers of the control logic unit E. The control logic unit also stores those parts of the program during which the tool has not departed from the programmed contour. It also forms the link between the normal NC section and the adaptive control attachment. Since the adaptive control system described eliminates the need for including the subdivision of the total metal removal into separate passes in the program it enables a considerable amount of programming effort to be avoided. Sample programs have shown that the length of the programs can be reduced by about 60% (210). In addition adaptive control offers further advantages in that it makes it unnecessary to make allowances for variations in the characteristics of the material and to some extent also for unforeseen tool wear since this produces a marked increase in the torque. Figure 209 shows an NC lathe that is equipped with an adaptive control attachment.

Since it is more difficult to make measurements on rotating tools than on stationary ones, and also since optimisation with a view to cost minimisation (machining time and tool costs) should be at least discussed in principle, as it is the most complicated case, the following section describes an application of the system to a milling machine.

8.4.2   *AC and NC techniques on a milling machine* (196, 199, 205)

The control system described below optimises the machining process on a portal milling machine and consists of a numerical control system to which an analogue adaptive control attachment has been added. The required optimisation target for this adaptive control system is the achievement of a high rate of metal removal together with minimum machining costs for the individual workpiece. Since the two parts of this target, namely 'high rate of metal removal' and 'minimum unit cost' are in conflict, the achievement of minimum cost was taken as the primary object. The conflict arises from the fact that although the unit costs are related to the machining time, any attempt to make a radical reduction in the machining time will increase the cost of the tools, since their life is considerably reduced due to the higher feed rates and excessive cutting speeds. To enable the machining process to be controlled automatically in such a way that the cost is a minimum, the values of the spindle speed and the feed rate must be changed as the machining cycle progresses in such a way that at any instant they lie on an empirical curve which is stored in the circuit; to enable this to be done these parameters must be measured and the readings must be evaluated.

Here again analogue methods are chosen for making and evaluating the measurements in order to reduce the cost.

The machining process on milling machines is also subject to varying conditions and to external influences. Five such factors are of primary importance:

> Variations in width of cut,
> Variations in depth of cut,
> Tool wear,
> Variations in hardness of material, and
> Interruptions in the cut due to the geometry of the workpiece.

Figure 210 shows the parameters that are measured and that are evaluated and utilised in the central control system. Apart from these parameters the programmed feed rate is also introduced into the control algorithm. Its value is given by the internal interpolator. An interesting feature is the determination of the tool deflection as a measure of the cutting force during

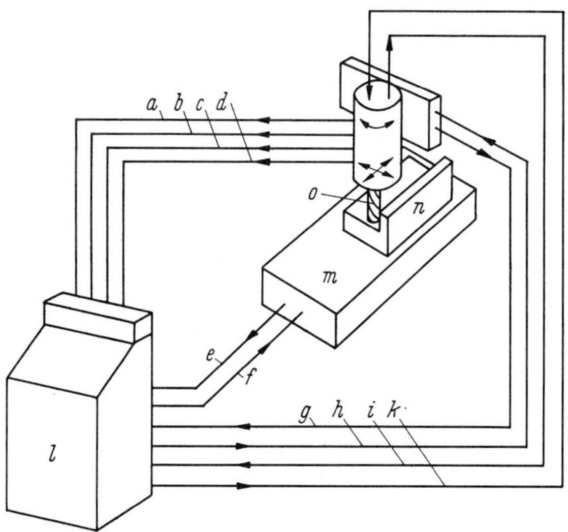

Fig. 210 Measurements and adjustments on a numerically controlled milling machine with adaptive control (199)

a Driving power. *b* Spindle torque. *c* Tool deflection *y*. *d* Tool deflection *x*. *e* Displacement measured in direction *x*. *f* Control of drive in direction *x*. *g* Displacement measured in direction *y*. *h* Control of drive in direction *y*. *i* Measurement of spindle speed. *k* Control of spindle speed. *l* Control console. *m* Machine slide. *n* Workpiece. *o* Cutter.

the machining process. Four pickups on the lower end of the spindle give the deflections in the X and Y directions. The limiting value for the deflection is selected so that on the one hand it is below the point at which the tool will break, while on the other hand it enables the required manufacturing tolerances to be maintained. The range within which control can be exercised is further limited by the permissible maximum torque value and by the maximum power setting.

Before the workpiece is machined it is necessary to supply the adaptive control system with the values of some parameters, such as the maximum permissible deflection, the ratio of cutter stiffness to spindle stiffness, the permissible range of spindle speeds, and a power index.

In the case of machine programming (Chapters 10 and 11) these values are calculated in a preprocessor for every combination of tool and material. If the tool wear exceeds a preset limiting value during the course of the machining operation, a pilot lamp lights up as an indication to the operator that the tool should be changed.

The sketches in Fig. 211 compare the feed behaviour of machines with and without adaptive control when various disturbances are present. Whereas in the case of NC machines without adaptive control the value of the feed is fixed to suit the most adverse cutting conditions, the adaptive control described will automatically adjust the feed to suit the cutting conditions. It reduces it at a constant rate if the width or depth of cut increases continuously, or reduces it instantaneously if the cutter hits a hard spot. Where the cutter is not required to remove metal, for example between two projections, the feed is increased to the rapid traverse value.

It took about seven years to develop the adaptive control attachment for this numerical control system (205). In view of the length of time that this

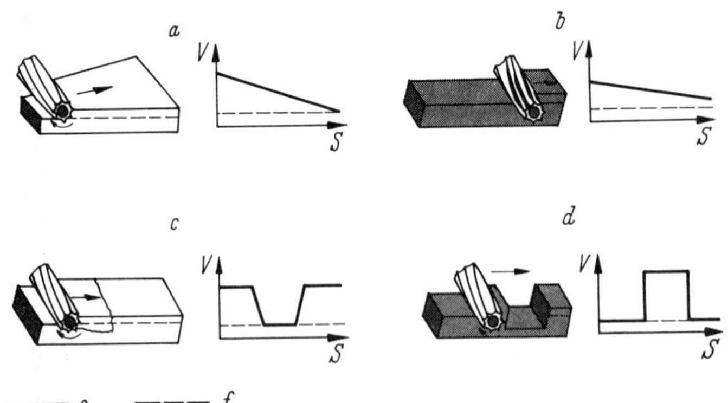

Fig. 211 Optimisation of feed in the presence of various disturbances (199). – Four disturbing factors compared.

a Variation in width and depth of cut. *b* Tool wear. *c* Variations in hardness of material. *d* Interruption in cut. *v* Feed rate. *s* Displacement.

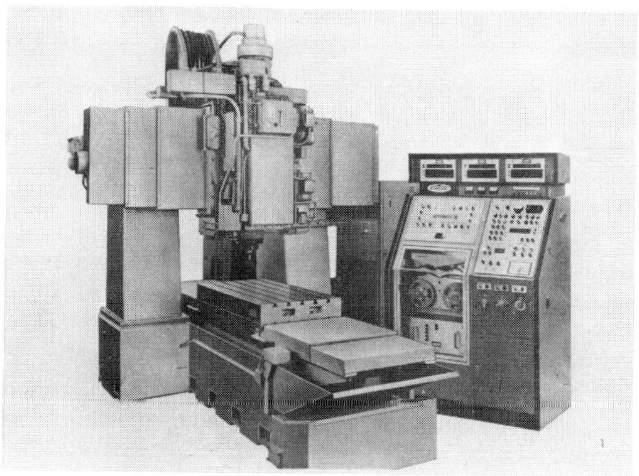

Fig. 212 Vertical milling machine with analogue control of the cutting process to result in minimum costs. (Photo courtesy of *Cincinnati*).

work took, it is not surprising that no technical details have yet been released of the sensory and the evaluation circuits. Presumably this will, however, change during the course of the next few years.

Figure 212 shows the portal milling machine described in the example together with the combined AC-NC system. At present it is being used mainly for metals that are difficult to machine and for non-homogeneous materials, since in these applications the efficiency of the process soon shows to advantage despite the high cost.

8.5 Design of computing circuits —
Miniature computers and magnetic stores for internal data processing

The digital computing circuits employed in the numerical control systems mentioned in Section 8.1 are used for performing arithmetical operations on data which they receive from other units in the control system. The individual stages of the calculations can be determined either by permanent wiring or by stored flexible programs. For economic reasons permanent wiring is at present preferred, but this position may change as further development work on small computers takes place. There is also a choice available for the storage media that are used in conjunction with these circuits. It would be beyond the scope of this book to go into details of the computer techniques, the digital circuits, or magnetic stores; interested readers must be referred to the technical literature that has already been mentioned. Nevertheless it seems advisable to make brief mention of the basic features and of some of the choices that are open.

All arithemetical operations in a digital computing circuit can normally be reduced to additions. Some of the computing circuits that are encountered in practice will be briefly described below. In these use is made of the symbol for an adder that is shown in Fig. 213A). From three binary input values, the two values that are to be added C and D, and the carry-over U, from a previous adding stage, two binary output values are produced, namely the sum S and the next carry-over U'. The input values C, D, and U, as well as the output values S and U', can all be only L or O. Their logical relationship is shown in the table in Fig. 213B). In addition to an adding stage a complete computing circuit will contain at least one store, known as the accumulator, which can hold intermediate results. Basically there are two possible types of computing circuit. One of these is built up from adders which supply at the output the sum of two values which are applied simultaneously to the inputs. The other circuit is based on the counter principle (Chapter 3). In this the two values that are to be added are represented by sequences of pulses which are applied in turn to the same input of a counter. The value on the counter then represents the sum. The counter principle is, however, used only very occasionally in positioning and straight-line control systems, and since it has already been mentioned in Chapter 3, it need not be considered further here.

The addition of multi-place figures can be effected in parallel or in series. Parallel processing requires a separate data transmission channel for each decimal place. Each time an adding or storage order is received all places of the numbers that are to be added are transmitted simultaneously through the appropriate channels.

For the serial processing of numbers, only one transmission channel is required, and the individual decimal places of the numbers are passed through

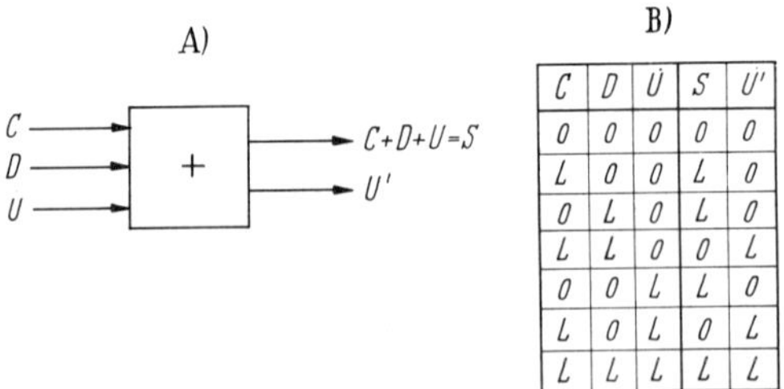

B)

C	D	$\ddot{U}$	S	$\ddot{U}'$
0	0	0	0	0
L	0	0	L	0
0	L	0	L	0
L	L	0	0	L
0	0	L	L	0
L	0	L	0	L
L	L	L	L	L

Fig. 213 Functional representation of a digital adder

A) Circuit symbol. B) Logic Table.
C,D Values to be added. S Sum. U,U' Carry-overs.

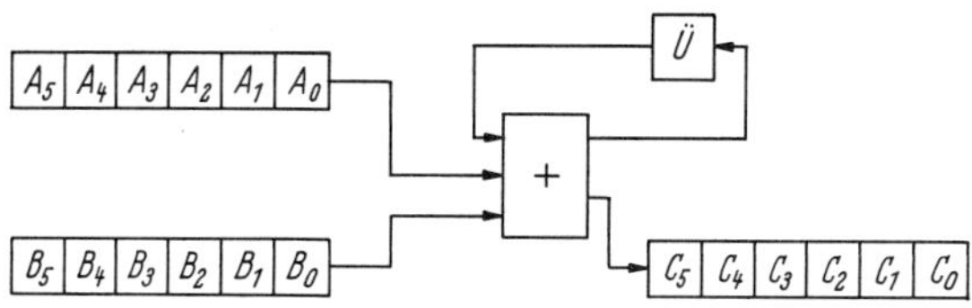

Fig. 214 Block circuit of a parallel adder for adding two six-place binary numbers

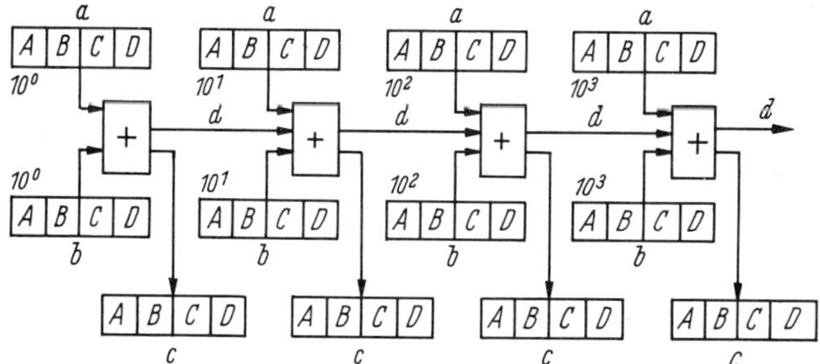

Fig. 215 The processing of binary-coded decimal numbers (Chapter 2). Four-symbol binary groups processed in parallel; decimal numbers processed in parallel

a 1st Number to be added. *b* 2nd Number to be added. *c* Sum. *d* Carry-over.

this in sequence (Fig. 214). This reduces the complexity of the arrangement. With serial addition it is only necessary to ensure that the equivalent decimal places of the two figures which have to be added are applied simultaneously to the inputs of the adding circuit. Any carry-over that arises is temporarily stored in U and is taken into account during the addition of the next higher place. In addition to computing circuits that employ strictly series or parallel methods, there are also mixed circuits which are frequently used for the processing of binary numbers and the corresponding binary four-symbol groups are processed in parallel (Fig. 215).

In addition the corrective or comparative calculations that have to be performed can be carried out either in several computing circuits arranged in series (Fig. 216) in which each computing circuit applies a different operand to the result of the preceding operation, or all calculations are performed in a central computing circuit in which the operands are connected in sequence (Fig. 216B). In this solution the intermediate result is passed back into the computing circuit after each stage, by way of a store and accumulator. The sequence of events is shown in the cycle diagram of Fig. 216 C.

The computing circuits described up to the present are non-selfcontained units within numerical control systems. The functional performance of the calculations is determined by the way in which the individual electrical

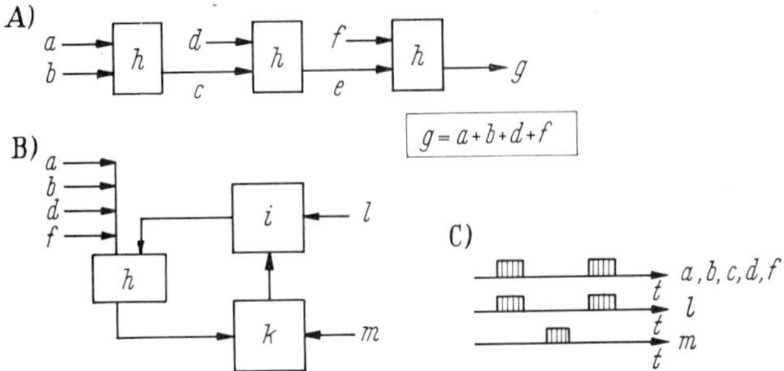

Fig. 216 Block circuits of computers for producing the sums of several numbers

A) Using several adders. B) Using only one adder. C) Cycle diagram.

a, b, d, f Numbers to be summed. *c,e* Intermediate sums. *g* Final sum. *h* Adders.
i Accumulator. *k* Store. *l* Accumulator cycle. *m* Store cycle.

components and groups are wired up. The complete circuits form small special computers which have been designed for one specific application and which work in accordance with a fixed algorithm. The internal interpolators that have been discussed previously have also been made on these principles up to the present. This permanently-wired combination of functional groups may require a considerable design and production effort for each individual control system.

The computer industry has put forward the idea of connecting the functional units flexibly rather than rigidly; this can be done by a flexibly stored program. Functional units of this type are therefore more accurately described as store-programmed small computers. They can perform the calculations in accordance with freely-programmable algorithms. All the logical and arithmetical functions that are required by a machine-tool control system can then be simply modified. Some manufacturers of control systems are attempting to take advantage of this feature, but the users' reactions are not yet clear. In these designs a freely-programmable small computer forms a central part of the machine-tool control system. Figure 217 shows an example of an actual numerically-controlled milling machine, the continuous-path control system of which has been equipped experimentally with a freely-programmable small computer. A teleprinter machine with a punched-tape reader can be used for the input of both the machine control programs and also of small external computing programs. In this design an effort has been made to make full use of the greatly increased computing speed and capcity of small computers for this type of application.

The typewriter can be used not only for the input of programs, but can also print-out reports on the work and the contents of stores. A control console near the machine tool is provided with special push-buttons and switches for the correction and manual input of machine programs of the type described in Chapter 9. The keyboard also contains an arrangement for

Fig. 217 Experimental version of a machine tool with a built-in small computer (Photo courtesy of *Kongsberg/Köping*).

a NC machine. *b* Freely programmable small computer. *c* Input and monitoring typewriter with punched tape input and output. *d* Control console for the numerical control system.

Fig. 218 NC machine with extended storage capacity for correction values (Photo courtesy of *Burkhardt/Sperry*).

displaying the numbers of parts of program sentences. An adaptor unit transforms the output information from the computer into command signals for the drives and control valves on the machine. The use of adaptor units of appropriate design at the input and output of the computer enables a control system of this type to be fitted to any machine tool with the programs altered as needed (218). It remains to be seen whether widespread use will be made of this possibility in the future in view of the development of direct numerical control systems (Chapter 9).

Another concept has been employed on the machine and control system illustrated in Fig. 218. In this instance the control system is equipped with a small magnetic drum store to enable large quantities of tool correction values to be accommodated. Each new tool that is loaded into the machine is moved into contact with a measuring plate, and the actual values of the tool dimensions are determined and automatically entered into the drum store; there it is available for use in the calculation of corrections as required. Since details of this arrangement have already been published it will not be described in full here (216, 217). Further discussion of these matters can conveniently be postponed to include the subject matter of Chapter 9.

8.6   Summary

The material covered in this Chapter can be summarised as follows:
1. The requirements for ease of operation and simpler programming led to the early adoption of simple computers even for positioning and straight-line control systems. One of their main advantages is that they enable zero-point and tool corrections to be performed at the machine while it is in operation.
2. The fact that with numerical continuous-path control systems it is necessary to maintain a functional relationship between the movements in the directions of the various machine axes means that a comparatively large computing capacity is required which replaces mechanical form stores.
3. Internal interpolators, no matter what their type, act as real-time computers and must enable at least two requirements to be met simultaneously:
   a)   The relative movements of the tool and workpiece must be controlled in several axes so that the required workpiece contour is attained.
   b)   The resultant feed speed must be kept constant, as nearly as possible at all points on the path.
4. The development and dimensioning of internal interpolators must be such as to keep the space occupied on the machine tool to a minimum while programming and the operation of the machine must be kept as simple as possible. At the present time these objects are met most closely by permanently-wired digital computing circuits.
5. In view of the limited resolution capacity of the associated digital/analogue converters and displacement measuring systems, interpolators that make use of analogue quantities have not been widely used in practice.

6. Digital internal interpolators can have digital-absolute and digital-incremental signal outputs for the commands to the subsequent control circuits. If digital-analogue converters are employed analogue values can also be produced at the signal output and analogue follow-on control circuits can then be used.

7. Digital linear interpolation can be performed comparatively simply to a high degree of accuracy, circular interpolation can be performed only approximately but with adequate accuracy. The internal interpolators currently employed in practice are for the most part capable of being switched over from linear to circular interpolation; parabolic interpolation is used only in special cases.

8. In current practice two methods are used for linear and circular interpolation; these are the DDA method or the direct calculation of the functions. Of these the DDA process is the more common. There is little to choose between these two methods; alternative methods would be possible.

9. The use of a thread cutting attachment for the internal interpolators on NC lathes results in a considerable extension of the field of applications of these machines and converts them into fully universal lathes.

10. During the course of the last six years the efforts that have been made to reduce machining and programming times have led to the development of optimisation circuits which, in the first instance, are of comparatively simple form (NC machines with adaptive control). In the first place these have been arranged either to keep certain predetermined characteristics of the machine tool (e.g. the power) constant at a required value, no matter what external factors are tending to vary this, or to vary automatically the programmed technological working conditions in accordance with a permanently-wired optimisation law as required by measured values.

    Up to the present practical experience has been gained and published for only a few adaptive control systems. Two of these are described to indicate the importance of the subject; to reduce the costs, the measurement and evaluation of the factors concerned has been by analogue means. It is probable that considerable progress will be made in this field during the course of the next few years.

11. With the exception of the adaptive control systems that are described all the functional cycles referred to in this chapter can be realised with the aid of digital computing circuits. In addition to the logic units that are described in Chapter 3, these require adding elements and registers. The electronic circuit elements can be joined together into computing circuits in accordance with a variety of different principles.

12. In place of computing circuits with permanently-wired functions of the type used up to the present it is also possible to employ freely-programmable small computers. These form a link with direct numerical control systems. It will be convenient to defer any

conclusions regarding the use of small computers until after this latter development has been discussed (Chapter 9).

13. The use of large storage systems for correction values must also be discussed at a later point against a wider background.

9   Input units and control consoles

9.1   Basic methods available for feeding-in work information

The input equipment provides the interface between the internal and external data processing, as defined in Chapter 2. In practice the boundaries between these two concepts are sometimes becoming a little blurred as a result of the current trend towards integrated data processing; but it nevertheless seems advisable to continue subdividing the data processing operations in this manner for various reasons which are discussed in Chapter 12. During the last two or three years the direct numerical control systems that are considered in this chapter have emphasised this tendency for the two types of data processing to merge. This development does not, however, make a discussion of the input units, control panels, and control consoles superfluous; on the contrary, these items continue to form an important means of communication between man and machine (20) and their arrangement is not merely of practical interest. The main methods of feeding-in information are shown in Table 14.

The punched card is an old-established and well-known medium for simplifying office work. One of its major advantages is that a strictly defined, if somewhat limited, amount of information can easily be sorted if it is punched onto cards and that the data can be processed mechanically while keeping the individual self-contained data carriers intact (e.g. a punched card for each employee for salary records, sets of punched cards for spare parts, customer records, and tools). In the manufacturing field it would be possible to include all the data required for drilling a particular hole (coordinates, tool details, speeds and feeds) on a single punched card.

A complete drilling programme would then comprise the appropriate punched cards, the sequence of which can be changed as required at any time. The situation is different where technological considerations require the operations to be carried out in a strictly defined sequence; getting the cards mixed would then have disastrous results. Since one can seldom afford the luxury of re-sorting the cards, even on an automatic sorter if this is located at some distance from the machine tool, punched tape, with its positive and unvarying sequence of perforations, offers much greater security in such cases. It must also be borne in mind that punched card readers are much more expensive than punched tape readers which again makes them less suitable for machine tool control systems. Punched card installations will therefore not be considered further in the discussions which follow.

Type of input for work information	Storage capacity of individual data carriers (6, 221)	Applications
Manual input Section 9.2		For machines with positioning and straight-line control systems and work-pieces (or small batches) where the amount of work information is small. May be of advantage for one-off jobs.
Input using data carrier – punched card used as information carrier.	approx. 7 bit/cm^2	Seldom used for machine tools; if used, then only on machines with positioning control systems (e.g. wiring machines)
Input using data carrier – punched tape used as information carrier. Section 9.3.	approx. 15 bit/cm^2	Used for machine tools with positioning, straight-line, and continuous-path control systems and for work-pieces (and batches) with medium to large amounts of work information. Punched tape is the data carrier that is currently most widely used.
Input using data carrier – magnetic tape used as information carrier. Section 9.4.	approx. 1250 bit/cm^2	Used for machine tools with continuous-path control systems in conjunction with external interpolators and workpieces containing very large amounts of work information.
Direct numerical control (DNC) Section 9.5.		Where several machines with positioning, straight-line, or continuous-path control systems are coupled to a central computer. The positions of the actual point at which cutting takes place between the central computer and the machine control systems can vary.

Table 14. **Methods used in practice for feeding work information into numerically-controlled machine tools.**

Manual methods of feeding-in work information, on the other hand, are by no means obsolete, and will, presumably, continue to occupy an important position for such applications as the production of one-off workpieces on NC machines, although in many cases they are likely to be an alternative to punched-tape control systems which are available on the same machine. For various reasons *magnetic tape control systems* are far less widely used than would appear likely on theoretical grounds; current practice is dominated by *punched-tape control systems*. It seems probable that they will continue to be of importance in the future for economic reasons.

9.2 The manual input of work information

The manual input of switching information by means of push buttons, hand levers (e.g. on gearboxes), rotary switches, handwheels, etc., is well-known, and no changes have been made to these methods where they have been adopted for manually operated numerical control systems. As far as "number-reading" machines are concerned, only displacement information is fed in numerical form. Two of the many possible arrangements have been found most convenient, namely decade switches and cross-bar distributors. Both have the advantage that the numbers that are fed in by these means are at the same time automatically stored and are available at all times for internal data processing.

Fig. 219 Example of a manual input control panel for numerical displacement information on a radial drilling machine. Fig. 296 in Appendix (Photo courtesy of *GSP*).

a 5-place decade switches for dimensions in the *x*-axis. *b* 5-place decade switches for dimensions in the *y*-axis. *c* Job card with programme giving sequence of coordinate dimensions.

Figure 219 shows an example of a simple control panel for feeding in two 5-figure coordinate dimensions using decade switches. To simplify the work, the job card which contains a list of the required dimensions is placed in a holder directly beside the control panel and a straight-edge assists in reading it off line-by-line. The method of operation of the very simple NC machine will be readily understood if reference is made to the illustration of the complete radial drilling machine in the Appendix.

Figure 220 illustrates the principles of the well-known cross-bar distributor. With this device it is not only possible to set up complete

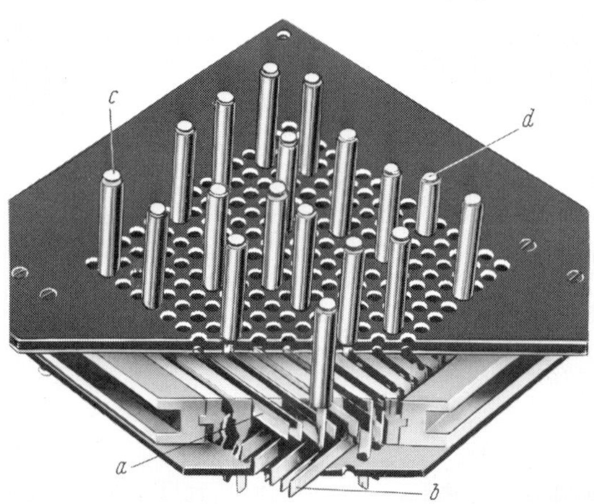

Fig. 220 Sectioned illustration of a cross-bar distributor (Courtesy of *Ghielmetti*)

a Row of contact bars 1. *b* Row of contact bars 2. *c* Plug with built-in diode (corresponds to Fig. 4). *d* Short-circuiting plug.

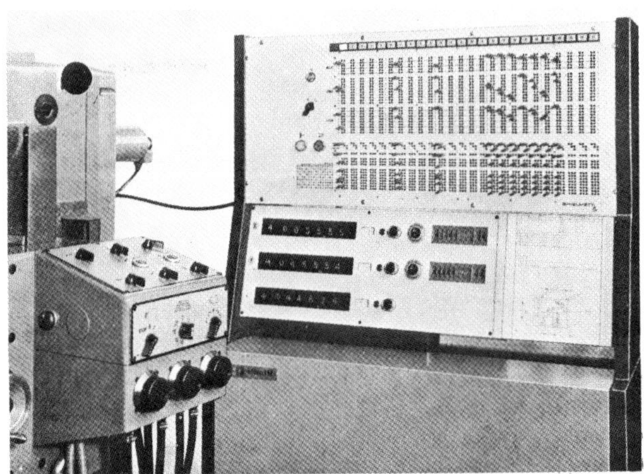

Fig. 221 Cross-bar panel for a horizontal boring mill (Photo courtesy of *Ghielmetti*)

programmes of switching information, but also to feed in manually the required values of large numbers of displacements; to save space it is, of course, advisable to encode the various decimal figures into the BCD code (Chapter 1). As the work progresses a stepping switch causes the individual lines to be read off in sequence and so to obtain the required switching and displacement information. The machines are then numerically controlled in accordance with this programme. The machine tool that is being controlled must clearly be equipped with displacement measuring systems and comparators or with directly positionable drive systems (Chapter 6) to enable it to meet the requirements for an NC machine. Figure 221 shows a complete program plug board of this type for a boring mill. The same effect is also produced by programming cylinders of all types, on which cams are arranged

A)

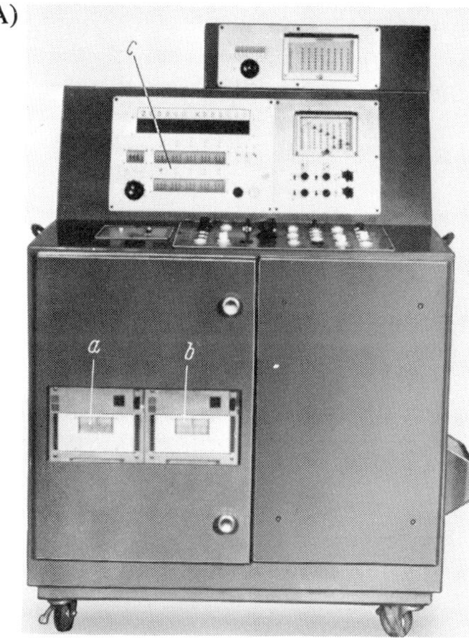

B)

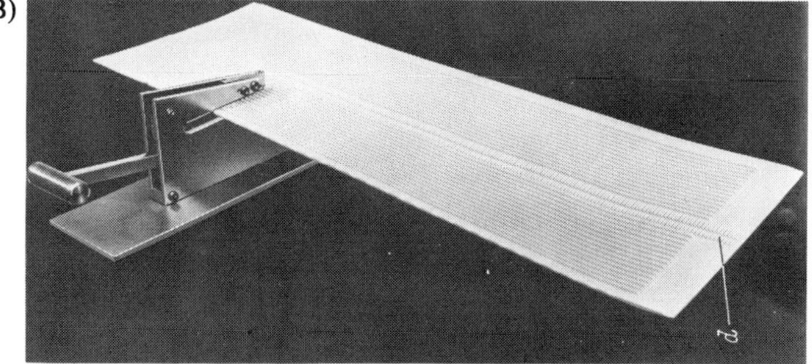

A) Control console. B) Plastic card with notcher.

a Program card for switching information. *b* Program card for displacement information (input accuracy 0.1 mm). *c* Decade switches for adjusting tool positions to an accuracy of 0.01 mm. *d* Feed indentations on commercially available plastic card.

Fig. 222 Control console for a Pittler Numeric installation, which uses notched plastic cards as information carriers, Fig. 296 in Appendix (Photo courtesy of *Pittler*).

in series in such a manner that all the switching and displacement information required for a given work stage is available along one line (223). The ratchet mechanism driving the camshaft then replaces the indexing device for tapping off the cross-bar plug board.

The plug-board of the cross-bar distributor on the turret lathe shown in Fig. 297 in the Appendix serves another purpose. To avoid inaccuracies during the presetting of the tools, all slide movements are manually programmed on the plug board when the machine is being set up, a trial run is then made and all deviations from the required finished dimensions of the workpiece are recorded. Any corrections needed are then made by changing the positions of some of the plugs on the machine. In this way it is possible to eliminate completely the need for accurate presetting of the tools (Fig. 161). Many manufacturers of turret lathes therefore still offer simple and cheap cross-bar distributors as an alternative to a punched-tape control system.

A similar line of reasoning led to the development of the machine illustrated in Fig. 296 of the Appendix. Since the size of the plug board is limited, and hence the number of machining steps that can be accommodated is comparatively small, two corrugated plastic programming cards, one for switching information and the other for displacement information, are notched in such a way that they can accommodate a comparatively large machining programme; the cards are also provided with indentations so that they can be passed through the reading units a step at a time (224). The smallest displacement that can be programmed in this way is 0.1 mm; smaller dimensions and corrections are entered into the machine manually by means of decade switches (Chapter 8). This simplifies the task of presetting the tools. Figure 222A) shows the control console of the machine and Fig 222B) shows a plastic card and the notcher. Programme cards of this type are used fairly frequently in automated process technology.

The design of the control system for the machine illustrated in Fig. 229 of the Appendix is based on a further development of these ideas. Since this is a pneumatic control system it seems advisable to postpone discussion of it until Section 9.3 where it will be dealt with in another context.

One special application of manual data input, which for technological reasons is similar to this method, will, however, be discussed here. This relates to the use of jig borers for toolmaking where an accuracy of up to ± 0.001 mm is required. A machine of this type together with its control system is shown in Fig. 301 of the Appendix; Fig. 223 shows the control cabinet by itself. It is interesting to note that the manufacturers have developed their own electro-optical control system to ensure that the required high accuracy is achieved under all conditions, including manual and automatic operation with workpieces weighing up to 1200 kg. All machines are equipped in principle with optical precision scales on each axis and these enable the operator to set the machine up by hand to any position, or to check its operation when it is under automatic control. To simplify the task of coarse positioning between 0.01 and 0.1 mm, a rotary position measuring system with magnetically-scanned code discs (similar to Fig. 50) is arranged in

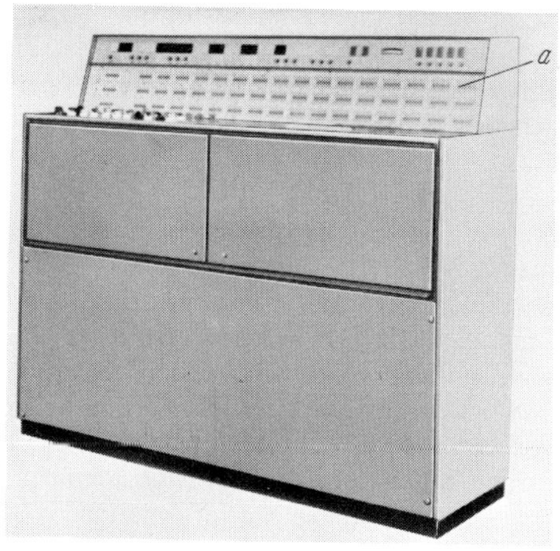

Fig. 223
Control console for a
precision jig borer as shown
in Fig. 301 of the Appendix.
Photo courtesy of *Dixi*).

a Bank of decade switches
for tool correction.

parallel by means of a rack and pinion. Fine adjustment using the optical linear scale is effected under automatic control with the aid of a TV camera tube (Vidicon-Patent DIXI) on each axis with automatic adjustment to within 0.001 mm of the required position. It is not possible to discuss the details of this remarkable positioning and straight-line control system here (225); in the present context it is, however, of interest to note the following features:

1. No automatic tool-changing, all tool changes being effected manually (Chapter 7).
2. Up to 48 tool-length or cutter-diameter corrections can be made, with five-place decade switches available for each.
3. Hydraulic cylinder-piston drives (Chapter 6), with the oil temperature stabilised at room temperature to prevent thermal expansions.
4. Photoelectric punched tape reader.
5. Complete programming keyboard which not only enables the machine to be controlled manually but at the same time operates a tape punch. The manufacture of the first workpiece can thus simultaneously produce on the machine a punched tape which is then available for machining repeat items.
6. Copy punched tapes can also be produced on the machine, so that no separate programming space is required.

Through these six main features the control system is very largely adaptable to the operations that are performed on precision machine tools of this type:

a) if one-off jobs predominate, as is usual in tool-making, the high-speed automatic positioning mechanism enables all the advantages of manual data input to be utilised;

b) there is every opportunity of using the machine for small-batch production where a number of repeat parts are to be made; no separate programming office is needed;

c) the provision of a tape punch makes the control system a repeatable one.

A detailed description of a control system of this type with magnetic drum stores has been given in the literature (7, 26) so that it is not necessary to go into the details here. However, other manufacturers of precision jig borers have also developed special numerical control systems to meet the requirements for high accuracy and to suit the specific features of their own machines; these are suitable both for the manual input of work information (rational one-off production) and for repetitive control (for subsequent batch production purposes). In principle there is no real need to provide a punched tape; if it is provided this is solely in order to enable the information to be transmitted to other machines in the plant.

9.3   Punched tape as a data carrier (information carrier)

In 1965, draft Standard DIN 44 300 included punched tape in the category of data carriers. Since the earlier expression 'information carrier' is still current in technical writing and in speech both terms will be considered here as having the same meaning.

Almost universal use is made on machine tools of 1 inch wide tape with internationally standardised hole spacings and hole sizes; the dimensions are such that it is possible to arrange for a maximum of eight rows of holes. As a result of the small amount of information that could be stored, punched tapes having a width of 11/16 inches with five rows of holes have become obsolete. these were derived from the equipment used for teleprinters, and at one time were fairly frequently used for NC control systems. For similar reasons a further argument is currently in progress. Although there is general agreement on the use of 1-inch tape with 8 tracks, the most satisfactory form of coding to adopt is not yet settled. Since this again raises the question of the number of characters required and of providing adequate security of the characters (Chapter 1) it seems desirable to discuss this subject in Chapter 10 in greater detail than is possible here. At this point it will be sufficient to consider only the physical dimensions of the tape and the resultant working conditions.

Figure 224 shows a portion of an internationally-standardised 8-track punched tape to DIN 66016 Sheet 2. So far as machine control systems are concerned it is possible to use either strong paper tape or plastic-laminated opaque tape. The material must be resistant to oil and must have a high tear strength to enable it to stand up to the severe conditions to which it will be subjected in a machine shop. The photo-electric readers which are at present very frequently employed make it necessary for the tape to be opaque, and it must remain so even if it comes into contact with oil. In general reliability in service should always be considered more important than cost.

No reference is made to coding in the Standard that has been quoted. In principle, it is possible to employ longitudinal or transverse coding, these two types being compared in Fig. 225. With longitudinal coding, the numerical work information is located on each track in binary code to save space. An electronic computer is required to produce a punched tape coded in this way. Longitudinal coding is very seldom used nowadays, and can be ignored in this context. In the old EIA (Electronic Industries Association) Recommendation RS-244, which was issued in 1961, definite recommendations for longitudinal coding were included, and these are illustrated in Fig. 225; the revised version of the Recommendation, RS-244-A, issued in 1967 no longer includes this type of coding and it has also not been included in the ISO Recommendations.

With transverse coding $2^8 - 1 = 255$ hole combinations are possible for each row of holes, if the combination 'no holes' is ignored as being ambiguous. If no check digit (Chapter 1) is employed there are 255 different characters that can be represented on each row of an 8-track tape. In practice, however, as mentioned in Chapter 10, a redundancy of 1 bit is always provided, so that the number of characters available for processing will in fact be only $2^7 - 1 = 127$ characters. All further discussions then hinge on the question whether this possible number of characters should be utilised in full or only in part; this question will also be considered in detail in Chapter 10,

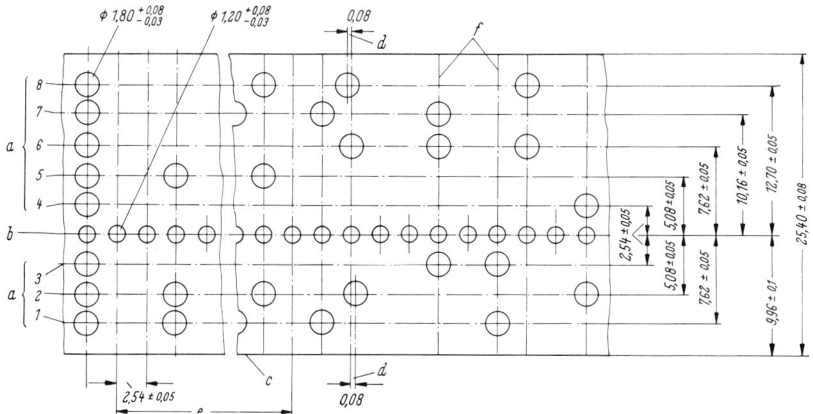

Fig. 224 Standard 1-inch punched tape to DIN 66016 Sheet 2

a Information tracks. *b* Feed holes and timing track. *c* Reference edge. *d* Permissible aberration in hole spacing *e* Permissible cumulative error ± 0.25 mm over 10 divisions and ± 0.90 mm over 50 divisions. *f* Rows (in transverse coding the combination of holes in a row represents one character).

since it is of importance for the future development of NC machines. Here only the effects of the physical dimensions of the punched tape, as shown in Fig. 224, on the scanning equipment will be considered.

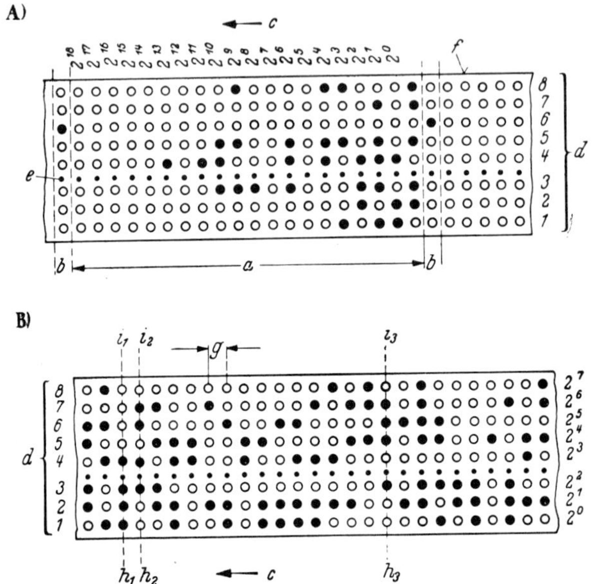

Fig. 225 Coding for punched tape.

A) Longitudinal coding (example). B) Transverse coding.

a Information block in binary code. *b* End-of-block symbol (Reader 'Stop'). *c* Direction of motion of tape (reading direction). *d* Tracks 1 to 8 (Track 1 = main spindle speed, track 2 = feed speeds, track 3 = coordinate dimensions for *X*-axis, track 4 = coordinate dimensions for *Y*-axis, track 5 = coordinate dimensions for *Z*-axis, track 6 = control of punched tape reader, track 7 = information of any additional calculations (e.g. interpolator control), track 8 = check symbol). *e* Row of feed holes. *f* Indication of sign. *g* Reader step. *h* to *i* Binary-coded symbols. ○ not punched ● punched.

9.3.1 Punched tape scanners

Series and parallel (block) scanning

Despite the transverse coding of the tape it is necessary to distinguish between the methods of operation of two forms of scanner in machine tool control systems:

a) Each combination of holes in the transverse coding is regarded as a character, and hence read as a unit, and all the displacement and switching information for a machining step forms a block of characters which are read off in sequence. This is referred to as series scanning of the block[1] and parallel scanning of the characters.

The individual blocks of information can be of various lengths, depending on the nature of the machining step (Chapter 10). A special combination of holes is placed at the end of each block in order to stop the scanner. The characters read off in sequence within a block, which follow each other rapidly and in the form of pulses are, in general, not directly suitable for a machine-tool control system. Instead the information must be held in suitable buffer stores until it is required by the machine. This process will be described in detail later, when an example will be given. In this context it is only necessary to point out that series-scanning and buffer stores belong together, and that the variable 'blocks' on the punched tape correspond to the 'program sentences' of Chapter 10.

b) Each block is taken as a unit and scanned. In this case the information required for a machining step remains available until the tape is moved on by one block when the next step is called up. Obviously under these conditions the buffer stores will not be required. In its method of functioning this type of control corresponds closely to the scanning method employed with the cross-bar distributor referred to earlier or to the method of operation of program cams of all types.

If these two scanning principles are compared it will be seen that each has advantages and disadvantages.

1. With series-scanning there will be the additional expense of the buffer stores, and this can be eliminated in parallel scanning. Since stores of all types are comparatively expensive it will always be necessary to check whether block-scanning might not be suitable if expense must be considered.

2. With parallel- or block-scanning, the maximum length of the block depends on the design of the scanner, whereas it is not limited in this way with series-scanning. In other words a parallel-scanner can, strictly speaking, be made only for a particular machine tool which has to perform a clearly defined task, since the length of the block, and hence the number of the contacts, is constant. This means that the length of the sentence in programming is also constant. Using parallel-scanning methods, several types of scanner having different

---

[1] This must not be confused with the series scanning of the individual holes in a combination as, for example, in the transmission of characters by a teleprinter. There each hole in turn is checked for punched/not punched and the resultant signals are used for telegraphy. This fundamental resolution of the characters is not required in machine control systems; the method of scanning used here is also sometimes referred to as series-parallel scanning.

numbers of contacts are thus required for the various types of machine tool. This is a major disadvantage, so that parallel-scanning, where it is used at all, is primarily employed with simple 2-D positioning control systems where, in the majority of cases, no switching information is introduced by way of the punched-tape scanner. It is also found occasionally with pneumatic control systems. It will therefore be convenient to discuss this method of scanning briefly later in connection with pneumatic control systems. Series scanners with their variable block and sentence lengths are not restricted to specific types of machine; they are therefore very suitable when employed as standard appliances for use with punched-tape control systems on all types of machine tool and so will be discussed in detail below.

Scanners for the series-scanning of information blocks

In principle there are three methods that can be employed for scanning punched tapes:

> Electro-mechanical scanning,
> Photoelectric scanning, and
> Electrostatic (dielectric) scanning.

In view of the adverse conditions encountered in service only the first two of these methods are of practical importance for machine tool control systems. Figure 226 shows the principles of some designs that have been tested in recent years. *Mechanical scanners* have scanning speeds of about 7 to a maximum of 120 characters/second, while the scanning speeds of *photoelectric scanners* are 20 to 500 characters/second and more. In general, recent experience has shown that any electro-mechanical scanning method will damage the punched tape to a greater or lesser extent, and that reliability is not very high due to the presence of mechanical contacts, especially at high scanning speeds. Figure 227 complements Fig. 226 by showing the principles of an improved electro-mechanical scanning system (smaller masses); even this system is, however, available only for speeds of up to 120 characters/second. Because of the frequent demand for higher working speeds and also their absence of wear, photoelectric scanners are becoming increasingly popular, and the examples to be described will all be of this type.

Reliability has become the most important criterion that a punched-tape scanner must meet, since this is the critical link on which the constant serviceability of a valuable item of plant depends. The trend is thus undoubtedly following two paths:

a) to make the process for feeding-in the data as reliable as possible;

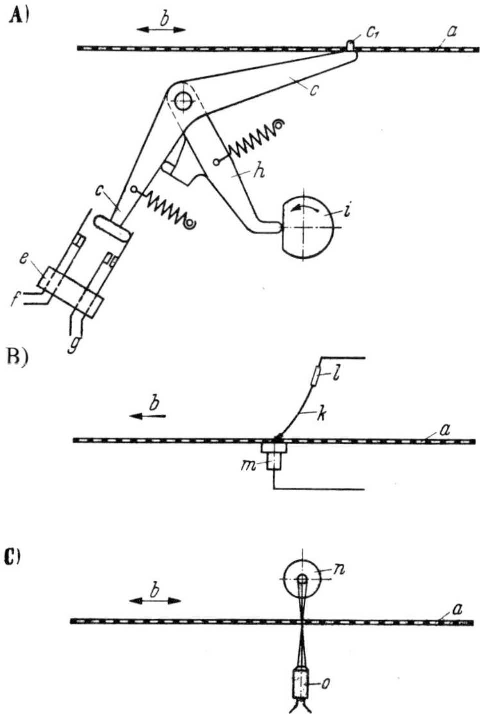

Fig. 226 Principles of operation of some punched-tape scanning systems (6)

A) Electro-mechanical scanning with moving detector pins. B) Electro-mechanical scanning with fixed brushes. C) Photoelectric scanning.

a Punched tape. *b* Possible directions of motion of punched tape. *c* Detector arm. c_1 Detector pin. *e* Contact block *f* Contact 1 (closes if hole present). *g* Contact 2 (closes if no hole present). (*f* and *g* provide the signals which indicate the presence or absence of a hole to the processing circuits). *h* Actuating link. *i* Motor-driven camshaft (the motor also drives the tape feed intermittently through a clutch; to simplify the figure this drive is not shown). *k* Spring-loaded contact brush. *l* Brush holder. *m* Fixed contact. *n* Light source with concentrated rays. *o* Photocell.

b) to eliminate the reader and the data-carrier completely, i.e. to take the work information directly from the computer (DNC).

Both these possibilities will be discussed in this chapter.

The most vital points in all high-speed punched tape scanners are:

the tape transport,
rapid, error-free reading, and
the ability to deal with long lengths of tape without causing wear of
the components.

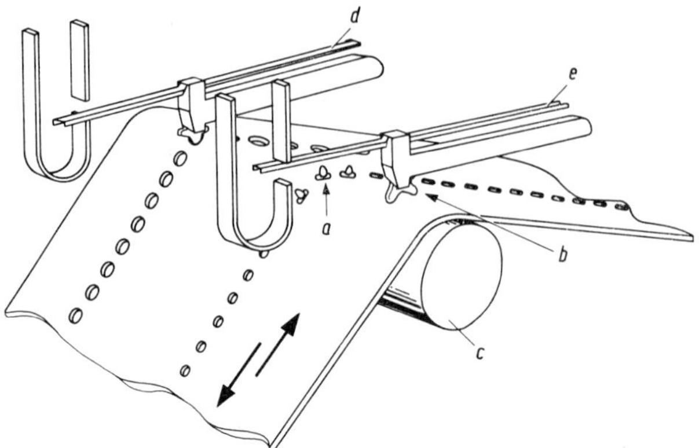

Fig. 227 Diagram of operation of an improved electro-mechanical punched tape scanning system for faster scanning (Courtesy of *Tally*)

a Perforated tape on feed sprocket. *b* Star wheel for making contact through perforations. *c* Feed roller (also completes electrical path to *b*). *d* Low-mass detector lever in hole-present position. *e* Low-mass detector lever in hole-not-present position.

Tape transport

Punched tape scanners have long been used in conjunction with teleprinters; Figs. 226 and 227 show the types of scanner that have been derived from this application. They use a feed sprocket that is connected to the drive motor, which rotates continuously, by means of a quick-acting electro-magnetic clutch. If a reversing mechanism is provided and two clutches are used it is possible to run the tape through the scanner in two directions. If a particular tape is frequently used this type of feed will clearly give rise to wear. For photoelectric scanning use is therefore made of feed rollers, the tape being gripped between a movable pressure roller and a fixed drive roller; here the pressure roller is controlled by a solenoid. Figure 228 shows a drive system without a drive sprocket of this type which is suitable for *one* reading direction. If both reading directions are to be available a transport roller must be fitted on either side of the reading head to ensure that the tape is under tension at all times. The feed track shown in Fig. 224, which was originally intended for use with mechanical scanning, then becomes a photoelectrically scanned *timing track* to achieve correct pulse scanning by means of the electronic system.

Both electro-magnetic couplings and magnetically-operated pressure rollers are components that possess mass and are subject to wear. A number of attempts have therefore been made to replace them by other components. In the punched-tape scanner shown in Fig. 229 the drive system employs an electric stepping motor, the principles of which have been discussed in

Fig. 228 Transport system for a high-speed photoelectric punched tape reader for one reading direction (400 characters/second) (Photo courtesy of *Remex*).

a Drive roller. *b* Pressure roller, controlled by a solenoid. *c* Interchangeable source of light. *d* Magnetic brake for punched tape.

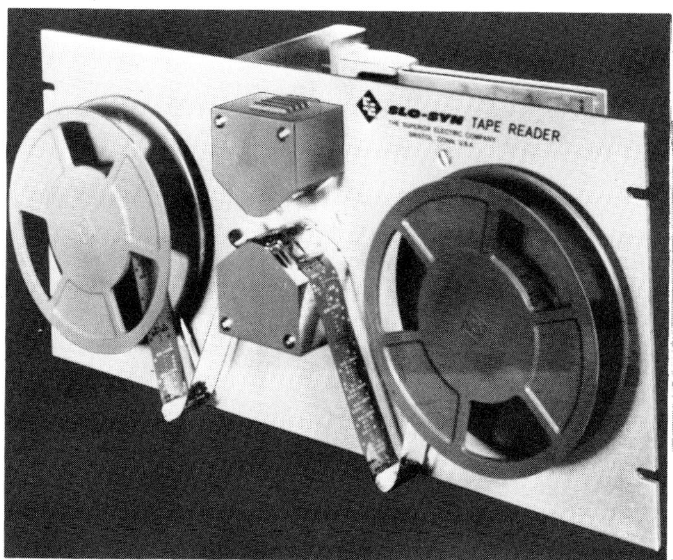

Fig. 229 Photoelectric punched tape scanner driven by an electric stepping motor for scanning speeds of 0 to 125 characters/second (254). (Photo courtesy of *Superior Electric*).

Chapter 6. For this particular application the low torque produced by an electric stepping motor is quite sufficient. The drive roller and the angle through which rotation occurs at each step must be matched in such a way as to ensure that each angular step of the motor corresponds to the spacing between successive characters on the punched tape (2.54 mm in Fig. 224). In

Fig. 230 Example of a photoelectric punched tape scanner with start-stop operation of a disc-armature motor (Servalco) for reading speeds up to 80 characters/second (Photo courtesy of *Infranor*)

the example shown in Fig. 230, on the other hand, a *Servalco* disc armature motor is used. This type of motor has also already been referred to in Chapter 6, but it was not powerful enough for driving machines, an application which was then under discussion. Disc armature motors can readily be used for the pulse operation of a punched tape scanner; the drawback is, however, that the electro-mechanical components previously described are now replaced by brushes and commutators, which require maintenance.

It is hardly possible to predict on theoretical grounds which of the drive elements that have been described will prove the most popular in the long run. It is probable, however, that further experience with numerical controls will result in a preference becoming apparent during the course of the next 2 or 3 years. Finally, if accurate start-stop operation is required, a reader that is equipped with pressure rollers must be provided with means of braking the punched tape as it passes rapidly through; this can be achieved without difficulty with the use of magnetic brakes.

Rapid, error-free reading

Very little difficulty is encountered with photoelectric equipment; the silicon photocells used nowadays present no difficulties as regards switching times, sensitivity and life, especially as the amplifiers and pulse production stages

Fig. 231 High-speed photoelectric punched tape reader for two reading directions with spooling capacity for long tapes (234) (Photo courtesy of *Remex*)

usually take the form of integrated circuits fitted in the readers close to the photocells. The practical difficulties arise rather in the light sources and the focusing of the rays of light. As is apparent from the illustration of a punched tape shown in Fig. 224, there is not much room for nine photocells side by side (8 information tracks, 1 timing track). To some extent conditions are similar to those obtaining in the optical angular step transmitters described in Chapter 3. The lamp must be powerful, have a long life, and be easily interchangeable. In addition the beam that is concentrated by the lens system should be 'cold' so as not to damage the heat-sensitive photocells.

Figure 231 shows a reader the design of which was based on similar considerations to those prevailing in the graduated scale scanner shown in Fig. 23.

The use of a glass-fibre optics system enables a sufficient distance between a powerful source of light (halogen bulb) and the photocells to be achieved; in addition, full use is made of the possibilities offered by integrated circuits (Chapter 3). The combination of measures of this type enables such a reader to produce useful signals even if the tape is 70 per cent transparent; the life of the quartz bulb amounts to 15 000 hours (234). The electronic system will work reliably at ambient temperatures ranging from 0 to 70 deg. C, and it is possible to read the tape while it is moving in either direction.

Ability to deal with long lengths of tape

When simple workpieces are being machined the punched tapes are usually not long; it is even possible to join their ends together to form a loop and to house them in a flat cassette so that they are protected from dirt (Fig. 232). If at all possible efforts should be made to keep the programming tapes short by arranging for repeated machining cycles being scanned by rewinding the tape; in such a case the reader must be capable of working in both directions of motion of the tape and must switch off when a particular symbol is reached (Chapter 10). When machining workpieces of very complex form (129) it will, however, not be possible to avoid the use of very long tapes. Table 14 shows that the storage capacity of a punched tape is in any event not particularly high, so that in some circumstances the limit is soon reached. In such a case provision must be made right from the beginning for adequate spooling capacity; this is nowadays available in various forms. In view of the desirability of rewinding which has already been mentioned — it is probably advisable to provide two spools of adequate capacity. These are now available with capacities of up to 600 metres of punched tape; Fig. 233 shows an example of such an arrangement. The technical requirements are similar to those for the well-known domestic tape recorders: Rapid forward and reverse winding, good precision braking device for the spooling device, avoidance of excessive tape loads, breakage etc. If a programme requires tapes that are longer than can be accommodated in the available spooling device, it is possible to overcome this difficulty by adapting the programme itself. Nearly every programme has places where it is technically feasible to interrupt it; arrangements can be made to stop at these points and to insert a new tape.

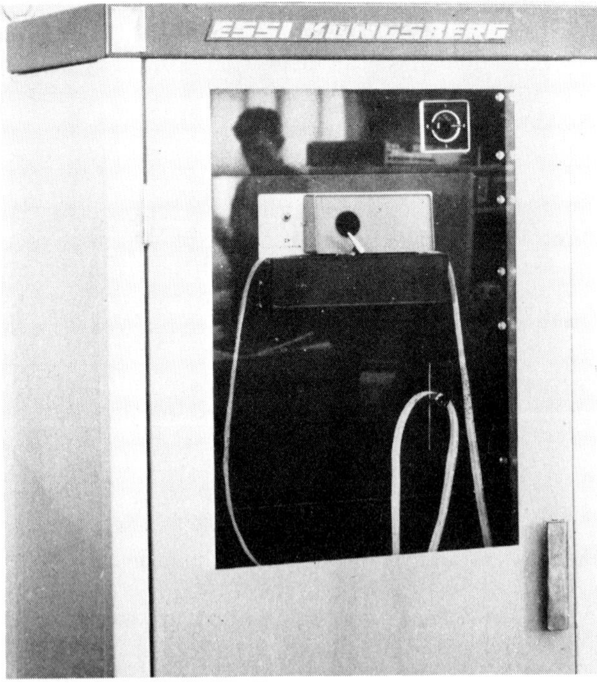

Fig. 232 Punched tape reader with a cabinet to protect the loop of punched tape (Photo courtesy of *Kongsberg*)

Fig. 233
Spooling device for very
high tape speeds (up to
500 characters/second)
and for tape lengths of
up to 300 metres.
Rewinding speed in both
directions 500 cm/s.
(Photo courtesy of
Remex)

9.3.2 Decoders and checking of characters

During the discussion on various types of code in Chapter 1 it was pointed out that it is necessary to provide for a certain amount of redundancy in automated data processing as a precaution against errors. Since this, however, is also a question of economics, which is closely connected with the reliability of components and groups of components, it was necessary to allow a certain amount of trial-and-error time during which the various proposals could be tested. Now it appears that, at least with NC, there is international agreement on the combination of 7 information bits and 1 check bit.

Code B in Table 4 (Chapter 1) is a code comprising 6 information bits and 1 check bit. The two codes, with 6 and with 7 information bits, will be discussed in detail in Chapter 10, since both forms are currently in use (EIA and ISO codes).

With the input equipment it is sufficient to take the 7-bit code as an example to show the principles of checking the characters. The additional 8th bit is used for checking, and is known as the *parity bit* or the *parity check*. Since 7 is an odd number, the parity bit is always introduced into the character when the number of holes is odd; i.e. it is used to make up an even number of holes, and the check consists of making sure that the number is in fact even. A check logic circuit of this type is shown in Fig. 234. In practice, these are nowadays almost invariably integrated circuits which are very little larger than those illustrated in Fig. 37. Whether the check circuit is to be incorporated in the tape reader itself or in the subsequent control system is a matter for the control system manufacturer. The readers illustrated in Figs. 229 to 233 can be fitted with this attachment or not as desired. There are various ways in which the error signal can be utilised:

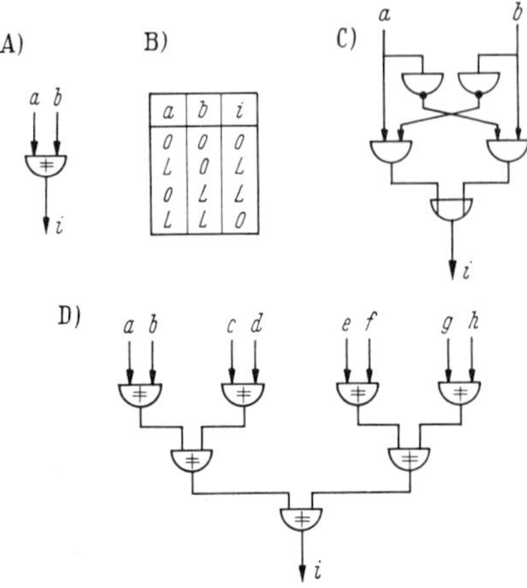

Fig. 234
Logic circuit for a parity check that there is an even number of holes per character.

A) Symbol for the logical relationship EXCLUSIVE OR.
B) Corresponding logic table.
C) Construction of EXCLUSIVE OR circuit from basic AND, OR and NOT units. D) Construction of check logic circuit for an 8-track punched tape.

a to *h* Input signals L or O. *i* Output signal L = correct O = error.

Either the reader, and hence the machine, is stopped immediately or the control system has buffer stores arranged in such a way that it always has 4 to 5 programme sentences in-hand. There is then sufficient time to rewind the tape and once again to read the new block containing the character that has been reported to be wrong. If the error is reported again the machine will have to be stopped; if it was merely a case of a reading error the machine can continue to work.

A matter quite apart from checking for errors is the decoding of the combination of holes. As has already been mentioned, with 7 bits it is possible to represent a total of 127 discrete characters and to output these onto 127 circuits as individual signals. With current NC machines this large number of signals is as yet not required. In practice efforts are made to make do with about 50 characters, (DIN 66024 and Chapter 10) and to add the characters that are not required ($127 - 50 = 77$) to the redundancy and hence to increase the reliability. The precise number depends on the machine tool and its control system. For this reason decoding is usually done in the control system and not directly in the reader, i.e. signals are usually present at 8 leads coming out of the reader simultaneously, and these are then passed on for further processing. To prevent this discussion becoming too complicated, Fig. 235 shows the logical circuit for a decoding and checking circuit for a $\left(\frac{5}{3}\right)$ code (Code E in Table 4 of Chapter 1).

9.3.3 *Buffer stores*

It has been mentioned on several occasions that buffer stores are required for series scanning, which is the system most widely used. These stores are selected, filled, and erased by the output signals W of the decoding circuit (e.g. that shown in Fig. 235). In practice, these stores often take the form of

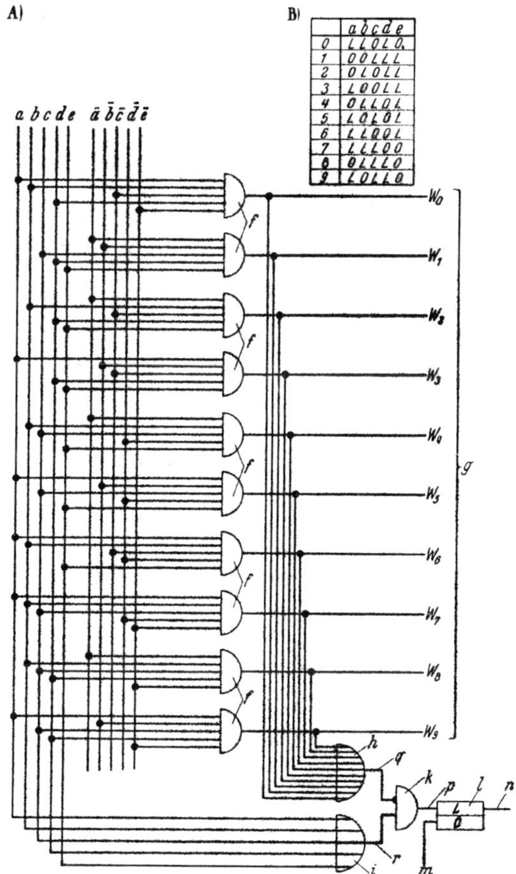

Fig. 235 Logic circuit for a decoding and checking circuit (5/3) corresponding to Code E of Table 4.

A) Logic circuit. B) Code table of a (5/3) code (Code E).

a b, c, d, e Places of elements of binary numbers. $\bar{a}, \bar{b}, \bar{c}, \bar{d}, \bar{e}$ Negations of corresponding values. *f* AND elements. *g* Output signals from circuit. *h* Ten-part OR element. *k* EXCLUSIVE OR element. *l* Binary store. *m* Erase signal. *n* Stored alarm signal. *p* Output signal (error signal) from check circuit. *q* Output signal of OR circuit *h*. *r* Output signal of OR circuit *i*.

flip-flops (Figure 33). The fact that these stores (especially when used for storing switching information) are often linked by logic couplings (interlocks) is of no consequence in this context. The selection of a store is also termed the 'information distribution' within a machine tool control system. Selection

can be achieved in two ways, depending on the programming technique
employed:

    a) by the address method, and
    b) by the stepping method (which corresponds to the tabulator process
       in programming, Chapter 10).

The two methods are compared in Fig. 236.

The address method

Here each store has a definite 'name' and is 'called up' by at least one discrete
signal from the decoder. Strictly speaking the number of discrete signals
should then be as large as the number of stores that are to be filled with
information. In complicated machine control systems this requirement would
lead to considerable complexity of decoding and checking equipment, and
would cause difficulties when representing the commands to the machine by
symbols for programming purposes. The stores are therefore gathered
together into a relatively small number of convenient groups.

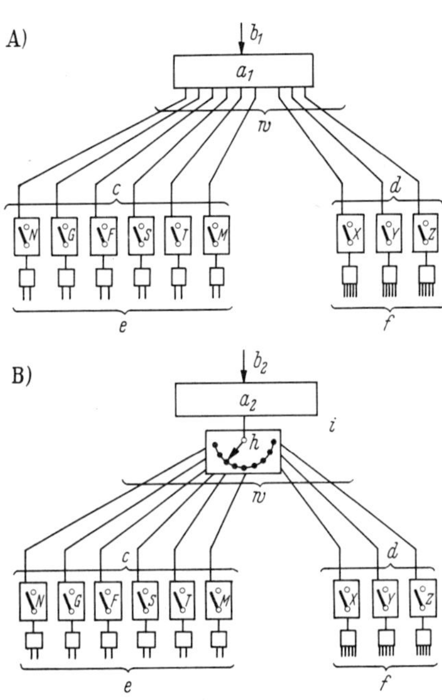

Fig. 236
Block diagrams of information distributors.

A) Address method.
B) Stepping method.
N,G,F,S,T,M Symbols for six switching-information stores (Table 12). X,Y,Z Symbols for three displacement information stores. a_1, a_2 Character checks (Fig. 234) and decoders (similar to Fig. 235) for the address process. b_1 Punched tape information with address details. b_2 Punched tape information without address details, but with TAB symbol as separating symbol (Chapter 10). c Switching information store. d Displacement information store. e Two-place shift store (in example). f Five-place shift store (in example). h Stepping switch or similar. i Control signal for stepping switch (derived from TAB symbol). w Address signal.

The method can best be explained by taking an example:
Consider the main spindle head of a milling machine, which has a spindle speed range of 18 steps. Each of the 18 spindle speeds can then be selected by a group of three signals:

> one signal S as the address for the 'main spindle' section of the control system,
> one tens signal,
> one units signal.

In Fig. 236A) the six principal storage groups c for the switching information are N, G, F, S, T, M (Table 12 in Chapter 7) and there are three storage groups d for the displacement information, X, Y, Z; a total of 9 addresses must therefore be provided. The contents of the subsequent shift stores e and f can then be defined by 10 to 12 further signals, e.g. the 10 decimal numbers and the two signs + and −. The presence of the two-place shift store e and the five-place shift store f in this example keeps the number of address letters required to an acceptable figure.

The stepping switch method

In Fig. 236b) item h is a component which can be a stepping switch or a shift register. Each time a stepping pulse i is emitted from the decoder a_2 (e.g. on the appearance of the TABULATOR symbol) the stepping switch will move by one further step and at each step it opens the input to another shift store; these registers can then be filled with numbers as previously described. After each new TAB impulse the stepping mechanism moves on to the next store. With this process no addresses are therefore needed, and it is only necessary to ensure that the sequence of the stores to be used is fixed. With this process, known in programming as the TAB process (Chapter 10) it is possible to save on address letters, but on the other hand a further possible source of failure in the form of the stepping mechanism is added to the system. This is probably the reason why the TAB process for programming, which is associated with the use of stepping switches, has lost considerable popularity; the current ISO proposals for Standards include only the address method or a mixed TAB-address method, although there are bound still to be many simple control systems in service which use stepping switch methods (TAB programming).

9.3.4 *Pneumatic scanning of punched tape and pneumatic logic elements*

Two pneumatic-hydraulic drive systems were described in Chapter 6 and the complete machines were illustrated in Figs. 123 and 126. Figure 237 shows the method of operation of a pneumatic punched tape reader for series

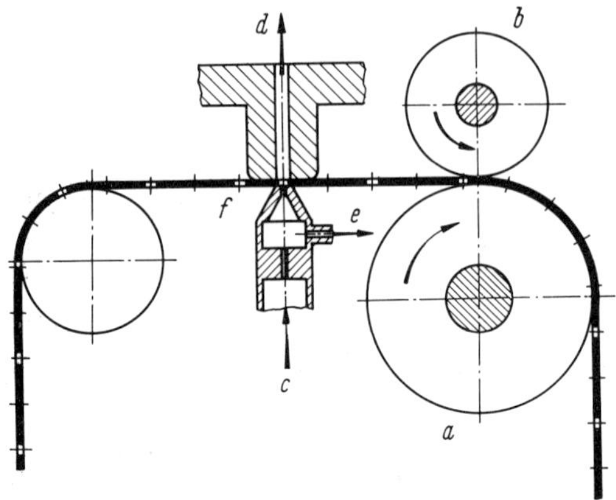

Fig. 237 Method of operation of a pneumatic punched tape reader for series scanning.

a Drive roller. *b* Pressure roller (can be controlled pneumatically). *c* Supply pressure. *d* Air exhaust. *f* Air nozzle. *e* Static pressure signal.

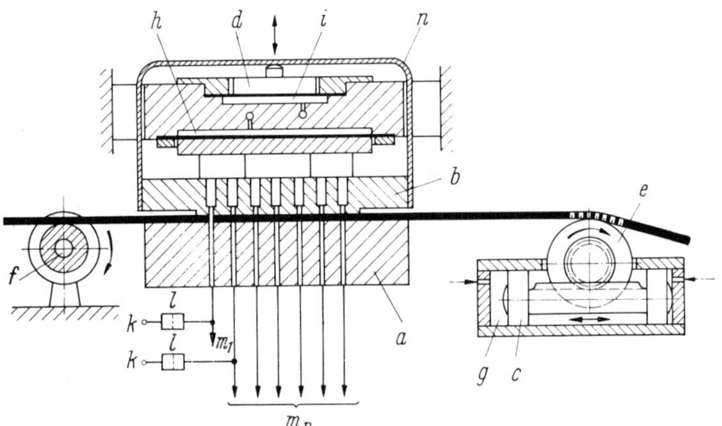

Fig. 238 Schematic diagram of operation of a pneumatic punched tape reader for the block scanning of a maximum of 25 8-track characters per block.

a Distributor block, viewed in direction along block; further rows of nozzles are arranged behind those shown, depending on the number of tracks on the tape. *b* Movable clamping plate with exhaust holes. *c* Pneumatically-operated feed piston. *d* Diaphragm-controlled plunger unit for raising casing *n* with clamping plate *b*. *e* Sprocket for feeding punched tape (driven by piston *c*). *f* Guide roller for punched tape. *g* Compressed-air cylinder (the length of the cylinder depends on the length of the block). *h* Diaphragm with pressure space for pressing down clamping plate *b*. *i* Diaphragm with pressure space for raising clamping plate *b*. *k* Connections for auxiliary air supplies. *l* Pressure reducers. m_1 to m_n Static pressure signals if no hole present at relevant position. *n* Casing for raising item *b*.

scanning of symbols of the type required for the control system shown in Fig. 126. Figure 238 shows a schematic diagram of a pneumatic punched-tape reader for block scanning. The principle on which the pneumatic scanning system operates is the same for both these punched tape readers. Air at the supply pressure c or k, throttled by l if necessary, is led to the nozzles f or a. If a hole is punched in the relevant position in the tape the air can escape through d or b; there will then be no pressure build-up at e or m. If, on the other hand, no hole is present, the nozzles will be blocked, and an appreciable pressure builds up at e or m. This can be processed by pneumatic logic elements (Figs. 240 and 241) or can be converted into electrical signals by pneumatic-electrical signal converters (similar to Fig. 240) which are then processed in the normal manner. Figure 238 also illustrates a pneumatic punched tape feed system.

Finally Fig. 239 shows more detail of the control console for the machine shown in Fig. 123. Interesting features here are that the control system contains:

a) pneumatic block reader a, and
b) pneumatic tape punch b.

Like the control system shown in Fig. 123, this control system can therefore act as a repeating unit for reproducing tape, so that it is not necessary to make use of the services of a programming office, and the machine can be used for both one-off jobs and batch production. For this reason it is also included in the Appendix (Fig. 299).

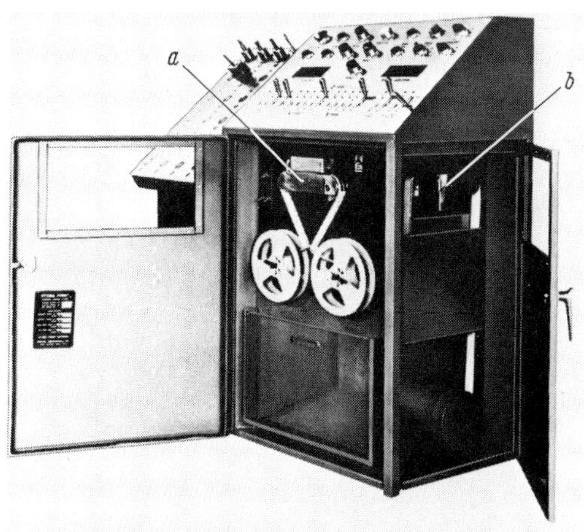

Fig. 239 Pneumatic NC console (Figs. 121, 122, 123, and Fig. 299 in Appendix) (Photo courtesy of *Moog*)

 a Pneumatic punched tape reader for block scanning. b Pneumatic tape punch.

Experience to-date indicates that pneumatic tape readers have the advantage that any particles of dirt, which are inevitably present in a workshop, and which could act as a disturbing factor, will be blown clear, so that a high degree of reliability is achieved. A new development of considerable promise, however, is pneumatic processing of digital signals; pneumatic components have long been widely used in analogue systems (3). Even though it is not possible at this point to discuss details of this new fluidics technique — the literature is now quite extensive (117, 235, 236), it nevertheless seems advisable to touch briefly on the principles underlying the major components concerned.

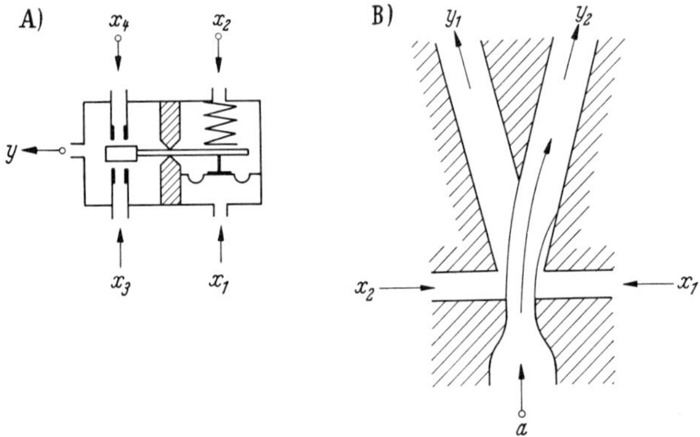

Fig. 240 Diagramatic sketches to illustrate the principles of types of pneumatic logic elements.

A) Versatile diaphragm relay. B) Bistable flow element (wall flow element) (117).

a Air inlet. x_1 to x_4 Input signals. y, y_1, y_2 Output signals.

In general it is necessary to distinguish between two principles of operation for these components, namely components with moving parts and components in which the flow itself performs the control function; these may be regarded as static and dynamic components respectively. Figure 240A) illustrates the basic principle of a typical diaphragm relay which can be used as the basic unit for a variety of logic circuits. Since it includes masses which have a certain inertia and mechanical moving parts, there must be a certain amount of wear, even though this is hardly detectable, and the switching times are slightly longer than for flow-controlled components. On the other hand the construction is such that it is easy to include electrical contacts, and similar components are used at the interface of all electro-pneumatic control systems.

Figure 240 B) illustrates the principle of a bistable wall flow element. The operation of this depends on the tendency of a flowing medium to adhere to the wall. This effect was discovered by Coanda about 30 years ago and has been named after him. In practice, the effect of this is that the air that flows in at a can flow out either at y_1 or y_2, but never at both. A short pulse of compressed air at x_1 or x_2 can 'tip over' the direction of flow, and it will again be stable in its new direction. The arrangement acts in precisely the same way as the electronic flip-flop shown in Fig. 33. It is, however, much simpler, and can easily be produced in plastics in large quantities. It forms a virtually ideal non-electrical component with storage capacity. Figure 241 shows an example of a fluidics control console which is suitable for many applications.

From the few examples which it has been possible to show, it should be apparent that in some applications it may well be worthwhile to consider alternatives to electronic methods. Both their low price and their high reliability under the adverse conditions obtaining in practice in workshops are advantages of the pneumatic tape readers which should not be underestimated. As against this, they have the disadvantage that pneumatic elements are slower in operation than electronic elements and that pneumatic control systems cannot be connected directly to an integrated data-processing system.

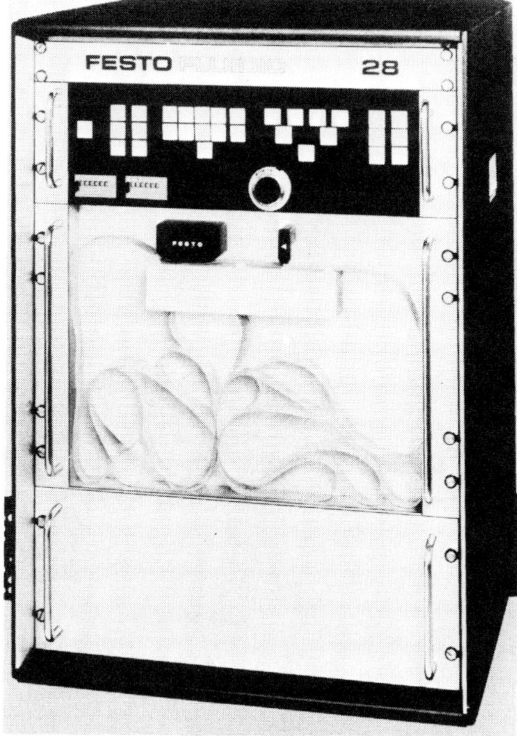

Fig. 241 Versatile control console with pneumatic punched tape readers (series scanning) and fluidic logic elements (Photo courtesy of *Festo*).

9.4   Magnetic tapes as information carriers on numerically-controlled machine tools

Compared with other types of information carrier the magnetic tape offers two major advantages:

  a) the density of information per unit area is very high (Table 14); and
  b) the information can easily be stored and also erased without damaging the tape.

For both these reasons magnetic tapes are widely used in general data-processing applications. Here, however, it is only the specific case of the control of machine tools by magnetic tape which is of interest. As has already been mentioned in Chapter 2, the high specific density of the information makes the magnetic tape particularly suitable for use as an information carrier in conjunction with external interpolators. In workshop service, on the other hand, the second of the storage factors, namely that the information can readily be erased, can be a nuisance. In a data-processing installation that has been installed with care and which is properly supervised at all times, and in which the magnetic tapes are always stored in special units in air-conditioned rooms, this does not prove a major difficulty. The technical objections to the use of magnetic tapes in conjunction with machine tools are based on three grounds:

  a) During its long journey from the computer, via the library, to the machine tool it is possible for the information on the magnetic tape to be damaged (partially erased) or completely erased, either deliberately (e.g. through sabotage), or accidentally (e.g. through carelessness). Punched tape cannot suffer undetected a damage of this type.
  b) With magnetic tape it is not possible to detect whether the information it carries has become defective without the aid of complex inspection devices. With punched tape it is necessary for fairly severe damage, which is easily recognisable, to occur (e.g. tape torn, additional holes punched, or holes covered by pasting on strips) to alter the information once this has been punched in by machine.
  c) The magnetic tape and its scanning devices are very sensitive to dust (especially if this contains particles of cast iron), so that compared with punched tape it is necessary to adopt more extensive precautions when it is used on metal cutting machines (e.g. use of cassettes, installation of control console in dust-proof rooms).

Both the organisational reasons, which will be discussed later (Chapter 11), and these technical arguments have together resulted in frequent objections being raised to the use of magnetic tapes as information carriers under the severe conditions prevailing in the workshop, so that punched tapes are often preferred despite the undoubted advantages of magnetic tapes. It is possible, however, for a change to occur in this attitude as progress is made in the development of tapes and the associated equipment, especially since the ½-in tapes offer the further advantage that they are fully compatible with the general data processing equipment (244). In the plant illustrated in Fig. 178A) full use is made of all the advantages mentioned and all the appropriate safety precautions are taken; the individual machines forming the Molins 24 system are controlled by magnetic tape and are 'administered' from air-conditioned central control rooms.

It is a basic feature where magnetic tapes are used in conjunction with external interpolators that the interpolated displacement information must be entered continuously. Intermittent operation and the widespread use of buffer stores, as used for punched-tape systems, are not employed but, on the other hand the control consoles are small and simple (see Fig. 243). Since the magnetic tape has to act as a continuous emitter of required dimensions, even this will have to be very long for complex workpieces (large tool displacements). To enable the spools to be kept relatively small — even though the programmes are extensive — it is desirable not to select too high a tape playback speed. There are technical limits to the extent to which this advice can be followed. Tape speeds employed in practice at the present time range from 9.5 cm/s to 38 cm/s.

Storage of digital and analogue guide values

The method by which the information is stored on the magnetic tape is a factor that is of major importance. The information always takes the form of the guide values for the internal control circuits in the machine, which can operate in either digital or analogue mode. Because of the amount of space required on the tapes and the resultant technical difficulties, it is not possible to adopt digital-absolute or analogue-absolute methods of storing information. In principle magnetic-tape control systems can therefore operate by digital-incremental methods using pulses or by using continuous analogue guide values and follow-up control loops (8, 51, 239); the most recent British installations employ an intermediate solution (246). Two possible methods of storing information on magnetic tape are shown schematically in Fig. 242. Figure 242A) shows a four-track magnetic tape having a ¼-in width; this corresponds to the tape dimensions normally used for domestic tape recorders. It is not advisable to accommodate more than four tracks on a narrow tape of this type; the interference is increased unduly as a result of which the reliability is reduced. In the example illustrated there are three tracks for the displacement information in three axes and one track for checking purposes. The number of impulses for each axis describes the displacement that is to be effected, while the frequency of the impulses gives the speed at which the machine slide is to move in the direction of the axis concerned. The direction of motion is determined by the polarity of the impulses. The information that is produced and stored in this way using a computer and an external interpolator can be fed in the form of required values into a digital comparator, similar to that shown in Fig. 43. The movement of the machine slide is then effected by means of digital follow-up circuits (9), the actual values being determined by means of measuring devices of the type shown in Fig. 24. As a precaution against faulty pulses the fourth track carries check pulses which meet the requirement $a = \dfrac{x + y + z}{4}$. In effect then one-quarter of the sum of the pulses is stored and is checked in the magnetic tape reader (40, 239). If suitable power amplifiers are employed, stepping motors, as shown in Fig. 120, can be operated directly from the pulses on the magnetic tape; a control system as shown in Fig. 1 will then be obtained, although to be fully effective this should be supplemented by a workpiece measurement control system as discussed in Chapter 6. Simple though the method using pulses stored on magnetic tape may appear, it nevertheless introduces difficulties owing to the presence of defective or 'blind' sections in the magnetic layer. This is probably one of the reasons why the method is no longer employed, having been replaced by a method that will be discussed later. Furthermore, it is not possible to accommodate on the narrow ¼-inch tape further tracks with switching information. If more information is to be accommodated it is necessary to go over to ½-inch tapes with seven or nine tracks, such as are usual for data-processing installations. Further displacement information for machine control systems can then be accommodated on five or six tracks. Figure 243 shows an example of a magnetic-tape control system for ¼-inch tapes, while a machine control

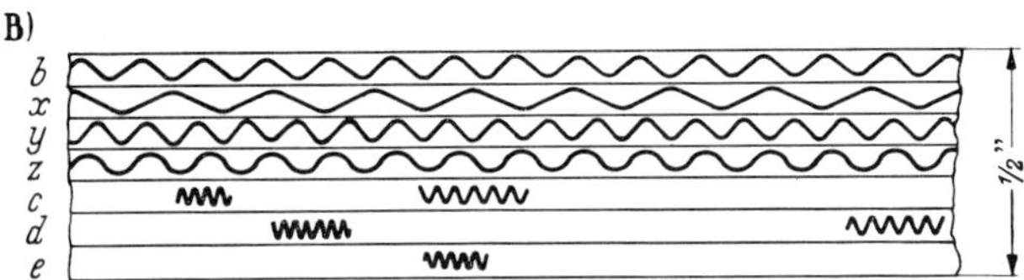

Fig. 242 Two methods of storing information on magnetic tapes (partially superseded by later developments).

A) Digital storage (pulse method). B) Analogue storage.

a Check track. *b* Master timing frequency. *c,d,e* Signals of various frequencies to provide switching information. *x,y,z* Fully-interpolated displacement information for the three machine axes. (Guide values for feeding into follow-up digital or analogue control circuits.)

Fig. 243 Production milling machine with 4 spindles equipped with a magnetic-tape control system for ¼-in tapes (Photo courtesy of *Ferranti*).

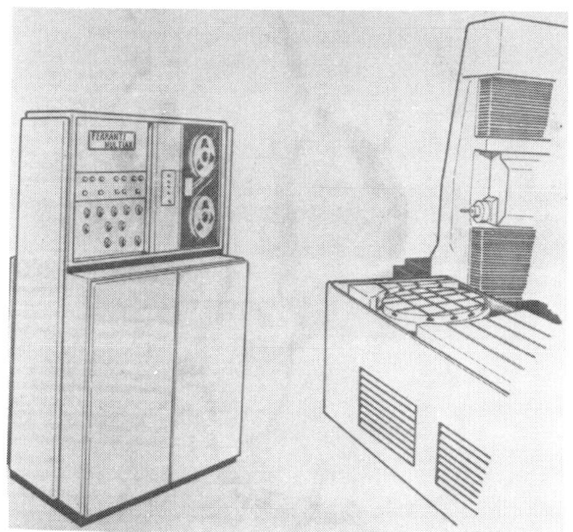

Fig. 244 Sketch of a MULTIAX control cabinet with magnetic-tape control system for ½-in tapes (see also plant illustrated in Fig. 178A) (Photo courtesy of *Ferranti*).

console for ½-in tapes is shown on the drawing Fig. 244. The machines illustrated in Fig. 178B) are to be equipped with control cabinets of this type and are to be joined together to form complete machining lines as shown in Fig. 178A); the control cabinets for the machines will be accommodated in a separate air-conditioned computing centre.

Figure 242B) is a representation of the technical features of magnetic tape control by analogue means (240). A seven-track ½-in magnetic tape is taken as an example. Four tracks show voltages of different frequencies and phase positions. It must be emphasised that the frequencies are, of course, not drawn on the tape itself like oscillogram figures (as shown for convenience on the sketch), but the sine curves are intended to represent the rhythmic variations in the induction of the magnetic tape. Let track *b* carry a master timing frequency of say 100 Hz, single-phase. When the tape is played back this can be used to produce a three-phase reference field within the machine tool by means of an electronic three-phase generator, and, as indicated in Fig. 61, any required number of synchros can be attached to this. Track *x* carries a single-phase voltage supplied by the external interpolator of a frequency and phase relationship that corresponds to the required speed and position of the slide in the *X* axis. The magnetic tape thus replaces a synchro transmitter. If the slide is coupled in the X-axis with, for example, a synchro (fed from the three-phase reference voltage) that detects the actual movements, the signal on the magnetic tape and the single-phase voltage of the synchro receiver can be compared by means of a phase-sensitive bridge circuit. It is thus comparatively simple to build up an analogue follow-on control loop using the magnetic-tape signal as the required value, the synchro signal as the actual value, and the phase-sensitive bridge as the comparator. The same principles are adopted for the other axes. It is obvious that a system of this type can be used for the simultaneous control of even very complicated slide movements (243). These control circuits must, of course, be very carefully dimensioned, but this also indicates the direction in which improvements can be expected.

The process offers two advantages:

1. Discrete pulses have been replaced by continuous magnetic variations. If small 'blind' spots occur on the magnetic tape their effects are outweighed by the inertias of the synchros, etc.
2. The tape speed is eliminated as a factor. The large machine tool is no longer dependent on the small tape drive mechanism. On the other hand the operator no longer has the opportunity of modifying feed rates to allow for any difficulties arising from the materials or the tools, in the way that he can where internal interpolators are used (Chapter 8).

Signals of various frequencies are shown on tracks, *c, d,* and *e* of the example. These can be used for feeding switching information into the machine. The various frequencies are then filtered out and the signals are transmitted to the remainder of the machine control system.

The analogue process that has been described has clearly not proved successful in practice; no plants operating on this system are known to have been produced in recent years. It seems probable that the greatest difficulties were encountered in the dimensioning of the control circuits and in the time constants of the drive systems. A different solution is therefore adopted in more modern machines (244, 245). The sinusoidal wave forms are replaced by square wave forms, the measuring system is of the type shown in Fig. 84 with a pulse formation stage (sinusoidal/square form), and phase comparison is no longer made by analogue means but digitally with high sensitivity and accuracy along the lines indicated in Fig. 86. The subsequent positioners and power drives must then clearly have very satisfactory time constants (Chapter 6).

This latest development appears to have overcome many of the technical difficulties previously encountered with magnetic-tape control systems, and their advantages (e.g. ease of coupling with a data processing system) have again become more attractive (244, 245). Before making a final comparison of the features of punched tapes and magnetic tapes for use as data carriers in NC machines it is, however, necessary to take into account the latest developments which are discussed in the next section and to make allowances for the organisational effects of the use of magnetic tapes on the external data processing arrangements. The latter point will be discussed in Chapter 11.

9.5   Direct numerical control (DNC, on-line control)

Frequent reference has been made in this book to the close relationship that exists between the numerical control systems of machine tools and electronic computer installations. It is not surprising, therefore, that with the rapid developments that have taken place in the computer field, an attempt should

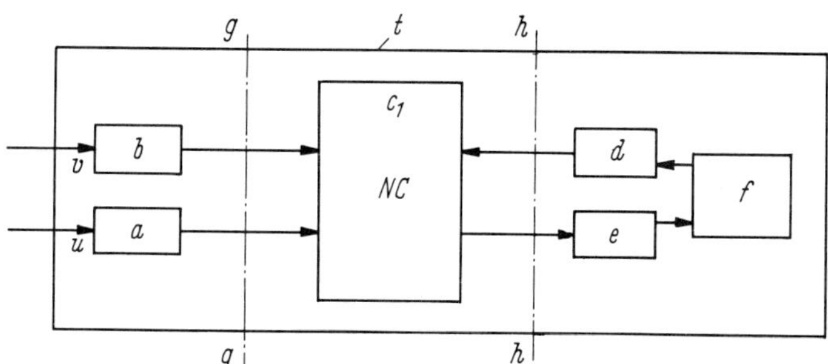

Fig. 245 Block diagram for a normal NC machine with punched-tape input and continuous-path control. To enable the illustrations below to be derived more readily, an arrangement which differs slightly from that employed previously has been adopted.

a Punched tape reader. *b* Control panel with correction switches. c_1 Normal NC combination with character check, decoder, intermediate stores, internal interpolator, comparator, matching unit for external groups *d* and *e*. *d* Displacement measuring systems. *e* Drives. *f* Machine slide. *g* Junction 1. *h* Junction 2. *t* Machine/control-system combination. *u* Punched tape data. *V* Manual input of correction data.

be made to couple these two components together more closely than in the past. As has been shown, magnetic-tape control is already a step in this direction. There are probably three main reasons why serious experiments into the coupling of several NC machine tools directly to a computer should have been made during the course of the last two or three years:

1. The lack of reliability of punched-tape and magnetic-tape readers.
2. The technical under-utilisation of the opportunities offered by

A)

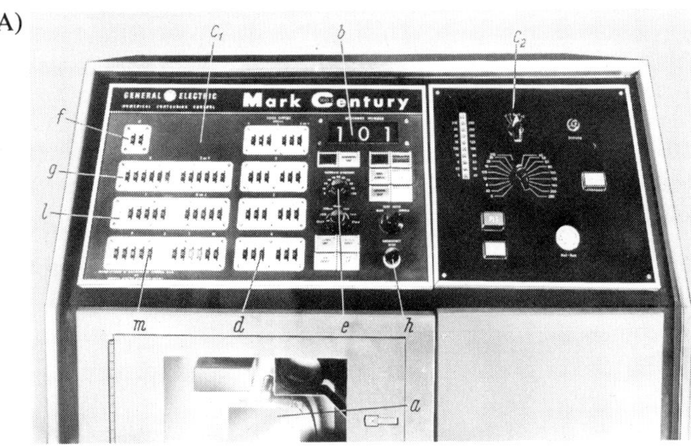

B)

Fig. 246
Example of a numerical 2-D continuous-path control system with punched tape input (Photo courtesy of *General Electric***).**

A) Control panel (corresponds to item *b* in Fig. 245).
B) Control console open.

a Punched tape reader.
b Sentence number indicator. c_1, c_2 Manual data input (c_1 introduced by control system manufacturer, c_2 considered necessary by machine manufacturer).
d Tool position indicator.
e Feed corrections. *f* Selector switch for G functions.
g Coordinate dimensions XY or XZ. *h* Emergency stop.
i Interpolator circuit boards.
k Power supply. *l* Interpolator inputs or pitch inputs for threads (Table 12 and Fig. 205). *m* Switching information.

extremely fast computers and the possibilities of using them as time-sharing computers.

3. The unnecessary increase in the cost of numerical continuous-path control systems due to internal interpolators.

It was predictable that while the direct connection of machine tools to computers would help to ease matters in these three areas it was almost bound to introduce new difficulties (246). All the installations of this nature that have been produced to date are to some extent of an experimental nature which can by no means be regarded as fully developed, quite apart from the fact that although a lot has been written on this subject (e.g. 248, 249, 250, 251, 252) the failures and the real know-how have not been publicised.

In view of this situation it is the intention to give only a brief description of the basic problems together with some sketches to illustrate the principles.

In Fig. 245 the signal flow between the principal units of a numerical continuous-path control system is represented in such a manner that it is possible to follow the further discussions without difficulty. Speaking very generally, it is possible to connect the computer output either:

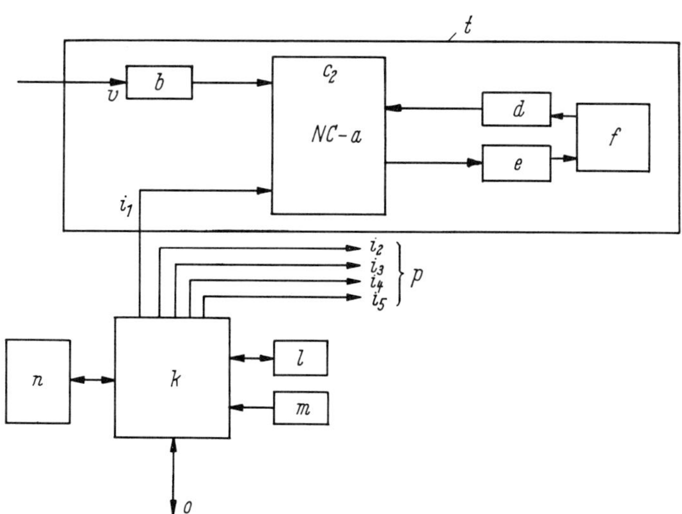

Fig. 247 Block diagram for a DNC system with the punched tape readers eliminated on the machines being controlled.

b Control panel with correction switches. c_2 NC equipment similar to c_1 in Fig. 245, without punched tape reader *a*, character check, and decoder. *d* Displacement measuring systems, *e* Drives. *f* Machine slide. i_1 to i_5 Data transmission connections. *k* Small process computer. *l* Typewriter. *m* Punched tape reader. *n* Disc or drum store. *o* Data link to and from next-largest plant computer. *p* Connections for further NC machines. *t* Machine/control system combination. *v* Manual input of correction data.

a) at the junction g and so in the first instance to bypass the punched tape reader; or
b) at the junction h and so to entrust the central computer also with the tasks performed by the internal interpolator (the comparator remains in the NC section)

To clarify matters, Fig. 246 illustrates a control console of the type required for conventional NC systems. Figure 246A) shows the control panel with the controls which, according to experience, are necessary to enable manual corrections to be made to the operation of the machine. In Fig. 246B) the control console is shown open; the circuit boards shown include the linear/circular internal interpolator for the 2-D continuous-path control system. With a above — computer output is connected directly after the punched tape reader — it is possible to eliminate only the punched tape reader (Block a in Fig. 245). This phase of modification is shown in greater detail in another block diagram in Fig. 247. In many cases, however, the punched tape reader would not be dismantled but retained in the control system (249) in order to:

a) make it possible to continue to process the punched tape even if the computer fails, or
b) make it possible to continue to use existing punched tapes if necessary.

The connection to the computer k is by data link i_1. Opinions differ as to the size of the computer k; this is, however, never the large central plant computer. It seems that it may be desirable to establish a certain hierarchy of computers to enable a sound and reliable mixture of centralisation and decentralisation to be achieved. (253) It may be correct to choose the process computer k so that it is of a suitable size to enable $8 - 10$ NC machines to be connected to it in this manner. To keep down the costs of the cables i (screened, multi-core leads, Section 9.3.3) the distance between the computer and the 10 or so machines that it controls should be kept within limits; the maximum distances currently considered are of the order of 50 or 60 metres.

These brief remarks are sufficient to show that even in the simplest cases the machine installation plans and the computers must be carefully matched. This is another reason why it may be desirable in a large plant to arrange the NC machines into groups and to place them under the control of a number of small process computers. A comparatively new idea is to provide the relatively small process computers with high-capacity peripheral stores (drum or disc stores) which can then accommodate several current programs for the NC machines in their entirety ready for calling up as required. New programs can be entered by means of the central punched tape reader m or by means of the data transmission link o from the next larger group or central computer. In other words, it is immaterial whence the programs originally emanate; it must merely be possible to draw them quickly from the peripheral store to

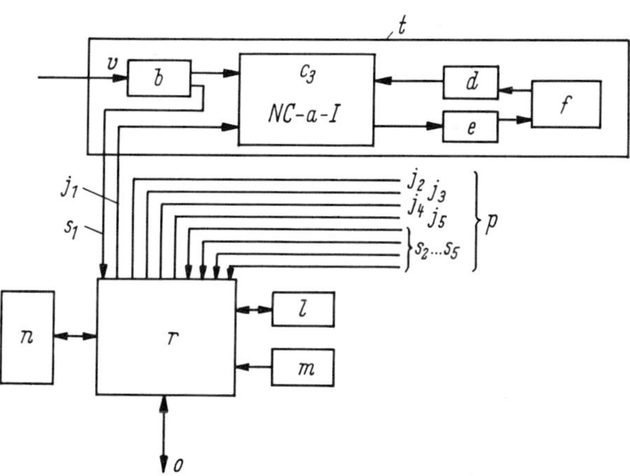

Fig. 248 Block diagram of a DNC installation with no punched tape readers or internal interpolators on the machines

b Control panel with correction switches. c_3 NC equipment without reader or interpolator, i.e. comprising only comparator and matching unit. *d* Displacement measuring systems. *e* Drives. *f* Machine slide. j_1 to j_5 Data links, may have to be suitable for high frequencies. *l* Typewriter. *m* Punched tape reader. *n* Drum or disc store. *o* Data link to and from larger computer. *p* Connections for further NC machines. *r* Time-sharing process computer. s_1 to s_5 Feedback of manual input data into process computer. *t* Machine/control-system combination. *v* Manual input of correction data.

enable the flexibility of the plant as a whole to be further increased. The machine control system itself will, of course, retain the capability for manual data input, as shown, for example, in Fig. 246A). In an emergency the individual machines can also work independently if the punched tape reader is left installed in the control console.

Much more far-reaching than this first step towards DNC is the second possibility mentioned, namely, transferring the tasks normally performed by internal interpolators to an external computer; this is sketched in Fig. 248. The absence of the internal interpolator now causes the size, weight, and cost of the control console to be reduced to a minimum. On the other hand the leads i_1, i_2, i_3 etc., must now be arranged to suit much higher frequencies, since the servo systems remaining in the machine control system (comparator, measuring systems, and drives) draw their guide values, which may be digital in form, from the computer r. This also has to be larger and more powerful than the computer k in Fig. 247; the latter acts mainly as a program distributor, but in this application the computer has to perform a much more extensive task and to calculate simultaneously the information required for the continuous-path control of 8 to 10 machines (e.g. in the time-sharing mode).

A computer of this type will certainly be less expensive than 8 or 10 internal interpolators, but experience will have to show whether, in the end, the additional costs of the installation, the provision of high-grade data links, etc., do not largely eliminate this saving. It must also be borne in mind that there is a tendency for the costs of internal interpolators to drop, while in experimental operations there have been cases where the possibility was soon recognised to use the high-grade time-sharing computer for other tasks than the interpolation calculations for several machine axes. Reference has already been made to these possibilities in Figs. 176, 177, and 178 in Chapter 7. This subject is to be discussed again against a wider background in Chapter 12. Here it can merely be concluded that the ideas of DNC operation lead into an area where considerable experimentation will be required and where a number of points will require classification. Under normal circumstances it would be possible to make progress in such a field only step by step. The experimental installation referred to in Chapter 8 (Fig. 218) also now appears in a new light.

9.6   Conclusions

Some interesting points on internal data processing have come to light in this last chapter:

1. With the increasing integration of the information flow through data processing systems the division of the production calculations into internal and external data processing that was made in Chapter 2 retains only limited validity. Nevertheless there are several reasons why it seems desirable to retain these terms, since they make it possible to make a clear distinction within a closed system in the case of systems analysis, namely that relating to the machine and that related to organisational factors. This applies in particular to automatic machines in which there is no extensive integration of the information flows. As will be mentioned in greater detail in Chapter 12, this includes cam-controlled machines, copying machines, cam-operated automatic lathes, NC machines with manual data input, and also conventional NC machines with punched tape readers.
2. Both for technological reasons (e.g. extremely high accuracy and simplification of tool pre-setting) and for organisational reasons (e.g. manufacture of one-off jobs without having to use a programming office) manual input of numerical data has proved very successful in a comparatively large number of practical cases. The most important components involved are the decade switch and the cross-bar distributor.

3. In punched tape control systems the use of eight-track punched tapes has become the standard practice throughout the world; the choice of codes is, however, still open to question. The decision as to whether the ISO or the EIA code should be adopted largely depends on organisational arrangements, and is discussed in detail in Chapter 11.

4. Photoelectric punched tape readers at present occupy a dominant position. These have scanning speeds of 50 to 500 characters per second. All development work is directed towards making the equipment more reliable than in the past; this applies both to rapid, error-free reading and to the feeding of the tape and the provision of arrangements for dealing with long lengths of tape.

5. Little use is made today of the block scanning of punched tapes; intermittent reading of blocks of variable lengths predominates.

6. Because of its greater reliability, the address method is preferred to the stepping method (TAB programming).

7. Pneumatic punched tape readers and control systems have proved to be practicable alternatives to electronic models; they offer certain advantages especially under the adverse conditions prevailing in machine shops. On the other hand, this type of control system is not suitable for linking directly to a computer.

8. Despite their widespread use in the data-processing field, magnetic tapes have not yet been used to any appreciable extent as data carriers for production purposes. Recent British designs of magnetic-tape control systems have, however, raised the hope that it may prove possible to achieve a low-cost linkage between computers and NC machines using seven or nine track standard ½-inch magnetic tapes. Owing to the effects that this will have on organisational matters this question will be further discussed from a different point of view in Chapter 11.

9 In view of difficulties in reliability, especially of punched tape readers, there is some inducement to avoid completely the use of data carriers in some cases, and instead to group a number of NC machines together and to have them controlled directly by computers. This direct numerical control (DNC) system would also enable a better utilisation of high-speed computers to be achieved. The analysis has shown, however, that progress in this field will have to be slow, and that the question of a suitable compromise between centralisation and decentralisation will have to be decided separately for each individual case. For economic reasons alone it seems likely that punched tape control systems using improved tape readers will continue to dominate the field for some years to come.

10. Even now it is possible to distinguish between two types of DNC installation. In the one group the main step that has been taken is the elimination of the punched tape reader; while a further increase in flexibility is achieved by providing rapid access to ready-to-use programs stored on magnetic discs; in this case the small process

computer employed is used basically only for distributing the programs rapidly and correctly to the 8 or 10 NC machines with which it is linked. In the other group not only is the punched tape reader eliminated, but the functions of a number of internal interpolators are taken over by a time-sharing computer. In both instances larger plant computers can be introduced to coordinate several installations. This leads to the provision of hierarchies of computers, i.e. to development trends in the field of external data processing which will be mainly discussed in Chapters 11 and 12.

III   External Data Processing

10   Manual and semi-automatic parts programming—
The use of small computers for machine-orientated
programming

10.1   The basic possibilities of programming

Mention has previously been made in this book of the "transfer of responsibility" from the workshop floor to the production planning and work preparation departments; what this implies and the extent to which this internal rearrangement of the work is gaining in importance will have become clearer from a study of Chapters 7 to 9. In the chapters that follow less emphasis is placed on the numerically-controlled 'numerate' machine tool and its many design details. The discussions are concerned rather with the best way of incorporating them in the factory organisation. Computers of all types and sizes will play a decisive role, both now and in the future. A first indication of the most suitable future arrangement was given in Chapter 9, where mention was made of the development of hierarchies of computers. The adoption of micro-miniaturisation and the constantly improving cost/performance ratio makes it likely that the remedy of available computers and computer systems, will exert a major influence on factory organisation and on workshops. It is impossible in this book to go into detail of the large number of machines and auxiliary devices available for office automation in general and for computer-aided production control in particular. The large number of publications in this field makes it necessary to limit discussion to basic factors and mention of a few selected literature references.

In this connection it is advisable to start with an adequate supply of characters for the direct control of machining processes on NC machines, and to illustrate the possibilities of the compression of data and its adaptation to the requirements of the problem on the one hand and the transition to higher degrees of automation using hierarchies of computers in four broad stages. For convenience. the discussion which follows is subdivided into the following four subject areas:

1. Machine-orientated parts programming[1] with the aid of typewriters

---

[1] Where the term 'machine-orientated' is used in this context it always relates to a work-performing machine, i.e. in this instance a machine tool, and not to a calculating machine (computer, data-processing installation) as is the usual practice in handbooks for data-processing installations (6, 57, 69).

and simple desk calculators; this method will be described as 'manual' parts programming.

2. Machine-orientated parts programming with the aid of miniature computers and simulation equipment; in such instances it would be appropriate to speak of 'semi-machine' programming.

3. Problem-orientated parts programming using miniature computers of various sizes; this method of operation could also be described as semi-machine' programming, since many decisions are left to the operator.

4. Problem-orientated parts programming with the aid of large computers and graphical data-processing equipment; in this instance the common term 'machine' programming is fully justified.

The term 'computer-aided programming' is also widely used for Groups 2 to 4; because of the large variety of different computer sizes it is desirable, however, to make some distinction, and it is important to make a reference at this stage to the man/machine problem, which has to be solved with all combinations of equipment (cf. Chapter 12). In any event it is desirable not to speak of 'automatic programming' whenever computers are used; this is not compatible with the definition of an 'automatic machine'. Since the use of problem-orientated languages and the employment of large computers must be considered against a broader background, the discussion of the problems associated with Groups 3 and 4 will be left to Chapter 11.

10.2   An adequate supply of characters

There is considerable argument as to what should be considered an 'adequate' supply of characters, and the decision on this must be based on experience in this field. When NC machines first began to be widely used, at the end of the 1950's, it was generally assumed in the machine-tool industry that a supply of 15 to 20 characters would be adequate to control the NC machine tools then developed [2].   For general data-processing, which also includes data transmission, it was even then necessary to provide between 50 and 60 characters. During the course of the 1960's agreement was reached in ISO/TC 97[3] on a proposal '6 and 7 Bit-Coded Character Sets for Information Processing Interchange', from which the 7-Bit Code was adopted as a German Standard (DIN 66 003) (259). Because of its basic importance this code is reproduced in Table 15. Since then there have been a number of international

---

[2] VDI Richlinie 3559, which is now obsolescent, was based on this assumption.

[3] ISO = International Standardization Organization/Technical Committee 97.

b7	b6	b5	b4	b3	b2	b1	Column / Line	0	1	2	3	4	5	6	7
								0	0	0	0	1	1	1	1
								0	0	1	1	0	0	1	1
								0	1	0	1	0	1	0	1
0	0	0	0	0	0	0	0	NUL	(TC7) DLE	SPA	0	\	P	@	p
0	0	0	0	0	0	1	1	(TC1) SOH	DC1	!	1	A	Q	a	q
0	0	0	0	0	1	0	2	(TC2) STX	DC2	"	2	B	R	b	r
0	0	0	0	1	1	1	3	(TC3) ETX	DC3	#	3	C	S	c	s
0	0	0	0	1	0	0	4	(TC4) EOT	DC4	$	4	D	T	d	t
0	0	0	0	1	0	1	5	(TC5) ENQ	(TC8) NAK	%	5	E	U	e	u
0	0	0	0	1	1	0	6	(TC6) ACK	(TC9) SYN	&	6	F	V	f	v
0	0	0	0	1	1	1	7	BEL	(TC10) ETB	,	7	G	W	g	w
0	0	0	1	0	0	0	8	FE0 (BS)	CAN	(	8	H	X	h	x
0	0	0	1	0	0	1	9	FE1 (HT)	EM	)	9	I	Y	i	y
0	0	0	1	0	1	0	10	FE2 (LF)	SS	*	:	J	Z	j	z
0	0	0	1	0	1	1	11	FE3 (VT)	ESC	+	;	K	[(Ä) *	k	{ (ä) *
0	0	0	1	1	0	0	12	FE4 (FF)	IS4 (FS)	,	<	L	~(Ö) *	l	─ (ö) *
0	0	0	1	1	0	1	13	FE5 (CR)	IS3 (GS)	─	=	M	] (Ü) *	m	} (ü) *
0	0	0	1	1	1	0	14	SO	IS2 (RS)	.	>	N	^	n	| (ß) *
0	0	0	1	1	1	1	15	SI	IS1 (US)	/	?	O	─	o	DEL

Table 15. ISO 7-Bit-Code from DIN 66 003 (Draft 1965)

proposals with a view to increasing the useful supply of characters to 8 or even 10 bits; whether international agreement will be reached on these proposals is at present uncertain. In the current state of development of numerical machine control systems, the supply of characters available with the 7-Bit-Code should be adequate for some considerable time to come, even if the development possibilities of DNC are borne in mind (cf. Chapter 9). In this case one is not restricted by the limitations of an 8-track punched tape. The control characters for the transmission of data are also included in the code. Even the 7-Bit Code includes control characters (e.g. SO, SI, SS, and ESC) which enable it to be extended so that further character meanings can be adopted if necessary. In this case, however, the code would no longer be unambiguous. The characters shown in Table 15 makes use of 7 information bits; the redundancy required for operational reasons is not included.

From the remarks on 8-track punched tapes in Chapter 9, it is an obvious step to select an adequate number of characters for the numerical punched-tape control of machine tools from the 120 distinct and unambiguous characters listed in Table 15; and to add an eighth symbol to the resultant combination of holes to enable a parity check to be made.

The 50 characters which have been selected for NV purposes are printed in bold type in Table 15. Since the code contains the odd number of 7

| | | Bit No. (P = test bit) | P | 7 | 6 | 5 | 4 | | 3 | 2 | 1 |
| | | Track No. (T = timing tk) | 8 | 7 | 6 | 5 | 4 | T | 3 | 2 | 1 |
No.	Symbol		Hole combination								
1	NUL							·			
2	BS		●				●	·			
3	HT						●	·			●
4	LF						●	·		●	
5	CR		●				●	·	●		●
6	SP		●		●			·			
7	(				●		●	·			
8	)		●		●		●	·			●
9	%		●		●			·		●	●
10	:				●	●	●	·		●	
11	/		●		●		●	·	●	●	●
12	+				●		●	·		●	●
13	−				●		●	·	●		●
14	0				●	●		·			
15	1		●		●	●		·			●
16	2		●		●	●		·		●	
17	3				●	●		·		●	●
18	4		●		●	●		·	●		
19	5				●	●		·	●		●
20	6				●	●		·	●	●	
21	7		●		●	●		·	●	●	●
22	8		●		●	●	●	·			
23	9				●	●	●	·			●
24	A			●				·			●
25	B			●				·		●	
26	C		●	●				·		●	●
27	D			●				·	●		
28	E		●	●				·	●		●
29	F		●	●				·	●	●	
30	G			●				·	●	●	●
31	H			●			●	·			
32	I		●	●			●	·			●
33	J		●	●			●	·		●	
34	K			●			●	·		●	●
35	L		●	●			●	·	●		
36	M			●			●	·	●		●
37	N			●			●	·	●	●	
38	O		●	●			●	·	●	●	●
39	P			●		●		·			
40	Q		●	●		●		·			●
41	R		●	●		●		·		●	
42	S			●		●		·		●	●
43	T		●	●		●		·	●		
44	U			●		●		·	●		●
45	V			●		●		·	●	●	
46	W		●	●		●		·	●	●	●
47	X		●	●		●	●	·			
48	Y			●		●	●	·			●
49	Z			●		●	●	·		●	
50	DEL		●	●	●	●	●	·	●	●	●
51											
52								·			

Fig. 249
Code for 8-track punched tapes to DIN 66 024 (Draft 1967); ISO Code as a subset of the ISO 7-Bit-Code given in DIN 66 003 (Table 15) for data processing and transmission of data.

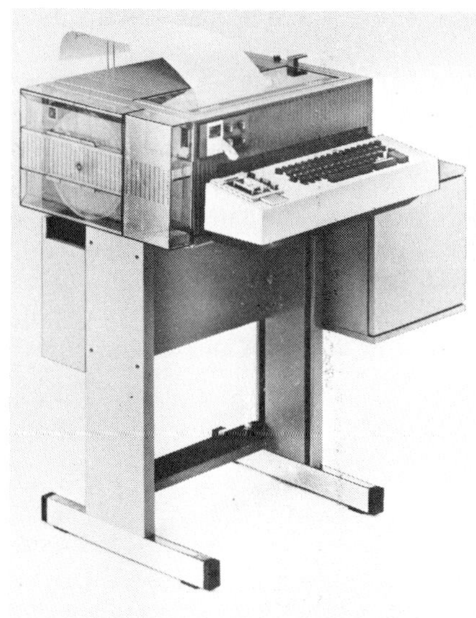

Fig. 250
Punched-tape controlled
typewriter for ISO Code.
This machine can also be
used as an input/output
device for computers and for
transmission of data (274).
(Photo courtesy of *Olivetti*).

information bits, the parity check that is to be given in the 8th track can only be used to make up an even number of holes, and the parity check has to be for an even number as shown in Fig. 234. Figure 249 shows the relationship between hole patterns and characters for this ISO Code. It has the advantage firstly that it is of logical construction and internationally recognised; and secondly that it is compatible with the internal coding of data-processing installations and with data transmission systems; it is thus likely to have a wide field of application.

On the other hand, it has the disadvantage that it is at present rarely used for NC machines, since the trend to integrated data processing is only just commencing. It is not necessary to change the code if direct computer control of machine tools is adopted (DNC, Chapter 9); it is already comparatively widely used in the USA, where it is know as ASCII (American Standard Code for Information Interchange). When comparing the two codes shown in Table 15 and in Fig. 249 it should be borne in mind that of the large number of characters shown in Table 15 only the capital letters, as shown in Fig. 249, have been adopted. This has certain consequences when using typewriters, which will be considered in more detail in the next section. Figure 250 shows a typewriter console which can be used as an input/output unit for computers and for the transmission of data as well as for coding punched tapes for NC machines.

10.3   Special punched-tape codes which cannot be derived from Table 15

Before the selection of characters for numerical controls for machine tools is discussed further two NC codes will be mentioned which appeared very advanced when first introduced, but seem to be becoming less important with the increasing integration of NC machines with data-processing installations. For some time to come, it will be necessary to use, in many practical applications, old and new codes side by side, since it is not possible to adapt either the programming arrangements or the machine control systems immediately unless an automatic code translator of the type shown in Fig. 253 is used.

The EIA Code[4]

This is the oldest — and the most widely used — punched-tape code; it was last revised in 1967 and issued as EIA Standard RS 244-A. Originally, this

Bit No. (P = test bit): 7 6 5 P 4   3 2 1
Track No. (T = timing tk): 8 7 6 5 4 T 3 2 1

No.	Symbol	8	7	6	5	4	T	3	2	1
1	1						·			●
2	2						·		●	
3	3				●		·		●	●
4	4						·	●		
5	5				●		·	●		●
6	6				●		·	●	●	
7	7						·	●	●	●
8	8					●	·			
9	9				●	●	·			●
10	0			●			·			
11	a		●	●			·			●
12	b		●	●			·		●	
13	c		●	●	●		·		●	●
14	d		●	●			·	●		
15	e		●	●	●		·	●		●
16	f		●	●	●		·	●	●	
17	g		●	●			·	●	●	●
18	h		●	●		●	·			
19	i		●	●	●	●	·			●
20	j		●		●		·			●
21	k		●		●		·		●	
22	l		●				·		●	●
23	m		●		●		·	●		
24	n		●				·	●		●
25	o		●				·	●	●	
26	p		●		●		·	●	●	●
27	q		●		●	●	·			
28	r		●			●	·			●
29	s			●	●		·		●	
30	t			●			·		●	●
31	u			●	●		·	●		●
32	v			●	●		·	●	●	
33	w			●			·	●	●	●
34	x			●	●		·	●	●	●
35	y			●	●	●	·			
36	z			●		●	·			●
37	.		●	●		●	·	●	●	
38	,		●	●		●	·	●	●	●
39	/		●	●	●	●	·			●
40	+		●				·			
41	−		●	●	●		·			
42	&					●	·	●	●	
43	%		●			●	·		●	●
44	Tab HT			●	●	●	·	●	●	
45	Cr or End of Block LF	●					·			
46	Delete DEL		●	●	●	●	·	●	●	●
47	End of Record					●	·		●	●
48	Space Sp					●	·			
49	Back Space BS			●		●	·		●	
50	Upper Case		●	●	●	●	·	●		
51	Lower Case		●	●	●	●	·		●	
52	Blank Tape NUL						·			

A)   B)

Fig. 251
Typewriter Code for 8-track punched tape to EIA RS 244-A (1967 Edition).

[4] EIA = Electronics Industries Association/USA

code was used for controlling punched-tape operated typewriters. Since this type of office equipment was already widely used in the USA in the 1950's and, because it included both punched-tape readers and punches, it was an obvious matter for the US machine-tool industry and the US manufacturers of control systems to use this existing equipment for the production of control tapes and for printing-out programme lists. The punching principle has already been described in Table 4 (Chapter 1) where it is referred to as Code B. Unlike the ISO code shown in Fig. 249, the EIA code contains only 6 information bits, and so provides in theory only 63 distinct characters. This is, however, fully adequate for controlling a typewriter. The 5th track carries the parity check (as in Code B, Table 4); it is thus interposed between the 6 information bits so that the external appearance is that of a 7-Bit-Code which can be parity checked. Figure 251 shows the relationship between the hole patterns and the characters; of the 64 possible characters, 52 are used[5]; Because of its origin as a typewriter code, the EIA code possesses two interesting features:

1. The 8th track is occupied by only a single character, the carriage return, which in the majority of punched-tape controlled typewriters is combined with the line feed. It was shown in Chapter 9 that the arrangement of the information in individual information blocks of variable length in punched-tape control systems led to intermittent operation. The common features of intermittent reading or reading a line at a time can be utilized for a manually-produced text. Each return of the typewriter carriage and start of a new line can be interpreted as 'Reader Stop' for the machine control system, and each programme sentence will automatically begin on a new line. The close relationship between typewriters and NC techniques not only led to the use of the term 'manual programming', but also accounts for the popularity of the EIA code. There is also the advantage that the NC machine operator can easily distinguish the individual blocks (programme sentences) directly on the tape, since the single symbol in the 8th track is readily identifiable.

   For economic data processing, it is waste of space to use one of the punched tape tracks for a single symbol only merely to enable the operator to distinguish the blocks. It is likely that with the penetration of integrated data processing into production shops the importance of EIA code will gradually diminish. Nowadays the majority of control systems can therefore be equipped either for EIA or ISO Code, depending on customer preferences; the user will have to decide whether and when to change over to the computer-orientated code form.

---

[5] For comparison, ISO code uses 50 of the 128 characters that are possible (since NUL is available as a character or a punched information pattern). The difference, i.e. $128 - 50 = 72$ characters, can be used as additional redundancy to increase the system reliability. In EIA code the excess number of characters is only $64 - 52 = 12$, and it is not possible to increase the number of characters beyond 64.

Fig. 252 Punched-tape controlled automatic letter-writer for EIA code.The machine can also be used to assist in the rationalisation of the internal organisation of the plant (264) (Photo courtesy of *Friden*).

2. The programme text contains both numbers and letters. To avoid having to keep changing over on the typewriter it is necessary to adopt the type of letter which is in line with the numbers and symbols on the typewriter hammers. In most instances these will be lower case letters, and it will be necessary to reach agreement with the typewriter manufacturer on the remainder of the symbols required (.,/+–&%). Figure 252 shows a special machine of this type; this is the FLEXOWRITER which is also used for general office automation (automatic letter writer); on this machine the figures and the lower-case letters are on the same line. In the data-processing field, on the other hand, it is usual to use the capital letters (cf. the programme languages in Chapter 11); with ISO code the use of capital letters is therefore specified in the code itself.

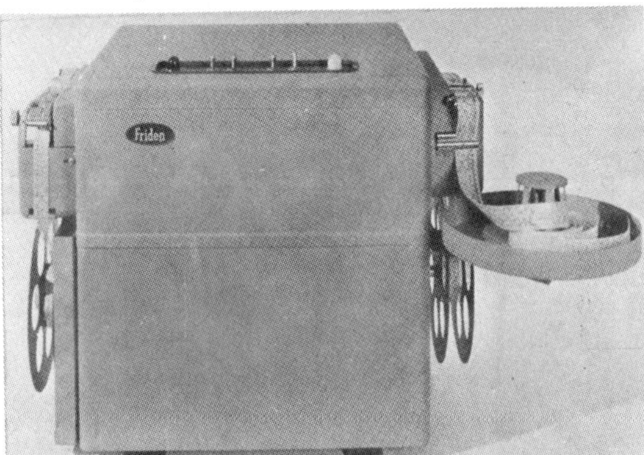

Fig. 253 Automatic code converter (Photo courtesy of *Friden*).

Conversion possibilities: 5 to 8 tracks and vice versa. ISO to EIA code and vice versa.

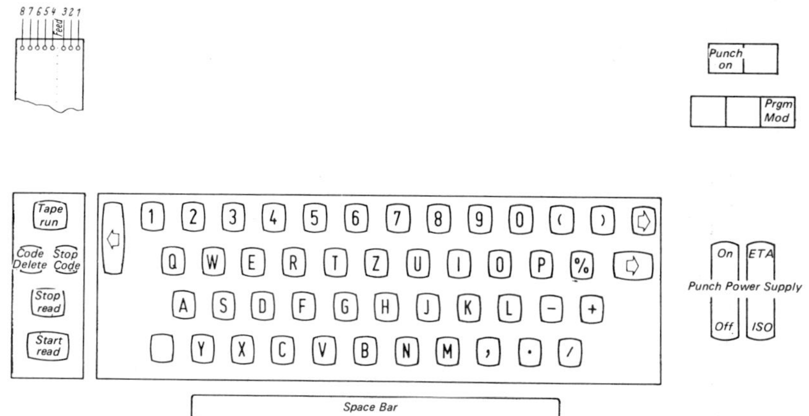

Fig. 254 Layout of keyboard of FLEXOWRITER (Type 2376) to serve as special coding station with EIA/ISO dual coding (264) (Courtesy of *Friden*).

Note: The external appearance of this machine is identical with that shown in Fig. 252, but it cannot be used as an automatic letter-writer.

Since manual programming, which includes the production of punched tapes by punched-tape controlled typewriters, will probably continue to be used for some considerable time owing to its simplicity, it is desirable to arrange the coding appliances so that they are capable of being switched over, at least for a transition period. Figure 254 shows the keyboard of a dual-purpose machine which can be switched over for either EIA or ISO code; in its external appearance the machine is identical to that shown in Fig. 252. Unfortunately, this machine cannot be used as an automatic letter-writer, since the code changeover mechanism is linked to the capital/lower case shift. This restriction may make it necessary to use a code converter similar to that shown in Fig. 253, or to use two types of coding machine; the alternative is to adhere rigorously to one type of code with no transition period. Questions of this nature are, however, more likely to be of importance in American firms than in European ones.

Programme Code 8 C (25)

This code has been developed in the German Democratic Republic to overcome any unreliability that may arise in service when using punched-tape control systems. This code was based on the work of Hamming (14); the theory of the self-correcting Hamming Code has already been discussed in

Chapter 1 (Code C of Table 4). This contains 4 information bits and 3 check bits; the number of characters that can be used is therefore only fifteen.

This is too few even for comparatively simple NC machines if the safe address method is to be used. If the number of useful characters is to be increased to 31, so that 5 information bits will be needed, there will have to be, according to Hamming, 4 check bits to produce a self-correcting code. Since only 8-track punched tape is available, some other method must be sought for extending the number of characters. This can, for example, be achieved by including a syntax prefix in the sequence of characters. In this code, as shown in Fig. 255, there are two groups of characters; 15 numbers with prefixes and operating characters for the typewriter and 15 letters. It is possible to use the Hamming test procedure (Chapter 1) directly for the first group; while for the second group it is first necessary to negate for example Track 5. The internal permanent wiring of the control system must then be arranged so that it can automatically recognise a character that is not a letter. If one of the programming rules states that letters are read as store addresses and that one of the characters 13, 14, 15, or 16 must appear before each letter, this condition can be met. If, in addition a sequence of decimal numbers having a prescribed number of digits must appear after each address

No.	B)	A)	8	7	6	5	4	T	3	2	1
1		1	●	●	●			·			●
2	2		●	●		●		·		●	
3		3			●	●		·		●	●
4	4		●		●	●		·	●		
5		5		●		●		·	●		●
6	6	.		●	●			·	●	●	
7		7	●					·	●	●	●
8	8			●	●	●	●	·			
9		9	●			●	●	·			●
10	0		●		●		●	·		●	
11		−		●			●	·		●	●
12	+		●	●			●	·	●		
13		TAB			●		●	·	●		●
14	LF, CR	< ≡				●	●	·	●	●	
15	DEL	IRR	●	●	●	●	●	·	●	●	●
16	SP	ZWR						·			
17		x	●	●	●	●		·			●
18	y			●	●	●		·	●	●	
19		z	●			●		·	●	●	●
20	r			●		●	●	·		●	●
21		k	●		●	●	●	·		●	
22	s				●	●	●	·	●		●
23		o	●	●		●	●	·	●		
24	i				●			·		●	●
25		v	●		●			·	●		
26	a		●	●				·		●	
27		h		●	●		●	·			
28	l		●				●	·			●
29		n		●				·	●		●
30	d						●	·	●	●	●
31		p	●	●	●		●	·	●	●	●
32											
33											
34											
35											
36											
37											
38											
39											
40											
41											
42											
43											
44											
45											
46											
47											
48											
49											
50											
51											
52											

Header of the table: Bit No. / Track No. (T = timing tk) — Hole combination, with bit columns 8 7 6 5 4 T 3 2 1.

Fig. 255
Code for 8-track punched tapes as given in TGL 28-210 (25)

Note: The special NC code is self-correcting for one error and will recognise the presence of two errors (275).

letter, after which one of the operating characters (e.g. space) is programmed, the cycle will be closed. The 8-track code has been made self-correcting for 31 simple character errors.

In view of the possible unreliability of punched-tape readers (Chapter 9), the idea of an error-correcting code with relatively large number of characters is very attractive. But since the amount of effort required for checking is relatively large — and faults in punched-tape readers and in NC machines are not all due to incorrectly read characters — this code has not been adopted internationally despite its good theoretical basis, especially since it is compatible with neither the ISO proposals nor the EIA Code. A simple parity check, together with improvements in the design of punched-tape readers, appears to be considered adequate throughout the world.

If one summarises what has been said in this Section it appears that even though the path followed by the ISO Code will no doubt meet with considerable resistance and suffer temporary difficulties, it is the most hopeful, since it is compatible with future development trends.

10.4   The meaning of characters and the construction of programmes

Standardised programming, i.e. an internationally accepted relationship between letters and numbers on the one hand and the functions of NC machines and the sequence in which they are fed in as defined control signals on the other, is of greater practical importance than the allocation of hole combinations to letters, numbers, and signs. Since these relationships are not dependent on whether programming is machine-orientated or problem-orientated, it is advisable to consider this important link between internal and external data processing first. The ISO discussions on this have not yet been concluded, and the German Standards based on these — DIN 66025 (168) — have remained in draft form since 1968/69, and may still be modified in detail. Nevertheless a brief survey will be given at this point; only the ISO Recommendations that are still to be issued and the final DIN Standards will, however, be binding.

The relationships between letters and machine functions have already been given in Table 12 (Chapter 7); these account for the use of the 25 letters of the alphabet — 0 being omitted — for address purposes. Dependent on the nature and extent of the control system, many letters can have more than one meaning (e.g. D, E, I, J, K, P, Q, R), which is an indication that the time will come when the number of letters in the alphabet will be insufficient to enable suitable single-letter symbols to be allotted to all the machine functions.

In addition to the letters, there are the decimal numbers and the two signs '+' (plus) and '−' (minus); these require no explanation. Some special symbols and control symbols are also required for building up the programme; they are listed in Table 16.

Symbol	Meaning
%	Start of programme, also unqualified stopping of rewind of punched tape.
:	Main sentence, also qualified stop of rewind of punched tape.
/	Sentence suppression
(	Start of note.
)	Conclusion of note.
NUL	No holes punched, apart from timing track; tape feed at maximum speed.
BS	Backspace
HT	Horizontal tabulator
LF	End of sentence, also line feed
CR	Carriage return
SP	Space
DEL	Delete

**Table 16. Meanings of special symbols and control symbols according to
DIN 66025 Sheet 1 (Draft 1968)**

The information given in Tables 12 and 16, which is based on DIN 66025, provides a means for describing the machining of a workpiece on a NC machine operation-by-operation in a kind of shorthand. Each operation produces a 'programme sentence'; each 'sentence' appears on the punched tape as a 'block'. The following remarks apply exclusively to this type of programme arrangement.

Basically, the term 'programming' as applied to a numerically controlled machine tool covers three distinct functions:

1. The arrangement of the form data and the technological information into a work schedule arranged in order of machining operations. The work information will include the values of the coordinates, the cutting speeds and feeds, as well as the auxiliary functions and possibly also the numbers of the cutting tools in the sequence in which they will be required in the machine. The information can be listed wholly or partially in plain text which can be universally understood (e.g. cutting speed 90 m/min, feed 40 mm/min, etc.). A skilled and experienced planning engineer is needed to produce this specification; in some cases it may even be necessary to use a team consisting of a designer and a production engineer. Experience to date shows that machine setters and foremen are often very suitable for this class of work; this provides many opportunities for internal training schemes.

2. These plain-language specifications must then be transformed, in accordance with definite rules, into an abbreviated form which will

enable the work information to be represented as compactly as possible by means of symbols; at the same time this abbreviated text must be capable of being read and understood without difficulty. The preparation of these abbreviated specifications and the application of the programming rules is an essential and novel aspect of the work planning process. Every effort must be made to make this operation as simple and as easy to learn as possible, because in the majority of cases it will not be possible to employ specialists for this work. The symbols and programme language must be simple enough to enable any good planning engineer, who is primarily concerned with the processes that take place in the machine, to learn it. Any proposed symbol language must be examined from this point of view. In addition the symbols and rules of a machine-orientated programming language of this type, and also the construction of the programme, should so far as possible be understood and accepted internationally. The work information that has been encoded in this manner is then either entered manually on programme sheets or programme lists or — with machine programming — it takes the form of the computer print-out.

3. With manual programming the programme sheets or lists, which should follow a standard layout, are then passed to the coding section. Here a clerical assistant can finally encode the work information, which is still in readable form, into the machine language in the form of punched tapes produced on suitable office equipment (card puncher, typewriter with tape perforating attachment, etc.). With machine programming, the punched tape is produced automatically by the computer together with the programme lists.

The extent of the operations described in paragraphs 1 and 2 above will be determined primarily by the machine tool and the workpiece. The time required will depend ultimately on the amount of work information that has to be calculated and written down.

A new look at displacement and switching information

The terms 'displacement information' and 'switching information' were introduced in Chapter 2; these, however, provide only a rough and ready means of subdividing the information, and now that the use of address letters makes it possible to break the work down in greater detail, it will be necessary to specify the items more closely.

1. The addresses listed in Table 17 form part of the definite displacement information:

X	Movement in the direction of the X-axis
Y	Movement in the direction of the Y-axis
Z	Movement in the direction of the Z-axis
U	Second movement parallel to X-axis
V	Second movement parallel to Y-axis
W	Second movement parallel to Z-axis
A	Rotation about X-axis
B	Rotation about Y-axis
C	Rotation about Z-axis

Table 17. Addresses for definite displacement information items. (Definitions of directions of axes, selection from Table 12).

In some of the illustrations given in Chapter 7 (e.g. Figs. 171 and 173) an attempt has already been made to illustrate the meanings of these definitions of directions of axes. These important factors will be discussed in greater detail in Section 10.5 which follows, and will be described in connection with the examples of machines that are given in the Appendix. The further letters D, E, P, Q, R that are included in Table 12 (Chapter 7) may also be displacement information associated with coordinate dimensions, but this is not imperative. In the address method of programming each displacement information word (coordinate word) thus consists of one of the letters listed above followed by a multi-figure number (coordinate dimension) which gives the end point for the next variation in the position of the machine slide concerned. This can be written with or without a sign, depending on the type of control system and the position of the zero point (origin). And it is given as an integral multiple of the smallest element Δs that can be programmed (cf. Chapter 1). In other words, no decimal point is used. In many control systems the initial zeros need not be written down for small numbers. Coordinate words can therefore basically be of the following forms:

X	25 000	
Y	000 30	or Y 30
Z	375	or Z 00375
W	007240	or W 7240
C	32570	
	etc.	

In case of doubt reference should be made to the latest edition of DIN 66025 Sheet 1. The numerical form of the coordinate information is the characteristic of numerical control techniques.

2. According to Table 12 (Chapter 7) the following addresses form part of the definite switching information:

F Feed
S Spindle speed
T Tool

If the same method as used for displacement information is to be employed here some further conventions must be agreed on.

The discussion in Chapter 7 will have made the tool information T clear; this information will to a large extent depend on the manner in which the tooling system is organised in the plant. In the simplest case, the position of a turret can be called up as a single-figure number; in the most complicated case, a three-to-five-figure tool number can be called up (cf. Figs. 150 and 169). If required, the tool numbers can be shown by illuminated figures on the control console (cf. Fig. 149); this is often used for calling up the next tool where manual tool-changing is employed.

Special conventions and rules apply to coding of feeds and speeds, they are contained in DIN 66025 Sheet 3. Before these are considered in more detail, however, reference will be made to a third group of possible addresses where doubts may arise as to whether they fit into the existing rough subdivision. These have only arisen in recent years as a result of the progress that has been made in machine design; these are the letters:

G Displacement conditions
M Auxiliary function
N Sentence numbers

The number of the individual machining operations should, preferably, equal the sentence numbers. With the G and M functions, matters are more complicated; again it is essential for international agreement to be reached to ensure that a uniform practice is followed. They are contained in DIN 66025 Sheet 2 (168).

The information given by the F, S, G, and M functions will now be discussed very briefly; further details can be obtained from the standards quoted.

The coding of feeds and spindle speeds to DIN 66025 Sheet 3

In addition to direct information concerning the values of feeds and speeds, arithmetical, geometrical, free, and time-reciprocal codings are admissible. Here reference will be made to geometrical and arithmetical codings only, since these have probably the widest actual or potential use:

Code number	Numerical value
00	Stationary
08	0.25
25	1.8
40	10
60	100
70	315
80	1000
95	5600
99	Fast traverse

Table 18. Examples of geometrical coding of F- and S- functions.

In *geometrical coding* the numerical values of the feeds and spindle speeds are given by two-figure numbers. The numerical values are based on the R 20 standard number series (see DIN 323 Sheet 1). Some examples are given in Table 18.

The range of numerical values covered by the coding is about 10^5 and this should suffice for most applications. If necessary, the numerical values may be shifted upwards or downwards by a power of ten. This change must be indicated in the machine specification.

In *arithmetical coding* it is usual to employ three-figure code numbers which are derived directly from the numerical values by applying certain rules. This code is therefore sometimes referred to in the United States as the 'magic three code'. The rules are roughly as follows:

1. The numerical value is rounded off to two figures which occupy the second and third places of the code number. The first place of the code number indicates the position of the decimal point in accordance with the following rule:
2. The code number for the position of the decimal point is obtained by allotting the code number 300 to the numerical value zero. This means:

a) If the numerical value is greater than or equal to 1, the code number will be 3 plus the number of figures to the left of the decimal point.

    (Examples   1250 rev/min =   713

                                   71 mm/min =   571)

b) If the numerical value is less than 1, the code number will be 3 minus the number of zeros to the right of the decimal point.

    (Examples   0.63   mm/min =   363

                                0.045 in/rev   =   245)

The dimensions for the arithmetical or geometrical code number are derived from the address

> S Spindle speeds in rev/min
> F Feeds in mm/min or mm/rev.

and from the machine specification.

Coding of G *and* M *functions to DIN 66025 Sheet 2*

Both functions are represented by two-figure codes. Since these codes have to be capable of defining a wide variety of operations it is not possible in practice to build them up logically in the same way as those for the S and F functions. The *displacement conditions* or *preparatory functions* are characterised by the address letter G. They represent additional information to the coordinate dimensions and can be considered as forming part of the displacement information. The items covered by the G functions are indicated by the partial list given in Table 19.

Code number	Meaning
$G\,00$	Positioning control
$G\,01$	Straight-line interpolation
$G\,02$	Circular interpolation, clockwise
$G\,03$	Circular interpolation, anti-clockwise
$G\,17$	Interpolation plane xy
$G\,18$	Interpolation plane xz
$G\,19$	Interpolation plane yz
$G\,33$	Thread-cutting, constant pitch
$G\,41$	Tool correction, left
$G\,42$	Tool correction, right
$G\,54$	Translatory displacement X
$G\,55$	Translatory displacement Y
$G\,56$	Translatory displacement Z
$G\,60$	Accurate stop
$G\,62$	Immediate stop
$G\,80{-}89$	Subprogrammes

Table 19. **Examples of** G **Functions (excerpt from DIN 66025, Sheet 2,**
        Draft 1968).

As can be seen from Table 19, many of the requirements referred to in Chapters 7 and 8 are programmed as part of the G functions; this provides a great deal of flexibility in the programming, especially as no fixed meanings have been allocated to 25 to 30 of the code numbers, so that these are freely

Code number	Meaning
M 00	Programmed stop
M 01	Optional stop
M 02	End of programme
M 06	Tool change
M 07	Coolant 2 on
M 08	Coolant 1 on
M 09	Coolant off
M 10	Clamp)
M 11	Release) tools
M 19	Spindle stop with specified end position
M 30	End of punched tape, possibly rewind
M 36	Feed range 1
M 37	Feed range 2
M 38	Spindle speed range 1
M 39	Spindle speed range 2
M 55	Translatory tool displacement, position 1
M 56	Translatory tool displacement, position 2
M 60	Workpiece change
M 61	Translatory workpiece displacement, position 1
M 62	Translatory workpiece displacement, position 2
M 68	Clamp workpiece
M 69	Release workpiece
M 78	Table clamp on
M 79	Table clamp off

Table 20. Examples of *M* Functions (excerpt from DIN 66025, Sheet 2, Draft 1968)

available to the manufacturers of the control systems and the users of NC machines for special applications.

The auxiliary functions with the *M* address letters (miscellaneous functions) are for the most part technological functions which cannot be accommodated under any of the other address letters; they can conveniently be considered part of the switching information. To illustrate this, some of the *M* functions to which specific meanings have been allocated are given in Table 20.

About 40 code numbers of the *M* functions still remain unallocated. The information given in Tables 19 and 20 and in DIN 66025 is in agreement with ISO Draft ISO/DR 1316 'Punched tape block formats for the numerical control of machines, coding of preparatory functions *G* and miscellaneous functions *M*.' Since this Draft is not yet approved as an ISO Recommendation, a check should be made before using these items.

Now that the most important address letters have been described it is possible to discuss the established sequence of the words within a programme sentence.

Sequence of words within programme sentences and hence of the input signals in NC machines (to DIN 66025, Sheet 1, 1968 Edition)

The start of the programme is indicated by the 'Programme start' character '%' (character No. 9 in Fig. 249 or No. 43 in Fig. 252). Any text can be included in the programme schedule and in the punched tape prior to the character %, provided this text does not itself include the character %; this text is then ignored by the control system. The end of the programme is indicated by the word *M* 02 (Table 20).

A sentence consists of several words and the character 'End of sentence' LF (character 4 in Fig. 249 or character 45 in Fig. 252). The words in a sentence should be arranged in the following sequence:

1. Word for sentence No. N . . . .
2. Word for displacement condition G . . . .
3. Coordinate words $X, Y, Z, U, V, W, A, B, C, I, J, K$, etc.
4. Word for feed F . . . .
5. Word for spindle speed S . . . .
6. Word for tool T . . . .
7. Word for auxiliary function M . . . .

The sentences that form a programme can be interrupted by instructions for the machine operator. These instructions are inserted in round brackets () and are then ignored by the control system. Further details are given in the Standard which goes fully into the matter.

Since, as noted in Chapter 9, each word input is associated with a storage function in the control system, the following simplifications of the programming work can be introduced.

1. A word that does not change in several successive sentences in a programme need be given once only and can be omitted in all the following sentences in which it is to be used unchanged.
2. There are therefore two types of sentence: 'complete' which include all the words, and 'incomplete' in which some words (e.g. coordinate dimensions, spindle speeds, etc, which are unchanged) are omitted. The complete sentences are known as 'main sentences' and are characterised by the inclusion of a colon ':' in front of the sentence number. If the control system is suitably arranged, the punched-tape reader can stop at this point on the rewind.
3. By adopting these measures it is possible to arrange the programme in sections. The first sentence in each section is the main sentence. This arrangement not only provides for a clear layout of the text but also provides the possibility of rewinding a tape back to the preceding main sentence and so enabling all the stores to be refilled. The only requirement is that the readers must be capable of recognising the characters : or % during the rewind.

Construction of the words

Here mention may be made of the side effects that different writing of words may have on the programming operation.

A word can be written in three ways:

a) Using an address letter and a sequence of numbers with or without a sign (address method).

b) By the character 'tabulator' (HT in ISO Code, TAB in EIA Code) and a sequence of numbers, with or without a sign (tabulator method, the character TAB or HT not being printed out; instead the typewriter carriage moves a preset amount so that the figures are tabulated).

c) The 'tabulator' character, an address letter, and also a sequence of figures with or without a sign (tabulator address method).

The method of word structuring to be adopted will depend on the control system and must be given in the specification. The tendency seems, however, to be to allow the TAB method to lapse slowly, and instead to make use of the other two methods of writing the words (cf. Chapter 9).

Now that a standard method of building up the words and of arranging the word sequence has been established it is possible to arrange for the programmes to be checked formally by means of permanently wired check routines. This is done with all semi-machine and machine programming systems.

Once the sequence of words had been agreed upon agreement had to be reached on the symbols for the control system and the corresponding programme structure. The proposals for this are also included in DIN 66025 Sheet 4. These require a control system and the corresponding sentence structure to be described in accordance with the following rules:

1. The abbreviated description of a word consists of the appropriate address letter and a figure giving the number of places contained in the subsequent sequence of figures.

2. The address letters are also written out when the tabulator method is used. Instead of the tabulator being operated, the character '.', which is printed out, is inserted between the abbreviated forms of the words.

3. In the abbreviated description of the words for the coordinates the address letters are usually followed by a two-figure number. The first figure gives the number of places to the left of the decimal point and the second figure the number of places to the right of the decimal point.

4. If both positive and negative absolute dimensions are acceptable in the control system this is indicated by inserting '+' after the address letters.

5. If incremental dimensional information is to be used by the control

system and in its programmes, this must be indicated by inserting D after the address letters.

6. If both absolute and incremental dimensional information can be employed 'L' must be inserted after the address letters.

7. The end of the sentence is indicated by the symbol '*' which can be printed out on a typewriter (in place of LF or CR).

Using the address-tabulator method it would be possible to describe the boring mill shown in Fig. 171 together with its control system somewhat as follows:

N3.G2.X3.2.Y3.2.Z3.2.W2.2.B3.1.F2.S2.T2.M2*

To describe a control system fully it is also necessary to add further information in abbreviated form, such as:

Minimum element that can be programmed $\Delta s = 0,01$
Indication of tool number,
Manual tool changing, etc.

If the machine symbols quoted in Chapter 7 are added a clear and complete indication of the plant as a whole will be obtained.

10.5   Coding of axis directions

The proposals made to date for coding of axis directions are given in VDI Code of Practice 3255 (169)[6], which is based on the provisional international agreements given in ISO Draft DR 1315 (1968). Unlike earlier proposals, the latest rules are based on the clockwise rotating coordinate system that is normally used in mathematics. This is to take account of the continuously increasing use that is being made of computers for programming and to ensure that a unified system of description is used for both manual and machine programming. It is necessary to accept that in some instances this will result in a certain loss of clarity in the case of manual programming; in special instances it may be necessary to permit deviations from the rules, but these must then be stated in the machine specification.

The relationships between the directions of the axes and the rotations about the axes (or parallel to them) are illustrated in Fig. 256; reference

---

[6] In practice it is always advisable to comply with the latest editions of Codes of Practice or Standards, since full agreement has not as yet been reached on the international proposals. The information quoted here dates from 1968.

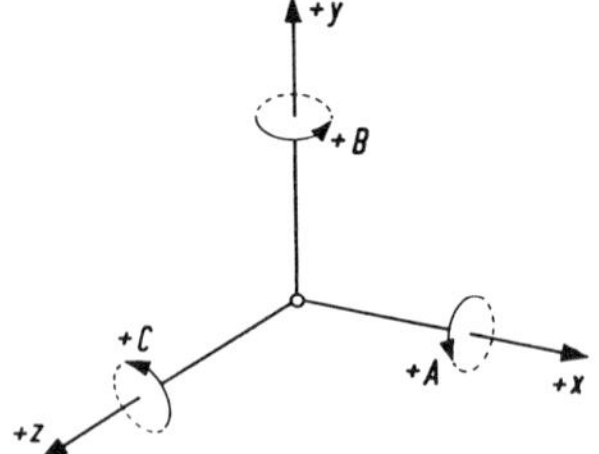

Fig. 256
Conventions adopted for the coordinates in ISO Draft DR 1315 (1968) and VDI 3255 (1968)/169/.

should be made to the definitions in Table 12. Four basic rules should be observed, especially for the selection of the signs:

1. The coordinates, which are perpendicular to each other, are referred primarily to the workpiece, and the directions of motion of the machine tool are derived from them.
2. The relationship of the coordinate axes to the workpiece is retained for all machining phases (e.g. swivelling of workpiece); a coordinate system is assigned to each machining phase (side of workpiece). This rule is of particular importance with machining centres, and is illustrated in Fig. 257.
3. The references to angles on drawings (e.g. of inclined surfaces) should follow the mathematical direction of rotation. A simple example of this is given in Fig. 258.

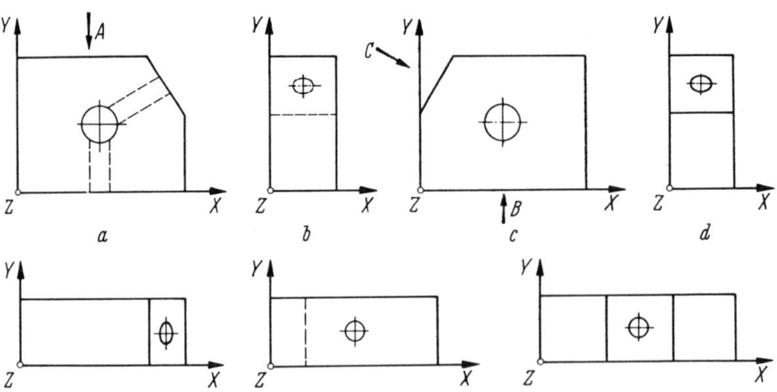

Fig. 257 Arrangement of various machining planes in the coordinate system (to VDI 3255).

a View on machining plane 1. *b* View on machining plane 2. *c* View on machining plane 3. *d* View on machining plane 4. *e* View on machining plane 5 (viewed in direction A). *f* View on machining plane 6 (viewed in direction B). *g* View on machining plane 7 (viewed in direction C).

4. The coordinate origin can lie outside the workpiece or precisely on an edge or corner. If it is carefully selected only positive coordinates will result (single-quadrant operation). The coordinate origin can, however, also be located within the boundaries of the workpiece; both positive and negative coordinate values will then occur (multi-quadrant operation).

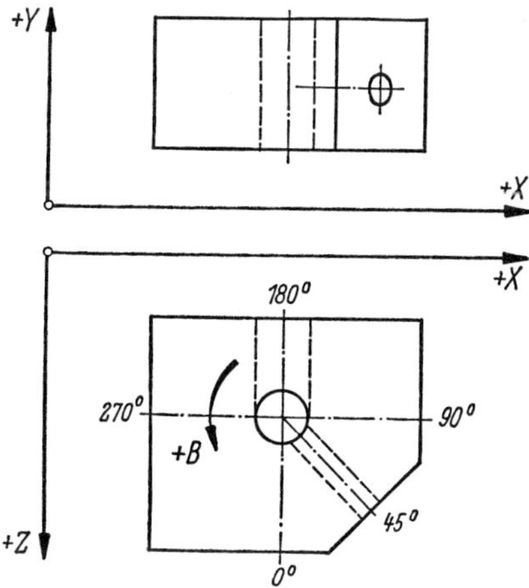

Fig. 258
Example showing angles inserted on a drawing where the axis of rotation is parallel to the Y-axis (to VDI 3255)

As has already been mentioned, the directions of motion for the machine tools are derived from these rules which are established relative to the workpiece. If there is conflict between the mathematically logical rules and normal engineering practice, the programming rules provided by the manufacturers of the machines and the control systems should always be followed. The variations mainly relate to the allocation of the sign to avoid excessive use of negative coordinate dimensions. It must always be the aim to apply a uniform set of rules within any one plant, even if the machines are made by different manufacturers. The most important relationships given in VDI 3255 can be summarised as follows:

a) The Z-axis is perpendicular to the XY-plane and always refers to a driven spindle. With rotating tools (e.g. on drilling and milling machines) the positive coordinate direction is from the workpiece towards the machine spindle. This convention follows from the use of a clockwise coordinate system. With rotating workpieces (e.g. lathes) the Z coordinate dimensions that are of positive sign increase from the clamping point towards the cutting tools.

b) If the cutting tool is moved, the axis direction corresponds to the direction of movement. The positive directional vectors are in the same direction and are designated X, Y, and Z.

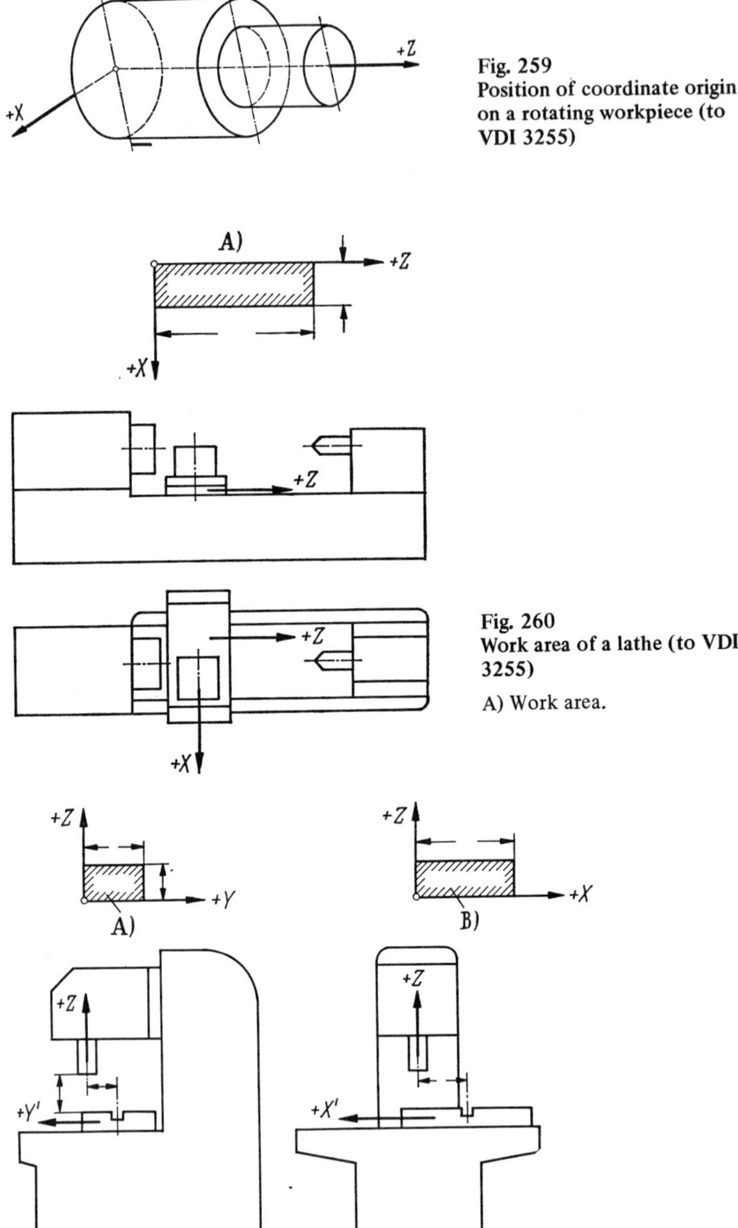

Fig. 259
Position of coordinate origin
on a rotating workpiece (to
VDI 3255)

Fig. 260
Work area of a lathe (to VDI
3255)

A) Work area.

Fig. 261 Work area of a vertical drilling or milling machine (to VDI 3255)
A) Work area in *YZ*-plane. B) Work area in *XZ*-plane.

c) If the workpiece is moved, the axis direction and the direction of motion are opposed to each other. These directions of motion are designated X', Y', and Z'.

d) The relationships for superimposed longitudinal motions (e.g. U, V, W) and for rotations (e.g. A, B, C) are similar.

e) Programming sketches are required of each machine tool. These must show the directions of motion, the work area, and the position of the origin. Sketches suitable for this purpose are illustrated in Figs. 260 and 261.

10.6   Example of programming

Many of the problems encountered in programming and work scheduling, to which reference has been made, will become clearer from an example of programming which contains positioning, straight-line, and continuous-path requirements, as follows:

> 3 mm deep grooves as illustrated in the drawing in Fig. 262 are to be machined in a flat cast iron (e.g. GG 22) plate, using a cylindrical end mill 10 mm dia. The geometrical reference points are numbered in the sequence of the machining operations; for clarity the corresponding coordinate dimensions (X- and Y-axis, I and K information) are tabulated. The in-feed of the cutter to the workpiece takes place in the direction of the Z axis. The clamping fixture is assumed to be designed in such a manner that the cutter can move freely in a plane 1 mm above the workpiece surface (Z = 16 mm).

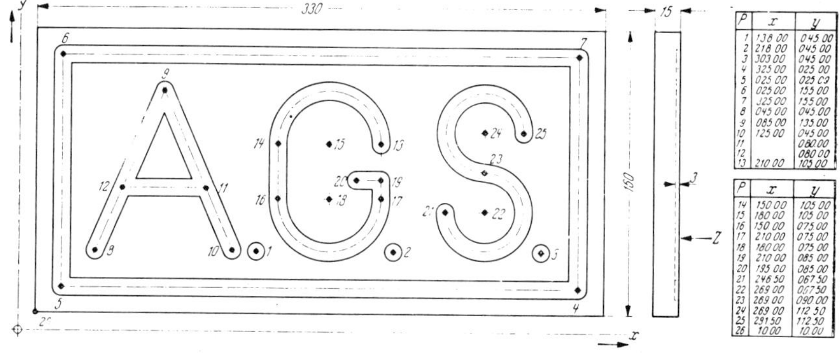

P	x	y
1	138 00	045 00
2	218 00	045 00
3	303 00	045 00
4	325 00	025 00
5	025 00	025 C3
6	025 00	155 00
7	325 00	155 00
8	045 00	045 00
9	085 00	135 00
10	125 00	045 00
11		080 00
12		080 00
13	210 00	105 00

P	x	y
14	150 00	105 00
15	180 00	105 00
16	150 00	075 00
17	210 00	075 00
18	180 00	075 00
19	210 00	085 00
20	195 00	085 00
21	246 50	067 50
22	269 00	067 50
23	269 00	090 00
24	269 00	112 50
25	291 50	112 50
26	10 00	10 00

Fig. 262 Example of programming. Drawing and tabulated dimensions for a simple workpiece.

Since it is not necessary to change the cutting tool the amount of switching information required is small. The displacement information predominates during programming. It is assumed that the work is to be performed on a vertical milling machine with absolute measurement of displacement (measurement of position). The control system must be suitable for positioning control (points 1, 2, 3), straight-line control (movement parallel to axes between points 4-5-6-7-4), and continuous-path control (for movements between points 8 to 12, 13 to 20 and 21). The internal interpolator is assumed to be suitable for all these functions, i.e. it can be changed over from linear to circular interpolation and vice versa, and it can be used for multi-quadrant working (cf. Chapter 8)[7].

The programme schedule given in Table 21 has been drawn up with these requirements in mind. The TAB-address method is used (the decimal point is not programmed), and it is further assumed that each word (in each column) is stored until a fresh word appears in the same column. As a result the amount of information that has to be inserted for each sentence decreases appreciably[8]. The 32 sentences on the programme schedule cover the whole of the machining process on the milling machine, so that it need not be repeated here in plain language. A newcomer to this subject is, however, advised to work through the example in detail. The technological data has been chosen arbitrarily and makes no pretence to represent ideal machining conditions.

When the programme in this example is evaluated, the following conclusions become apparent:

1. When the workpiece is clamped on the machine table it must be located in an accurately prescribed position to maintain the mathematical relationship to the programme, which is numerically fixed. In this particular simple example it is sufficient to locate point 26 or the coordinate origin and to meet the requirement that the edges of the workpiece are parallel with the axes. For most workpieces these simple alignment requirements would not suffice, and fixtures would have to be made to enable these initial conditions to be met without difficulty. Alternatively, the machine control system must incorporate means of adjustment to enable a reference point on the workpiece to be brought into alignment with a reference point on the machine (cf. Chapters 3, 4, 5, 7, and 8, Fig. 192). A close interrelationship exists between the provision of fixtures and the possibility of making adjustments by some auxiliary

---

[7] If an internal interpolator is used in which circular interpolation is restricted to one quadrant at a time, the circular arcs will have to be subdivided, and more intermediate reference points will be required.

[8] The TAB-address method is gaining in popularity. It produces a clear verbal statement of the programme; can be prepared by means of a typewriter or the printer of the data processor; and at the same time provides the reliability of the address method even without the use of a stepping switch (cf. Figure 236).

	N	G	X	Y	I	K	Z	F	S	T	M	Remarks
%	N 01	G 00	X 13 800	Y 04 500			Z 1600	F 99	S 78	T 5	M 04	(Point 1)
	N 02	G 05					Z 1200	F 40				
	N 03						Z 1600	F 99				
	N 04	G 00	X 21 800									(Point 2)
	N 05						Z 1200	F 40				
	N 06						Z 1600	F 99				
	N 07		X 30 300				Z 1200	F 40				(Point 3)
	N 08						Z 1600	F 99				
	N 09		X 32 500	Y 02 500								(Point 4)
	N 10						Z 1200	F 40				
	N 11	G 01	X 02 500									(Point 5)
	N 12			Y 15 500				F 48				(Point 6)
	N 13		X 32 500									(Point 7)
	N 14			Y 02 500								(Point 4)
	N 15						Z 1600	F 99				
	N 16	G 00	X 04 500	Y 04 500								(Point 8)
	N 17						Z 1200	F 40				
	N 18	G 01	X 08 500	Y 13 500				F 48				(Point 9)
	N 19		X 12 500	Y 04 500								(Point 10)
	N 20		X 10 944	Y 08 000				F 60				(Point 11 calculated)
	N 21		X 06056					F 48				(Point 12 calculated)
	N 22						Z 1600	F 99				
	N 23	G 00	X 21 000	Y 10 500								(Point 13)
	N 24						Z 1200	F 40				
	N 25	G 03	X 15 000		I 18 000	K 10 500		F 48				(Point 14)
	N 26	G 01		Y 07 500								(Point 16)
	N 27	G 03	X 21 000		I 18 000	K 07 500						(Point 17)
	N 28	G 01		Y 08 500								(Point 19)
	N 29		X 19 500									(Point 20)
	N 30						Z 1600	F 99				
	N 31	G 00	X 24 650	Y 06 750								(Point 21)
	N 32						Z 1200	F 40				
	N 33	G 03	X 26 900	Y 09 000	I 26 900	K 06 750		F 48				(Point 23)
	N 34	G 02	X 29 150	Y 11 250		K 11 250						(Point 25)
	N 35						Z 1600	F 99				
	N 36	G 00	X 01 000	Y 01 000								
	N 37										M 05 M 02	(Point 26)

Table 21. Programming sheet corresponding to Fig. 262
Type of machine: CAM-332
Type of control system: N2. G2. X32. Y32. I32. K32. Z22. F2. S2. T1.M*
Descriptive remarks: Vertical milling machine; work area 1000 x 600 x 500 mm; manual tool changing; minimum element that can be programmed 0.02 mm; linear and circular interpolation in the XY-plane for all quadrants; G and M functions in accordance with DIN 66025, Sheet 2; F and S functions in accordance with DIN 66025 Sheet 3; rapid traverse 10 m/min.

feature of the control system. This is of importance with regard to the setting-up times that are still a factor where numerical control is employed, and with methods of reducing them (cf. Chapter 7).

2. The example chosen is deliberately of a simple nature. The majority of workpieces encountered in practice are likely to be much more complicated. Although it is assumed that the machine and control system to be used are such that many of the calculations are performed by the internal interpolator, the programmer is still left with the task of listing every detail of the machining process. If the degree of complexity exceeds a certain level it may be uneconomic to do this manually and it will then be essential to make use of programming aids of all types.

In view of the wide variety of the numerically controlled machine tools now available and of the workpieces that are produced on them there is little point in giving further examples of programmes here. The number of publications and reports produced on this subject in recent years is large and the reader is therefore referred to these (e.g. 25, 124, 261, 262, 263, 268, 272). It is much more important to bring out the relationships between the machine tool and its control system on the one hand and its direct machine-orientated programming on the other hand, and to point out the urgent necessity of producing rules of procedure that will be accepted internationally as widely as possible.

The next problem area is to find equipment and methods that will make it easier to perform these new tasks. In principle, written programmes of the type shown in Table 21 will always be required in great variety; they provide in an abbreviated form, which is still intelligible, all the form information and the technological details, as called for in Chapters 1 and 2. The method of producing this abbreviated form is the next subject to be discussed; here again the computer can help in many ways.

10.7   Arrangement of programming area and the use of miniature computers

10.7.1 Use of typewriters

Mention has already been made of the simplest method of coding production data and of producing punched tapes with the aid of typewriters. The use of these machines on their own led to the use of the term 'manual programming' (6, 276, 277). This covers only the final phase of an extensive amount of data processing which is carried out by human operators and which, in the simplest case, consists of the preparation of drawings, card indexes, lists of cutting speeds for various materials, standard times, etc. Here again a

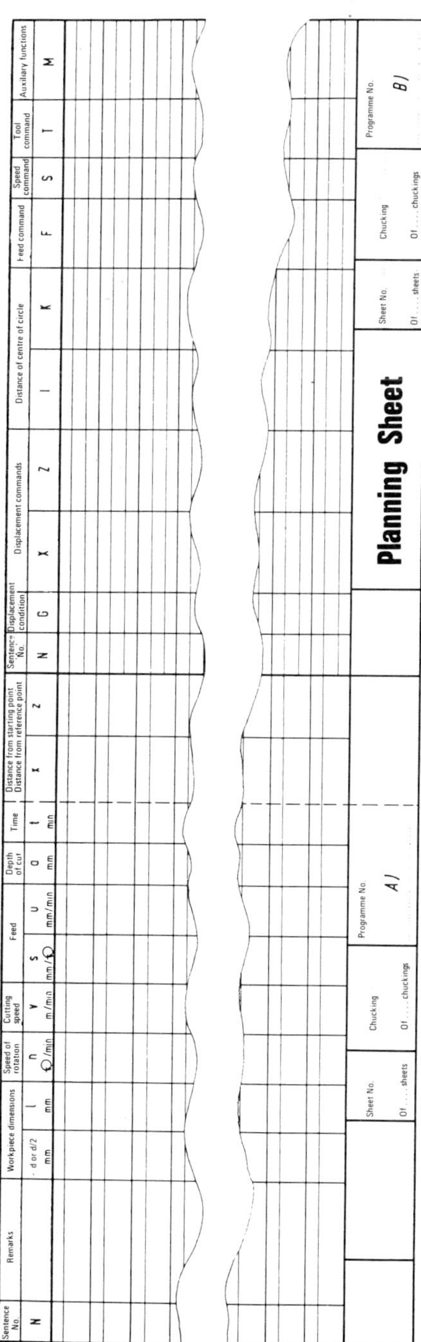

Fig. 263
Example of a planning sheet for lathes (262)

A) Plain-language entries made by planner for spindle speeds, cutting speeds, feeds, depths of cut, machining times, etc. B) Coded data provided by programmer in accordance with DIN 66025. Three lines are available for each programme sentence; one for the hand-written entries made by the programmer, one for the type-written entries made when the initial punched tape is prepared, one for a check print-out when the punched tape is duplicated.

considerable amount of information has been published and there are also instructions drawn up for the various machines (e.g. 124, 256, 262, 263, 268, 272, 273). Figure 262 shows, as an example of the documentation required, a proposed planning sheet for a lathe. This would be supplemented by setting sheets (instructions for the machine-tool setter regarding tools and fixtures required, tool corrections, etc), and tool and fixture index cards (262, 263). Because of their basic importance some examples of these are shown in Fig. 264. Since these card indexes are required for all forms of part programming, and they also prove a useful aid for conventional machine tools, it is advisable to make use of them as soon as possible, or at least to take the necessary administrative measures to introduce them (10, 124). Machine-tool manufacturers are always willing to advise and make recommendations, so that there is no need to go into detail regarding such planning aids here. An interesting feature of the form shown in Fig. 263 is the gradual transition to coding in the machine language when passing from section A) to section B) of the form. Section A) is for use by the planner who completes it entirely in plain language; section B) on the other hand contains solely the NC-orientated symbols as given in DIN 66025, although some of these are entered by hand. These symbols can then be converted into punched tapes in the appropriate codes by a clerk using a coding machine. At the same time a print-out for checking is produced on the same sheet; this can be reproduced each time the punched tape is run through, and so enables a check to be made on the punched tape. For this reason three lines are provided for each programming sentence; one is for hand-written entries, and the other two for typewritten entries that contain the same information. Every coding machine that is at all widely used will include a punched tape reader and a punch, so that the machines can be used both for reproducing punched tapes and for producing check print-outs.

Any of the typewriter type coding machines (e.g. those shown in Figs. 250 or 252) can also be used as an automatic letter-writing machine or for general document duplication in the works. Where the machines are to be used for multiple purposes such as this, it is only necessary to make sure that the codes employed are compatible with those employed on other typewriters in use in the plant that are controlled by punched tapes. This problem has only arisen with the gradual introduction of the ISO Code; it will no doubt retard the general adoption of this code in practice, but in view of the advantages that the code offers it is unlikely to prove a serious obstacle in the long term.

The use of typewriters as coding machines suffers, however, from three basic disadvantages:

1. Filling in the planning sheets by hand (cf. Figure 256) is a time-consuming task. With programmes that contain more than 100 to 200 sentences the amount of time, and the consequent costs, become a major factor (10, 265).
2. The need to make frequent reference to a wide variety of card indexes, lists, tables of cutting speeds, drawings, etc. is tiring and is likely to introduce errors.

3. The whole process is bound to a specific machine tool and its particular control system. Despite intensive efforts to achieve standardisation it is unlikely that any major improvements will be made in this respect.

10.7.2 *The use of auxiliary calculators*

A wide range of auxiliary devices is now available to assist in overcoming the effects of these three disadvantages, ranging from simple desk calculators to large computers. In view of the variety of the equipment and the continuous progress that is being made in developing new solutions, it will again be necessary to restrict discussion to the essentials. In Section 10.4 it was already proposed that the uses of a computer be divided into three groups. In many instances the whole of the effort will be directed to reducing the amount of data as much as possible so as to reduce the work that has to be manually performed. It will be obvious that the use of storage systems of all types can prove of considerable benefit in this respect. These systems, however, are particularly costly. For the further analysis of the problem it is, therefore, convenient first to identify four specific applications:
   a) Performance of calculations (small storage capacity required)
   b) Performing allocations (large storage capacity required as replaces card-index)
   c) Preparation of check routines (small storage capacity required)
   d) Automatic preparation of punched tape (no storage capacity required).

No discussion of allocation problems need be given here, since only simple computer systems to be employed as auxiliary devices are discussed in this chapter, and for reasons of cost it is frequently desirable to continue to employ machine-orientated programming for simple workpieces. These problems include, for example, the automatic determination of cutting speeds, tool selection, etc. On the other hand, a high proportion of the simpler calculations and programme checks can now be performed on very small computers that possess no significant storage capacity. As a result the work of programming can be greatly simplified. This line of thought is

The following caption refers to the figures on pages 410, 411 and 412.

Fig. 264 Examples of setting sheets and card-index entries for lathes (262, 268)

A) Setting sheet for use by machine operator. Gives details of zero-point (origin) shift, reference point, starting point, speed range, feed range, coolant, chuck, tools, and card-index entries, etc. B) Entry for tool card-index with main dimensions, radius of cutting edge, setting data, etc. C) Entry for fixture card-index; example shows power chuck.

A)

	Setting Sheet		Chucking of	Programme No.	
			Date	Name	Checked

Machine Type		Workpiece	
Control system		Drawing	
Machine No.		Material	
		Blank	

	X	Z	Sketch of fixture
Zero-point shift			
Reference point			
Starting point			
Speed range			
Feed relationship			
Coolant			
Fixture			
Serial No.			
Chucking dia.			
Length of bar feed			
Position			

Tool description	Tool serial No.	Tool comnd	Tool locn	Cutter material	Cutter radius	Remarks

Sentence No.	Required dimension	Tolerance A_o	A_u	Tool correction	Remarks

B)

Cutting tool	Machine	Date/name	Serial No.
Copy-turning tool	VDE turret lathe	6. 2. 68	11

Cutting tool type		L 170.5 - 4225
Tool bit		Tungsten carbide, clamped
Cutter material	Application group	
	Grade	
Approved angle	$\varkappa$	63°
Rake angle	γ	+ 6°
Clearance angle	α	
Front clearance	λ	
Chip breaker	b x t	
min. int. turning dia.		
pushing cut		pulling cut
Tool holder No.		11.9471.0085 - 01
Setting values	A	480
	B	540
Swing circle radius		275

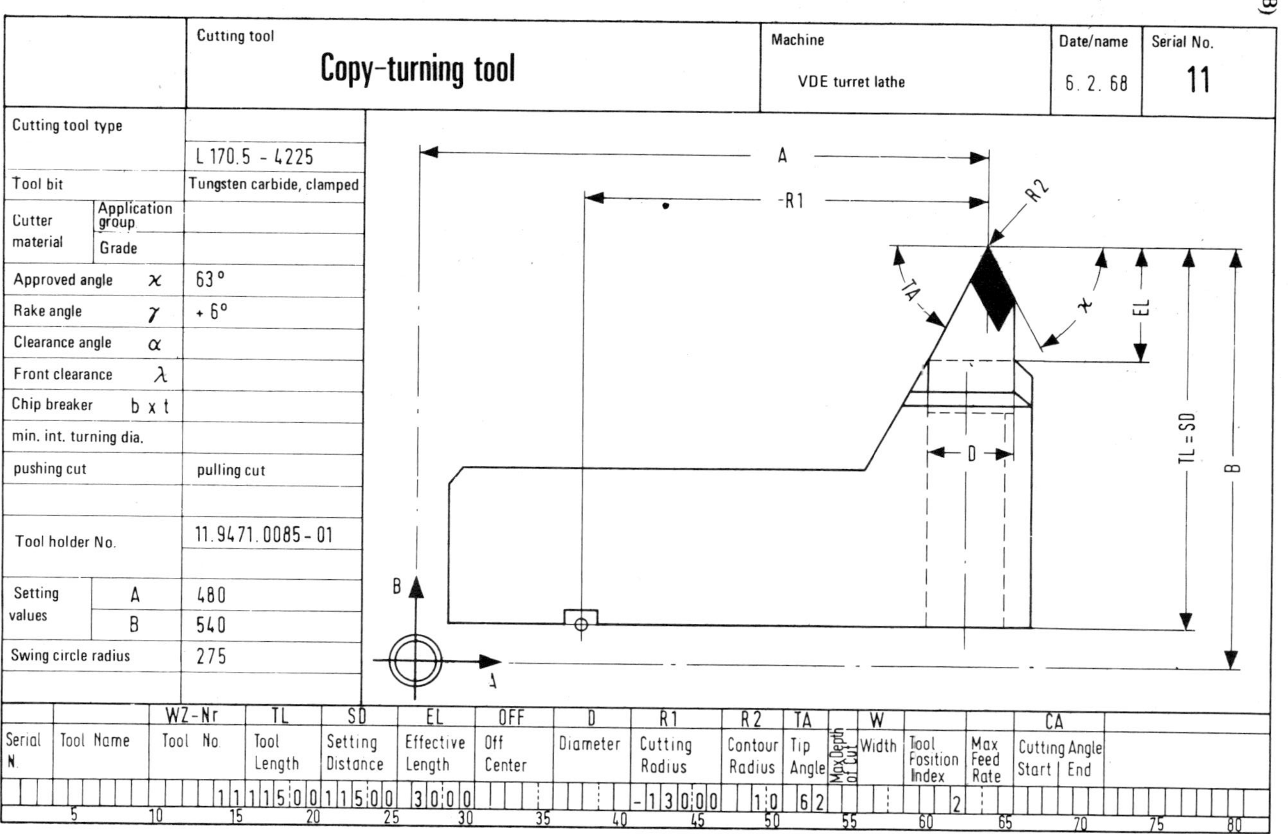

		WZ-Nr	TL	SD	EL	OFF	D	R1	R2	TA		W			CA	
Serial N.	Tool Name	Tool No	Tool Length	Setting Distance	Effective Length	Off Center	Diameter	Cutting Radius	Contour Radius	Tip Angle	Max Depth of Cut	Width	Tool Position Index	Max Feed Rate	Cutting Angle Start	End
			1 1 1 1500 0	1500 0	300 0			-1 300 0	1 0	6 2				2		

C)

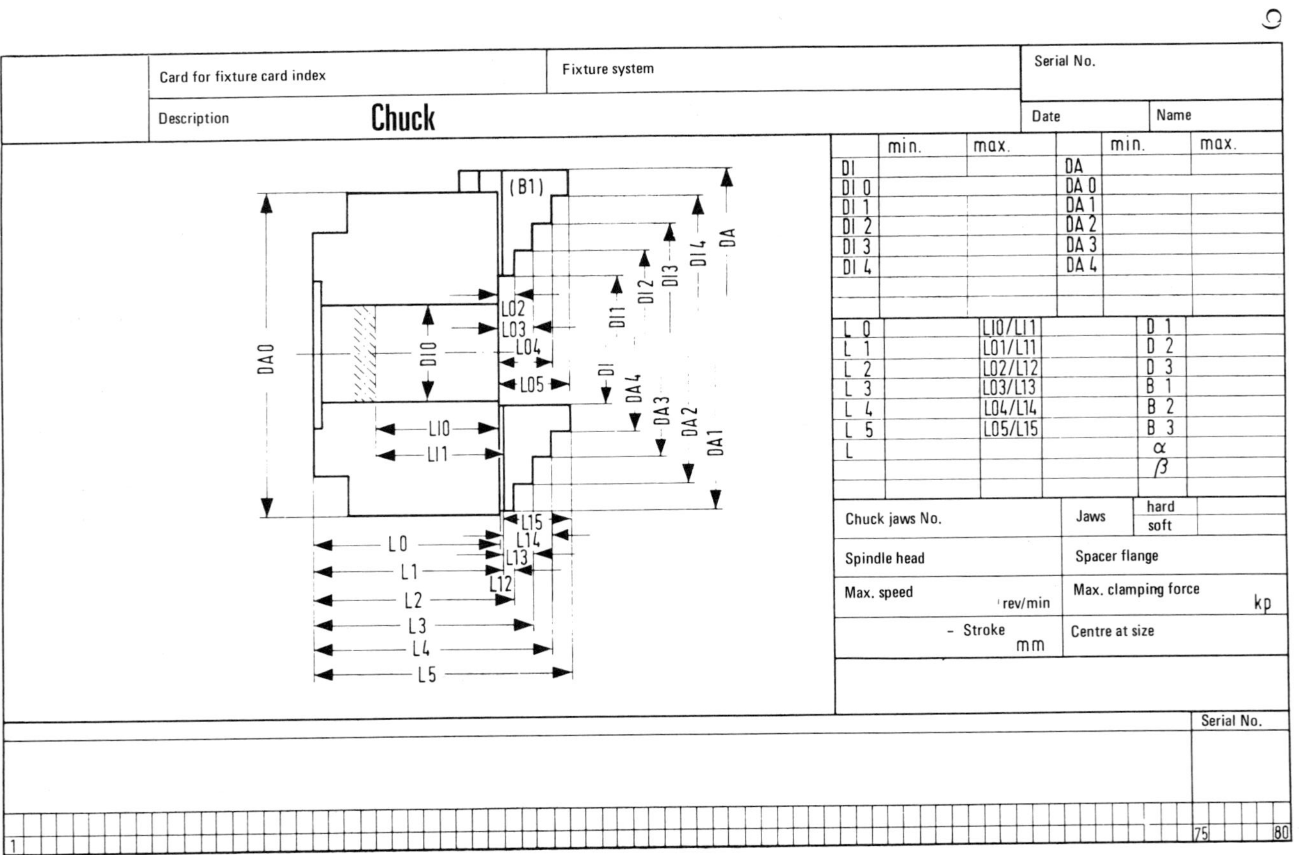

Card for fixture card index	Fixture system		Serial No.	
Description	**Chuck**		Date	Name

	min.	max.		min.	max.
DI			DA		
DI 0			DA 0		
DI 1			DA 1		
DI 2			DA 2		
DI 3			DA 3		
DI 4			DA 4		

L 0	L10/L11	D 1
L 1	L01/L11	D 2
L 2	L02/L12	D 3
L 3	L03/L13	B 1
L 4	L04/L14	B 2
L 5	L05/L15	B 3
L		α
		β

Chuck jaws No.	Jaws	hard
		soft
Spindle head	Spacer flange	
Max. speed rev/min	Max. clamping force kp	
− Stroke mm	Centre at size	

Serial No.	

continued below with the aid of some examples. Figure 265 shows in outline some of the problems encountered in drilling operations. In A) of this figure there is no obvious pattern to the positions of the holes or their diameters. So far as the form information is concerned, there is no simple means of reducing the data to any extent; in most instances, hole patterns of this type are governed by functional considerations (e.g. in gearboxes) and can be included in programmes only as part of a computer-aided design process. If an exclusively geometrically orientated programme language is used, machine programming will produce little improvement (cf. Chapter 11).

In Part B) reference is made to a common way of dimensioning holes arranged on a pitch circle; these dimensions are usually expressed in polar coordinates even when the remainder of the drawing is dimensioned in Cartesian coordinates. For NC machine programming the hole spacings must be expressed in Cartesian coordinates, i.e. the programmer must often make conversions involving angular functions for which slide-rule accuracy is not sufficient. This problem can easily be solved with adequate accuracy with the help of a small desk calculator.

In C) the hole pattern is symmetrical about the line $x_1 - y'$. The programmer need work out the dimensions for half the holes only and the task of completing the hole pattern is left to a miniature computer which has been given a 'mirror image' instruction. It is also possible to design control systems in such a manner that they will solve the same problem as part of the internal data processing (cf. Fig. 217), in which case the 'mirror image' could be called up as one of the G functions. In view of the remarks (made in Chapters 8 and 9) regarding the most effective and least costly means of

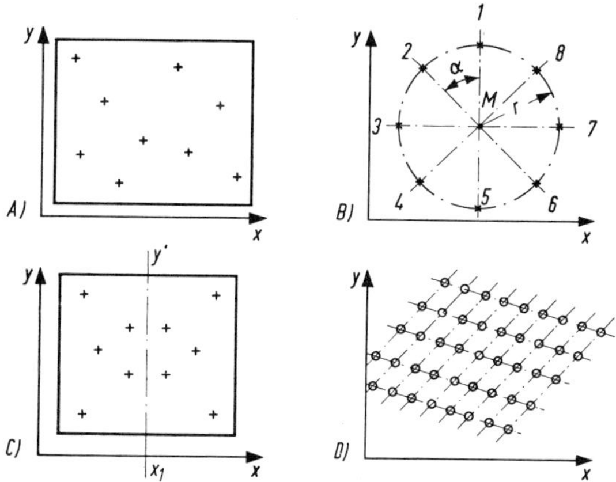

Fig. 265 Sketches illustrating some typical hole patterns

A) Irregular hole arrangement with no obvious pattern. B) Holes uniformly distributed on a pitch circle: positions dimensioned in polar coordinates. C) Symmetrical hole pattern. Pattern is symmetrical about line x_1-y'. D) Regular hole pattern to readily discernible formula.

dividing the work between the internal and external data processing systems, the methods of doing this cannot be pursued here.

Figure 265 D) shows a portion of a hole pattern which is built up in accordance with a formula that can easily be defined geometrically. Hole patterns of this type are often met in applications such as heat exchangers, and it is only the number of holes which have to be provided in a tube plate that determines whether a small or a larger computer will be needed to work out the pattern (cf. Fig. 180) (154). Figures 266 and 267 illustrate problems in machining. 'Thread-cutting' provides a suitable example because at the very least it comprises the following separate operations: 'Centre', 'Drill minor diameter', 'Countersink', and 'Tap'; these are then repeated from position to

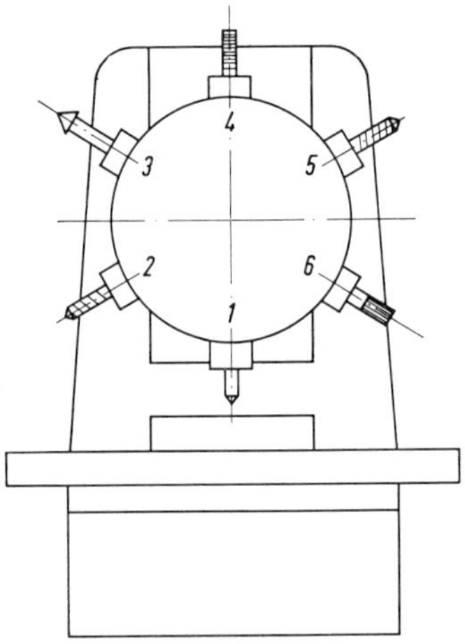

Fig. 266
Basic arrangement of a six-spindle turret drilling machine, set up for two operations: Tap, and Ream-to-size.

Position 1, Centre. Position 2, Drill minor diameter. Position 3, Countersink. Position 4, Tap. Position 5, Drill. Position 6, Ream to size.

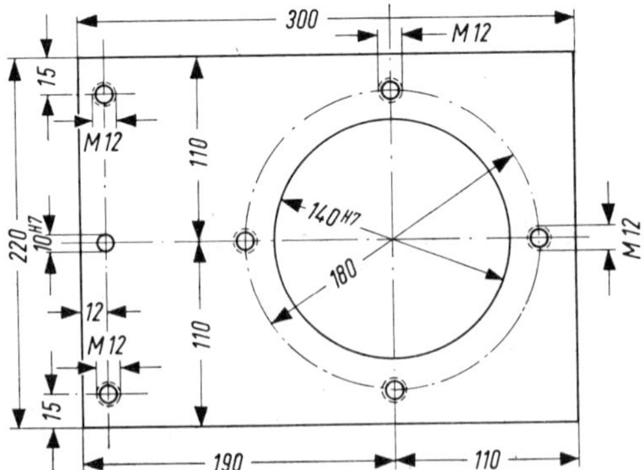

Fig. 267 Sketch of workpiece containing tapped holes (may be machined using 'thread tapping machining cycle'; decision depends on workpiece size and machine characteristics.

position. The machine control system can be arranged with either free or fixed sub-programmes for this purpose (cf. Fig. 217) and these operations can then be called up by, for example, the *G* functions *G* 80 to *G* 89 (168) as part of the internal data processing. The same problem can, however, also be solved as part of the external data processing using a small computer; with the advantage that this can also make the decision whether a full cycle of operations is to be performed at each position in turn, or whether one tool is to be used at all the positions before the turret is indexed. This decision will be based on workpiece size, the number of holes, and certain of the machine parameters, such as the times required for the machine slide and the turret to move into new positions. Even from the point of view of only achieving the maximum possible utilisation of the computer, it is desirable to perform as many as possible of the external data processing tasks on

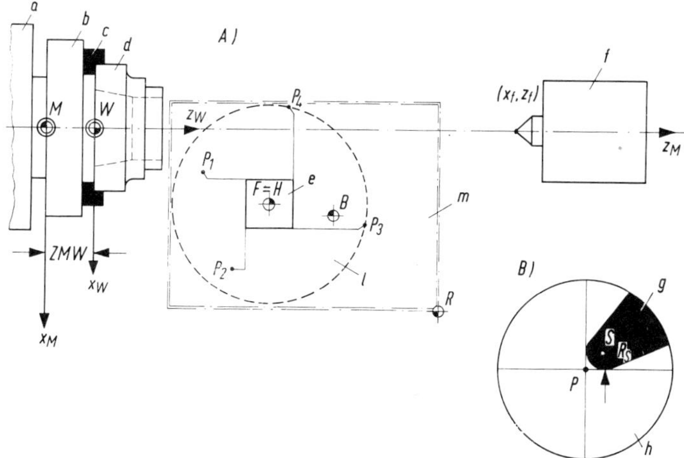

Fig. 268 Diagrammatic sketch of a lathe, arranged for chucking work and equipped with a turret.

A) The lathe, its work area, and its main reference points. B) Field of view of the appropriate tool pre-setting microscope (for example, as shown in Fig. 166 C or Fig. 167).

a Headstock. *b* Power-operated chuck (e.g. as shown on card-index card Fig. 264 C). *c* Replaceable chuck jaws. *d* Workpiece. *e* Turret. *f* Tailstock (retracted, since not required for this work). *g* Tip of turning tool against crosswires of presetting microscope (similar to instrument shown in Fig. 166 C or 167). *h* Field of view of presetting microscope. *l* Maximum swing of tools in turret *e*. *m* Work area. X_M, Z_M Coordinate axes of lathe. X_W, Z_W Coordinate axes of workpiece. *P*, Tip point of tool. (Setting point for tool in presetting instrument; point of intersection of tangents = centre of crosswires). *S*, Centre of tool tip radius (Centre point of tool slide from which locations of other points on tool holder are measured; usually lies on axis of rotation of tool turret). *R*, Reference point (for incremental control systems).*B*, Starting point (position of slide reference point *F*, when programme commences). *H*, Tool changing position (the position occupied by point *F*, when tool changing (turret indexing) occurs. The slide is drawn with point *F* at point *H*).

freely-programmable small computers. This enables the control consoles to be simplified and hence makes it possible to reduce the overall cost of the NC system. The discussion which follows of the conditions relating to lathes, points in the same direction.

Figure 268 summarises, in very simplified form, some of the programming problems that arise with lathes. In particular, there are some special programming features which make an early use of a computer desirable; furthermore, there is at present a very rapid increase in the use of NC lathes (266) — very probably because voluntary cooperation for standardising the programmes was established at an early date between thirteen of the major lathe manufacturers (262, 267, 268). Part A) of the figure shows in diagrammatic form the machine-specific aspects of the problem. Without going into detail at this point (268), it will be apparent that large numbers of coordinate transformations are required before it is possible to produce the proper coordinate words of the type shown in Fig. 263 or Table 21. In Fig. 268 B) reference is also made to the relationships that exist between the presetting of the tooling, the dimensional accuracy of the tooling, and the accuracy of the workpiece. In the long run the tolerances that can be maintained on the workpiece depend solely on the dimensional accuracy of the cutter contour radius R_S and the accurate location of the points P and S in the work area of the machine. As a result — for maximum accuracy — with lathe control systems in particular, it is necessary to provide at least 2×4 decade switches for the correction of the tool position (cf. Fig. 246 A). It is then necessary to arrive at a compromise between the measurement and control of the slide movement and that of the workpiece. Further details are given in (268).

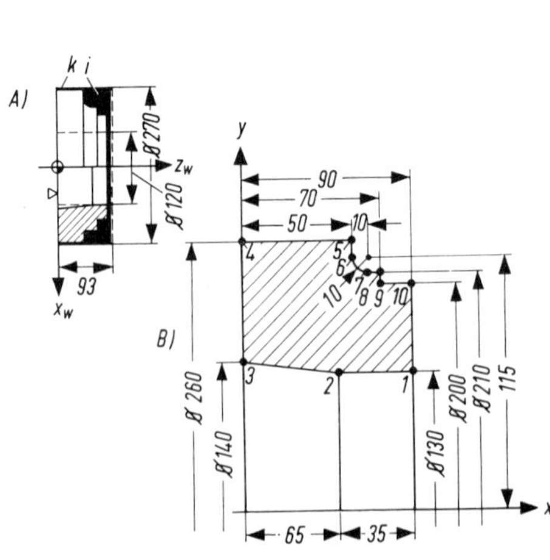

Fig. 269
Sketches of workpieces from Fig. 268

A) Dimensions of workpiece blank and finished contour in machine-orientated *XZ* coordinate system.
B) Finished contour of workpiece in computer-orientated *XY* coordinate system; sketch for use as basis for machine programming, e.g. using EXAPT 2 (cf. Chapter 11).

i Volume of material to be removed in the first clamping position of workpiece at *k*. *k* Residual volume of material to be removed on finish-machined face in second clamping position.

Figure 269 further develops the ideas illustrated in Fig. 268, primarily to make it easier to understand the transition to machine programming which is discussed in Chapter 11. Both parts of this figure show the workpiece illustrated in Fig. 268 A) from different aspects. In Fig. 269 A) the workpiece is again shown in the machine-orientated XZ coordinate system. The amount of material which has to be removed to produce the finished part from the blank is clearly shown. From the difference in the contours before and after machining, the part programmer has to determine the theoretical optimum amount of material to be removed with each cut, and the times required. At the same time he has to take into account the machine capacity, the life of the tools, and the physical properties of the workpiece material at the same time allowing a margin of safety in the calculations and the metal removal rate which must be below the maximum capacity of the machine. Factors that used to be under the control of the skilled machine operator, and which determined the amount of bonus that he would earn, are now fixed within close limits on the basis of tabulated values; only in an emergency is the machine operator allowed to take advantage of the opportunity to vary the feeds manually as shown in Fig. 246 A). Manual intervention is often omitted as a matter of principle, and with magnetic-tape control systems it is pointless (cf. Chapter 9). In this context the importance of adaptive control systems, as discussed in Chapter 8, becomes clear. Depending on the future costs of adaptive control systems in internal data processing, it may eventually be possible to eliminate the need for the external data processing equipment (cf. Chapter 11) determining, more or less theoretically, the distribution of the amount of metal to be removed by the various cuts.

10.7.3 *The VDF Autoprogrammer and Simulator as an example*

In the previous sections reference has been made on several occasions to the relatively high proportion of simple numerical calculations that are required for manual programming; at the present state of development and cost level of small electronic computers it is possible to use them to perform the majority of these calculations easily and economically. There is also another reason why it is desirable to improve the equipment which is available at the programming station; it is certainly not just a coincidence that proposals of this nature have originated from the makers and users of lathes (269, 270, 271, 272). As is shown in Fig. 268 the limited amount of working space available on lathes makes the question of interference or collisions of major importance. Collisions in this context are unintentional contacts between tool, tool holder, or tool carrier on the one hand, and the workpiece, chuck, tailstock, or other tool stores on the other hand (see Fig. 268 A). Most serious are the possibilities of collision that occur during tool changing (i.e. when the turret indexes) or during machining (268). It is very difficult, and time consuming, to determine all the possible interference paths and to store the information in its entirety in a large computer. The movements are largely two-dimensional, and simulating them on a drafting machine adjacent to the programming station may prove much simpler, clearer, and cheaper than using a large computer (cf. Chapter 11). Figure 270 illustrates a work station that includes a drafting machine of this type. The chuck holding the workpiece is drawn on the left-hand side of the drawing. The pivoting head *a*

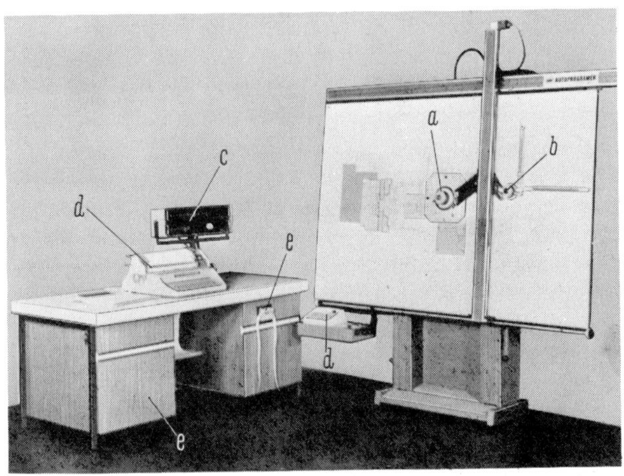

Fig. 270 Programming station for semi-machine programming using a small computer and simulator (Photo courtesy of *VDF-Autoprogrammer*). (269, 270)

a Swivel head for simulating turret. *b* Drafting machine for auxiliary drawings. *c* Digital readout for coordinate dimensions. *d* Typewriter for data input and check printout. *e* Miniature computer with punched-tape output.

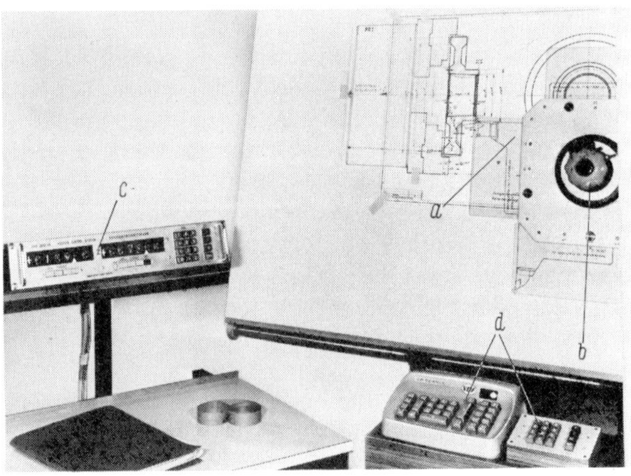

Fig. 271 Part view of programming station shown in Fig. 270

a Interchangeable tool templates. *b* Swivel head for turret indexing. *c* Digital readout for coordinate dimensions. *d* Keyboard for small computer.

carries interchangeable templates which are shaped to represent the tools contained in the card index (see Fig. 264); the pivoting head *a* can thus readily be used to represent the turret of a lathe and by moving the whole assembly longitudinally and transversely the movement of the cutting edge of the tool over the workpiece can be simulated. Any dangers of collision will then be clearly visible, especially if the ordinary drafting head *b* is used to sketch the other boundaries of the work area. All the points *P, S, F, R, B,* and *H* (Fig. 268) required for programming can be easily located at the correct positions within the work area. It is, of course, possible to use any suitably modified drafting machine for making these visual checks. It is, however, also possible to go a stage further, and to determine the longitudinal and transverse movements of the turret head *a* with sufficient accuracy by two digital-incremental displacement measuring systems (cf. Chapter 3) and counters. And, then, at least, to display the positions of the head *a* in numerical form and to transfer them into the storage system of the computer by pressing a key. Figure 271 shows an enlarged view of a work area of this type. The next stages in the development are simple and obvious. If all the other data given in Fig. 268 is available as machine constants and is first entered into a small computer by means of a keyboard, there is now nothing to prevent the automatic calculation of the coordinate words required for producing the programme being calculated automatically.

Additional programme simplifications and check routines will usually be added, to enable the fullest possible use to be made of the small computer. Some examples of these include:

> determination of the coordinates of the circle centre points *I* and *K*;
> determination of the code numbers for the *F* and *S* words in accordance with DIN 66025 from the speed and feed values which are provided in plain language (Fig. 263 A);
> checking that the sentence and word structure is correct where the machine specification as discussed in Section 10.4 is known;
> determination of the machining times, etc.

The typewriter *d* can be used to enter the following limiting values for the machine into the computer:

> minimum and maximum spindle speeds;
> maximum slide movements (work area);
> maximum torque of main spindle or maximum forces that the workpiece or tool can withstand;
> specific cutting force on workpiece;
> maximum permissible cutting speed in m/min:
> range of adjustment for feed (in %) available at machine control console;
> works identification number of control system employed if several types of control systems are in use;
> required type of code for punched tape (EIA or ISO); etc.

All values introduced by the programmer as part of the programme can then be checked against these limiting values to determine whether they are acceptable, and any errors can be reported by the computer. The production of the programme thus remains machine-orientated, but the work is greatly simplified and more reliable through the use of the computer. In addition, an elegant solution is provided to the problem of determining the type of coding that is required, and this is likely to prove especially convenient during the period of transition from EIA to ISO Code. The drafting machine as an aid to visually showing the danger of collisions is especially useful for beginners in the programming field during their training; whether it would be fully utilised by experienced personnel is another question. To arrive at the correct decision regarding the acquisition policy to be pursued, a small computer divorced from the drafting machine should be considered as an alternative; computer installations of both types, with or without a simulator, are commercially available, but machine-orientated programming will continue to find a very wide field of application. And in many cases its adoption will avoid the need for a larger computer for production purposes. At least it will postpone the need for such a machine until a soundly based decision can be made regarding the type and size of computer that is to be acquired or until the personnel concerned have gained adequate experience (cf. Chapter 12).

10.8   Summary

The seven sections that comprise this Chapter give the information on all the additional items needed, beyond the requirements for internal data processing together with the descriptions of the equipment that were given in Chapter 9, to enable external data processing to be dealt with. In a sense these two chapters are complementary, since they deal with a very important transition area from two different points of view. This chapter has however ranged beyond the actual transition area to include the simplest forms of programming station layout and the use of computers with particular reference to the machine tool and its control system. The analysis had led to the following conclusions:

1. The introduction of small electronic computers having a good cost performance ratio has shown that the older subdivision into 'manual' and 'machine' programming is too coarse; on the other hand, it is undesirable to increase the number of categories of programming methods to such an extent as to complicate the issue unduly. It is therefore proposed that one further category, namely 'semi-machine' programming, be introduced because it is possible for semi-machine programming with small computers to be either machine-orientated or problem-orientated. In this chapter only machine-orientated programming is however discussed.

2. During the course of the last 5 to 6 years the trend towards integrated data processing has become increasingly marked, also in the production field. This has been taken into account by the International Standards Committees, and a 7-Bit Code, based on the American ASCII, has been developed for information processing and transmission, and the numerical control of machine tools. The arrangement of the 120 hole patterns and the symbols (capital and lower-case letters, numbers, and other signs) have been incorporated into the German Standards as DIN 66003, and this forms the basis for further Standards in the field of data processing. Since all these Standards are currently available only in draft form, it is advisable to obtain the latest versions from the DNA before basing any definite decisions on them.

3. A subset of 50 characters was initially found adequate for the control of machine tools, and from this an eight-track punched tape was derived by adding a check bit to the seven information bits. The punched-tape code which has been developed for NC machine tools — known as the ISO Code — is fully compatible with data-processing installations and, with the increasing degree to which integration is taking place, is also to be preferred for the transmission of information within the plant.

4. In the 1950's, the EIA Code was developed in the USA based on typewriter techniques; this exerted a considerable influence on the development of NC machines. It is very widely used in the United States with the result that considerable opposition to the adoption of the ISO Code exists there. It may be assumed, however, that the wider technical scope that is offered by the ISO Code will ensure that this is eventually also adopted in the USA, even though there may well be a long transition period. The self-correcting PC-8C special Code, which has a supply of only 31 characters, is also likely to give way to the ISO Code.

5. Much more important than the allocation of the individual hole patterns to the various characters is the use of these characters to indicate internal processes within the machine. Here again international agreement has been arrived at by the ISO; the results are summarised in DIN 66025 Sheets 1 to 4.

6. The standard method of programme construction which is given in DIN 66025 Sheet 1 can be used directly for machine-orientated programming; however it also forms the output format for problem-orientated machine programming.

7. International proposals have been made for designating the directions of the axes of NC machine tools; they are at present given in VDI Code of Procedure 3255, but have not as yet been incorporated in a German Standard.

8. In view of the current state of the art, the simplest programming stations are at present based on typewriters; the programme schedules

are first filled in by hand to the rules given in DIN 66025, and a check print-out is produced when the tapes are punched on the coding machine. If necessary, the punched tape can be duplicated by the typewriter. This method of working is known as 'manual programming'.

9. Large numbers of arithmetical calculations have to be made when geometrical programme data is being prepared. These can readily be delegated to small computers, which — with automatic check routines and limiting value stores — can considerably ease the task of the part programmer. Furthermore, simulators based on drafting machines can be employed, especially for spotting collision risks. The process has been developed initially for the programming of lathes. The method described as an example can be designated as a machine-orientated semi-machine method of programming. It will however be considered further, in the light of the discussions on problem-orientated semi-machine programming in Chapter 11.

11 The use of computers and machine parts programming— Problem-orientated languages for production processes

Together with the development of nuclear energy and space travel, the automatic processing of information is one of the most remarkable phenomena of this highly technological age. The first steps in this field were taken barely thirty years ago, it is too early to appreciate to the full the far-reaching effects that this technique will have; especially since the rate at which progress is being made is such that the latest models have a useful life of only 5 to 8 years before they are superseded by newer and still better systems (300). It is not surprising to find that production technology is also being affected to an increasing extent by this development, the effects being both beneficial and detrimental. The previous chapters have shown clearly how important it is to link machine tools with this new technique of data processing from both the theoretical and the practical points of view. To gain a full understanding of the wider implications the reader is advised to make a detailed study of both the lay-out and the programming of electronic data-processing equipment about which a considerable amount of suitable introductory literature has been published (e.g. 2, 6, 57, 185, 186, 230, 279). Here it will be sufficient to discuss only the most essential aspects in simplified form.

11.1 Fundamentals of electronic computers

Whenever the term electronic computers is employed here it refers always to digital computers. Figure 272 shows a rough sub-division of these machines into the purposes for which they are employed. As has been shown in Chapters 3, 4, and 5, one of the primary advantages of digital techniques is that the accuracy is limited only by the number of decimal places employed in the numerical calculation of a problem and, in theory at least, the number of decimal places to be used can be made as large as desired. A second advantage of digital techniques, which is just as important, is the ease with which digital values can be stored: for example, the punched card or punched tape with the binary statements 'hole' or 'no hole'. Examples of components used in the construction of digital stores include self-holding relays; bistable multi-vibrator circuits as shown in Fig. 33; magnetic cores made of ferrites which have almost rectangular hysteresis loops; magnetic tapes, and magnetic drums or discs (rapidly rotating drums or discs having a thin magnetisable coating) (6, 230).

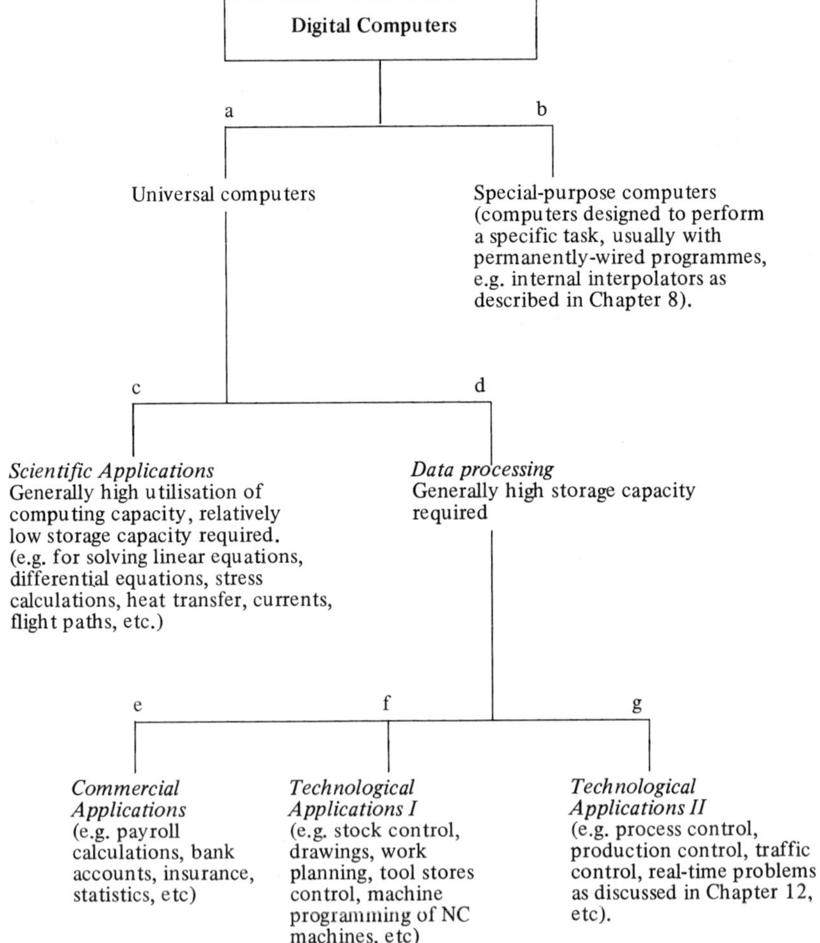

Fig. 272 Computers subdivided according to the purposes for which they are used, based on reference (230)

If, in addition, electronic components are used to process the digital signals there is the third advantage that very short computing times are achieved with compact equipment (cf. Chapter 3 and the remarks it contains regarding integrated circuits).

These three characteristics taken together result in digital data processing equipment possessing yet another major advantage. This is their extreme flexibility, i.e. the ease with which they can be adapted to perform a new task by the programme being changed (282). When using digital techniques for carrying out automatic computing operations it is necessary to bear three points in mind:

1. Basically, digital computers are capable of performing only very rapid additions and logic operations (e.g. AND or OR operations, comparisons). Multiplication is performed by repeated addition and division by subtraction[1].
2. To a certain extent some departure from the methods of pure mathematics is necessary. The concepts 'infinitely small' and 'infinitely large' do not occur in numerical mathematics. The development and use of highly convergent series is thus one of the more important means for making a problem capable of solution by a digital computer. The series can be terminated when the solution is 'accurate enough'. Differential equations become equations in differentials, and integrals must first be replaced by sums of finite terms. Examples of this type have already been given in Chapter 8 (digital internal interpolators as special-purpose computers).
3. For automatic calculation the necessary computation steps must be specified at the outset, i.e. the computer must be 'programmed'. While the machine itself is capable of taking logical decisions, the method that is to be followed for the computation must be specified in advance. The machine can determine whether the rate of growth of the sum of a series of terms for an independent variable has become sufficiently small for the series to be terminated.

Figure 272 shows the basic construction of a digital computer, in so far as this is necessary for the understanding of its method of operation and for following the discussions presented later in this chapter.

The command unit c controls the flow of data between the other units in accordance with the programmes with which it is supplied; it also determines the manner in which the data is manipulated in the computing unit b.

---

[1] Strictly speaking digital processing equipment can perform only elementary logic steps. The use of these logic steps for mathematical operations is only one of many possible applications. The computer is thus only a special case of general data processing equipment. This fact should be borne in mind where, for simplicity, the term 'computer' is used in the following discussion (282).

The store a contains raw data or partially-computed data arranged so that it can be retrieved within a given access time; for economic reasons alone the stores have to be arranged in a certain hierarchical order, as indicated in Fig. 273. Each call of the store has its own address. The capacity of the stores may be expressed in bits, but more usually in 'machine words' (numbers of a specified length, e.g. 8 binary places = 1 Byte (283), 10 decimal digits, 36 binary places (287)).[2]

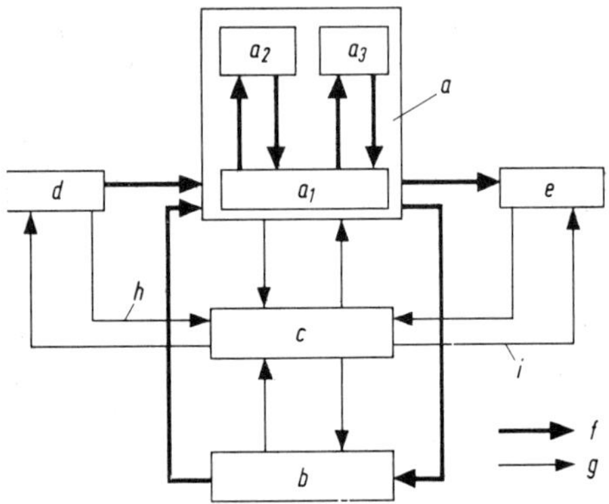

Fig. 273 Schematic diagram of a computer (data processor) (230)

a Stores (Memory). a_1 Quick-access store (working store, internal store) usually built up from magnetic cores (access time 1 μs and less). a_2 Large-capacity store having medium access times, e.g. magnetic drum store (access time 10 ms), and magnetic disc store (access time 20 to 100 ms). a_3 External large-capacity store e.g. magnetic tape store (access time 5 s and longer), several of these often being connected to the computer, b Computing unit (cycle frequency if integrated circuits are used about 10^6 to 10^7/s) also contains small working stores, known as registers, built up from flip-flops (cf. Chapter 3) having very short access times (about 0.1 μs). c Command unit. d Input equipment (e.g. typewriters, punched-card or punched-tape reader). e Output unit (e.g. typewriter, card or tape punch, line printer). f Data flows. g Control signals. The most important characteristics of a computer are the core store capacity and the cycle frequency or cycle time

The access time to the words is determined by the nature of the store (e.g. core store a_1, magnetic drum or magnetic disc store a_2, or magnetic tape store a_3. This time is of the order of micro-seconds (or less) for core stores, and up to several seconds for magnetic tape stores. The storage costs per bit become higher the shorter the access times.

---

[2] When giving the store capacities, it is customary to describe the number $2^{10} = 1024$ machine words with the letter K. A core store having 32 K storage cells can thus accommodate 32×2^{10} machine words, it is still necessary to define how large the machine words are (e.g. 24 bits). To standardise the terminology in this book, all capacities are quoted in Bytes = 8 bits, or in K Byte.)

The four basic arithmetical functions and the logical operations are performed in the computing unit *b* using numbers which, as controlled by the command unit, are either produced by the input unit *d* or are called up from a store. The results of the computation are supplied either to the output unit *e* or, provided with a new address, into a store.

The computer communicates with the outside world by means of the input and output units (these two functions may be performed by separate units). For this reason it is often necessary to distinguish between an internal and external code to ensure that communication between the man and the machine is unhindered (20). The larger the amount of data that has to be stored, the greater the importance of the external large-capacity stores (currently primarily magnetic tape stores, cf. Table 14). Large data banks can only be established with their aid; although where these are in use there is some deterioration in direct man-machine communication. Where this is of high importance increasing use is being made of graphic devices (288, 331, 332, 333).

In general, the input and output units are the parts of the installation that are most likely to develop faults and that require the greatest maintenance; this is because these units are primarily of electro-mechanical construction. The mechanical nature of the input and output units makes them the slowest links in the whole machine complex. Together with the limitations imposed by man, who cannot acquire information at a rate in excess of about 16 bit/s[3] (20), there is a resultant mis-match between the speed at which the electronic processor works and the speed with which it can communicate with its surroundings; this is the primary reason for the development of multi-programming and time-sharing systems. The disc, drum, and tape stores, which also contain mechanical components are, however, also parts of the installation that can easily become disrupted.

In recent years, the various different requirements and the numbers of manufacturers active in this field have led to a rapid increase in the number of computer types available. Only recently has there been any sign of a reduction to a manageable number of basic types. In general, it is more economical to use one large computer rather than several smaller units. This principle cannot always be adopted, however, and for economic reasons it often proves essential to match the machine to a particular application. Here the adoption of the unit-construction system can prove of advantage just as can the development of hierarchies of computers including small computers; this has already been suggested as a desirable development in Chapters 8 and 10. (207, 253, 350).

This short survey of the method of operation of computers shows the surprisingly large extent to which in many ways it resembles that of machine tools, as described in the earlier chapters. The relationship will become even more obvious in the Sections that follow.

---

[3] If anything unexpected occurs, human reaction time may be much longer.

11.2   General discussion of problem-orientated programme languages

The increasing sizes of computer installations, the ever-widening scope of their applications, the wide variety of computer types, and the relatively large number of manufacturers has already resulted, over a brief period of time, in a situation arising where the direct machine-orientated programming of computers has become difficult, time-consuming, and uneconomical; while the programmes that are produced cannot be transferred from one type of computer to another.

At a comparatively early stage (from about 1951) attempts were made to avoid being bound to a particular computer when programming and rather to develop a symbolic language that would be suitable for all programming operations. The most important factor here is the problem to be solved. The methods that make use of these languages are therefore referred to as problem-orientated programming as compared with the machine-orientated programming of computers which has been mentioned earlier.

In the main, there are three requirements which a symbolic language of this kind must meet if it is to serve as a means of communication between man and machine (249, 279, 282):

1. It must be possible to describe the problem fully by means of symbols which are known or at least can easily be learned.

2. All the logical steps required for the solution of the problem must be capable of being represented in abbreviated form, and in correct chronological sequence, by means of the chosen symbols, the construction of the words and the use of definite rules of syntax.

3. All the symbols must be capable of being reproduced on standard commercially-available typewriters.

The first fully-developed language of this type was introduced by IBM in the United States in about 1956; this language is known as FORTRAN (from FORmula TRANslation) (289). In this, as in all the symbolic languages developed subsequently, the operation of translating the generally valid problem-orientated version into the specific machine language is always the same. The 'dictionary' and the rules governing the syntax of the symbolic language are first fed into the computer, together with the interpretation rules for the computer concerned; this forms a translation programme or compiler. The compiler reads the individual instructions of the problem-orientated programme, decodes them, checks them for errors, and produces a sequence of machine instructions. This sequence of machine instructions — the machine-orientated programme — is stored either in the core store or in an external data carrier. Not until the translation has been completed can the machine-orientated programme be run so that the machining operations specified by the programmer can be performed[4].

This programme, which is expressed in special machine language (machine code) is in many ways very similar to the programme that would otherwise have had to be produced laboriously by hand. The problem can now be handled by a computer, and the result (e.g. the values for a table of logarithms) can be printed out (304).

The advantages of the method are obvious, but in practice certain difficulties arise, for example:

a) Where extensive problems have to be dealt with the dictionary becomes very voluminous and a considerable amount of valuable storage space is required to store the many instructions (sub-programmes). Small machines with limited storage capacities are unable to accommodate the complete dictionary. It has to be subdivided (in which case the translation has to be performed laboriously in several stages) or curtailed, in which case the full range of problems can no longer be solved.

b) It is very time-consuming and expensive to work out a translation programme. FORTRAN language, in its various stages of development and extension, is now widely used for scientific and technical purposes, and appropriate compilers are now available for most modern computers (284, 292). The latest versions to become available are FORTRAN IV, (293), 294) and the US Standard USASI[5] -FORTRAN, ASA-FORTRAN.

During 1957/59 the universal symbolic language known as ALGOL (abbreviation for ALGOrithmetic Language) was developed by various organisations in the United States and in Europe. Translation programmes for ALGOL (or at least for parts of ALGOL) are also available for many American and European computers. This language is particularly suitable for the mathematical-scientific field (29, 295, 304), and an international group of scientists is constantly at work extending and improving it.

The programming language COBOL has been developed by IBM for commercial purposes. The compilers for this are available for nearly all IBM computers, and COBOL can also be used on many other computers (283, 286, 296).

There is a great temptation to produce a programming language that is as widespread as possible in its application; the latest attempt to achieve this is PL/1 (programming language one) which has been proposed by IBM (283, 285, 297). To complete the picture, mention will be made of process

---

[4] Conditions are somewhat different for the programme languages used in production engineering, as will be described in Section 11.3, because there the data processing operation is always separated from the further processing in the machine tool.

[5] USASI = *United States of America Standardization Institute, formerly ASA.*

computer systems (or real-time computers) and their special, programming requirements (already mentioned in Chapter 8). (298, 299) The programmes for these machines are still often produced in the assembler, since this enables the most effective programmes, i.e. the shortest computing times, to be achieved. The assembler transforms the machine commands, which are introduced in symbolic form, into their final coded form, which is usually numerical. The programmer is thus relieved of the task, which is laborious and also easily subject to error, of having to determine the core store addresses for his commands and his data. Compared with programme-orientated languages, this enables valuable storage space to be saved, simplifies programming relative to the machine language, still remains orientated towards a specific computer, and may, in some cases, be readily adaptable to specific problems.

If these brief remarks regarding computer programming are compared with the discussions presented in Chapters 7, 8, and 10, many points of similarity will be found between the problems and the possible solutions. With machine tools increasing difficulties are also found with parts programming as the form of the workpiece becomes more complicated; the number of machine tool makers and control system manufacturers in this branch of industry is, however, so large that due to the large number of people engaged in this work it may be much more difficult to simplify matters by reaching agreement on the standards that should be adopted rather than by using computer technology. Fortunately the machine language for NC machines has now for the most part been established; this was discussed in detail in Chapter 10. The next problem is to become at least to some extent independent of the machine tool and the particular computer for programming machining operations, and instead to develop programming languages suited to the production problems. The basic ideas and suggested solutions can be described only very briefly in this context.

11.3   Programme languages for production engineering (381)

Soon after numerically-controlled machine tools were first introduced into American factories (about 1955/56), the need became apparent for the experience that had been gained in programming computers to be applied to the programming of NC machines. From the start one basic difference between this work and the familiar task of computer programming was recognised: In production engineering it is desirable to use one computer to supply the largest possible number of different types of machine tool (different manufacturers and different designs) with control tapes, and so to remain problem orientated and to work in general terms for as long as possible. In addition, for organisational reasons it is necessary to delay the decision as to which specific machine is to be used for machining a particular

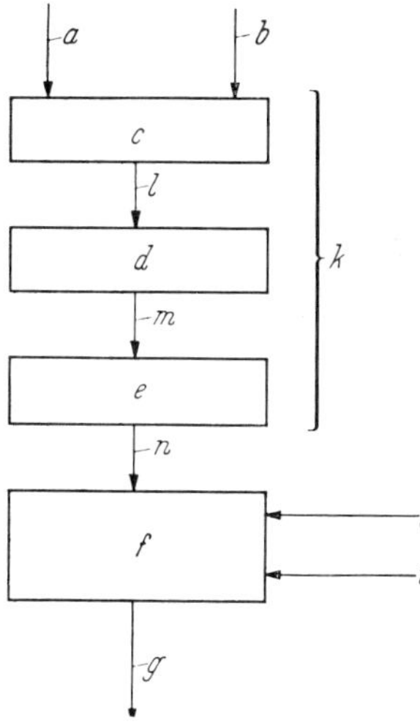

Fig. 274
Information flow in machine part programming of machine tools

a Form data. *b* Technological data. *c* Combination of these two to form work information and preparation of a plain-language work schedule. *d* Coding in accordance with the rules of a production programming language and production of a programme list (source programme). *e* Coding position (usually card punching machine). *f* Universal computer (see Fig. 272). *g* Information carrier for controlling a machine tool (usually punched tape, see Chapters 9 and 10), *h*, *i* Translation programme 'Dictionary and syntax rules' of the programming language, compiler. *h* Processor section (may be divided into a geometric and a technological processor). *i* Machine-tool adaption section (post-processor).*k* Programming section. *l* Plain-language work schedule, source programme in problem-orientated programming language). *n* Stack of punched cards (infrequently punched tape or magnetic tape)

component until the latest possible stage. The basic principles to be employed were established at an early stage at MIT[6], and these have mostly been retained. The translation programme (compiler) was divided into two sections: the processor and the post-processor. In the processor the problem is dealt with in the most general manner, a fictional tool path for a given tool is determined and its values are stored in an external store (e.g. magnetic tape). If necessary, this information can also be read-out and stored on another data carrier (punched tape or punched cards). This type of output is known as CL tape (cutter location tape). The data specific to the machine tool is gathered together for each machine in the appropriate post-processor. When it is decided which particular machine tool is to be used, the post-processor for that particular machine is loaded into the computer. The data from the CL tape is read and then converted into the final form to suit the machine tool. The result of this conversion is produced in a punched tape. The information flow and the equipment concerned is shown in principle in Fig. 274. The methods employed have led to the use of the expression 'computer-assisted programming' of NC machines; the term 'machine parts programming' is also employed. At first sight the method seems a little complicated, but it largely meets the requirements of the production engineer and provides flexibility in the allocation of machines to the various tasks.

---

[6] Massachusetts Institute of Technology, Cambridge, Mass., USA.

The first production engineering programming language to be developed on this basis has become known under the name of APT (Automatically Programmed Tool). This will now be discussed in greater detail as the first example of this type of language.

11.3.1 The APT programming system (311, 314, 316, 343)

In about 1955 the MIT was awarded a contract by the US Air Force Department to develop a problem-orientated programming language for production engineering purposes. The initial problem was to develop a means for the rational machining of aircraft components which could no longer be produced on conventional milling machines owing to their size and complicated shape. In many instances it was also necessary to machine these components out of solid metal to enable them to meet the increasing demands for structural strength. This initial object of the system, and also the date at which the work was commissioned (1955, at the time of the second generation of computers) must be borne in mind if an objective impression is to be gained of the APT system of today. Over the years, as progress was made, several versions of APT have been produced; currently the most widely used one is APT III. The next task was to build up the APT compiler in such a manner that it could readily be processed by various types of computer. Agreement was reached at USASI[5] that all APT programmes and sub-programmes should be written in FORTRAN USASI on the assumption that all modern computers would at least be provided with a FORTRAN compiler. This decision caused the core store capacity needed for APT III to increase considerably but on the other hand virtually no firm is now excluded from operating this system in principle (166). This is both a strength and a weakness of the APT system; many of the subsequent developments can only be understood in the context of this technical-scientific relationship.

The minimum storage capacity required for the economical use of the APT III system is rather high, and currently amounts to:

> 256 K Bytes in the core store, supported by a disc or drum system with about 3×10^6 Bytes and a 700-metre magnetic tape with about 3×10^6 Bytes (cf. Fig. 273).

Since 1961/62, the further scientific development, programme support and administration of the APT system has been in the hands of a private cooperative research group at IITRI[7].

---

[7] IITRI = Illinois Institute of Technology Research Institute, Chicago, USA.

This group is financed by user subscriptions. Detailed handbooks and programming instructions (314) are supplied only to members of the users group or through some of the computer manufacturers. An excerpt is contained in (311,312, 313) from which the following examples are taken.

Symbols and words

All capital letters and the decimal numbers are available as symbols, together with 10 special symbols (Chapter 10). Letters and numbers can be combined to form words, but the words must not contain more than 6 symbols. A certain number of words, based on the English language, are designated APT words and are contained in a dictionary: at present there are 350 to 400 of these words (vocabulary) to which definite meanings have been assigned. The programmer has to learn these words. The vocabulary is divided into main words and auxiliary words (modifiers).

Examples of main words include:

CALL, CIRCLE, CUTTER, GO, GOTO, LINE, POINT, TOOL

while examples of auxiliary words (modifiers) are

ARC, CENTER, LEFT, PARLEL, TANTO

Main and auxiliary words are separated by an oblique stroke, as, for example, in

CIRCLE/CENTER, P1, TANTO, L1[8]

In addition the programmer has the freedom to 'invent' any desired number of 'names' or 'symbols' (which may be in his own language) to provide abbreviated versions of any expressions that occur frequently, and so to save programming time.

It is possible to distinguish between three main forms of statement:

1. Definitions (Definition statements)
2. Executive statements
3. Instructions for the computer.

In the following sections examples will be given of all these three forms of statement. The examples given earlier represent definition statements.

---

[8] In this example a circle is defined by means of its centre P1 and its tangent L1. It is of course also necessary at the same time to define P1 and L1.

Description of tool paths

To prevent the discussions in this section from becoming too complicated they will be restricted to simple three-dimensional bodies. Tool paths can be defined as the geometric locators of a reference point on the tool. A close sequence of such points represents the required three-dimensional curve. It can, for example, be definitely fixed by means of the lines of intersection of surfaces. The relationships are illustrated in Fig. 275. For simplicity, assume that the base FLD is a horizontal plane area. It is intersected by a series of vertical surfaces FLA-FLB-FLC that stand on it. These surfaces govern the path taken by a tool, such as an end milling cutter. The tool path is considered as a combination of individual path elements. If the tool path formed by three path elements illustrated in Fig. 275 is to be described, the following is apparent:

1. All the path elements that form the tool path are in contact with the common area FLD.
2. The tool must first pass along surface FLA until it touches the boundary line between this and surface FLB.
3. The tool must then pass along surface FLB until it touches the boundary line between this and surface FLC.
4. The tool must then pass along surface FLC.

The surface FLD can be defined as a 'boundary surface' and the other three surfaces as 'tool path surfaces', and for this simple case it is apparent that the tool path surface differs for each path element, while the boundary surface remains the same for several path elements. Using these definitions all surfaces can be described; to complete the technical description of the milling process it is, however, necessary to define at least some minimum features of the tool geometry.

An APT tool definition is constructed as follows:

CUTTER/Definition details.

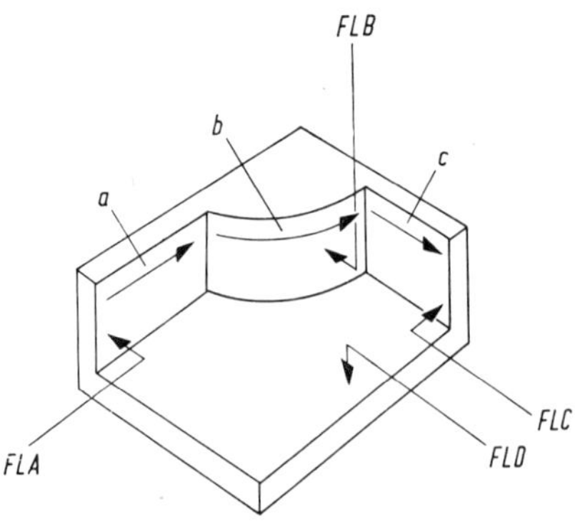

Fig. 275
Tool path as a combination of individual path elements

a Path element 1. *b* Path element 2. *c* Path element 3. FLA, FLB, FLC tool path surfaces. FLD boundary surface

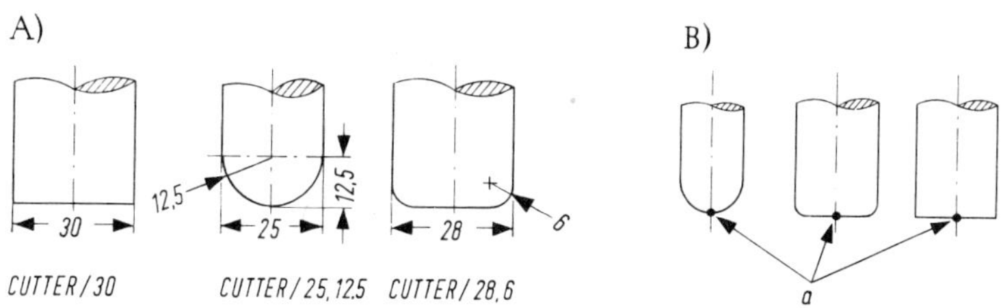

Fig. 276 Tool details
A) Tool dimensions. B) Tool reference points

Figure 276 shows the most important forms and dimensional information required for milling on numerically-controlled machine tools:
Using the geometric data for the workpiece defined in the form of the surfaces and the tool dimensions shown in Fig. 276, it is possible to calculate the required tool paths provided that the tool reference point is defined as shown in Fig. 276 B). The tool path then moves at a certain distance from the surfaces of the workpiece, this distance depending on the curvature of the path, so that it is not constant. The computer then represents the tool path by a series of points for which it computes the three-dimensional coordinates. If the milling machine is equipped only with an internal interpolator suitable solely for linear interpolation the sequence of points calculated by the APT computer must be close enough together to ensure that the permissible dimensional tolerances on the workpiece are not exceeded at any point on a curve (see also Fig. 203 in Chapter 8). The tolerances can be given by the APT words INTOL/ti and OUTOL/to where ti and to are numerical values.

Movement instructions for tools

The simplest way of visualising and deriving the movement instructions is for the programmer to imagine himself as moving with the tool. FROM/P1 means starting from the initial point P1; GO is the APT word for the start of a movement, and this has to be supplemented by a modifier such as TO, ON or PAST. The meanings of these three modifiers are shown in Fig. 277, so that a sequence of instructions might be:

    CUTTER/25
    FROM/P1
    GO/TO, FLA

The path along a running surface is described by executive statements which start with one of the APT words GORGT (go to the right), GOLFT (go to the left), GOFWD (go forwards) followed by an oblique stroke. The directional instructions 'right', 'left', etc. are with reference to the direction of the preceding path element. The position of the cutter relative to the plane must be distinguished from the directional instructions (GOLFT,

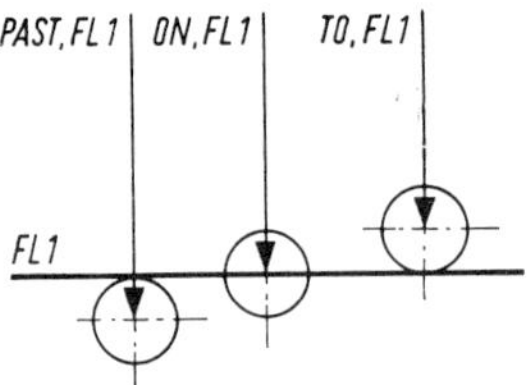

Fig. 277
Description of the approach of a tool to a surface

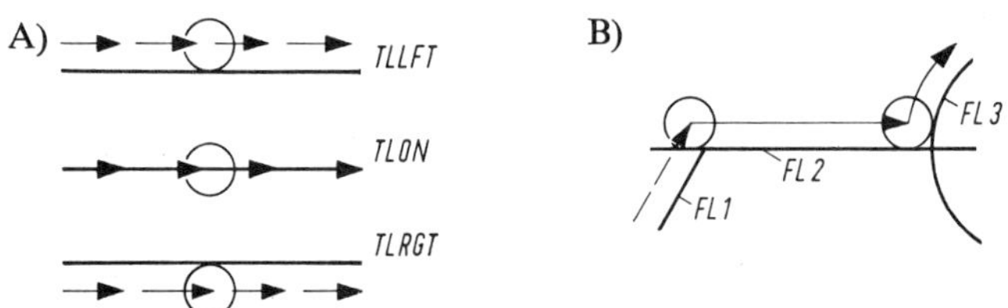

Fig. 278 Description of the movement of a tool along a surface (or line)
A) Determination of tool off-set. B) Example

GOFWD, etc); the former is indicated by TLLFT, TLRGT, or TLON, in which case the computer can calculate the magnitude and direction of the tool off set (Fig. 278 A). An example showing the relationships is given in Fig. 278 B). Using the APT rules, the description of this takes the form of the executive statements

    TLLFT, GORGT/FL2
    GOLFT/FL3

where FL1, FL2, and FL3 have previously been defined.

Production of sub-programmes

In many instances the task of programming can be considerably simplified by the provision of sub-programmes (for example, for machining processes that are frequently repeated). They can be defined by the APT words

    Sub-programme name = MACRO/ . . . .
                          . . . .
                          TERMAC (to indicate end of macro
                          statement)

and these are inserted at a suitable point in the programme by means of

    CALL/sub-programme name.

Using APT it is possible to perform all the calculations for the geometry of three-dimensional bodies. The proportion of words available for the solution of mathematical operations is therefore relatively high (e.g. LOGF, EXPF, ATANF, in the form of computer statements). The end of a complete APT programme is indicated by the APT word FINI. It is not the object of this section to train APT programmers; these brief remarks may, however, suffice to illustrate some of the basic principles, which will also be encountered in other languages. For a more detailed study reference should be made to the published literature (311, 314, 315, 330).

Summarising, it is possible to arrive at the following general conclusions: The advantages of the APT System are:

- the versatility with which even complicated geometrical configurations of workpieces can be described,
- the relatively widespread use that is made of this system, and
- the continuous updating of the language that is performed by IITRI.

The disadvantages of APT are:

- its complexity when it is used to describe simple forming problems,
- the small number of technological description possibilities[9], and
- the need for large computers with a high storage capacity.

These disadvantages of APT led at an early stage to slimming operations designed to enable specific problems to be solved by methods based on APT but less complicated in form (325). Examples of such languages are ADAPT (IBM) (319, 320), SYMPAC (UNIVAC) (317, 322), IFAPT (318, 367), NEL 2 C,L (323, 366), and more recently MINIAPT (330, 340). These can be described generally as APT-like languages (345).

Encouraged by APT many firms have developed their own special languages which are, however, not compatible in syntax or vocabulary with APT; these include:

AUTOPROMT (IBM), AUTOSPOT (IBM) (324), AUTOMAP (IBM) COMPAC (Bendix), SPLIT (Sundstrand), and many others. It has been estimated that by 1964 no fewer than 40 production engineering languages, some of which differed considerably, had been developed in the USA; but these have gained only limited acceptance since they were developed by individual firms specifically for their own use (381).

A further major encouragement to the further development of APT originated in Europe in 1964. The lack of means for expressing technological terms, the necessity of using large computers, and the close dependence on IITRI were found particularly disadvantageous. In Europe, NC techniques had gradually become widely used in mechanical engineering in general, and

---

[9] The technological instructions are almost all 'exiled' to the post-processors, although many of them could be included in generalised form in the processor section.

the aeronautical and astronautical industries, with their specific machining problems, did not play the same major role that they do in the USA. In 1964 at a machine-tool exhibition in Hanover PITTLER and IBM/Germany introduced for the first time a programming language which they had developed cooperatively, known as AUTOPIT[10]; this also included technological components (326).

Originally this language was not compatible with APT and was too closely associated with *one* lathe manufacturer and *one* computer manufacturer for it to become widely used; nevertheless, a promising start had been made, and it only required putting on a more general basis. The most important development of APT in this connection is probably EXAPT (EXtended subset of APT). The basic principles on which this is based are briefly described below.

11.3.2   The EXAPT programming system (307, 308, 309, 310, 327, 328) and MINIAPT (330) as examples of APT-like languages.

The basic specification of the EXAPT system can be summarised in four paragraphs.

1. The language is to be both geometrically and technologically orientated and capable of being extended so that the task of work preparation for conventional machine tools is simplified (automatic tool selection and determination of cutting values, computer-assisted machine selection, etc).
2. The language is to be built up from APT elements and is to use APT syntax rules, and is to have the effect of extending APT in the technological field; it is not to be a competitor to APT with its versatile geometric expressions for complicated three-dimensional bodies.
3. Since there are three main machining processes with very different technological aspects (drilling, turning, milling) it seemed advisable to split the language into three parts to prevent the computers needed from becoming excessively large and to avoid making the language unnecessarily difficult for a technician to learn.
4. The development work was to be performed a cooperatively by as many firms as possible (machine tool *and* computer manufacturers) under the chairmanship of a neutral body.

After some preliminary work had been undertaken by the VDI Committee on 'Automation of Production — Information Processing Subcommittee' in

---

[10] AUTOPIT = AUTOmatic Programming Inclusive Technology; from about 1969 it became closely akin to EXAPT 2 in the form of AUTOPIT 2 (326).

1963/64, theoretical work was started in 1964/65: it was shared between the Technische Hochschule Aachen (*Opitz*) and the Technical University Berlin (*Simon*), and was financed by the DFG[11].

The interest expressed by the West European industry in this work expanded so rapidly, that in 1965/66 two further University Institutes were added to the team (*Spur*, Berlin, and *Stute*, Stuttgart).

Figure 279 shows the structure of the three-part EXAPT System and its relationship with APT. The first stages of sections EXAPT 1 and EXAPT 2 have been completed since 1969, and have passed through their initial industrial test phases (307, 308, 309, 310, 327, 328). Maintenance of the system and advice on its use is available from the cooperative EXAPT Association (328), membership of which is completely open, and is not

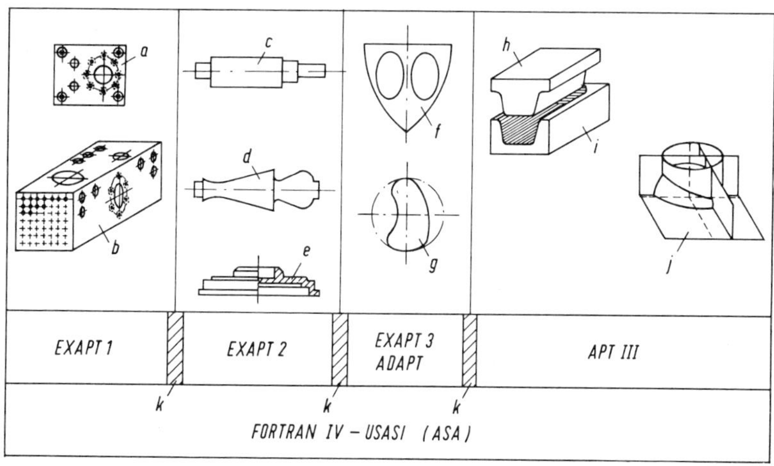

Fig. 279 Relationships between APT – EXAPT – FORTRAN

EXAPT 1 – Drilling and milling operations with positioning and straight-line control systems. EXAPT 2 – Turning operations with straight-line and continuous-path control systems. EXAPT 3 – Milling operations with straight-line and continuous-path control systems up to 2½-D (similar solutions within the APT family available using ADAPT, IFAPT, 2-CL). APT III – Mainly for multi-dimensional milling operations. FORTRAN IV is common basis language

a 2-D positioning control example. *b* 3-D positioning and straight-line control example. *c* Straight-line control example of turned workpiece. *d* Continuous-path control example of turned workpiece. *e* Straight-line and continuous-path control example of turned workpiece. *f* 2-D continuous-path control of flame-cutting machines. *g* 2-D continuous-path control example for milling cams. *h* 3-D continuous-path control example of milling an upper press tool. *i* Ditto for lower press tool. *j* Complicated profile-machining example. *k* Overlapping areas

---

[11] DFG = Deutsche Forschungsgemeinschaft, Bad Godesberg.

restricted to any national or international groups of firms. The individual parts of the language will now be described so far as space permits.

A common feature of all three parts of this language is that the processor consists of two sections: the geometric processing programme and the technological processing programme, the technological programme in its fully-developed form requiring properly established catalogue cards for tools, clamps and chucks, materials, and machine tools. If the EXAPT System is to be fully used it is essential that these catalogues be available in the form of punched cards to enable many of the routine operations associated with the work preparation (looking up tabulated values, selection of tools, etc.) to be delegated to the computer. Since the processor is split into two parts two CL tapes will be obtained: CLTAPE 1 from the geometric processing and CLTAPE 2 from the technological processing (309, 329). The post-processor for the specific machine tool is entered for the third pass through the computer only after the two first passes through the computer have been completed. As a result the final punched control tape can be obtained (344). Because of the general nature of processor technology, post-processors will in most cases be small and cheap; with the advantage that this brings to manufacturers and users of machine tools.

EXAPT 1 (327, 273)

The first part of the language is suitable for programming positioning and straight-line control problems; it is mainly used, however, for programming all types of drilling operation. These include such operations as centering, reaming, tapping, countersinking and counterboring, boring with a boring bar, etc. EXAPT 1 can also be used for programming punching process, spot welders, and wire-working machines. Geometric definition statements for points and circles are built up in just the same manner as with APT. The statement

ZSURF/30

defines a plane which is parallel to the XY plane and is located 30 mm from it. Once the Z plane is defined this remains valid for the subsequent point and point pattern definitions until it is replaced by a new ZSURF statement.

Since there is no difference between the geometric definitions of EXAPT 1 and those of APT, there is no need to consider this point any further. Characteristic of the technological part of EXAPT 1 are definitions for describing the machining of the workpiece which go further than those available with APT. These include the selection of tools at those positions which have previously been defined by geometric statements. All the technological instructions also include the option of either entering the values needed for machining (e.g. cutting speeds) separately, or calculating them on the computer automatically in accordance with preselected laws. During the development of EXAPT 1 provision was thus made for the data which is usually determined by the machine to be fed in by the programmer if required. This makes it possible to adapt the system to suit a particular firm's special practice or to allow for varying requirements (due, for example, to the adoption of new materials).

With EXAPT 1 the following can be determined by the machine:
1. Tool paths.
2. Feeds and cutting speeds (as given by the material and tool catalogue cards).
3. Spindle retractions when drilling deep holes (swarf removal).
4. Tool selection (from tool card index).
5. Sequences of operations (in so far as these can be automated as technological routines).

Card indexes that contain technological data are important aids for programming with all parts of EXAPT. With EXAPT 1 the clamping and chucking fixture card index is not of the same importance as it is for turning operations covered by EXAPT 2 (see also Chapter 10, semi-machine programming).

Furthermore, the computer will also carry out collision course calculations where EXAPT 1 is used. In drilling operations, for example, a check is made as to whether the effective length of the tool (as given in the card index) is sufficient for the programmed operation (depth of hole), and a collision-free tool change is determined.

It is not possible to discuss all the technological statements that are available, but they can be studied in reference (309) or (327). As an example, however, some of the EXAPT words that are not yet included in APT are listed:

EXAPT WORDS		Explanations
Modifiers of	MATERL, b	Material No. b
APT words	UNMACH	Solid Material
PART/ or PARTNO	SEMI	Semi-machined
	CORED	precast
	SMOOTH	smooth surface
	ROUGH	rough surface
EXAPT main	SAFPOS/x, y, z	Tool change position
words	CDRILL	Centre-drill
	SINK	Counter-bore
	SISINK	Countersink
Modifiers to	DIAMET, d	Tool diameter d
above	DEPTH, t	Machining depth t
	FEED, s	Feed s
	SPEED, v	Cutting speed
	SPIRET, g	Spindle return g
	SO	Single operation

Table 22. Some examples of EXAPT 1 words

An important extension to APT is either to combine technologically linked operations and call them up by a single main word, or to call up the operations individually by the modifier SO (single operation). If, for example, reaming a hole to standard limits is to be performed as a single operation, the instruction will be:

Symbol = REAM/SO, DIAMET, d, DEPTH, t, SPIRET, g.

In the description of a work cycle, the end point of a machining operation only is specified; the preliminary work and the machining operation are automatically determined by sub-programmes in the computer. The following work cycles are defined and permitted in EXAPT 1:

DRILL, REAM, SINK, SISINK, TAP.

If the modifier TOLPO is added, a centre-drilling operation is added to the work cycle.

Example: Symbol = REAM/DIAMET, d, DEPTH, t, TOLPO.

The cycle that is called up will then include the following operations:

1. Centre drill
2. Drill, using suitable pilot drill(s)
3. Countersink
4. Ream.

The movement statements (e.g. FROM, GOTO) have a wider meaning than they do with APT. Not only is the tool moved to the point indicated, as with APT, but this is followed by the preselected machining operation, e.g. tapping.

It is also possible to include computer statements in the EXAPT 1 system. Because of the simplified geometric statements compared with APT, it is possible to restrict oneself to the four basic arithmetical operations and some angular functions. The symbols and syntax employed are identical to those used in APT.

Since the work is limited to drilling problems, the EXAPT 1 processor at present contains 108 words only (main words, modifiers, and computer statements) even though some technological words have been added, compared with the 350 to 400 words included in APT. As a result, the storage requirement is reduced to about 50K Bytes in the core store and about 1×10^6 Bytes on a disc system. As has been shown in (273) the core store can be considerably reduced, even beyond this by segmentation of the processor and the use of assembler programmes.

EXAPT 2 (308, 309, 328, 334, 339)

In turning operations both the geometric and the technological conditions are very different from those encountered in drilling. Some of the special features are apparent in Fig. 268 and 269 of Chapter 10.

1. Since the workpieces that are to be machined are axially symmetrical their contour description can be restricted to one half expressed in a two-dimensional coordinate system.
2. .The sum of all contour elements including the centre line, must result in a closed loop.
3. In the final version of the programming language the computer should be able to determine all the intermediate phases with a minimum of instructions, from the data for the unmachined contour and the finish-machined contour. It is, therefore, essential for using EXAPT 2 (309, 361), to have tool and material catalogue cards available. This suggests that the technological processes for EXAPT 2 will be considerably larger than the geometric processor.
4. Normally, four coordinate systems have to be brought into agreement (cf. Fig. 268)
   a)   The workpiece coordinate system,
   b)   The clamp (chuck) coordinate system,
   c)   The machine coordinate system,
   d)   The tool coordinate system.

The two-dimensional contour description for the computer input is given in an *XY* system and the conversion to the *XZ* machine coordinate system specified by ISO R 841 and VDI 3255, together with all the coordinate transformations, is performed automatically by the computer (see Figs. 269 and 280).

The main ideas and extensions for APT and EXAPT 1 can be considered under the following headings:

    Contour descriptions and contour reference points
    Technological definitions
    Executive statements
    Machine-determination of technological data in the external data-processing system
    Description of any desired work sequences
    Programming examples

Contour descriptions and contour reference points

Points, straight lines, and circles may be used as the geometric elements of contours. Three types of point definition, seven types of straight-line definition, and four types of circle definition are available; they have been adopted from the APT system. The contour description is built up as shown in the following table:

Description of blank	Description of machined part
CONTUR/BLANCO	CONTUR/PARTCO (EXAPT words)
.	.
.	.
.	.
Contour description	Contour description
.	.
.	.
. .	.
TERMCO	TERMCO (EXAPT word for end of contour description

To enable the material side to be identified in the contour description it is necessary for the description to follow an agreed direction. As clockwise rotation was selected, the material side is always on the right of the described contour element. The starting point of the contour description and the initial direction are designated by the main word BEGIN/ ... (EXAPT word), and the following points are defined by numbered reference points. A simple example to illustrate this is given in Fig. 280. The corresponding verbal programme is given in the title of this figure; the auxiliary words (modifiers) used have the following meanings:

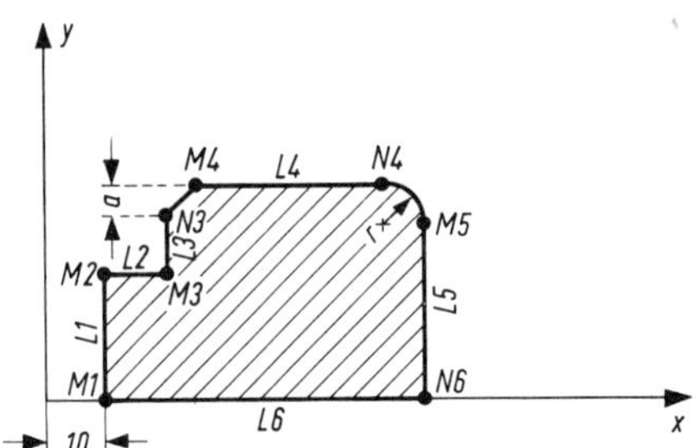

Fig. 280 **Example of an EXAPT 2 contour description for a turned component**

    A) Sketch of workpiece contour
    M1 to M5 Main contour reference points
    N3, N4, N6 Auxiliary contour reference points

    B) Programme
    CONTUR/PARTCO
    M1, BEGIN/10,0,YLARGE,PLAN,10
    M2, RGT/L2
    M3,N3,LFT/L3,BEVEL,a
    M4,N4,RGT/L4,ROUND, r
    M5, RGT/L5
    N6, RGT/L6
    TERMCO

BEVEL, *a*	Straight chamfer of width *a* (EXAPT word)
ROUND, *r*	Radius corner with radius *r* (EXAPT word)
YLARGE	Y value increasing (APT word)
PLAN	Faced-surface of a turned component (EXAPT word).

In the description of the machined part, it is possible to allot technological information to some of the contour elements, for example:

Surface finish
Limits required
Engagement of tool position correction switches (cf. Chapters 8, 9, 10).

The definition possibilities of these are given in the next section.

Technological Definitions

A characteristic feature of the technological part of EXAPT 2 is the use of the normal English technical terms to describe the workpiece technology and the organisation of the work in turning operations. The statements include all the data necessary to define the turning work. They include information about the workpiece, its machinability, the tools, and to some extent also the machine tool itself, unless the machine-tool details are so specific that they are included in the post-processor. Part of the data *must* always be supplied by the programmer; other parts *can* either be produced by the machine from stored tabular values or else determined by the programmer. Since lathes can also be used for drilling operations, some of the elements from EXAPT 1 are also used.

In the current stage of development of EXAPT 2, the following items are among those that can be determined by machine:

Turning operations	*Drilling operations*
Tool paths	Tool paths
(subdivision of total	Feeds and cutting speeds
metal removal into	Spindle withdrawals when
individual cuts)	drilling deep holes (see
Depth of cut	EXAPT 1)
Feeds and cutting speeds.	

The data processing equipment will also investigate the possibilities of collision. Here the computer checks whether the tool geometry is suitable for the operation (from tool card index). (361) All possibilities of collision between the tool and the workpiece or the tool and the chuck or clamps are eliminated when the tool paths are calculated. The decisions left to the programmer in 'semi-machine' programming, as described in Chapter 10, are now performed by the computer.

In addition to the drilling operations listed in Table 2 EXAPT 22 contains four main words for turning operations:

TURN	Turn (parallel to axis) The subdivision of the metal removal among the various cuts for the TURN type of operation is illustrated in Fig. 281 A); in general, there are many contours which cannot be generated completely by this method.
CONT	Contour turn (continuous-path controlled profile turning). Two methods of operations are possible: One, as illustrated in Fig. 281 B) (specific feed direction preselected), the other as illustrated in Fig. 281 C) (no specific feed direction programmed).
GROOV	Groove (see Fig. 281 D)
THREAD	Screw-cutting on the lathe.

There are also a number of modifiers, such as

SO	Single operation (See EXAPT 1)
LONG	Longitudinal turning
CROSS	Facing
ARANGL	Taper turn at specified angle.

Figure 281 shows some of the ways in which the total metal removal can be subdivided among the cuts for the TURN and CONT methods of operation.

Before the first machining definition is given the following information on the workpiece PART/ . . . must be given by using modifiers:

Material . . . /MATERL, b (see EXAPT 1);
cutting value corrections
machining allowances
surface finish required.

The stability of the workpiece cannot be taken into account during the calculation of the cutting values by the computer, it is necessary to make provision for the programmer to alter these values by stating that they are to be a percentage of those determined by the machine.

Example: CORREC, 80, 120
where   80 = 80% reduction in the feed speed as determined by the machine
           120 = Increase in cutting speed to 120%

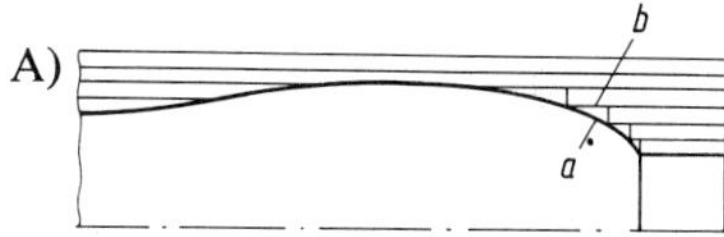

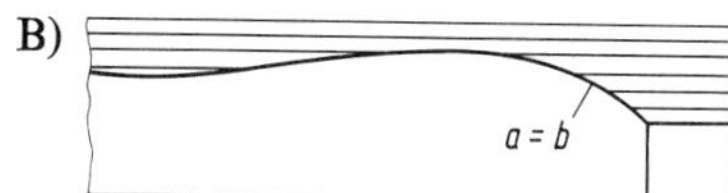

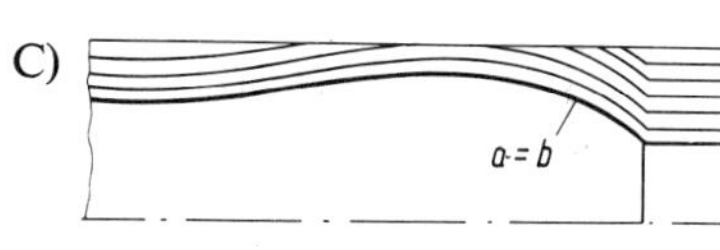

Fig. 281
Possible subdivisions of total metal removal for EXAPT 2 (from 339)

A) For basic machining operation TURN. B) For basic machining operation CONT with specified feed direction. C) For basic machining operation CONT without specified feed direction. D) For basic machining operation GROOV

a Required contour. *b* Actual contour

Machining allowances can be specified by

OVSIZE/FIN,cl,FINE,c2

and surface finishes by

SURFIN/FIN,c1,FINE,c2

where

FIN	=	Finish turn
FINE	=	Precision turn
cl and c2 =		equidistant oversizes of workpieces in mm or roughness values in mm.

After the modifiers FIN and FINE it is possible to specify

. . . . , FIT,hm . . . .

to indicate that a particular element of the contour is to have a given toleranced finish over a length of hm mm. To ensure that the required quality is achieved it is possible to include certain tool-position correction switches (see Chapters 7, 8, 9) in the written programme by placing

        . . . . OSETNO,no,. . . . .

after the modifier, where no is the number of the correction switch concerned. This can be done, however, only if it is known which particular lathe is to be used; and instructions of this type could also be included in the post-processor.

    Executive statements for EXAPT 2

The way in which the machining process is performed is determined by the executive statements. They are used to call up the previously defined methods of machining and coordinating them with the appropriate contour sections. The main form of statement is

        WORK/Symbol 1 of a machining definition
            Symbol 2 of a machining definition
                        .
                        .
                        .

and finally
        WORK/NOMORE

Machining is performed in the sequence in which the symbols are given. It is therefore possible to describe any desired follow-on operation. After each machining operation has been called up it is necessary to state at which points it is to be performed. The position for the machining operation is given by

        CUT/ . . . . (EXAPT word)

and may, for example, take the following form:

        CUT/M1,TO,M2 .

where

        M1 = Starting reference point for machining position
        M2 = Finishing reference point for machining position.

        Machine-determination of technological data in the external data-processing system (361, 370)

The possibility of determining depths of cut, feeds, and cutting speeds by machine is included in the design of all the versions of EXAPT; it is left to the user whether he takes advantage of these possibilities or not. Of great assistance in determining cutting values by means of a computer are the previously mentioned three card indexes:

Tool index   Chuck or clamp index   Materials index

In addition, there is, for a higher stage of automation of the work preparation, a *Machine index,* which may be used to simplify the preparation of machine loading schedules. The total metal removal is divided among the individual cuts in different manners according to the type of machining (i.e. roughing, finish-machining, etc.). Without question the introduction of adaption units, as described in Chapter 8, has resulted in at least the roughing operations being solved in a technically superior manner as part of the internal data processing; the main question is what the additional cost of these adaption units will be during the next few years when the teething difficulties have been overcome. In any event, this new trend will undoubtedly have some effect on the problem of determining automatically as part of the external data processing system, the machining speeds and feeds. Since adaptive systems are only in the initial stages of development and the question of cost will inevitably play a major part, these questions will be discussed against a broader background in Chapter 12.

Description of any desired sequence of operations

A sequence of operations that cannot be programmed by means of the machining definitions and machining position indications described, or that is considered unsuitable for programming in this way, can be described using separate technological data and tool movement information (in a similar manner to APT).

Among the items of technological data are:

Tool instructions	TOOLNO/	(APT word)
	. . . . /OSETNO, no	Number of correction switch (EXAPT word)
	. . . . /SETANG, w	Clearance angle of turning tool (EXAPT word)
	. . . . /FEDRAT/ . . .	Feed rate (APT word)
	. . . . /SPINDL/ . . .	Spindle speed (APT word)
	. . . . /CLW	Directions of rotation for spindle
	. . . . /CCLW	(APT words)
Spindle stopped	OFF	(APT word)
Rapid traverse	RAPID	(APT word)

For tool movements the same words and forms of statement are used as for APT, i.e.

$$\text{FROM}/x,y \quad \text{GOTO}/x,y \quad \text{GOLFT, GORGT, etc.}$$

If individual instructions are being given, starting and stopping the coolant flow must also be individually programmed, for example:

$$\text{COOLNT}/q \quad q = \text{Number of coolant}$$

and $\text{COOLNT}/ \quad < \begin{array}{l} \text{ON} \\ \text{OFF} \end{array}$

Both these instructions are APT words, just as is the use of

$$\text{AUXFUN}/tr \quad tr = \text{Number of a G or M function} \\ \text{(cf. Chapter 10)}$$

for calling up an auxiliary function.

The programmer is thus provided with further opportunities for introducing values based on his own experience and so in a sense to perform a combination of manual and machine programming.

Programming examples

To illustrate the remarks that have been made regarding EXAPT 2, a typical sample programme has been drawn up and is reproduced in Table 23. Table 24 is a problem-orientated description in EXAPT 2 of the turned component illustrated in Fig. 269 (363).

Finally, a few details may be given to show the extent of the vocabulary and the storage capacity required for the EXAPT 2 processor.

At the current stage of development of this programming language the numbers of words are as follows:

> 80 main words (49 of them from APT)
> 60 modifiers (31 of them from APT)
> 5 functions (Calculation statements, all 5 from APT)
> ———
> a total of 145 words.

The storage capacity required is larger than for EXAPT 1, mainly due to the large number of technological statements and the automatic determination of the machining values. At present the storage capacity required amounts to about

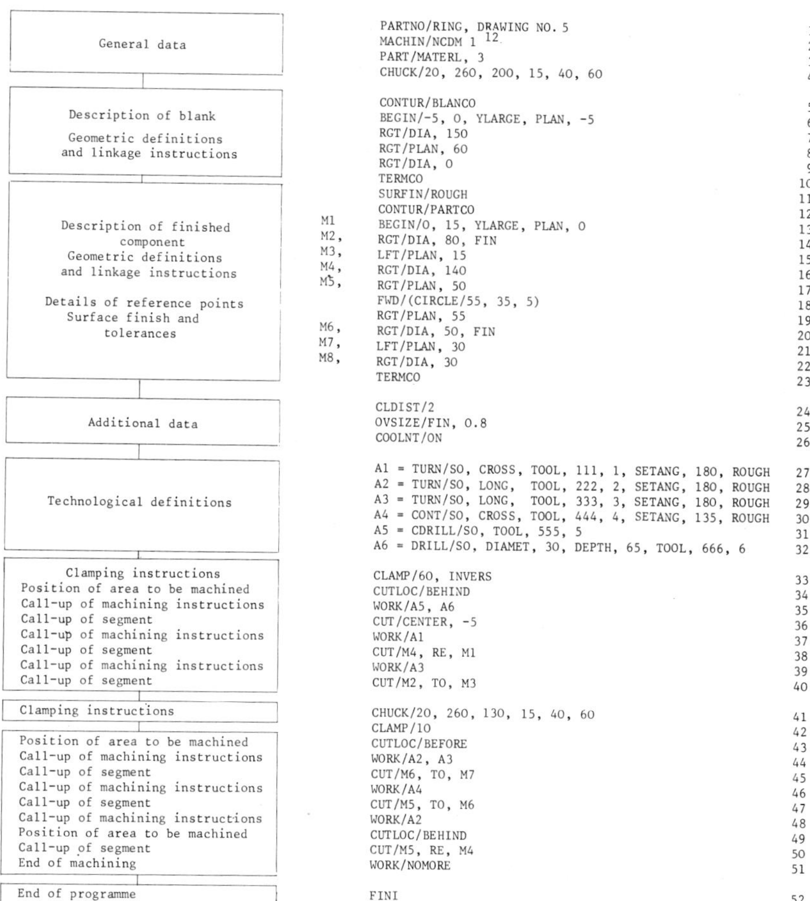

General data	PARTNO/RING, DRAWING NO. 5	1
	MACHIN/NCDM 1 [12]	2
	PART/MATERL, 3	3
	CHUCK/20, 260, 200, 15, 40, 60	4
Description of blank Geometric definitions and linkage instructions	CONTUR/BLANCO	5
	BEGIN/-5, 0, YLARGE, PLAN, -5	6
	RGT/DIA, 150	7
	RGT/PLAN, 60	8
	RGT/DIA, 0	9
	TERMCO	10
	SURFIN/ROUGH	11
Description of finished component Geometric definitions and linkage instructions Details of reference points Surface finish and tolerances	CONTUR/PARTCO	12
M1	BEGIN/0, 15, YLARGE, PLAN, 0	13
M2,	RGT/DIA, 80, FIN	14
M3,	LFT/PLAN, 15	15
M4,	RGT/DIA, 140	16
M5,	RGT/PLAN, 50	17
	FWD/(CIRCLE/55, 35, 5)	18
	RGT/PLAN, 55	19
M6,	RGT/DIA, 50, FIN	20
M7,	LFT/PLAN, 30	21
M8,	RGT/DIA, 30	22
	TERMCO	23
Additional data	CLDIST/2	24
	OVSIZE/FIN, 0.8	25
	COOLNT/ON	26
Technological definitions	A1 = TURN/SO, CROSS, TOOL, 111, 1, SETANG, 180, ROUGH	27
	A2 = TURN/SO, LONG, TOOL, 222, 2, SETANG, 180, ROUGH	28
	A3 = TURN/SO, LONG, TOOL, 333, 3, SETANG, 180, ROUGH	29
	A4 = CONT/SO, CROSS, TOOL, 444, 4, SETANG, 135, ROUGH	30
	A5 = CDRILL/SO, TOOL, 555, 5	31
	A6 = DRILL/SO, DIAMET, 30, DEPTH, 65, TOOL, 666, 6	32
Clamping instructions Position of area to be machined Call-up of machining instructions Call-up of segment Call-up of machining instructions Call-up of segment Call-up of machining instructions Call-up of segment	CLAMP/60, INVERS	33
	CUTLOC/BEHIND	34
	WORK/A5, A6	35
	CUT/CENTER, -5	36
	WORK/A1	37
	CUT/M4, RE, M1	38
	WORK/A3	39
	CUT/M2, TO, M3	40
Clamping instructions	CHUCK/20, 260, 130, 15, 40, 60	41
	CLAMP/10	42
Position of area to be machined Call-up of machining instructions Call-up of segment Call-up of machining instructions Call-up of segment Call-up of machining instructions Position of area to be machined Call-up of segment End of machining	CUTLOC/BEFORE	43
	WORK/A2, A3	44
	CUT/M6, TO, M7	45
	WORK/A4	46
	CUT/M5, TO, M6	47
	WORK/A2	48
	CUTLOC/BEHIND	49
	CUT/M5, RE, M4	50
	WORK/NOMORE	51
End of programme	FINI	52

Table 23. Typical example of EXAPT 2 programming from (339)

[12] Much of the information in the technological processor is associated with specific machines by the card indexes, so that it is necessary to specify the machine that is to be used in the source programme. If the machine is not specified until later — as often happens where APT is used — this information must appear, at the latest, in the post-processor.

PARTNO/DREHTEIL NACH BILD 269	(APT)
NOPOST (kein Postprocessor)	(APT)
REMARK/ROHTEILBESCHREIBUNG	(APT)
CONTUR/BLANCO	(EXAPT)
BEGIN/93, (120/2), XSMALL, DIA, 120	(EXAPT)
RGT/PLAN, 0	(EXAPT)
RGT/DIA, 270	(EXAPT)
RGT/PLAN, 93	(EXAPT)
TERMCO	(EXAPT)
REMARK/GEOMETRISCHE DEFINITIONEN	(APT)
P 1 = POINT/90, (130/2)	(APT)
REMARK/FERTIGTEILBESCHREIBUNG	(APT)
SURFIN/FIN, 0.025	(EXAPT)
CONTUR/PARTCO	(EXAPT)
M1, BEGIN/P1, X SMALL, DIA, 130	
M2, FWD/L 1	
M3, RGT/PLAN, 0	
M4, RGT/DIA, 260	
M5, M6, RGT/PLAN, 50, ROUND, 10	
M7, LFT/DIA, 210	
M8, RGT/PLAN, 70	
M9, LFT/DIA, 200	
M10, RGT/PLAN, 90	
TERMCO	(EXAPT)
FINI	(APT)

Table 24. **EXAPT 2 programming example for Fig. 269 (Geometrical section).**

    210K Bytes in the core store (segmented)
  2×10^6 K Bytes in the disc store
  1×10^6 K Bytes on each of two magnetic tape units.

The technological section is larger and more space-consuming than the geometric section, and it is the part that determines the storage capacity required. The manner in which the information stored is divided between the external disc and tape stores depends on the computer configuration; the figures quoted above are therefore only approximate.

EXAPT 3 (309, 310)

The type of work for which EXAPT 3 is intended was outlined in Fig. 279 and the way in which it differs from APT in its applications was indicated. EXAPT 3 is designed primarily for economic programming of drilling and milling operations with up to 2½-D continuous-path control systems; such as are encountered in machining centres where there is a wide variety of work (cf. Chapter 7). For this reason, EXAPT 3 has been so arranged as to include the whole of the vocabulary of EXAPT 1. The work at the Universities on the development of EXAPT 3 was expected to be concluded by the autumn of 1970.

Figure 282 illustrates some simple milled components which fall within the field covered by EXAPT 3; the many drilling operations that are possible overlap these machining problems and will be considered separately at this point.

The geometric definitions used in EXAPT 3 are for the most part a subset of those used in APT III, and are included in similar form in the programming languages ADAPT (319, 320), IFAPT (318, 367), and 2 C,L (323, 366). Extensive efforts are therefore currently being made to gather together all

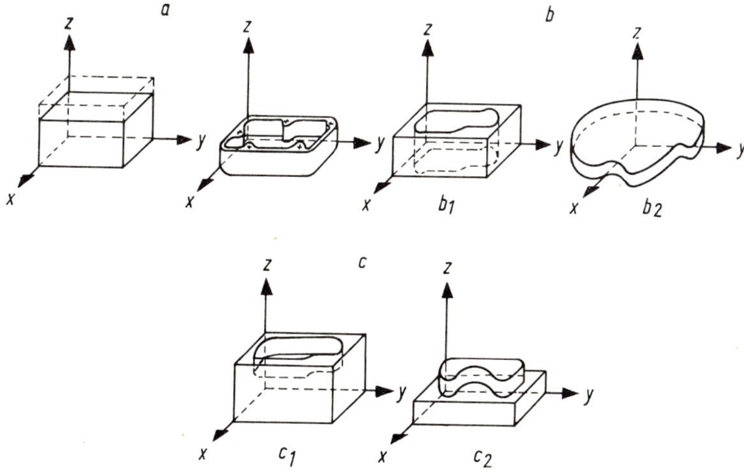

Fig. 282 Machining possibilities using EXAPT 3 (from 309)

a Plane milling. *b* Cylindrical contour milling. b_1 Internal contouring. b_2 External contouring. *c* Pocket milling. c_1 Internal contouring. c_2 External contouring

these APT derivatives into a common programme packet under the auspices of ISO (338). With EXAPT 3, as with EXAPT 1 and 2, it is, however, also possible to determine technological data automatically, i.e. by means of the data-processing equipment, and to feed it into the machining process. In the final stage of development it is intended that both the cutting data for milling operations, and the data for the milling cutters, be determined by the machine. This proposed stage of development is based more on the wish for as many as possible of the tasks performed in the work preparation department to be carried out by means of a computer, rather than on the direct needs of the machine programming of NC machines taking account of geometric and technological components; it has therefore been postponed for the present. It will be sufficient to point out the main differences compared with EXAPT 1 and EXAPT 2 (310).

Geometry with EXAPT 3

The geometric descriptions employed in EXAPT 1 and EXAPT 2 are, in the main, restricted to definitions of points, straight lines, and circles based on APT, and these are extended by the adoption of a few special words derived from the technology concerned. With EXAPT 3 it is in the main only the APT words

TABCYL
POLGON
ELLIPS
HYPERB
PARAB

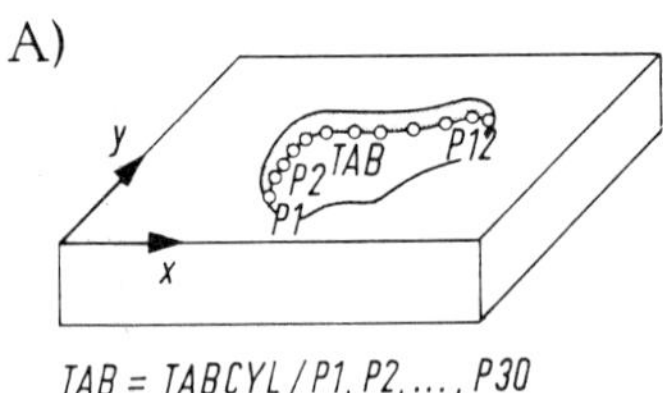

A)

$TAB = TABCYL / P1, P2, ..., P30$

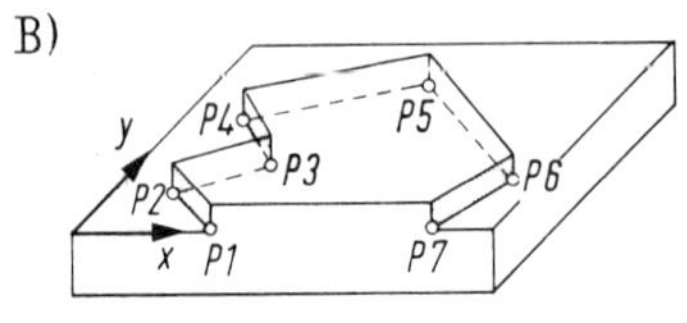

B)

$POL = POLGON / P1, P2, P3, P4, P5, P6, P7$

Fig. 283
Examples of extended contour descriptions using EXAPT 3 (adopted from APT) (310)

A) TABCYL statement (TABulated CYLinder) of the type TAB = $TABCYL/P_1, P_2, P_3 \ldots P_{30}$

B) POLGON statement (POLyGON) of the type POL = $POLGON/P_1, P_2, P_3 \ldots P_7$

which have been taken over from the APT system. The most important of these are probably TABCYL and POLGON. In the TABCYL definition a series of points are specified between which a continuous curve is interpolated (330); in the POLGON statement the points are joined by straight lines, so that a polygon is produced (Fig. 283). Whether the APT words for ellipses, parabola, and hyperbola are required frequently enough for it to be worthwhile making the necessary provision in the computer can only be answered by a thorough statistical frequency investigation. It is an area where an opportunity of simplifying the system may arise; in any event it should prove possible to programme the manufacture of cam discs easily (cf. Chapter 12) (365).

Technology with EXAPT 3

As with parts 1 and 2 of the language, a machining-value model is produced in the computer for determining speeds and feeds; this is inevitably more complicated than the one required solely for drilling and turning. The automatic selection of the tools is of lesser importance and can be postponed.

The determination of the required cutter paths when milling surfaces, milling out pockets, etc., can be performed automatically with the aid of complete machining definitions; they are called up by means of CONMIL[13] and FACMIL. Provision must be made for the overrun of the cutters, the overlaps of the cuts, and the method of working (e.g. reciprocal milling) selected by the programmer.

Closely connected with the calculation of the cutter paths as part of the machining definition is the calculation of the number of cuts that will be required. If the available tool data is used, it is possible to calculate both the number of cuts and the machining time that will be needed from the description of the Z value of the cylinder which is to be machined off, and hence the volume of metal that has to be removed. As with APT, ADAPT, and the other two parts of EXAPT, it is possible to programme EXAPT 3 using *individual* movement instructions. For simple operations this is less complicated than describing the contour with complete machining definitions. When individual instructions are used the cutting values are not determined by the machine, and the programmer has to supply the cutting data. The possibilities of collisions occurring between the tool, workpiece, and clamp or chuck are also worked out by the computer. With EXAPT 3, however, unlike EXAPT 2, the full contours of the complete workpiece are not described; to simplify programming only those areas are described in which machining actually takes place. Any edges of the workpiece that might cause a collision danger, and clamps or chucks — which are not included, or

---

[13] Typical applications of CONMIL include: milling of oil grooves, fitted keyways, cam discs, frame milling of gearcases, milling round precast cut-outs, etc.

not completely included, in the contour being handled — can be defined by a supplementary contour description, thus enabling the collision calculations to be made comprehensive. The principles on which an EXAPT 3 part programme is drawn up are illustrated in Table 25.

General information	
Part Number	PARTNO/...
Machine tool (see Footnote 11)	MACHIN/...
Material	MATERL/...
Remarks	REMARK/...
etc.	
Geometric information	
Geometrical elements	LINE, CIRCLE, TABCYL, etc.
Contour description	CONTUR
Technological information	
Technological details of workpiece	
Machining definitions	CONMIL, FACMIL
Call-up of type of machining	WORK/...
Call-up of machining area	CUT/...
Movement instructions	GOTO, etc (as for APT)

Table 25. Basis of an EXAPT 3 parts programme (310)

Like EXAPT 1 and 2, the programme is divided into three sections: General information, geometric information, and technological information; the contents are, however, somewhat different. The basic structure of all three parts of this language is the same.

Coverage of geometry and technology for EXAPT 3.

All the machining definitions used in EXAPT 1 are available for defining the type of machining. There are also two further machining definitions for milling operations:

CONMIL/ ...	Contour milling
FACMIL/ ...	Surface milling.

Both can be defined more closely by means of a number of modifiers, for example:

. . . ./SO	Single operation
. . . ./MEANDR	Meander type milling
. . . ./ZIGZAG	Zig-zag or reciprocal milling
. . . ./ZIG	Reciprocal milling with quick return.

The machining definitions (which are characterised by a freely-selectable symbol) are called up by means of the word WORK/symbol. Allocation to a particular machining area or contour is effected by the command CU/ followed by the symbol for the machining area or contour concerned. The arrangements are thus similar in nature to those employed for EXAPT 2.

Contour description for EXAPT 3

Only the basic form of statement will be given here.

```
C11   =   CONTOUR/CONNEC,        OPEN        Z1
 /          \     /                \           \
Symbol    Contour comprising      open on     Depth of cut
          several elements        one side    in Z direction
```

This could represent an example similar to that shown in Fig. 275. This can be followed by a more precise description of the straight lines and circles that form the contour if, as in EXAPT 2, a starting point and a starting direction have first been defined by means of BEGIN. It is not possible to go into further detail at this point, because

1. A description of the EXAPT 3 language is shortly to be published, and
2. The international discussions on methods of bringing the APT-like languages (IFAPT, 2C,L, EXAPT 3, and ADAPT) closer into agreement, at least in this field, to which reference has already been made, are currently in progress.

Extent of EXAPT 3 language

The number of words provided and also the storage capacity required, have increased, as a result of the adoption of additional APT words, the provision of contour details and of machining definitions with many modifiers, as well as the inclusion of the technological instructions. At present the number of words is

97 Main words (73 of them from APT)
106 Modifiers (44 of them from APT)
5 Functions (all 5 from APT)

———

a total of 208 words (122 of them from APT);

while, in addition, the proportion of EXAPT 1 words must be allowed for.

The storage capacity must therefore be at least as large as for EXAPT 2. If the entire EXAPT system is to be used in a plant, based on Fig. 273, the minimum configuration will be roughly as follows:

256K Byte in the core store,
2 x 10^6 Byte in a disc system, and
3 magnetic tapes each with 1 x 10^6 Byte (one tape for
each part of the language).

By further segmenting the processor into several passes through the computer, it is possible to reduce the required core storage capacity (273, 330, 340). If a comparison is made with the details of the required storage capacity given for the APT system, it will be seen that both APT and the whole of the EXAPT require about the same storage capacity. In other words, the reduction in the geometric expressions is virtually counter-balanced by the addition of technological instructions. This is not a very satisfactory solution for a very large number of European plants with only a small computing capacity. To supplement these discussions on EXAPT, a brief reference will therefore be made to the efforts of a private programming group, which were published in 1969 under the name MINIAPT.

MINIAPT (330, 340, 341)

The basic ideas on which MINIAPT is based are very simple and appropriate to the task in hand:

1. APT is at present the most widely used programming language for production engineering; any variations should be largely based on APT.
2. The high storage capacity required for APT is due to the use of a relatively small group of very powerful words which initiate a large number of automatic operations in the computer. This small group of words is, however, needed for only 15 to 20 per cent of all the workpieces encountered in practice, and these emanate for the most part from the aero space field.
3. If these, and also some associated words, are eliminated, and use is made of the possibilities opened up by suitable segmentation of the

programme (several passes through the computer) the required storage capacity is much reduced.

4. Among the words that have been cancelled are:
SPHERE
ELLIPS
HYPERB
PARAB
MULTAX.
In this way a true subset of APT has been produced, which can be described as MINIAPT, and which is sufficient for dealing with 80 to 85 per cent of all workpieces encountered in practice.

5. No technological executive statements are included in MINIAPT. As with APT, machining instructions and cycles can be included only in the post-processor.

6. All the other APT rules are maintained, so that the CL tapes for APT and MINIAPT are similar.

7. Twelve mathematical functions can also be called up using MINIAPT.

The total reduction in the vocabulary amounts to 201 words from 380, i.e. by 53 per cent, and so is of the same order of magnitude as for EXAPT 3. Combined with a segmentation over 5 to 8 computer passes for the processor, it is possible to use the MINIAPT processor on computers having a core storage capacity of 32K Bytes or higher, especially if a high-capacity disc system is available. If larger computers are used, the computing time will be considerably reduced. Furthermore, the lower core storage capacity that is required simplifies the multi-programming operation of large installations.

11.3.3 SYMAP and AUTOPROG as examples of languages that are not related to APT

Both these language systems were first made public in 1967/68 (25, 346) and both provide useful examples for discussion; their nature and their methods of use differ to such an extent, however, that they must be considered separately.

SYMAP (25)

The word SYMAP is an abbreviation of SYmbolic languages for MAchine Programming of NC machines. It was developed by a group drawn from the VEB Carl Zeiss Jena, the Institute for Machine Tools and the Institute for Production Technology Karl-Marx-Stadt, and the Institute for Mathematics at the Technische Hochschule at Karl-Marx-Stadt. There is to some extent an unmistakable similarity to APT, and especially with EXAPT; considerable changes have, however, been made in the principal structure of the language compared with the APT-like languages. Like all programming languages designed for production engineering use, SYMAP is subject to constant development. While the machine itself is capable of taking logical decisions, the method to be followed for the computation must be specified in advance, so that it is only possible to describe the situation as it is known at present.

The main similarities are:
1. The subdivision into a general processor section and a post-processor section specific to the machine tool was adopted from the APT system, as was the restriction of the lengths of the words to six characters.
2. The system is subdivided into several part systems to suit the most important machining processes in a similar way as EXAPT, and technological components are included in the processors. To enable the smaller computers to be employed, the subdivision of the language system has been taken further than with EXAPT. At the present time the subdivision is roughly as follows:

SYMAP(P) for positioning control systems

SYMAP(S) for straight-line control systems

These can be combined as SYMAP(PS) and are then approximately equivalent to EXAPT 1 with a slightly greater emphasis on straight-line control systems for simple milling operations.

SYMAP(DS) for lathes with straight-line control systems.

SYMAP(DB) for lathes with straight-line and continuous-path control systems.

SYMAP(DS) and (DS) together correspond roughly to EXAPT 2.

SYMAP(B) for 2½-D continuous-path control systems corresponds roughly to EXAPT 3.

The formation of what are termed complex programmes enables further detailed arrangements to be produced efficiently for special purposes, for example, the turning of grooved rolls.

The deviations from the APT-like languages are also easily recognisable in SYMAP and are principally as follows:
1. The vocabulary is based on German.
2. Even though SYMAP statements can, in principle be written in free style, there is a definite preference for one particular format for SYMAP input statements. These are written on a form, the headings of which are shown in Table 27.

Serial No.	Command word	Variable	Definition word	List of parameters (1),(2),(3),(4),(5),(6)

Table 26. Headings for a SYMAP form

3. The number of characters — especially the special characters — is somewhat smaller than for APT. The characters comprise:

Letters: A, B, C, D . . . . Z (capital and lower-case letters are equally permissible)

Decimal numbers: 0, 1, 2, . . . 0

Special characters: Signs — (+ is not written)
.   Decimal point (as for APT)
,   or TAB as equally valid word separating signs
$\leqq$ Carriage return and line advance, line separation sign (as used in telex).

4. Character combinations containing 8 permissible characters are termed literals; these include words, descriptions of variables, and numbers.
5. The words used in the symbol language include those of
Command
Definition
Modification and
Markers.

Command words are, for example,

PROGR	Programme
TECH	Technological instruction; the variable T is used for this.
INRIP	In direction of point; the variable P is used for this.
WELKS	Tool left
KUEEIN	Coolant on
etc.	

A *definition word* always describes an operator that is used for processing the definition of an element (e.g. a circle, straight line, point, a tool, or a technological process. Examples of definition words include

PXY	Point, defined by coordinates
KXYR	Circle, defined by radius and coordinates of centre,
WSONFR	Special tool for milling
LIFRAN	Line milling to finished
etc.	dimension

Modification words are used for a more detailed description of a definition.

Examples:

XGR	X larger
XKL	X smaller
RTS	right

For each definition word it is laid down where a modification word of a given sub-group should be placed in the list of parameters.

Markers are 'words' that consist of a single letter only. They are used for marking figures or arithmetic variables. The following rules should be observed: The marker is placed in front of the item that is being marked, is separated from it by a comma, and is entered together with the item in one sector of the parameter list.

To clarify the difference between what has just been described and the methods adopted in the APT-like languages, the same subject matter is given below both in APT and in SYMAP (Table 27).

In APT:
K1 = CIRCLE/100, 80, 20
P1 = POINT/10, 50
G1 = LINE/P1, RIGHT, TANTO, K1
COOLNT/ON

In SYMAP

Serial No.	Command word	Variable	Definition word	List of parameters (1),(2),(3),(4),(5),(6)
.				
.				
.				
66	DEF	K1	KXYR	100.0 80.0 20.0
67	DEF	P1	PXY	10.0 50.0
68	DEF	G1	GPK	P1 K1 RTS
69	KUE EIN			

Table 27. Comparison of APT and SYMAP instructions

6. Each line of a programme is a 'sentence' and can be punched as required either onto a card or onto a telex tape to enable the entire system to be accommodated in an existing data transmission system.
7. A block consists of a sequence of sentences. It covers a section of a programme that can be completed in its entirety by what is termed a complex programme. A complex programme is defined as a computer

programme by which it is possible to machine programme a type of control system for performing a certain production process. In (25) three complex programmes for SYMAP (B), which were produced in 1968, are listed:

KOPGK   Point — straight line — circle
PFOLGE   Sequence of points
KUME 1   Cam mechanisms 1, cam discs.

Each block commences with a sentence that contains the command word

BLOCK

a definition word to identify one of the complex programmes, and the appropriate values of the parameters. Each block is concluded with the command word

BLENDE

These short notes will have made clear the principal common features and differences between the two language systems. For closer study the reader is referred to the literature quoted (25) and to the distribution and administrative centre for the SYMAP language system[14].

Finally, Table 28 shows the range of the vocabularies for the language components (complex programmes); this makes possible a comparison with the other languages that have been discussed previously.

SYMAP language part	Applications	Versions	No. of words
B	Milling	Complex programme 2½-D continuous-path control KOB 2	184
DB	Turning	Complex programme Any turned components KODRBS	137
		Complex programme Form turning of rolls KOWADR	20
PS	Drilling and straight-line milling		95

Table 28. **Extent of vocabularies for 3 SYMAP components (complex programmes)**

---

[14] VEB Carl Zeiss Jena, Department 'Technische Vorbereitung'

Compared with those for the three EXAPT parts, the numbers of words for the corresponding SYMAP parts are slightly smaller; the storage capacity required could therefore be a little less than for EXAPT. It is unlikely, however, that it will be possible to get below the minimum core store capacity of 32K Byte required for MINIAPT — especially since the technological material is in part included in SYMAP or is to be worked into it. Just as with EXAPT, the tables for the automatic determination of cutting values, on the basis of the tool and material data stored in the computer, require a comparatively large amount of storage space. There is, however, the possibility of writing all the processors in a rational manner in assembler languages if the variety of computers to be employed is not large; this is known to reduce the required (273) storage capacity.

AUTOPROG (346)

Unlike the programming methods described up to the present, this system was initiated in the USSR. The background to the system is the graphical method of data processing, which is at present still in the very early stages of development (331, 332, 333, 363), and the elimination of all language difficulties, even in international operation, by the adoption of a form of 'picture writing', similar to the ancient hieroglyphs. It is proposed that the engineering drawing should be used as a standard means of communication, since this is produced to more or less standardised rules in all industrialised countries, and will probably continue to be adopted by developing countries as they become industrialised. The problem is to develop sufficient order among the extremely wide variety of workpiece shapes to enable these to be described by an acceptable number of 'ideograms'. The large amount of work that has already been carried out on the formation of families of components, (group technology) has led to the evolution of various methods for doing this (347, 348, 349); fundamentally these are all based on a statistical survey of variety of workpieces in existence and a subsequent data reduction by means of an ordering system based to a greater or lesser extent on logic. One particular example that may be taken as a model is the procedure described in (347) which formed the theoretical basis for the AUTOPROG system. The first stage of grouping is to divide all types of workpiece (100%) into four categories:

Rotationally symmetrical	55%
Flat	13%
Box-shaped	6%
and others	26%
	100%

This distribution of all workpieces among these four groups, which is based on Czecho-Slovak statistics, is very interesting. Since, according to these

Czecho-Slovak investigations, the bulk of the fourth group (others) consists of unmachined components, they can to a large extent be ignored. It is highly probable that figures for the engineering industries of other highly-developed industrial nations would be similar to the Czecho-Slovak statistics, so that it is possible to regard the figures quoted as being of relatively general validity, at least so far as this discussion is concerned. The most important fact to emerge is that the great majority of components are rotationally symmetrical, so that the decision to develop a new programming method initially for this class of workpiece is fully justified. This task has been undertaken by the Research Institute for Machine Tools and Machining (VUOSO Prague); whether the same methods can be adapted for flat and box-shaped components is at the moment not clear. The basic solution is illustrated in Fig. 284 (346). A catalogue showing the various basic forms of rotationally-symmetrical parts is also required. The dimensions of the workpiece need only be entered on the appropriate sheets of the catalogue to produce a simple tabular basis for the input punched tape of a data processing system. Since all dimensions and surface finishes are also obtained simultaneously from the drawing, it is only necessary to enter the material, the characteristics of the machine tool, and the tools that are available (together with their most important features) into the data processing system to enable a control punched tape for a specific NC machine tool to be produced complete in every detail. This method can be linked with computer graphics and with computer-aided design, as briefly discussed in Chapter 12, and this underlines the importance of the system for integrated data processing (288, 331, 332, 333).

There will, no doubt, be many difficulties in introducing this system generally; the basic ideas from which it has been developed are, however, so simple and relevant to the problem that the system has great possibilities.

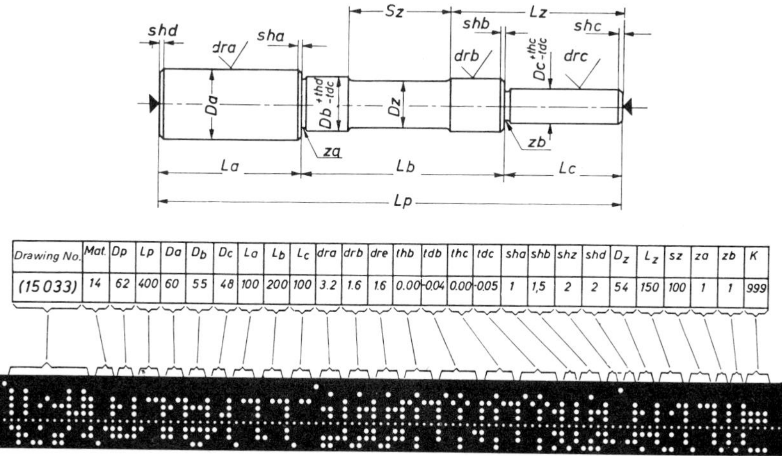

Drawing No.	Mat.	Dp	Lp	Da	Db	Dc	La	Lb	Lc	dra	drb	dre	thb	tdb	thc	tdc	sha	shb	shz	shd	D_z	L_z	sz	za	zb	K
(15 033)	14	62	400	60	55	48	100	200	100	3.2	1.6	1.6	0.00	-0.04	0.00	-0.05	1	1,5	2	2	54	150	100	1	1	999

Fig. 284 Complex workpiece represented in the AUTOPROG system

Undoubtedly it will become more successful, the more it proves possible to classify the remaining half of all workpiece shapes in a similar manner.

11.3.4 Symbol language for small computers

Since the market has been dominated by the manufacturers of large computers, the importance of the development of high-performance small computers was not realised for a long time; it is only in the last two to three years that developments have become known which fall into this category. The possibilities of using small computers were discussed in Chapters 8 and 10; here the interest is concentrated on the possibility of using production-engineering symbol languages on these machines, and thus to provide an alternative to machine programming of NC machines. The main openings for small computers lie mainly in three special sectors:

1. Owing to their favourable price/performance ratio, small computers are used primarily where only a small number of special programmes are frequently employed; in other words they can be used more easily for an assembler operation, and so are more quickly and cheaply adaptable to special problems.
2. In some instances it is possible to use permanently wired, and hence inexpensive, programme stores instead of freely-programmable core stores; this reduces the cost and improves the reliability. It is only necessary to have a good knowledge of the frequency with which the various routines are required, to enable the correct allocation to be made between 'fixed stores' and 'live stores'. If this analysis is correctly performed it is possible to use small data-processing units, having very satisfactory cost performance ratios, for production-engineering languages.
3. The cost — currently about £12,000 — appeals to a much wider range of users, and so introduces the possibility of large-batch production of this type of equipment, which would lead to further cost reductions. What is termed 'medium-sized data processing' thus gains considerably in technical and economic importance (208, 256).

It is not possible at this point to discuss in detail the many specific proposals that have been made for the use of small computers; reference must again be made to the manufacturers' literature (e.g. 350, 351, 352). Nevertheless, to give a clear picture of the possibilities of this new means of information processing, which will undoubtedly be widely used in the future as an alternative to the large computer, some of the basic features will be considered; there are, however, considerable differences between manufacturers in this respect.

The PDPS small computer made by Digital Equipment Corporation/USA has already been mentioned as part of the experimental arrangement illustrated in Fig. 217, where it was used as a freely-programmable internal interpolator. The manufacturers have developed a simple symbol language for solving drilling problems for use on this computer, known as the QUICKPOINT 8 system (350). The core store required must have a capacity of 4K words of 12 bits. The language is used in the same way as APT with a processor that is universally valid and various post-processors for adaptation to specific drilling machines. The vocabulary for the processor section comprises only

Fig. 285
Small-computer program-
ming station for NC
machines employing the
EASYPROG language
system (computer type
PDP 8) (Photo courtesy of
Max Muller)

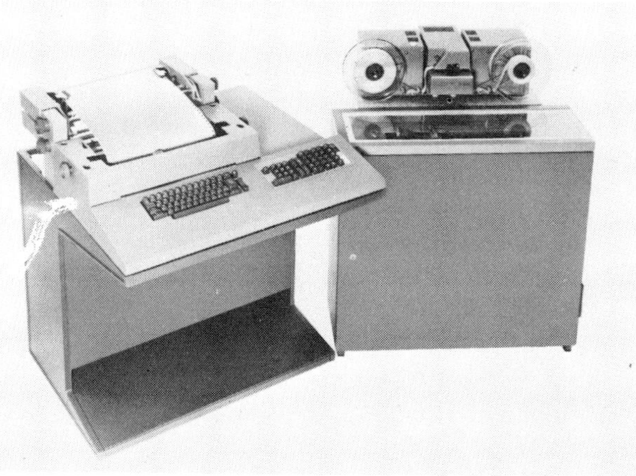

Fig. 286 Small computer for NC machine programming (computer type Nixdorf 820)
(Photo courtesy of *Nixdorf*)

23 three-part words
10 two-part words
    individual letters, and
    some special symbols.

The system is designed specifically for determining point patterns and can cope with a wide range of combinations of even complicated hole patterns. The small vocabulary inevitably limits the descriptions that can be produced to the most important applications, with the result that the storage capacity required for the processing programmes is appreciably reduced. The processor and post-processors are entered by means of punched tapes and the teleprinter associated with the computer using ASCII, i.e. virtually in ISO Code (see Chapter 10). The output of the control tape for the NC drilling machine is by means of the same teleprinter, but this is in EIA Code, which is currently more widely used. Further details will be found in (350).

The same small computer is also used for the EASYPROG system (351). In the form in which it is currently available this programming system can be used for the machine programming of lathes with straight-line and continuous-path control systems; it is, however, also to be extended to other types of machine. The processor will handle geometric and technological vocabularies. The total number of main words and modifiers combined is 44; the length of the words is restricted to 4 characters. Unlike the QUICKPOINT 8 system, with the larger APT-like languages used in EASYPROG the output takes the form of a CL tape. Since the equipment used with the computer is the same as for the QUICKPOINT 8 system, all inputs and outputs are by means of punched tapes, with the teleprinter acting as the printer. Figure 285 shows an EASYPROG computer; the QUICKPOINT computer is similar in appearance.

Figure 286 shows a Nixdorf Type 820 computer with a punched-tape reader and punch in use as an NC programming station. To achieve maximum flexibility and adaptability, both the computer itself and the programmes are built up from basic units so that it is possible to use unit construction for both the hardware and the software. This will increase the utility of the small computer (207). Among examples of possible combinations are:

a) The programming of NC machines using simple programming languages.
b) Programme-orientated calculation of workpiece contours.
c) The use of the machine as an automatic document writer which offers many opportunities for simplifying office work, especially in technical offices.
d) Desk calculator for the accurate and rapid calculation of any intermediate results, particularly those involving angle functions and logarithms.

Since we are here primarily interested in the computer as an aid to programming particular attention will be paid to examples a) and b) above, although they must not be considered in isolation from the others if high utilisation of the equipment is to be achieved.

The use of a small computer for machine programming of NC machines

Since the requirements differ greatly when lathes, drilling machines, and milling machines are being programmed, three specific programming languages have been developed; to this extent there is an unmistakable affinity with the ideas underlying EXAPT. The symbolic languages have been simplified in particular by adding an identifying letter to all numerical data as it is fed in, thus indicating the application of the number; to some extent this approaches the free-style address method employed for manual programming (cf. Chapters 9 and 10). Identifying letters of this type for a drilling machine programme (XY plane numerically controlled, Z axis cam-controlled) include:

X, Y	for the coordinate axes
A	Starting angle (for pitch angle descriptions)
B	Number of increments on a pitch circle
D	Pitch circle diameter
H	Increment spacing along a straight line
L	Stopping of drilling spindle for tool change
N	Spindle speed in rev/min
S	Cam No. in Z-direction
T	Call-up of Z movement
V	Feed in Z-direction in mm/min
F	End of programme.

The end of an input sentence is indicated by the character ◇. The use of individual identifying letters in place of word abbreviations containing 3, 4 or even 6 characters makes it possible to produce a machining programme that requires very little storage capacity. The numerical values for speeds and feeds, which are entered in plain language, are automatically converted to the nearest coded numbers to DIN 66025 (cf. Chapter 10) which are then punched onto the tape. If the width of the paper employed in the typewriter is suitable the print-out of the control programme for the NC machine, which corresponds closely to the format for a manually-produced programme as shown in Chapter 10, can be produced beside the source programme in the symbolised language. This is possible since each input sentence is immediately worked out completely in the processor and the post-processor to the point where a complete and error-free control sentence is produced. Moreover for each control programme the manufacturing time is determined, the machining times and the idle times being calculated separately, and this also forms part of the output.

The general contour calculating programme (352)

This programme is built up as a self-contained unit, so that it can either be used alone or if required can be incorporated in drilling or milling machine programmes. As in the large programming languages, many practical definition possibilities are provided for straight lines and circles as well as combinations of these elements. To save storage space, many of the configurations of displacement elements that are frequently encountered are allotted identifying numbers. As an interesting feature, this machining programme can take part in a man-machine dialogue; the computer continually informs the user, by means of the typewriter, of the information that has to be provided next, and so automatically provides a guide to the correct method of solving the problem.

The automatic typewriter programme (207)

As has been emphasised in Chapters 10 and 11, it is very convenient to use an automatic typewriter with a punched-tape input and output; when using a large computer it is also often more economical to use an automatic typewriter rather than the large computer for duplicating and correcting the control tapes. Used in conjunction with the small computer and the typewriter, the self-contained automatic typewriter programme enables many existing and new clerical tasks to be simplified, such as:

- Unrestricted choice of typewriter tabulation settings to suit various form layouts.
- The production of punched control tapes in any desired code.
- Code conversion when duplicating (cf. Chapter 10). etc.

The desk computer programme

Trigonometric functions or logarithms are employed in many technical calculations beyond the capabilities of ordinary desk calculators. To meet the requirements of a very varied range of users, a special small computer programme has been developed for use in technical offices. The small computer shown in Fig. 286 is used as a programming station for NC machines. As well as the typewriter, the operator has a numerical keyboard with some additional functional keys; this can be used for calling up the machining programmes referred to above. Extensive use is made of the mixture of fixed stores and freely programmable core stores. In its largest configuration, the core store can store 1024 (1 K) 16-digit decimal numbers to accommodate the variable data; for NC use the normal version has a capacity of 128 16-digit decimal numbers.

The fixed store, which is used for accommodating the standard programmes (automatic typewriter, desk computer, contour calculation, processors for drilling and milling and lathe languages) which do not vary, can accommodate a maximum of 16K commands each of 18 bits. The fixed store, which costs only about one-fifth compared with the freely-programmable core store, is designed so that any correction required can be performed easily even after it has been programmed.

The small computer offers many users of machine programming methods a genuine alternative to the large languages that have previously been described; but development in the field of small computers is only just beginning, and considerable further progress is likely to occur. The small computers mentioned in Chapter 10 in connection with semi-machine programming (e.g. for use with the AUTOPROGRAMER) are of the same order of size as the small computers being considered here. Since in this instance a symbolic language is used, instead of the NC machine being programmed directly, the use of the term 'semi-machine programming' is not appropriate in this connection.

11.4   External interpolators and magnetic-tape control systems

In Fig. 8 (Chapter 2) a system was shown and discussed in which the entire computing process, of the type which has to be performed for continuous-path control systems, is located in the external data processor. The great advantage of this arrangement is that the expensive computer installation has to be used once only for a considerable number of machine tools and the outlay on internal data processing is greatly reduced[15] (cf. also the alternative shown in Fig. 248).

Mention was made of a disadvantage of this form of organisation in Chapter 9; it is no longer possible to interrupt the machining process simply to apply corrections (e.g. to alter the feed speed to allow for unforeseen contingencies, or to correct for the diameter of a milling cutter when this has been sharpened, etc), as can be done when internal interpolators are used. There must also be doubts on the use of magnetic tapes under machine shop conditions, as has been mentioned in Chapter 9. Nevertheless, the organisational possibilities introduced by this process will be considered rather more closely in this section, since it is quite possible that developments will take place during the course of the next few years which will enable these difficulties to be overcome (244, 245).

---

[15] According to Ferranti, upwards of 50 machine tools can be supplied with control tapes by a central installation of this type, since the computing and storage processes are performed at high speeds independently of the machine tool (358).

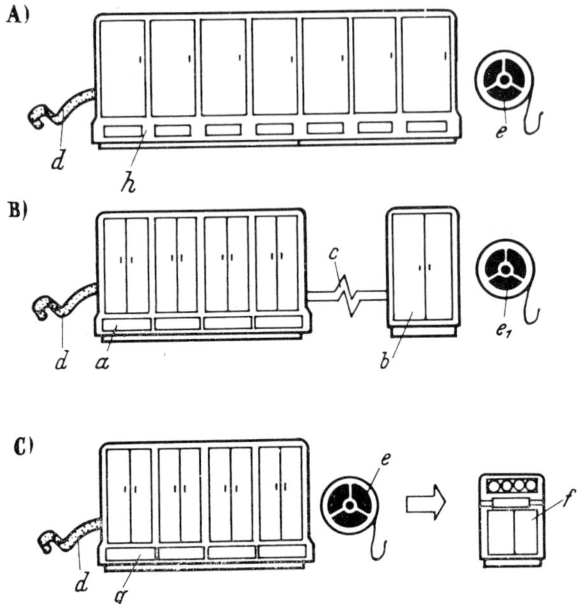

Fig. 287 The principal ways in which the continuous-path control computing processes can be divided up in the external data processing in conjunction with magnetic tape control systems (359, 244, 245).

A) Special computer with punched-tape input of geometric data and magnetic-tape output of completely interpolated guide values (cf. Figs. 242, 243, 244). B) All-purpose computer with curve generator permanently attached (external interpolator for complete interpolation). C) All-purpose computer with incomplete curve generation (pre-interpolation), magnetic tape as intermediate information carrier, and final interpolation in remotely situated customer service centres

a All-purpose digital computer. *b* External interpolator (curve generator). *c* Electrical connection. *d* Punched tape as input information carrier. *e* Magnetic tape as output information carrier. *f* Digital final interpolator in customer service centre. *g* All-purpose digital computer with magnetic-tape output. *h* Special digital computer with magnetic-tape output

There are three main ways in which the basic idea could be realised in practice, and these are shown schematically in Fig. 287; they introduce organisational problems rather than technical problems.

Example A.   Use of a special computer

If during the design of the digital computer, a decision is reached that it should not be used for other purposes associated with the running of the plant (e.g. pay-roll calculations, book-keeping, technical and scientific calculations, etc), and if it is arranged so that it is capable of computing complicated curves with the points spaced so closely together that subsequent internal interpolation is not necessary, a special computer solely for machine-tool control systems will be obtained (see Fig. 272). The data input can either be on punched cards or on punched tape; the output will always be at high speed on magnetic tape. The magnetic tapes can be used directly for controlling machines, but will then run at comparatively low tape speeds (Chapter 9). One of the advantages of this method is the complete separation of the fast electronic data-processing operation from the slow movements of the machine slides (358).

One of the main economic disadvantages of this arrangement is that the high-speed computer will be fully used only if it supplies the largest possible number of machine tools with control tapes. Since it is a special-purpose computer it cannot be used for other tasks and if it is not fully loaded idle periods will inevitably result and the installation as a whole becomes uneconomical. This type of installation is no longer made.

Example B. Use of an all-purpose computer with an auxiliary curve generator (external interpolator) located beside it.

A more economical arrangement than that shown in Example A) will be achieved if the computing process for the external data processing is divided in such a way that the greater part is performed in an all-purpose computer while the special task of calculating the curves or pure interpolation is allocated to a special machine. The overall function is the same as in the previous example. When it is not required for this work the all-purpose computer can be used for other tasks associated with the running of the plant, and only the comparatively small curve generator is liable to stand idle. The interpolation can be performed digitally to any desired degree of accuracy (e.g. by using cubic parabolas with closely-spaced reference points) without the consequent higher capital costs becoming significant, since they occur once only in the external data processing system. The control data can also be produced in analogue form. The cost of the control equipment on the machine tools will then be low since, as in Example A), no internal interpolator is required. The all-purpose computer is then programmed completely in accordance with the principles put forward in Section 11.3

With this organisational arrangement it is also possible to obtain certain design data and mathematical equations, which have to be determined experimentally (e.g. flow profiles from wind tunnel tests) directly by calculation from the measured values without drawings first having to be prepared (360). The only disadvantage of the arrangement is that a comparatively close network of service stations equipped with large computers has to be built up to take full advantage of numerical control e.g. its flexibility when a programme is changed. If delivery of the control tapes were to take a long time it would run counter to this principle, especially if the tapes were not ready for immediate use, and any programme errors had first to be detected and corrected. In a fairly small country (such as Britain) it would be possible to build up a system of this type comparatively easily and to maintain it at full efficiency, but in a country such as the United States it would be difficult to operate. These operational difficulties probably account for the fact that magnetic-tape control systems, sponsored by Ferranti, whose service agencies are relatively closely spaced, are to be found comparatively frequently in Britain, but are infrequent in the USA. Any really effective spread of magnetic-tape control systems of Example B) in Continental Europe thus appears likely only if the major manufacturers of NC magnetic-tape units, such as Ferranti, of Edinburgh, were able to establish an adequate network of suitable computing centres. This is more an organisational and financial problem than a technical one. There seems a good case, therefore, for basing the establishment of service centres on a different principle, as described in Example C).

Example C. Use of an all-purpose computer with magnetic-tape output and divided computing process using several geographically remote service centres.

If magnetic tapes are used as intermediate information carriers between a main unit and a large number of service centres, the interpolation process can be subdivided in a similar manner to that described in Example B), but there is much greater freedom to select the locations of the units. In such instances it would also probably prove worth while to supplement the service centres and their curve generators with NC drafting machines (Fig. 189) to enable the programmes to be tested at the computer centre. Whether it is worthwhile to adopt this method on a wide scale appears to be not only a question of organisation and finance, but also of the competition that will arise as further progress is made in the development of reliable punched-tape control systems and of direct numerical control (DNC, cf. Chapter 9). This is the case because, despite all the technical and organisational improvements that have been made, there still remain two disadvantages in the use of magnetic tapes:

1. Difficulties due to the principles employed, which result from the assignment of the entire responsibility for the production process to the external data-processing element. These are increased by the distances that separate the various units (e.g. no simple method is available at present for correcting the programme at the machine tool itself).

2. The close economic, technical, and organisational ties that must be forged with what are at present only a small number of manufacturers of magnetic-tape systems. Once management has decided for 'one

system' it has to adhere to this decision, and all the machine tools must employ the same control systems. This situation may, however, change in the next few years.

11.5 Summary

1. At the beginning of this chapter a brief introduction was given to the construction of electronic computers and of the ways in which they can be programmed to enable the reader to understand the interaction between computers and NC machines in manufacturing plants. This introduction had to be kept brief, but the references quoted should provide sufficient material for those who wish to study these problems in greater depth.

2. The most important feature of computers for these purposes was found to be the structure of the store hierarchies. The most suitable designs of these can be determined from a study of two characteristics — 'access time' and 'capacity'; to enable direct comparisons to be made, storage capacities are given in 'K Bytes'.

3. The following are required for a correct decision on the choice of suitable programming languages for production engineering:

a) as accurate as possible knowledge of the parts spectra encountered in the manufacturing plant; and

b) a clear decision as to the extent to which automation is to be introduced into the work planning and work preparation activities, and the extent to which it is possible to introduce such automation in view of the costs.

4. The APT programming language was developed in the USA for complicated machining processes; this language is administered in the United States and is constantly being extended. In this system the technical features and difficulties are solved in the post-processor, which is specific to the type of machine under consideration. Disadvantage of APT for many potential users is the high storage capacity required, i.e. expensive computers must be employed.

5. To reduce these economic difficulties, several APT-like languages have been developed, based on analyses of parts spectra. To some extent these are more or less extensive subsets of APT which maintain the original principle of dealing with special technological features in the post-processor.

6. Still using the basic principles of APT, but fully aware of its disadvantages, the idea developed in Europe in 1963/64 to make use of the possibilities offered by computers for solving various technological work planning problems, both large and small. The only economical way of undertaking this work was to make a rough

subdivision of the various machining tasks into the three groups of turning, milling, and drilling; in this way it is possible to reduce general technological routine tasks (e.g. the determination of suitable cutting values) in the processors and hence to make them cheaper to perform.

7. The three-part EXAPT system, which is closely allied to APT, was produced in this way; while SYMAP is a four or five part system that is less closely based on APT. At the same time, the subdivisions were designed to ensure that even computers with small storage systems could be used for production engineering applications.

8. A completely different approach was followed with the AUTOPROG system which makes use of the methods of group technology and of the formation of part families; in this system the basis is engineering drawing — which is an international means of communication. Up to the present it has been developed only for turned components; adapting it to other types of workpiece is likely to run into considerable difficulties.

9. New ideas for machine programming, which in some cases are completely divorced from the ideas underlying the APT system, were not developed until the introduction of small computers in about 1968. The development of the equipment and programmes for this purpose has, however, only just started seriously, so that it is not at present possible to come to any definite conclusions. However, for many purposes this development offers suitable alternatives, and it appears to be acting as a counterweight to the strong trend towards centralisation that is associated with large computers.

IV   Summary

12   The integration of NC machines and computers in the production process

At least three points that should form the basis for further work have emerged from the discussions of the preceding chapters:

1. NC machines are most conveniently considered as a specialised section of the general data-processing equipment.

2. The effect of developments in the field of computers has been to make confusingly large the number of possible combinations of equipment and methods for automating manufacture.

3. System operation in the workshop or at the individual machine is no longer the only sufficient criterion. The effects of computer technology make it necessary to include the entire plant in the survey if this is to produce sufficient accuracy as far as the effects of any changeover on personnel or economic factors are concerned; in any such survey the NC machine will only be one factor among many.

The system-based approach to the subject, Fig. 8, has proved to be satisfactory not only for the complex relationships to be analysed clearly, thereby making them easier to study, but also of building up imaginary models in such a way that various 'systems' can be run through, if necessary with the aid of data processing equipment.

To go into this system-based technique in detail is beyond the scope of a basic text book of this type, but the fundamental ideas underlying it have been published in an earlier study (1969)[1].

The results from two earlier publications (253, 372) have also been included to complete the picture of the flow of materials and the control of production. In this way it should be possible to build up a theoretical analysis of the complete operation of a plant and to use this as a basis for calculation.

---

[1] The study was accompanied by a questionnaire, to enable as many users as possible to give their views. In so far as these contributions have become available within a year of the earlier study being published, they have been incorporated in this chapter.

The most important factor in all these considerations is the 'system' as defined in DIN 1926 (12).[2]

A_t System output (e.g. finished and inspected parts, produced during the total manufacturing time t_d)
E System input (e.g. manufacturing order). F Internal data processing. G External data processing. M_1 Raw material. M_2 Tools. M_3 Clamps. W_2 Tool magazine. a_1 Form data for positioning and straight-line controlled machine tools. a_2 Form data for continuous-path controlled machine tools. b_1 Technological data for a_1. b_2 Technological data for a_2. c_2 Programming station for data-processing installation. d All-purpose data-processing installation (e.g. computing centre located 300 miles away). f_1 and g_1 Compiler for a production-engineering programme language. f_1 Processor section. g_1 Post-processor section. i_1 Internal interpolator for continuous-path controlled machine. j_1 Workpiece quality control (samples). k Interface between internal and external data processing. p_1 Punched-tape reader for positioning and straight-line controlled machine. p_2 Punched-tape reader for continuous-path controlled machine. r Switching information. t_2 Input for correction values. t_{f_1} and t_{f_2} Data transmission times if computer d is located in a remote computing centre. t_{f_1} Transmission time for source programmes. t_{f_2} Transmission time for control punched tape. t_d Total time for passage through system. t_v Preparation time (partly controlled by organisational factors). t_m Processing time (controlled by machine and transport factors). X, Y Coordinate axes

1. Displacement measuring systems.
2. Comparators. 3. Drives. 4. Machine slides

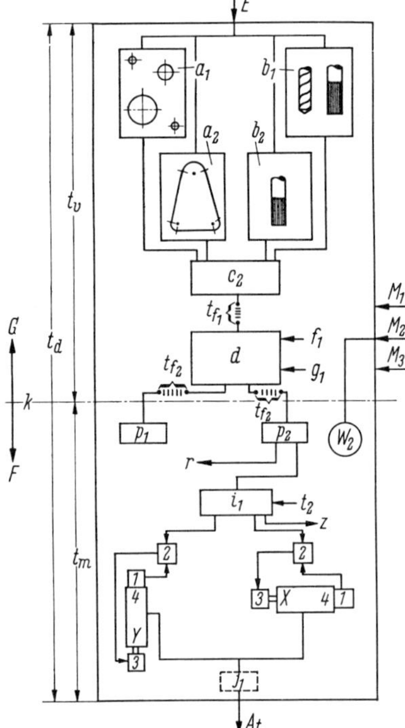

Fig. 288 Data flow sheet for a production engineering basic system

[2] In this Standard a system is a defined arrangement of elements that interact on each other. Such elements can be either material objects or thought processes and their results (e.g. forms of organisation, mathematical methods, programming languages, etc.). This arrangement is separated from its surroundings by a real or imaginary boundary envelope. The connections that join the system to its surroundings pass through the boundary envelope. The properties and conditions that are carried by these connections are the values, the interrelationships of which produce the behaviour that is characteristic of the system. Larger or smaller systems can be produced by joining up or splitting the original systems in a suitable manner.

Figure 8 shows such an arrangement in the sense of the term as used in this Standard; in particular it will be noted that it is possible either to divide the arrangement or to join it to others to form larger units as required. Extensive use has been made of the possibilities of subdivision in the earlier chapters. Only a few additions are needed to make such definable systems, as specified in DIN 19226, capable of being manipulated and joined together to form larger arrangements. A system with its boundaries as shown in Fig. 8 can thus be described as a production engineering basic system. Figure 288 shows a more complex production engineering basic system with suitable additions; its character remains that of a data flow sheet (384). The 'values' that represent the connections to the outside world and to other systems are in the first place the input E and the output A. In this limiting case these are:

E = Manufacturing order for (initially) *one* workpiece drawn from a parts spectrum, and

A = the finished, inspected part.

There are also the main material flows (shown here only as inputs):

M1 = Raw material (input only, since output is at A);

M2 = tools (supply and return, since they are stored in auxiliary stores for re-use);

M3 = clamping equipment (supply and return, since they are stored in auxiliary stores for re-use).

The power required for the manufacturing process can be neglected in this context.

The most important characteristic of the system is the total time t_d required for passage through it. As a first approximation this can be subdivided into the preparation time t_v and the processing time t_m; a more detailed sub-division may be required, and this will be introduced later. The total time t_d is a realistic measure of the flexibility of the system, to which frequent reference has been made; considerable future effort will have to be devoted to determining the precise relationships between the internal structure of the systems, the information content of the workpieces that are to be machined (129), and the total time t_d required for passage through the system. From the control aspect, the total passage time is also an important characteristic of the transmission behaviour of a production engineering system.

12.1   Arrangement of basic systems according to degree of automation

If one starts from the proposition that every automatic system is characterised by having at least one information store, the features of which, together with its location within the system, largely determine the behaviour of the latter, it is possible as a first approximation to draw up a table of the type shown in Table 29. In this several of the features of the system can be coded in binary terms, either of the symbols O or L being inserted according to whether the feature is present or not.

In the first instance 10 features that can be readily identified will suffice to indicate the degree of automation of a system. It is possible to argue about the relative importance of the features, i.e. the order in which they are listed, and the precise words by which they are defined; up to the present, however, all comments received by the author have been favourable, or at least no counter proposals have been made. The detailed text relating to Table 29 will be found in (10); however, after the earlier chapters of this book have been studied the following notes should be readily understandable, despite their brevity.

Listing order 1

Computer operation of a basic system covers both the use of a computer for the production of data carriers and also DNC operation.

Listing Order	Feature	Binary coding	
		O	L
1	Computer operation	no	yes
2	Type of operation	off-line	on-line
3	Displacement value processing (determination of actual values or comparator, or both)	non-automatic	automatic
4	Type of displacement information store	rigid	flexible
5	Location where store is produced or filled	internal	external
6	Processing of switching information	non-automatic	automatic
7	Adaptive control	no	yes
8	Nature of control system	Positioning, straight-line	continuous path
9	Tool change	non-automatic	automatic
10	Workpiece change	non-automatic	automatic

Note:
If neither of the binary codings can sensibly be applied to a feature, this should be designated O.

Table 29. Provisional list of binary features used to describe a system

Listing order 2

The type of operation gives information on the use (e.g. punched tape or magnetic tape) or non-use (DNC operation) of data carriers.

Listing order 3

A machine tool is considered to possess automatic displacement value processing if it is equipped at least with some device for automatic determination of actual or required values (displacement measuring systems or an analogue required-value pick-up) of the slide positions to enable a subsequent comparison to be performed between them. These can be, for example, numerical displays associated with displacement measuring systems, analogue follow-up control circuits in conjunction with copying devices.

Listing order 4

The machine tool is considered to be equipped with a 'rigid' displacement information store if the displacement information to be processed is called up from a mechanical programme store such as the following:

Cam discs, cam drums, templates, patterns, non-adjustable gearing, rigid crank and slider mechanisms, form-storing tools (form cutters, or fixed sets of tools), drilling jigs. That some of these rigid stores can be produced more rapidly with the aid of NC machines is another factor which will be considered again later.

The machine is considered to be equipped with a 'flexible' displacement information store if the displacement information is called up from a programme store such as the following: Cam strips with adjustable cams, adjustable crank and slider drives for machine slide movements, programmable drum cams, cross-bar distributors, programme cards for plug boards, decade switches, punched tape, punched cards, magnetic tape, electronic stores (including magnetic stores for reproducing control systems (226).

Listing order 5

'Internal' production of the stores is said to occur when the displacement information store is produced (Figs. 233, 229) or filled (Fig. 218) in the internal data processing system; 'external' production of the stores, on the other hand, is said to occur if the store is produced or charged in the external data processing system.

Listing order 6

A machine tool is said to have 'automatic' processing of switching information if it is at least equipped with some means of automatic detection (e.g. punched tape, cams) and processing (e.g. interlocks) of switching information. If one of these is not performed automatically, the machine tool is said to have 'non-automatic' processing of switching information.

Listing order 7

A machine tool is said to possess adaptive control if the values of 'speed' and 'feed' are automatically adjusted in order to optimise the cutting conditions as a result of measured data (cf. Chapter 9).

Listing order 8

A machine tool has positional or straight-line control if in principle it requires no interpolator; if it has a continuous-path control system, an interpolator will be required at some point in the system.

Listing order 9

A machine tool is said to have 'automatic' tool change if it is equipped with some device for automatic tool changing (turret, magazine, cf. Chapter 7) and/or simultaneous use of several tools (multi-spindle or multi-cutter operation cf. Fig. 310 (133, 134). Otherwise the machine has 'non-automatic' tool change.

Listing order 10

A machine tool is said to possess 'automatic' workpiece change if it is equipped with a device for automatic loading, clamping, and unloading of workpieces; otherwise it has 'non-automatic' workpiece change.

(Note: Movement of workpieces to other machining locations and the automation of transport and handling does not form part of the basic system as defined, but part of the materials flow and production control systems. It is more convenient to regard NC machine lines with integrated production control by means of a computer as shown in Figs. 176, 177, and 178 as self-contained basic systems, and these are discussed later).

The use of binary-coded features raises the possibility of regarding the values assigned to each feature as binary numbers, and of taking the listing order given in the table as the digit number for a binary figure. The degree of

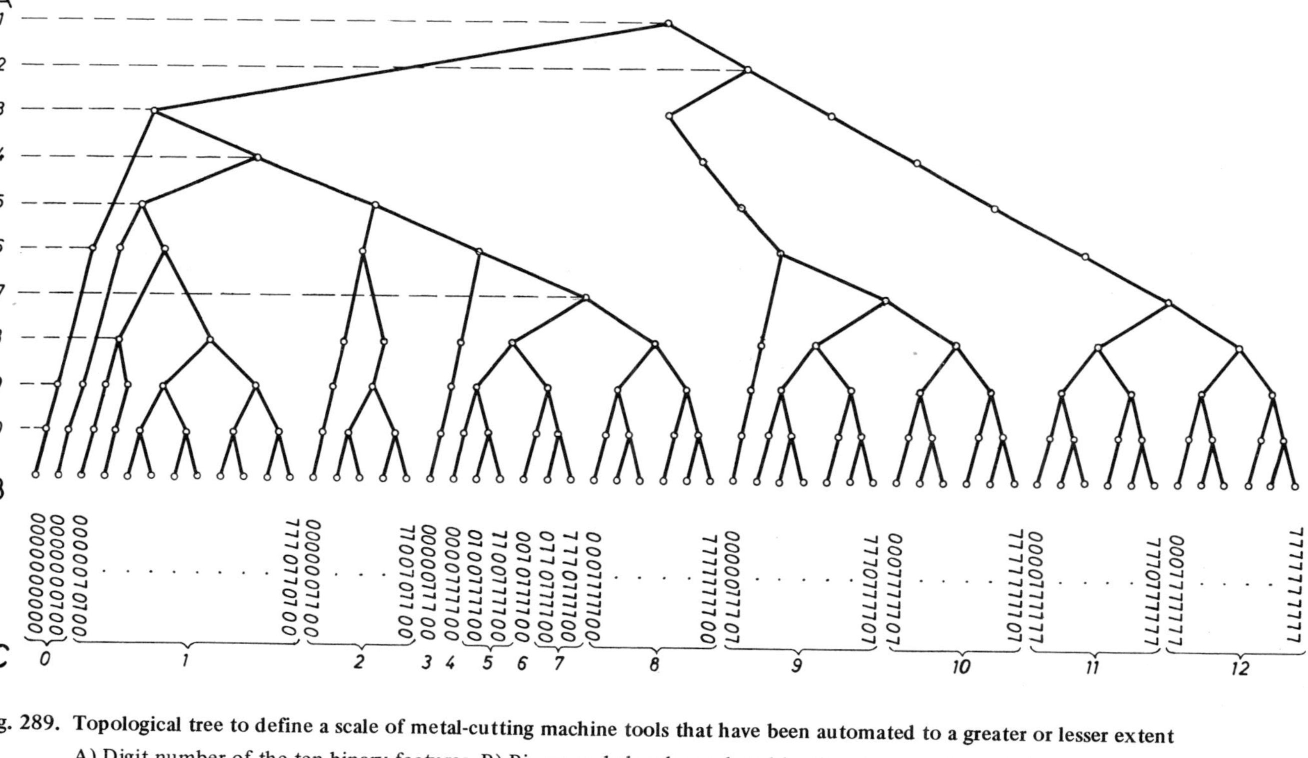

Fig. 289. Topological tree to define a scale of metal-cutting machine tools that have been automated to a greater or lesser extent

A) Digit number of the ten binary features. B) Binary-coded scale produced by the binary coding and the digit number of the features
C) Arrangement of groups for simplified verbal description in accordance with Table 31

Stage	Corresponding binary numbers	Description	Examples of machines in Figs. No.
0	OOOOOOOOOO . . . . OOLOOOOOOO	Manually-operated machines, i.e. machine tools with no appreciable storage capacity for displacement or switching information, without automatic tool or workpiece change but possibly with numerical position indicator	76, 42 numerical read-out
1	OOLOLOOOOO . . . . OOLOLLOLLL	Machine tools with rigid displacement information stores obtained from external data processing for analogue positioning, straight-line, or continuous-path control systems, possibly with automatic processing of switching information and automatic tool and workpiece changing	e.g. automatic lathes (133, 134) Copying machines (72)
2	OOLLOOOOOO . . . . OOLLOLOOLL	Machine tools with flexible manually-adjustable displacement information stores within the internal data processing devices for analogue or numerical positioning and straight-line control systems, possibly also with automatic switching information processing and automatic tool and workpiece changing	e.g. cam-controlled machines (29), machines with repeating controls (226) 223, 296, 297
3	OOLLLOOOOO	Machine tools with flexible displacement information stores obtained from external data processing for numerical positioning and straight-line control systems, without automatic switching information processing and without automatic tool and workpiece changing, mainly 2-D NC machines with cam controls in Z-axis	126, 179
4	OOLLLLOOOO	Machine tools with flexible displacement information stores obtained from external data processing for numerical positioning and straight-line control systems with automatic switching information processing, but without automatic tool and workpiece changing.	123, 87, 218, 141A, 148A, 149B, 171, 180, 303

Table 30. 12-part Stage scale for automated metal cutting machine tools, in accordance with Fig. 288. Summary giving brief description of main stages

Stage	Corresponding binary numbers	Description	Examples of machines in Figs. No.
5	OOLLLLOOLO ············ OOLLLLOOLL	Machine tool as for Stage 4 but with automatic tool changing and possibly also automatic workpiece changing	134, 143, 153, 154 155, 160
6	OOLLLLOLOO	Machine tools with flexible displacement information stores obtained from external data processing for numerical continuous-path control systems, with automatic switching information processing, but without automatic tool and workpiece changing	172, 183, 184, 187B, 188, 305
7	OOLLLLOLLO ············ OOLLLLOLLL	Machine tools as for Stage 6, but with automatic tool changing and possibly also automatic workpiece changing	217, 135, 139A, 140B, 144, 146, 147, 156, 158, 159.
8	OOLLLLLOOO ············ OOLLLLLLLL	Machine tools as in Stages 4 to 7 but with adaptive control	212, 209
9	LOLLLOOOOO ············ ············ LOLLLLOLLL	Computer centre for 'off-line' operation of machine tools of Stages 2 to 7, but without adaptive control of the machines	173, 175A, 308
10	LOLLLLLOOO ············ ············ LOLLLLLLLL	Computer centre as in Stage 9, but with adaptive control of the machine tools	
11	LLLLLLOOOO ············ LLLLLLOLLL	Computer centre for 'on-line' operation of machine tools, but without adaptive control of the machine tools	177, 247, 248
12	LLLLLLLOOO ············ LLLLLLLLLL	Process computer system for 'on-line' operation of machine tools of Stages 4 to 8 with adaptive controls	

Table 30. (Continued)

automation can then be expressed by means of a ten-digit binary figure. Listing order 1 is then at the start (left-hand end) of the binary figure

This sub-division can, of course, be extended if necessary, both to distinguish more clearly between various conventional automated machines (single-spindle automatics, multi-spindle automatics, various types of copying machine, etc), and also to allow for the details of the AC and DNC techniques that are on the point of being introduced. In general, however, with the $2^{10} = 1024$ variations of degree of automation possible with the system as proposed, a high proportion of the many different designs that have been introduced can be defined in terms of a small number of basic elements. The results are shown in the form of a topological tree in Fig. 289. Its elements are:

1. Nodes which represent a junction at which the binary coding for the feature concerned is allotted to the appropriate digit number (the absence of a node is equivalent to the value 0 appearing at the position concerned).
2. Lines, whose angle of inclination represents the binary code for the feature concerned; lines inclined to the left correspond to 0 and lines inclined to the right correspond to L.

This form of coding may go into too much detail for many practical applications, and for practical purposes a coarser subdivision could be introduced. This is shown as stages 0 to 12 on Fig. 289. The result is summarised in Table 30, and this is supplemented by examples of machines that are illustrated in this book; although it is not always possible to derive information about the external data processing system from the description of the machine tool. The state of current development is such that the bulk of present-day examples of machine tools will fall into Stages 4 to 9.

An interesting feature of this method of coding is that it gives no information about the nature or the size of the machines, and very little regarding their kinematics; for this type of information additional coding will be needed. At this stage the intention is merely to enable a clear numerical definition of the term (degree of automation) to be obtained. In practice, a machine shop will contain a mixture of machine tools of differing degrees of automation; the optimum mixture will depend on the parts spectrum that is to be manufactured and in practice is achieved only very rarely. By defining the degree of automation in terms of 10 binary features it is possible to analyse an existing stock of machine tools fairly rapidly (376, 377), and then to improve it gradually by establishing a suitable target for future acquisitions (378, 379).

The decisive advantage of this method of grading machine tools appears, however, to be that it provides an effective means of including, at least in broad terms, possible future developments (such as direct numerical control and adaptive control) in the scheme of operation of the plant as a whole, and hence in the necessary planning exercises. This simplifies the drawing up of long-term investment plans. It is of course possible to convert the unfamiliar 10-digit binary numbers into 3 or 4 digit decimal numbers without any loss of information. A continuous-path controlled machining centre with manual programming of the type shown in Fig. 292 would then have a degree of automation corresponding to:

> the binary coefficient   LOLLLLOLLO or the
> decimal coefficient   758.

The difficulty is that the logical way in which the coefficient is built up is completely masked when a decimal number is used; to emphasise the clear logical relationships it is better to retain the binary number. At least this method of grading provides a means of covering the whole range of automation from the manually-operated machine to the DNC machine in numerical form, thus providing a suitable basis for further use of data-processing equipment.

12.2   Detailed break-down of total time for passage through the system t_d

It was shown in Fig. 228 that another very important characteristic of the system is the total time, t_d required for a workpiece to pass through it, and it was found that dividing this into the preparation time t_v and the machining time t_m was too coarse and did not meet the requirements of the changed conditions prevailing in the manufacturing plant. Table 31 shows one method of subdividing the total time for passage, in this case into nine time intervals. Even this shows clearly the most important alterations that are produced by employing a refined method of system analysis. At the same time all the options for further automation are left open, since the time intervals are arranged to cover clearly definable sections of the work. For initial investigations, this nine-fold subdivision should prove fully adequate. Detailed comments on the various intervals are as follows:

t_1 This is a major bottleneck in most plants, and many attempts are being made to shorten this time interval, in particular, by using data-processing equipment. These attempts have become known under such names as 'computer graphics', 'computer-aided design', (375, 331, 332, 333, 363, 364, 303, 380). Time, t_1, is affected by the complexity of the workpieces (129), and for a given state of equipment of the design office, it will be found to be a parameter of the parts spectrum. Although t_1 forms part of the basic system, it is beyond the scope of this book to deal with it. Reference should be

Time interval	Operation performed	Symbol in Figs. 291, 292, 293
t_1	Time for determining the geometry of the workpiece (time required in design office)..	G
t_2	Time for determining the production technology of the workpiece (time required in planning and work preparation departments).	T
t_3	Time for listing tables of dimensions or for manual or semi-machine programming; or writing programme in a production-engineering programming language from source programmes, or production of cams, templates, patterns, etc. in auxiliary workshops	L P H
t_4	Time for transmission of data from source programmes to remote computing centres.	tf_1 in Fig. 288
t_5	Time for coding by means of manual programming; or operator activity in the data-processing centre (e.g. loading compilers).	K O
t_6	Computing time in data-processing centre for processing the source programme, and automatic production of the punched control tape.	D
t_7	Time for transmission of data from the remote computing centre back to the plant.	tf_2 in Fig. 288
t_8	Time for machining on machine tools, production time (can continue to be subdivided, as in the past, into setting times, main times, and auxiliary times).	*X, Y, Z*
t_9	Time for quality control, inspection	Q

Table 31. **Proposals for a detailed break-down of total time for passage through the** system, t_d, (based on 10).

made to the literature in this field, since this is a very important area for future developments.

t_2 The use of a computer to reduce the time required is still in its initial stages; information on this is given in the discussion of the programming languages in Chapter 11 (369, 351, 347, 339, 328). Such factors as:
automatic selection of machines and full utilisation of their capacity (361)

automatic selection of tools (328)
automatic determination of cutting parameters (370)
should be distinguished as clearly as possible from the mainly geometric tasks associated with the choice of the form of the workpiece, which occupy time t_1, even though they are related.

t_3 Basically, always the same task has to be performed here. The form data has somehow to be stored in a manner to suit the machines. In a simple case this can be done for an NC coordinate table by listing the xy values; in more complicated cases, manual or semi-machine programming, which must be machine-orientated, may be necessary (Chapter 10). The tasks can be shown by the symbol L in Figs. 291 to 293. The writing of a source programme in APT, ADAPT, or AUTOSPOT or another primarily geometrically-orientated programme language (Chapter 11) also falls into this time category, where it is denoted by the symbol P, as does the production in the tool room of cams and templates for mechanical automatic machines (symbol H). Here differences arise which depend on the system employed and which may have a major effect on the total time of passage t_d; there may be optimum batch sizes for these various possibilities. Since in this analysis of t_3 at least three clearly-defined separate activities are being dealt with, they are designated by the symbols mentioned, the meanings of which are apparent from Figs. 291, 292, and 293. It is important to note that in case H it may be possible to reduce the time t_3 by using NC machines in the auxiliary workshops (toolrooms) (see, for example, Figs. 172, 175, 179). The machines shown in Figs. 183 and 188 also enable the manufacture of tools and fixtures to be improved and speeded up, since this work is not confined to metal cutting machine tools. Auxiliary workshops equipped with NC machines can also have an indirect effect upon improving the productivity of forming operations and the production of plastic components since they can produce press tools and dies more rapidly. The same applies to the computer-aided production of cam discs, cam drums, etc., which can considerably extend the areas of application of automatic lathes and similar machines.

t_4 In theory, the use of long-distance data transmission techniques (371)
and t_7 makes it possible to gain access to a network of large computers so that even small firms can gain advantage from using the large production-orientated languages (such as APT or EXAPT, Chapter 11). At present this procedure is, however, unlikely to be widely employed in practice. There are two main reasons for this:
    1.   Transmitting the source programme and the control programme to and from the remote data-processing centre not only occupies time (which may be difficult to spare), but also may prove very expensive — depending on the postal charges.

Costs thus arise which under correct accounting procedures should be allotted to the individual cost centres, and which may become appreciable in one-off or small batch production, where the full cost has to be borne by a single workpiece or a small number of workpieces.

2.   Even if a processor for the programming language selected is available at the computing centre, it will, in many cases, be doubtful whether the post-processor will be available to suit the particular machines that are to be used. It requires a highly organised and experienced organisation, together with extensive standardisation of the technical equipment (Chapters 7, 8, 9, 10, and 11) to make an arrangement of this type work successfully and to keep it competitive. Although at present it is not clear just what the future developments will be, the small computer will undoubtedly play an important part in this area, where its use will often counteract centralisation.

t_5 Since the work covered by this time interval is performed at different centres, two symbols have again been introduced. The coding station K has been discussed in detail in Chapter 10; while the computer operator 0 in the data processing department also affects the time required for a job to pass through the system.

t_6 The actual computing time is very short, especially where large computers are employed. It enters into the cost of the operation only when combined with the hourly charge for the computer; compared with the other comparatively lengthy times t_6 has no appreciable effect on the total time t_d.

t_8 This covers the actual machining times; if the workpieces are ideally suited to the process and if suitable NC machines are used, these can be reduced to 10 to 15 per cent of the machining times required using conventional machines. This is the area where the reduction in the total processing time is at present most clearly apparent. Furthermore, there is the rapid changeover of NC machines to a different programme, due especially to the elimination of setting and marking-off times. Table 31 shows clearly that various other departments in the plant have to be reorganised to enable the overall value of t_d to be reduced. Another criterion is the 'constant availability of the NC machine'. Certain organisational measures must be taken to assure this; these include proper training of the staff, introduction of a preventive maintenance system, provision of an adequate stock of spare parts, an effective production control system. (10) In view of their importance, production control organisation and control of materials will be discussed in greater detail in Section 12.4.

t_9 As a result of built-in measuring systems and the elimination of the human fatigue factor, the scrap proportion is often considerably reduced when NC machines are used; inspection need be performed on a sampling basis or on the first component only to be produced. In Fig. 288 the quality control department is, therefore, shown dotted to indicate that it is used only occasionally. On the other hand, all the possibilities for automating the quality control function (for example, by the use of inspection machines, Figs. 89, 91, 92) have been left open; if these are employed either t_9 will be reduced or the quality control function will be improved.

Adding the various times for these individual functions will give the total time t_d. While it is important that this be reduced, inspection of the nine time subdivisions of which it is built up will show that the NC machine itself can have only a limited effect in doing this. It may, however, act as a form of 'pacemaker' to the plant as a whole; the major changes required in various parts of the organisation will then follow. Efforts should next be made to reduce the times t_1 and t_2, in particular. Further analysis of this aspect is beyond the scope of this book. Subdividing the total time in this way has thrown a great deal of light on the subject of 'integrated data processing'; and it also has another advantage; it provides a better indication of the changed cost structures and the whole operation of the plant becomes clearer. This idea will be followed up in the next section.

12.3   Introduction into the method of simultaneous examination

So far the subject has been considered solely from a technical point of view. Every automation problem has, however, also economic and social aspects. Only if all three areas are considered together and are paid the same degree of attention, is it possible to form an adequately accurate impression of the subject as a whole (383); Fig. 290 is intended to illustrate this point.

The total time t_d having been subdivided into clearly differentiated time intervals, the next step is to allot a cost factor C to each recognisable work area, and to select, say, one hour as the time unit. In the first instance some of the cost factors can only be given average figures based on the statistical records from previous years — for example, 'cost of 1 hour of design office time', 'cost of 1 hour of planning office time', etc. Not until computers become more widely used in these fields of work will it be necessary to adopt more refined methods of calculation, differentiating between machine activities and human activities. The current uncertainties of the use of small computers in technical offices, or the connection of these offices to large time-sharing systems (252), make it advisable to progress gradually towards

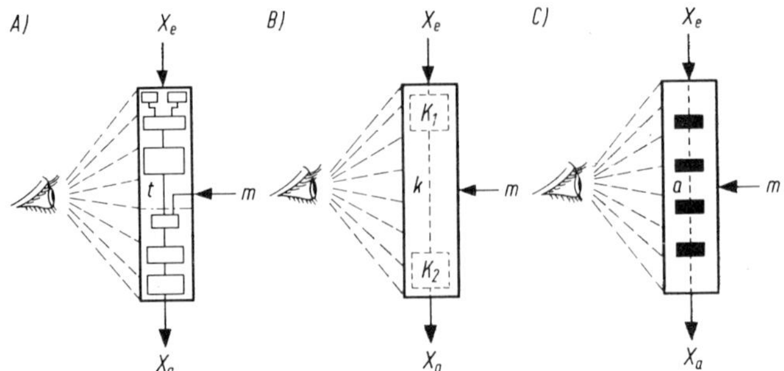

Fig. 290 Principal methods of examining the system as automation is extended as a part of the integrated data processing (383)

A) Technical method of examination based on times required (the internal structure can, for example, be as shown in Fig. 288). B) Economic method of examination based on costs. C) Industrial sociological method of examination based on the structure of the workplace (mixed man-machine systems).

t Time structure. x_e Input value (e.g. manufacturing order). x_a Output value (e.g. finish-machined workpiece). a Workplace structures. k Cost structures. k_1 (Data processing installation) and k_2 (e.g. numerically controlled machining centre) as particularly cost-intensive automatic installations. m Material input values (e.g. raw material, tools, fixtures)

Note: The symbols chosen for the values x_e and x_a are based on control system conventions to emphasise the close relationship that exists with the cybernetic approach

a condition that corresponds to the plant conditions. In any event, it is necessary to give early thought to the methods to be adopted for establishing a charge basis for the capital-intensive equipment used in technical offices. For this reason personnel costs and equipment costs are considered separately in the examples that follow to illustrate the problem. Comparatively little difficulty is presented by times t_6 and t_8, which have been familiar concepts for a long time. It is usually simple to obtain a figure for the 'computer cost per hour' for a hired computer; even if the computer has been purchased it will be possible to agree on a standard basis for the calculation of the hourly charge. Even simpler and more familiar is the use of machine costs per hour in the production field. On the cost side, the time t_8 may, however, cover three different cost factors, namely a $C8.1$ for the operator, $C8.2$ for the machine setter, who may be required by the features of the system to be in attendance for short periods, and $C8.3$ for the machine which is automated to a greater or lesser extent. In such an instance it will be found very useful to adopt a standard machine-hour rate as proposed in (374).

After covering the technical and economic factors it is possible to go a step further. In principle, dividing up the total time into time intervals also makes it possible to describe 'work areas' for which differing qualifications are required. Unfortunately nearly all the descriptions and definitions of work places produced up to the present have been concerned with operatives and tradesmen; the new and evolving technical jobs, such as part programming, the work of computer operators, design involving 'computer-assisted design', and so on have been completely neglected. This opens up another field of activity for inter-disciplinary research. The problems are not primarily those of remuneration, but rather of the mental work content and psychological factors, together with the training opportunities for the novel man-machine communication problems and their relationships with other activities. It is beyond the scope of this book to list in detail these problems in the field of work study and industrial sociology; the intention is merely to point out one possible way in which the outstanding questions can be dealt with in a unified manner.

To save space only three examples of systems will be given here; further discussions can be found in (10).

Figure 291 shows a three-part system diagram for the prototype of a group of conventional automatic machines. The binary code numbers in the figure cover the range

OOLOLOOOOO to
OOLOLLOLLL.

These machines are shown as Stage 1 in Table 30. Their main common features are:

a) No direct use made of a computer in production. The machines are not numerically controlled and are provided with rigid displacement information stores, which must be manufactured in an auxiliary workshop, H. All the remaining operations can be performed largely automatically; the workpiece can also be changed automatically.

b) The rigid displacement information stores can take the form of cam discs, cam drums, patterns or templates which control a follower, press tools, etc. As secondary effect, the production of these in the auxiliary workshop, H, may be speeded up by NC machines.

The machines under consideration include the following types:

Single-spindle automatics, multi-spindle automatics, copying lathes, copy-millers, punches and presses, plastics presses. The principal time intervals are likely to be t_3 and t_8, it being necessary to distinguish between two cases in t_3, namely the production of a new store and the provision of an existing store in the event of a repeat job. The machining time t_8 is usually comparatively long in metal-cutting operations, and short in metal or plastic presses; while in most instances t_3 will be quite long (new production necessary).

The cost diagrams V are drawn to a logarithmic scale, since the cost ranges encountered can extend over three decades. The cost factors are subdivided in both time interval t_3 and time interval t_8. In the first case, $C3.1$ relates to the cost of making a new pattern or cam; while $C3.2$ represents the storage costs for the patterns, etc. used for parts that are repeated. $C8.1$ represents the wages of the machine operator (including extras and contingencies; while $C8.2$ represents the wages of the machine setter who may be required for short periods, depending on the class of machine; $C8.3$ is the hourly cost of the machine used.

Conditions become more critical in the workplace diagram W. The principal difficulty is that there is as yet no standard and comparable scale of qualifications for office work and manual work. The small number of existing investigations into this subject so far (e.g. 382) do not give sufficient information. The problem is examined here as part of the system and possible solutions are taken into account. A considerable amount of research will be required, together with an exchange of experience and discussions with both management and labour and much goodwill, if answers satisfactory to all sides are to be obtained. To continue with these examples, and to enable the items for which solutions have yet to be found to be pin-pointed, an arbitrary 12-point scale has been assumed, and a rough estimate has been made of the features of the different jobs and of the corresponding degrees of difficulty. With Fig. 292, the field of NC machines has been reached. Apart from the decision as to whether or not automatic workpiece changing is to be adopted, the stages have been considerably narrowed down due to:

a) Automatic displacement data processing, the use of punched tape, automatic processing of switching information, continuous-path control, and automatic tool changing, but no adaptive control. This could be a modern machining centre of the type shown in Figs. 154, 156, 159, 308, or 310.

b) The external data processing, which in this instance has deliberately been made simple; programming is of either the manual or the semi-machine variety (Chapter 10); no computer for the 'complex' programming languages is therefore required.

A sub-division of cost structures has been made in $C8$. $C8.1$ represents the wages of the operator and $C8.2$ the wages of the machine setter who may be required because of the use of a more complex machine. Opinions differ on this subject; some production engineers consider the best solution is to employ a responsible skilled man as the operator, in which case $C8.2$ is not required, but $C8.1$ will be higher. Others state that in their experience a conscientious operator, who need be only semi-skilled, will be sufficient ($C8.1$); he will, however, need the advice and assistance of a highly-qualified man for setting the machine and in the case of an emergency ($C8.2$, part-time). But there has as yet been too little experience gained, especially with continuous-path controlled machining centres, for providing any generally valid answer to this type of question. Here attention shall merely be drawn to the nature of the problems that arise. The hourly machine costs

C8.3 for the complex machine will be high, and, in particular, they will be higher than the wage costs. The assumed values have been inserted in the workplace, diagram W.

Finally, Fig. 293 shows a basic system which covers a comparatively wide spectrum of usage of automatic machines (binary codes from LOLLLOOOOO to LOLLLLOLL). Characteristics of this are:

a) Employment of a computer for production engineering purposes in conjunction with data carriers (punched tape or magnetic tape); continuous-path control without adaption attachments but possibly automatic tool and workpiece changing (the version with automatic workpiece changing is not shown in the figure).

b) A large computer without remote transmission of data for use in the external data processing system, employing processors and post-processors for production engineering programme languages (e.g. APT or EXAPT).

c) The only additional automation stages that are still missing are adaptive control and direct numerical control (Chapters 8 and 9).

Since a large computer and machining centres are employed, it is assumed that in this plant the design office also makes some use of computing and drawing aids (380); in the cost analysis, $C1$, a subdivision has therefore been made into personnel costs, $C1.1$, and equipment costs, $C1.2$, the latter being the higher. It is possible either to apportion part of the costs of the central computer, operated on a time-sharing basis, to the design office, or the technical department can have its own computer. In either event the system illustrated by these diagrams is of a comparatively advanced standard for 1970.

The combinations that are shown still further to the right on the topological tree of Fig. 289 are to some extent future predictions. Combinations of this type are technically possible, but it is not known exactly when they will prove economic and be put into practice, or for what purposes they can be used effectively.

It is possible to make some qualitative predictions of the changes that will take place in the workplace structure as a result of the automation of production processes with the aid of NC machines, computers, and similar appliances (376, 377, 378). The guidelines for long-term investment programmes can also be determined more readily if it proves possible to simulate, in advance, the whole of the future operation of the plant (covering technical, economic, and labour requirements) with adequate accuracy. This is quite apart from the fact that it provides a good method of obtaining economic comparisons for the use of NC machines. The best method to adopt – if reliable predictions are to be obtained – is to sum the partial products 1 to 9 for the passage of several workpieces through two or three basic systems for different combinations of machines and appliances and for different workpieces:

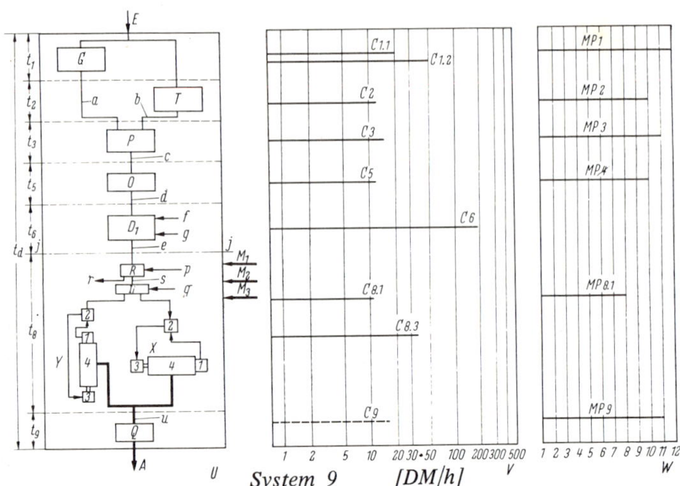

Fig. 291 Example of a system comprising a non-numerically controlled, automatic machine with rigid information stores (cam-controlled automatic lathe, copy-milling machine, etc.)

A	System output, inspected finish-machined workpieces
B	Central process computer for 'on-line' control of several machine tools (Stages 11 and 12)
C	Costs/hour, where appropriate sub-divided into wages, storage costs, and machine costs
$C3.1$	Information store production
$C3.2$	Information store storage
$C8.1$	Operator
$C8.2$	Setter
$C8.3$	Hourly machine cost
D_1	In-plant data-processing installation (computer)
E	System input, production order for workpiece or batch
G	Determination of geometric details of workpiece, design, possibly with aid of D_1.
H	Auxiliary workshop, tool room for production of information stores (cam discs, cam drums, templates, patterns, drilling jigs, etc.).
I_1	Internal interpolator
K	Coding station (automatic typewriter, etc)
L	Manual programming station, possibly with desk calculator assistance, semi-automatic programming (Chapter 10).
M_1	Raw material, semi-finished products
M_2	Tools
M_3	Jigs and tools, especially clamps
MP	Man power
N	Machine tools with automatic tool changing
O	Computer operator position
P	Computer programming position
Q	Quality control, inspection
R	Punched tape reader and control console for an NC machine (Chapter 9)
T	Determination of technological details for production of the workpiece. Work planning, possibly with aid of D_1

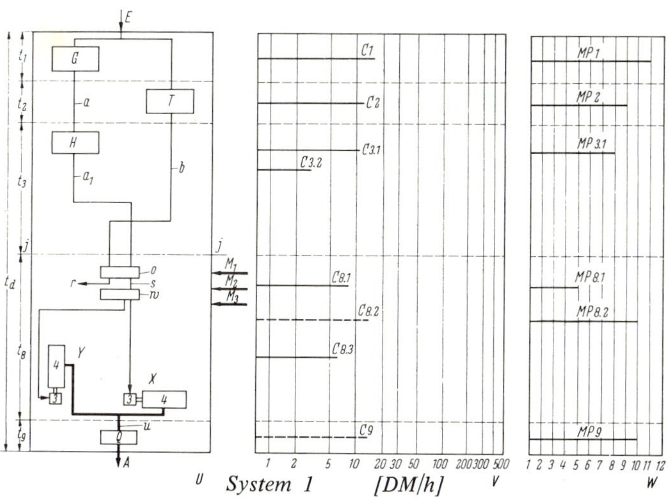

Fig. 292 Example of a system comprising an NC machining centre with continuous-path control and manual programming (e.g. as shown in Fig. 308)

U	Technical diagram, data flow diagram
V	Cost diagram for costs per unit time (hour), cost factors, charges
W	Workplace diagram, grading by difficulty of tasks
X,Y,Z	Coordinates axes
a	Design drawing, information store for geometric details
a_1	Geometric details held in rigid information stores (e.g. patterns)
b	Planning department instructions, information store for technological data
c	Programme schedule, source programme
d	Stack of punched cards, punched tape 1 for computer
e	Control punched tape, punched tape 2 for machine tools
f	Processor ⎫ Compiler for production engineering
g	Post-processor ⎭ programming languages
j-j	Boundary between internal and external data processing
o	Manual input station for all work information, the sum of all controls on the machine
p	Manual input of corrections at control console
q	Manual input of corrections in internal interpolator, feed corrections
r	Switching information in general (e.g. gearbox shifts, apply and release clamps, switch on coolant supply)
s, s_1, s_2, s_3	Displacement information (desired values for slide positions)
$t_1 .. t_9$	Partial times for passage through system, time intervals for individual processing stations
t_d	Total time for passage through system, system-specific time
u	Finish-machined workpiece
w	Rigid information store (from H) fitted to machine with analogue follower (e.g. copying by tracer, cams)
1	Displacement measuring systems
2	Comparator
3	Drive systems (stepped or infinitely variable)
4	Machine slides (including guideways, etc).

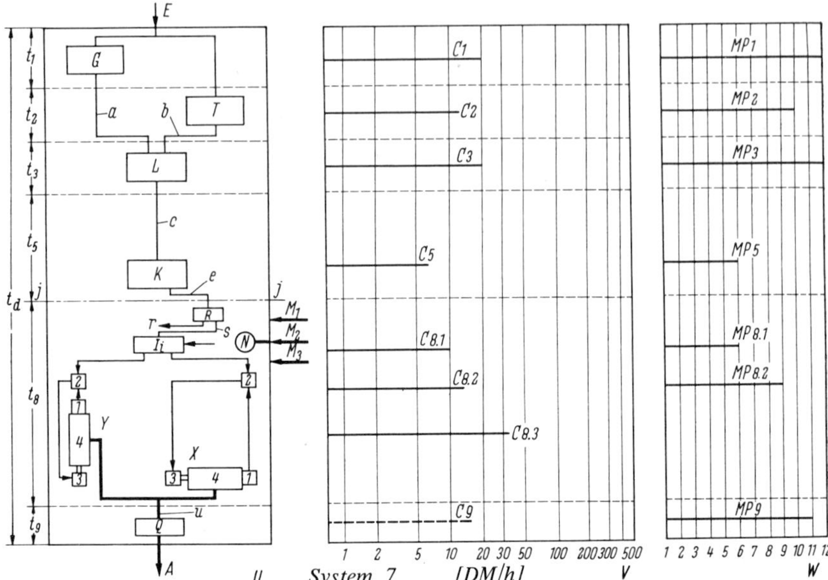

Fig. 293 **Example of a system comprising a continuous-path NC machine (e.g. lathe as shown in Fig. 305) used in conjunction with machine programming**

$$\sum_{i=1}^{i=9} t_i \cdot C_i = t_1 \cdot C_1 + t_2 \cdot C_2 + \ldots t_9 \cdot C_9.$$

It is possible to make certain simplifying assumptions, and so obtain approximate solutions; because of the large number of parameters, it will, however, not be possible to avoid completely the use of a computer if the results are to be reasonably reliable (379).

It is not sufficient to consider the basic systems in isolation; an attempt should be made to estimate, at least in broad outline, the effects that the changes in the material flow pattern will produce. The ideas involved will be analysed in the last section, even though they are not directly a part of the numerical control techniques that form the main subject of this book.

12.4   Higher order systems

As the preceding remarks have shown, the idea of gathering together into one basic system the whole industrial process from the initial concept (design instruction) to the finished workpiece ready for service is clearly very fruitful.

Even the rough subdivision of basic systems into the internal and the external data processing, which was introduced in 1960, has proved very useful (18). It also offers the opportunity of associating one defined external data processing system, which represents a possible form of organisation, with several production machines having various degrees of automation, and so closely simulating the actual practical situation in a factory.

Another question remains however completely open in principle. In a factory the problem is not merely to produce individual components efficiently, but also to transport them, to store them, to marshall them, to build them into sub-assemblies, to store these, to assemble these into complete assemblies, and in the end to sell the finished products. In short, a much more complicated multi-stage production process, rather than a single-stage production, such as has been considered up to the present, is almost invariably the problem in practice. In the long run, it is only by surveying a more complex system, such as a factory, a company, or even a group of companies, that it is possible to decide whether a new investment, for example in an NC machine, represents progress or not. An attempt is made below to derive a usable criterion for solving this problem by developing the system-based method of examination. The definition of the term 'system' in DIN 19226 (12) makes provision for systems to be joined together as well as broken down. Basically, it is desirable to regard the production process as a control loop, as shown in Fig. 294 (253).

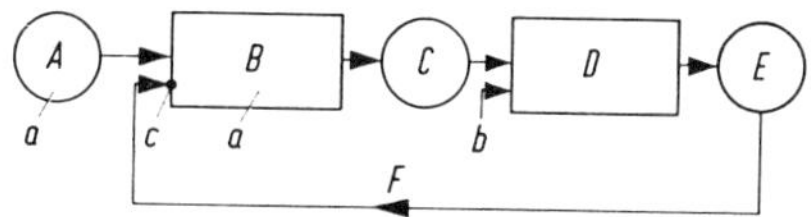

Fig. 294 Block diagram showing the principle of production control (253)

A Desired production programme. *B* Planned timetable — based on available capacities. *C* Timetable. *D* Production — manufacture of orders included in timetable. *E* Actual production results. *F* Feedback

a Causes of error 1 (e.g. state of ordering wrongly estimated). *b* Causes of error 2 = production difficulties (scrap, tool breakage, machine breakdown, etc). *c* Correction point

In Fig. 295 an attempt is made to subdivide the behaviour of the plant into systems to such an extent that the derivation of a usable mathematical expression is possible. Since the systems discussed earlier were referred to as basic systems, which could also be defined as 1st order systems, it is now necessary, in order to preserve the analogy, to speak of 2nd order, 3rd order, and higher order systems; there is no theoretical, merely a practical upper limit to the number of orders of systems. Assume that a group of n basic systems B is arranged organisationally and spatially in such a way that, for example, it could be used to produce all the components of a machine assembly. The machine tools available must then comprise B1, B2, . . . Bn lathes, milling machines, drilling machines, etc. installed in a mixed arrangement in such a manner that they can produce for example, all the parts required for assembly $e1$ with a minimum of transport distances; the machines employed may be NC machines, but this is not essential. A parts store g of adequate capacity is also needed to act as a buffer store to ensure that assembly area $f1$ and the individual machine tools can work continuously. The assembled components are placed in a further buffer store $h1$, and are ready to be drawn out as required for the final assembly stage c.

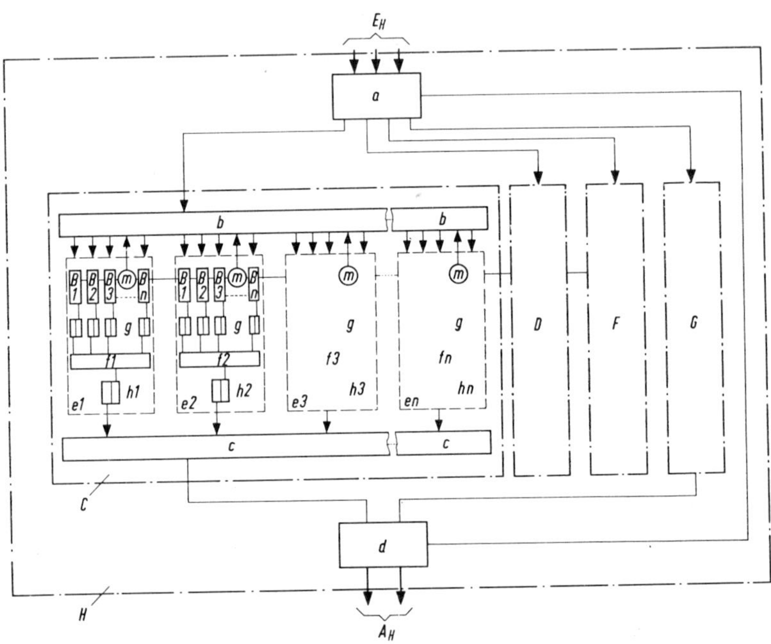

Fig. 295 Proposal for the use of systems engineering in an industrial application based on the definition of the term system in DIN 19226 (372)

A_H	Output values from system H (3rd Order system)
B	Manufacturing basic systems (1st Order systems, similar to Fig. 291, etc.).
C	Total manufacturing system (2nd Order system)
D,F,G	Examples of different departments of the company, e.g. purchasing, personnel, wages office, financial and commercial departments, etc.
H	Total company as a 3rd Order system
a	Company management
b	Plant management
c	Final assembly
d	Sales department
$e_1, e_2 .. e_n$	Product system 1,2,.. n (e.g. finished assemblies of various types)
$f_1, f_2 .. f_n$	Assembly areas for assemblies 1,2,..n
g	Parts store
$h_1, h_2 ... h_n$	Finished stores for assemblies 1,2..n
m	Measuring stations in the plant to provide information for production control system (data on finished parts, scrap, machine damage, etc.)

The same procedure can take place with assemblies e_2 to e_n, so that at certain time intervals finished product of type p_1 can be supplied to the despatch department d. It is also possible to solve this problem in a different way. It is, for example, possible to produce all the turned components produced throughout the plant in the first group of basic systems, making use of the most modern automatic machines of all types (single-spindle automatics, multi-spindle automatics, copy-turning lathes, NC lathes with turrets, NC lathes with magazines, etc) to produce the parts as economically as possible; these are then stored in the buffer stores g from which they are distributed to the various assembly areas $f1, f2.. fn$. All box-shaped parts, for instance, could then be made economically in the next group of basic systems, upon which they are stored in suitable buffer stores g and also distributed to the assembly areas. A similar procedure would be followed with flat parts which only require to be drilled. A typical feature of this type of production lay-out is that transport problems for individual parts have to be solved in a completely different manner than for the type of organisation described above. Important features common to both types of organisation are:

a) The advisability of forming groups of production units.
b) The central issuing of production orders by the plant management b
c) The necessity of monitoring each group by a measuring station m to provide information on the utilisation of the available capacity, queues, delivery dates, fault reports (machine break-down, illness, scrap, etc.).

All three functions can be delegated to computers, and the basic question again arises as to whether the problem should be solved centrally for all production groups using a large computer; or whether the work should be decentralised and several small computers employed, the work of which can be coordinated by means of a central process computer (253). Technically, the arrangement of the measuring stations (124) becomes easier the larger the number of NC machines in a production group, since the information on the current state of the machine can be picked up more easily on NC machines. The different batch sizes and the variety of parts that make up a parts spectrum, make it likely that in practice only a certain percentage of NC machines will be in use in a plant. For this reason, different solutions have to be found to enable production control systems (in the control engineering sense, with permanent feedback) to be built up. Details of work on this have been published in sufficient quantity elsewhere (124, 253, 227, etc).

Here a different approach may be of value. It is possible to regard the described manufacture of assemblies as a 2nd order system; the machine layouts illustrated in Figs. 175, 177, and 178 point in this direction. Here, however, a wider view will be taken, and the whole of the plant is defined as a 2nd order system. It includes now the whole of the production system which may be automated to a greater or lesser extent, and which extends from the technical plant management to the final assembly of the products, which may also vary. The parts stores g and h are now additional parameters for the cost minimisation calculations. By using process computers in the plant management h (decentralised or centralised (253) it is possible to take rapid steps by automatically altering the flow of materials) for overcoming any difficulties (machine breakdowns, scrap, absence of workers, etc).

The most important technical-economic problem then becomes the minimisation of stocks (tied-up capital) and

the machine waiting times, as well as

the effectiveness of the control loops that cover the whole plant and which are an organisational man-machine problem.

It is necessary to draw attention here to a consequence of the economic calculations where NC machines are in use. A significant reduction in the cost of all three items may be achieved even if a few NC machines only are in use. The increased flexibility (reduced total passage time t_d) enables unnecessary intermediate stores to be abolished, and at the same time the flexible automatic machines can form an 'emergency reserve' for use if the more inflexible machines should be out of action for technical or personnel reasons. These potentially cost-reducing factors could not be recognised when 1st order systems (basic systems) were being examined, and certainly could not be put into effect. When discussing the economics of NC machines it will, therefore, be necessary to distinguish clearly between

> 1st order cost minima, and
> 2nd order cost minima.

The problems that arise are then much more complex. The whole of the plant organisation must be re-planned and the possibility of using process computers for production control must be taken into consideration. If the plant is to remain financially healthy, these extra costs will have to be balanced by the savings produced. Here again it ought to be possible to make use of simulation techniques to check the capacity of the proposed plant. And to predict the effects that planned investments (including the use of larger or smaller process computers) and re-organisation (for example, arrangements for producing assemblies with a minimum of transport times) would have.

Finally, it is possible to carry the idea of creating systems still further. Up to the present the technical departments only have been considered as production factors. But a plant, a company, or a group of companies, also contains other departments, such as purchasing, personnel, wages office, accounts, commercial department for bought-out components. These make no use of NC machines, but they do require effective computers for stock control, etc., together with skilled and experienced staff. All those components that are located between the company management and the sales department can therefore be incorporated into a 3rd order system H. In this manner a further factor affecting the economic calculations of NC machines has come to light, which has hitherto not been apparent. The more flexible and trouble-free the combination of the basic systems with the material flows, the more scope there is for the larger system H, and it is possible to make products which cannot be supplied by less well-organised systems. All aspects of plan fulfilment (e.g. meeting delivery dates) are then simplified; this is an indirect accomplishment of NC machines which is, admittedly, not immediately apparent, and which is not directly reflected in an improvement of the cash flow situation.

V   Appendix

Examples of numerically-controlled machine tools

These examples are intended both to supplement some of the remarks made in the earlier chapters by showing examples of complete machines, and to illustrate means of arranging the widely differing types of machine in a generally-acceptable pattern, along the lines indicated in Chapter 12. Depending on the construction of the machine and its control system, there will be very different 'degrees of linkage' between the various NC machines and the remainder of the plant (10, 384). The higher the 'degree of linkage', the more it is necessary to make allowances for the existence of numerically-controlled machines when deciding on the plant organisation that is to be adopted, or, in other words, the greater the effects that the organisation adopted has on the efficient use of the numerically-controlled machines. An attempt has been made to illustrate this line of thought, which somewhat simplifies the ideas proposed in Chapter 12, by the sequence in which Figs. 296 to 316 are presented. In broad, general terms, it is possible to distinguish between four main groups of machines and their prospective users.

Group I, Figs. 296 to 301

There is virtually no programming in the sense in which this term is used in Chapters 10 and 11. Acquisition of these machines necessitates virtually no change in the plant organisation; for technical reasons there is no possibility of combining them with computers.

Group II. Figures 302 to 306

These five machines can be taken as being representative of the great majority of standard installations of NC machines. The linkage with the remainder of the plant organisation is quite marked (tool pre-setting, tool card index, maintenance service, workpiece card index, perhaps machine programming, etc.) (10).

Group III. Figures 307 to 310

This group covers machines for more advanced applications, and the plant has to be organised on NC-orientated lines if economic success is to be achieved. (10)

Group IV. Figures 311 to 315

The purpose of these five figures is to illustrate that NC machines are indispensable for special purposes; it was the existence of these problems in 1950/55 that led to the need for developing NC machines in the first instance. Their number is, however, likely to be restricted to certain specific applications.

The last machine shown in Fig. 316 is beginning to move away from the direct utilisation of machine tools into the sphere of integrated data processing. In a certain sense it is also possible to speak of a step towards 'geometric data processing'; here an attempt is made to derive a punched control tape for press tools directly from a prototype, without using any programming language. This is a development which is currently coming into use in the motor car industry.

Whereas the machines shown in Groups I and IV are to some extent special cases, with the control systems being produced in many instances directly under the control of the machine manufacturer, the majority of NC machine tools will fall into Groups II and III, and can be equipped with control systems to suit the customers' requirements.

The illustrations can cover only arbitrary selection of machine tools, and cannot represent complete coverage. The choice of machines is also not to be taken as a comment on their quality; the intention is merely to choose examples that illustrate some current designs and applications.

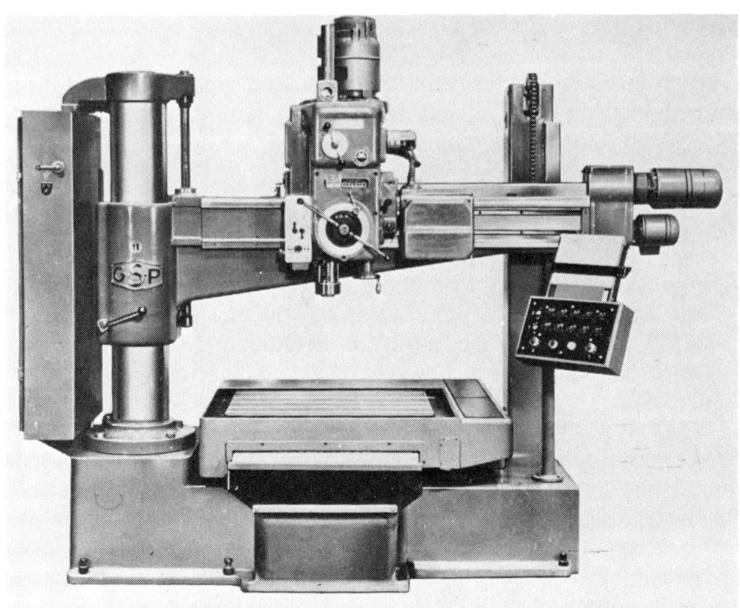

Fig. 296 Simple coordinate drilling machine with manual input of all machining information. Two axes are numerically controlled and can be programmed in sentences (using coordinate words only) by means of decade switches (cf. Fig. 219).

Manufacturer of machine:	*Ateliers GSP* *Guillemins, Sergot, et Pegard*
Address:	F-92, Courbevoie/Seine, France.
Type of Model:	Coordinate drilling machine 405 P8N
Capacity	Mounting surface 1030 x 800 mm
Slide movements:	*X*-axis, 600 mm numerical *Y*-axis, 800 mm numerical *Z*-axis, 325 mm fine adjustment of drilling spindle using hand-wheel. *W*-axis, 585 mm coarse adjustment of transverse slide by hand along column.
Number of numerically-controlled movements:	2 (*X-Y* plant)
Type of control:	Positioning control, absolutely direct.
Manufacturer of control system:	*GSP*

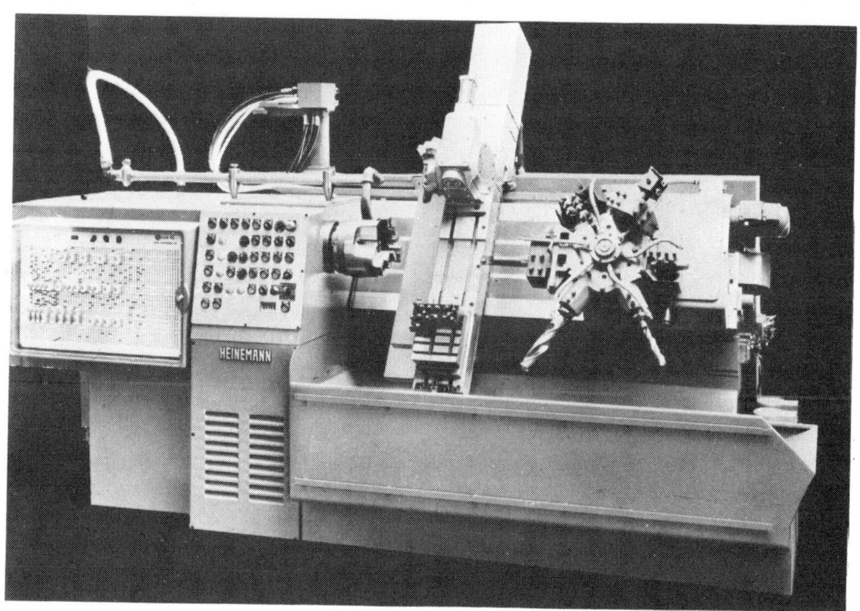

Fig. 297 Turret lathe with manual input of all machining information, numerical-control
for three axes and capable of being programmed for a complete work cycle by
means of a cross-bar distributor (cf. Figs. 136, 220).

Manufacturer of machine:	*Gebr. Heinemann AG*
Address:	D-7742 St. Georgen/Schwarzwald
Type or Model:	RAN 63 NC lathe
Capacity:	Spindle bore 63 mm
Slide movements:	Longitudinal slide, 645 mm numerical (*Z*-axis) Transverse slide, 300 mm numerical (*X*-axis) Turret slide, 660 mm numerical (*W*-axis) Tool corrections by decade switch
Type of control:	Straight-line control with stepping motors
Manufacturer of control system:	*Siemens AG*

Fig. 298 Turret lathe with flat table turret, numerical straight-line control in two axes and simplified programming of work cycles (can be performed at machine) with aid of plastic cards (cf. Figs. 131, 222).

Manufacturer of machine:	*Pittler Maschinenfabrik AG*
Address:	D-6070 Langen/Hessen
Type or Model:	PINUMAT
Slide movements:	Longitudinal slide, 460 mm numerical (Z-axis)
	Transverse slide, 560 mm numerical (X-axis)
	Max. workpiece dia. 315 mm
	No. of tools: 4 (with double cutters)
	Tool correction by decade switches.
Type of control:	Straight-line control
Manufacturer of control system:	*Pittler AG*

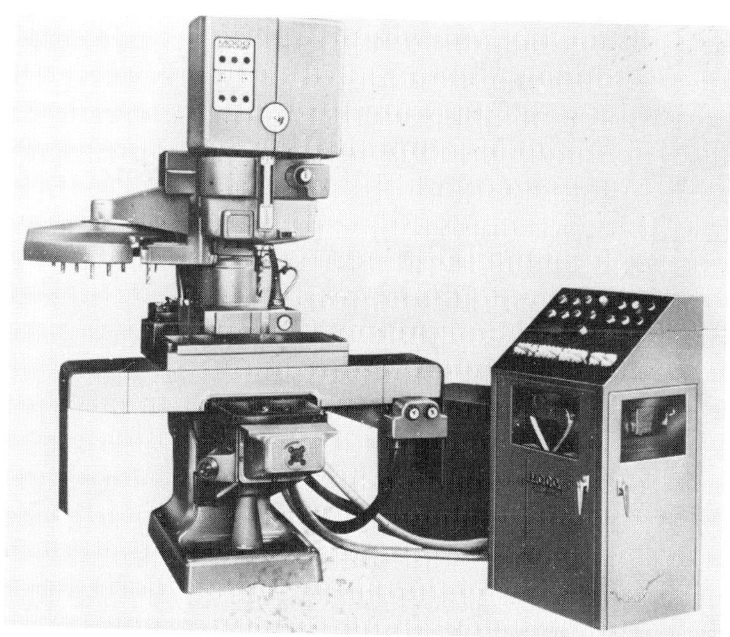

Fig. 299 Pneumo-hydraulically controlled machining centre with manual data input and preparation of punched tape in control console (cf. Figs. 121, 122, 123, 239).

Manufacturer of machine:	*Bridgeport-Moog Hydra-point*
Address:	Moog Ltd., Hydra-point Division, Cheltenham, England, P.O. Box 8.
Type or Model:	Model 1000 MC, Machining Centre
Capacity:	Table size 230 x 760 mm
Slide movements:	500 mm (X-axis) 250 mm (Y-axis) numerical 125 mm (Z-axis) The machine is also available without a tool magazine, and without automatic tool changing (cf. Fig. 123).
Type of control:	Straight-line control (pneumo-hydraulic)
Manufacturer of control system:	*Moog Inc., Hydra-Point Division*

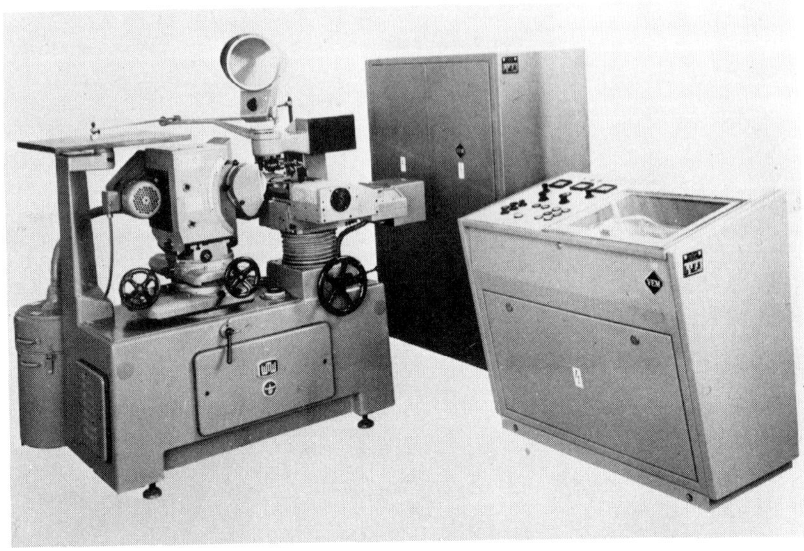

Fig. 300 Optical profile grinding machine with unrestricted manual control using projector and with punched tape control system (cf. Fig. 183)

Manufacturer of machine: *VEB Werkzeugmaschinenkombinat*
                             'Fritz Heckert', Karl-Marx-Stadt,
                             Betrieb Mikromat Dresden
Address: DDR-8036 Dresden
Type or Model: SWPO 80 NC
Capacity: Grinding wheel dia. 160 mm
Area within direct reach: 250 x 70 mm numerically-controlled
                             10 x 10 mm manually controlled
                             with pantograph
Type of control: 2-D continuous-path control,
                             Code as in Fig. 255.

Manufacturer of
control system: *VEB Starkstrom-Anlagenbau*
                             Karl-Marx-Stadt

Fig. 301 Large horizontal precision jig borer with optical and numerical controls for one-off and small-batch production (cf. Fig. 223). Accuracy to within 0.001 mm or ±1 second of arc.

Manufacturer of machine:	*DIXI S.A./Usine 2*
Address:	CH-2400 Le Locle/Switzerland
Type of Model:	DIXI-5S optical jig borer
Capacity:	Table dimensions 1250 x 1700 mm
	Max. table loading 5000 kg
Slide movements for large	Transverse 2000 mm (X-axis)
Type 5S with hydraulic	Longitudinal 1000 mm (Z-axis) ⎫ manual or
motors and recirculating	Vertical 1400 mm (Y-axis) ⎪ numerical
ball spindles; all smaller	Turntable rotation (B-axis) ⎬
types with hydraulic	360° ⎭
cylinders):	Table optical systems adjustable
	to within 0.001 mm
Type of control:	4-axis straight-line control,
	punched tape for repeat jobs
	prepared in control console.
Manufacturer of control system	*DIXI, S.A. – VIDIMATIC*

Fig. 302 Multi-spindle plate drilling machine for the production of heat-exchanger end
plates (condensers, cf. Fig. 180)

Manufacturer of machine:	*Herm. Kolb, Maschinenfabrik*
Address:	D-5000 Cologne 30
Type or Model:	MPN8 multi-spindle plate drilling machine
Capacity:	Power per spindle 38/46 kW Max. drilling force per spindle 2000 kp
Slide movements:	Numerical bridge travel min. 5000 mm (X-axis) Numerical drilling head traverse 3500/5300 mm (X-axis) Cam-controlled spindle stroke 350 mm (Z-axis)
Type of control:	Positioning control
Manufacturer of control system:	*AEG-Telefunken*
Programming:	The large number of holes (approx. 4000 to 7000) can be programmed economically only by machine.

Fig. 303 Horizontal boring and milling machine.

Manufacturer of machine:	*Collet & Engelhard,* *Maschinenfabrik GmbH.*
Address:	D-6050 Offenbach/Main
Type of Model:	Collet-Minor
Capacity	Table size 700 x 800 mm
Slide movements:	X-axis 800 mm ⎫ Y-axis 700 mm ⎬ numerical Z-axis 700 mm ⎭ Axial adjustment of spindle: 300 mm. If required can be fitted with 22-position tool magazine
Type of control:	Straight line control
Manufacturer of control system:	To customers' requirements As illustrated, *Siemens A G.*
Programming:	Machine or manual.

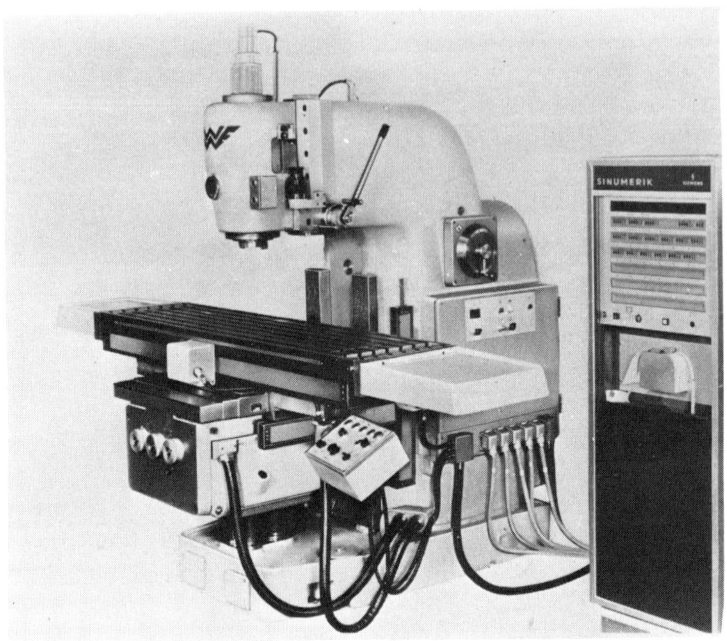

Fig. 304 Knee-type milling machine with three-dimensional straight-line control

Manufacturer of machine	*Wanderer-Werke AG*
Address:	D-8013 Haar bei Munchen
Type of Model	KFV 401 NC
Capacity:	Table surface 1600 (2000) x 400 (450 mm)
Slide movements:	X-axis 1230 Y-axis 380 mm } numerical Z-axis 465 mm
Type of control:	3-D straight-line control
Manufacturer of control system:	To customers' requirements, as illustrated, *Siemens AG*
Programming:	Machine or manual

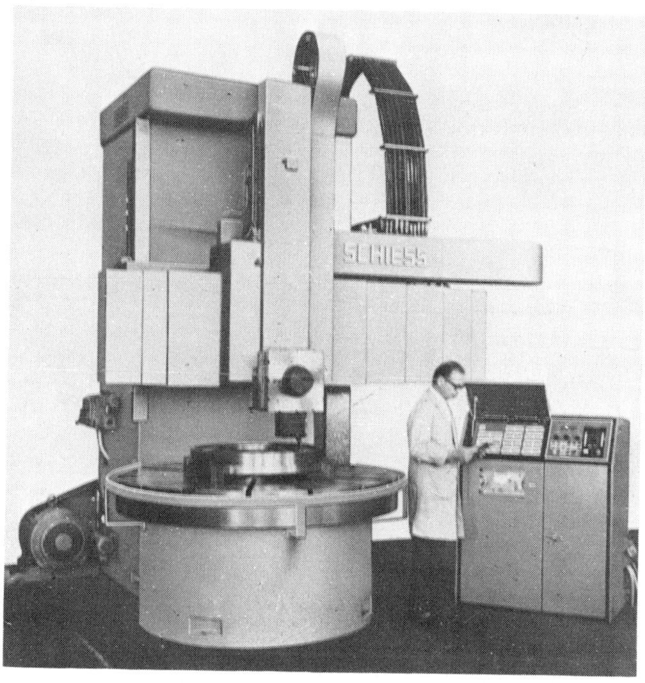

Fig. 305 Medium-size vertical lathe with continuous-path control, Type 20 DKE 180 NC/C

Manufacturer of machine:	*Schiess AG*
Address:	D-4000 Dusseldorf-Oberkassel
Type or Model:	DKE NC/C Five sizes from
	14 DKE NC/C
	to
	30 DKE NC/C
Capacity	Turntable dia. 800 to 3000 mm
	Turntable loading 8 to 20 tons
	Max. torque 2100 to 5200 mkp
	Max. ext. dia. 1400 to 3000 mm
	Max. machining height 975 to 2500 mm
Type of control:	Continuous-path control. Straight-line control also available.
Manufacturer of control system:	To customers' requirements. As illustrated, *General Electric*.
Programming:	Machine or manual.

Fig. 306 Lathe with a 9-position tool magazine (cf. Figs. 145, 147, 166).

Manufacturer of machine:	*Vereinigte Drehbankfabriken e. V.* *Werk Gebr. Boehringer GmbH.*
Address:	D-7320 Göppingen
Type or Model:	P 800 NC
Capacity:	Power (main spindle) 30kW Swing over bed 560 mm dia. Swing over saddle 270 mm dia.
Slide movements:	Longitudinal, 500 to 1500 mm (Z axis) numerical Transverse, 560 mm (X-axis) numerical The machine can be equipped as required with a four-way turret or a 9-position tool magazine.
Type of control:	Continuous-path
Manufacturer of control system:	To customers' requirements
Programming:	Machine or manual

Fig. 307 Magnetic-tape controlled 3-D profile milling machine (cf. Figs. 83, 242, 243).

Manufacturer of machine:	*Droop & Rein,,Werkzeugmaschinen-fabrik*
Address:	D-4800 Bielefeld.
Type of Model:	FS 130 g NB
Capacity:	Milling spindle dia. 130 mm Table loading 6 000 kg Table area 2400 x 1000 mm
Slide movements:	X-axis 1900 mm ⎱ simultaneously Y-axis 950 mm ⎬ numerically Z-axis 800 mm ⎰ controlled
Type of control:	3-D continuous-path control with external interpolator
Manufacturer of control system:	*Ferranti, Dalkeith, UK.*
Programming:	Machine by manufacturer of control system.

Fig. 308 Example of a machining centre with 2 x 20-position tool magazine (cf. Figs. 146, 176)

Manufacturer of machine:	*Marwin Machine Tools Ltd.*
Address:	Anstey, Leicester, UK.
Type or Model:	MIN-E-CENTER MEC 3
Capacity (143):	Spindle power 16 hp Spindle dia. 128 mm Speed range 20 – 400 rev/min Table area 20 x 20 inches (500 x 500 mm), double Max. table loading 1000 kg Separate chucking and machining stations
Slide movements:	X-axis 84 in (2100 mm) numerical Y axis 21¼ in (530 mm) numerical Z-axis 18 in (450 mm) numerical D-axis 0 to 6 rev/min continuously variable or indexing in 5° steps.
Manufacturer of control system:	To customers' requirements. As illustrated, *Plessey.*
Programming:	Machine and manual

**Fig. 309 Linkable machining centre normally equipped with 50-position tool magazine
(cf. Figs. 158, 165)**

Manufacturer of machine:	*Ludwigsburger Maschinenbau GmbH (Burr)*
Address:	D-7140 Ludwigsburg
Type or Model:	Transfer Center TC 22
Capacity (147):	Table area 1200 x 1200 mm Table loading 5000 kg Turntable dia. 1200 mm Tool changing time 10 s
Slide movements:	X-axis 1600 mm⎫ Y-axis 1200 mm ⎬ numerical Z-axis 1430 mm⎭ B-axis, turntable (numerical) or indexing table.
Type of control:	Continuous-path control recommended.
Manufacturer of control system:	To customers' requirements
Programming	Machine and manual

Fig. 310 Linkable machining centre for the simultaneous machining of several faces (cf. Fig. 154), equipped with loading stations and automatic feeding devices

Manufacturer of machine:	*Burkhardt & Weber KG, Werkzeugmaschinenfabrik.*
Address:	D-7410 Reutlingen
Type or model:	MT 3 A
Capacity (145):	Machining area 765 x 508 x 610 mm
	2 automatic loading stations
	Capacity of No. 1 tool magazine
	30 tools
	Tool changing time approx 2 s
	Capacity of No. 2 tool magazine
	14 tools
Slide movements:	X-axis 765 mm ⎫
	Y-axis 508 mm ⎬ numerical
	Z-axis 610 mm ⎭
	Dia. of turntable 508 mm
Type of control:	Straight-line control or
	continuous-path control
Manufacturer of control system:	Normally *Siemens AG* or *Bendix*
Programming:	Machine and manual

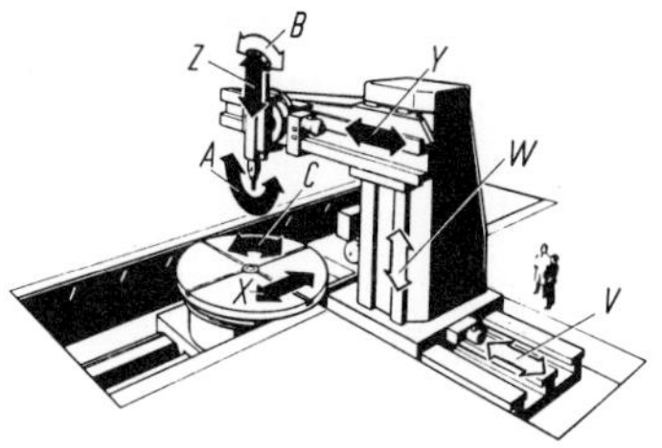

Fig. 311A Sketch to illustrate
movements in Fig.311.

**Fig. 311 Universal vertical milling and turning machine with numerical 5-axis continuous-
path control system for the US aerospace industry.**

Manufacturer of machine:	*Maschinenfabrik Froriep GmbH*
Address:	D-4070 Rheydt
Type or Model:	FRORIEP-SPHEROMILL
Capacity:	Table dia. 3700 mm Max. workpiece dia. 10000 mm Max. height of workpiece 3350 mm
Slide movements:	Column (V-axis) 2750 mm manual Arm (W-axis) 2900 mm manual Tool slide, horiz. (Y-axis) 3000 mm numerical Tool slide, vert. (Z-axis) 1500 mm numerical Table, longit. (X-axis) 6400 mm numerical Table rotation (C-axis) 360° numerical Milling head inclination (B-axis) ±120° numerical
Type of control:	5-axis continuous-path control
Manufacturer of control system:	*Bunker-Ramo,* USA
Programming:	Machine, using APT III.

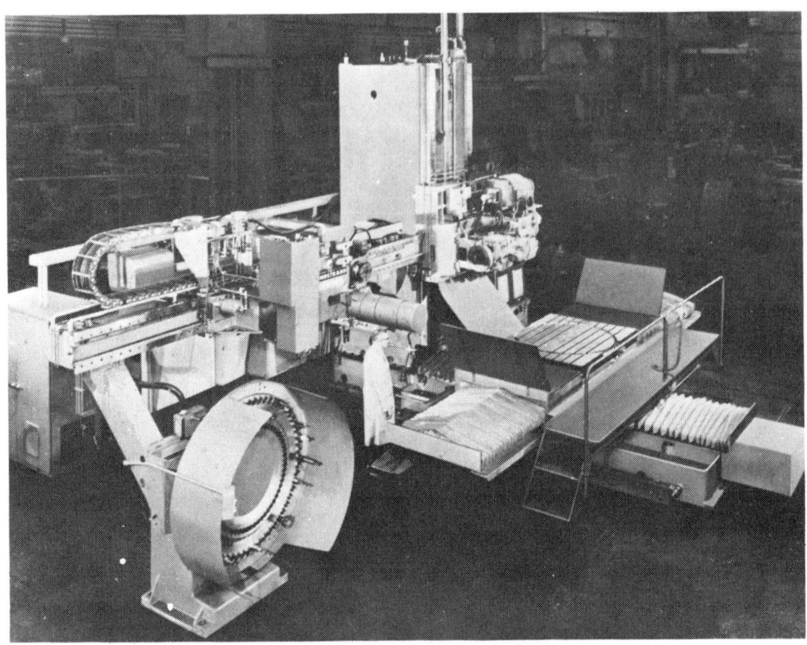

Fig. 312 5-axis machining centre with 60-position tool magazine (cf. Fig. 169)

Manufacturer of machine:	*Sundstrand Machine Tool*
Address:	Belvidere, Ill. USA.
Type or Model:	Model OM 3 Omnimill 5-axis NC Machining Center
Capacity:	Turntable dia. for mounting workpieces, 1050 mm
Slide movements:	X-axis 1200 mm Y-axis 1200 mm Z-axis 1200 mm } numerical Turntable rotation 3609 numerical (C-axis) continuously variable. Spindle head inclination 150° numerical (A-axis) continuously variable.
Type of control:	5-axis continuous-path control
Manufacturer of control system:	*Bunker-Ramo*, USA *Bendix*, USA *General Electric*, USA } to customers' requirements
Programming:	Machine, using APT III.

Fig. 313 Large profile milling machine with two gantries for large workpieces (*skin mill*)

Manufacturer of machine:	*Cincinnati Milacron Company* *Special Machine Division*
Address:	Cincinnati, Ohio 45209 USA
Type or Model:	Gantry Type Vertical NC Profiler (*skin mill*)
Capacity:	Special-purpose machine for machining highly-stressed large aircraft wing components from solid metal, including milling of recesses. Basic unit approx. 15 x 3 (4) m, can be extended in steps of 5 m to total length of about 110 m. Up to 5 gantries can be in use simultaneously, each equipped with 1 to 5 milling spindles. Up to 95% of solid material is machined away (166).
Type of control:	Continuous-path control, mainly 3-axis control.
Manufacturer of control system:	*Cincinnati-Acramatic*
Programming:	Machine, using APT III

Fig. 314 Heavy 5-axis profile milling machine (profiler)

Manufacturer of machine:	*Cincinnati Milacron Company* *Special Machine Division*
Address:	Cincinnati, Ohio 45 209, USA.
Type or Model:	Horizontal NC Profiler 5-Axis Table Type
Capacity:	Table area: Basic model 1200 x 4200 mm, can be increased to 1500 x 6600 mm Motor rating (main spindle) 30 hp
Slide movements:	X-axis 4300–6800 mm Y-axis 1300–1900 mm Z-axis max. 460 mm B-axis 21° to left 51° to right A-axis ± 22.5° } numerical
Type of control:	5-axis continuous-path control
Manufacturer of control system:	*Cincinnati-Acramatic*
Programming:	Machine, using APT III.

Fig. 315 Heavy vertical and horizontal profile milling machine for long, narrow workpieces (*spar mill*)

Manufacturer of machine:	*Cincinnati Milacron Company* *Special Machine Division*
Address:	Cincinnati, Ohio 45 209, USA.
Type or Model:	Vertical and Horizontal NC Profiler, Gantry Type (*spar mill*).
Capacity:	Width 600 to 900 mm Length up to 32 m Models available with up to 3 milling spindles having a total of 10 numerically controlled axes
Type of control:	Continuous-path control
Manufacturer of control system:	*Cincinnati-Acramatic*
Programming:	Machine, using APT III.

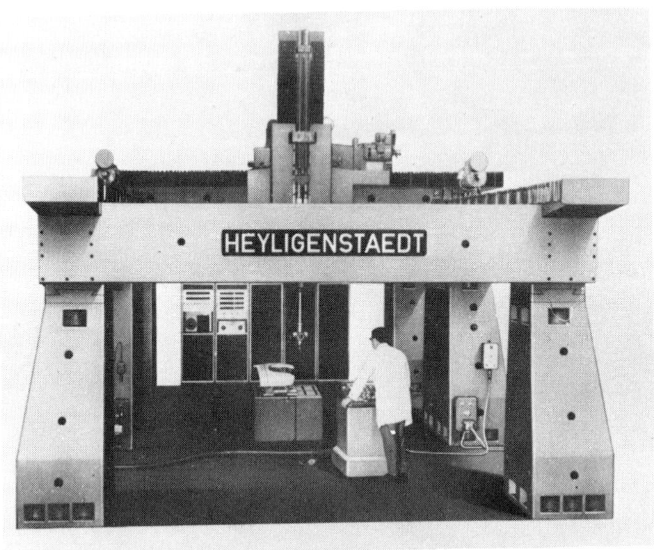

Fig. 316 Large 3-D measuring and scanning machine

Manufacturer of machine:	*Heyligenstaedt & Comp.* *Werkzeugmaschinenfabrik GmbH*
Address:	D-6300 Giessen
Type of Model:	Model TM measuring and scanning machine
Capacity:	Horizontal longitudinal displacement (X-axis) 3100 mm Horizontal transverse displacement (Y-axis) 2500 mm Vertical displacement of follower (Z-axis) 1600 mm

It is necessary to distinguish between two fields of application for this machine:

a)   Automatic measuring and recording of the dimensions of the workpiece to be checked (cf. Figs. 90, 91, 92).

b)   Transforming a prototype (styling model) into numerical values for subsequent processing in a data-processing installation, e.g. for producing punched control tapes for the production of press tools on NC machines.

Ref.:   Brochure: 'Heyligenstaedt Machine Tools in the Automotive-Tool and Die Industries'.

Bibliography

(Books are indicated with an asterisk *).

 1*   *Zemanek, H.:* Elementare Informationstheorie. Oldenbourg-Verlag, München 1959

 2*   *Neidhard, P.:* Informationstheorie und automatische Informationsverarbeitung. Berliner Union, Stuttgart 1964

 3*   *Oppelt, W.:* Kleines Handbuch technischer Regelvorgänge. Verlag Chemie, Weinheim, 4. Aufl. 1964

 4*   *Weyh, U.:* Elemente der Schaltungsalgebra. Oldenbourg-Verlag, München, 5. Aufl. 1968

 5*   *Koenigsberger, F.:* Berechnungen, Konstruktionsgrundlagen und Bauelemente spanender Werkzeugmaschinen. Springer-Verlag, Berlin 1961

 6*   *Steinbuch, K.:* Taschenbuch der Nachrichtenverarbeitung. Springer-Verlag, Berlin, 2. Aufl. 1967

 7*   *Saljé, E.:* Elemente der spanenden Werkzeugmaschinen. C. Hanser-Verlag, München 1968

 8*   *Schlesinger, G.:* Prüfbuch für Werkzeugmaschinen. Verlag G. W. den Boer, Middeburg, 7. Aufl. 1962

 9*   VDE-Buchreihe, Bd. 8: Digitale Signalverarbeitung in der Regelungstechnik. VDE-Verlag, Berlin 1962

10* *Simon, W.:* (Hrsg.): Produktivitätsverbesserungen mit NC-Maschinen und Computern – Investitionsentscheidungen im Rahmen der Beschaffung neuzeitlicher Fertigungsmittel. C. Hanser-Verlag, München 1969

11   *Lusser, G.:* Die Unzuverlässigkeit komplizierter Geräte. Flugkörper *1* (1959), H. 6

12   Normblatt DIN 19226: Regelungstechnik und Steuerungstechnik – Begriffe und Benennungen. Neufassung 1968 Beuth-Vertrieb Berlin-Köln

13   *Mitthof, F.:* Was brachte die 10. EWA Neues auf dem Gebiet der Numerik? TZ f. prakt. Metallbearb. *61* (1967), H. 12, S. 661/673

14   *Hamming, R. W.:* Error detecting and error correcting codes. Bell System Technical Journal Bd. *29* (1950), S. 147/160

15* *Flechtner, H.J.:* Grundbegriffe der Kybernetik. Wissenschaftliche Verlagsgesellschaft, Stuttgart 1966

16   Normblatt DIN 1319: Grundbegriffe der Meßtechnik. Ausgabe Januar 1962, Beuth-Vertrieb, Berlin–Köln

17   VDI-Richtlinie 3254 Blatt 1: Genauigkeitsangaben bei numerisch gesteuerten Werkzeugmaschinen – Begriffe und Kenngrößen. Beuth-Vertrieb, Berlin–Köln

18   *Simon, W.:* Die Werkzeugmaschine als Glied einer datenverarbeitenden Kette. Z-VDI *102* (1960), H. 25, S. 1171/1177

19   *Shannon, C.E.:* A mathematical theory of communication. Bell System Technical Journal Bd. *27*, S. 379/423 und 623/656

20* *Steinbuch, K.:* Automat und Mensch. Springer-Verlag, Berlin, 3. Aufl. 1965

21* *Caldwell, S.H.:* Der logische Entwurf von Schaltkreisen. Oldenbourg-Verlag, München/Wien 1964

22* *Föllinger, O.* und *W. Weber:* Methoden der Schaltalgebra. Oldenbourg-Verlag, München/Wien 1967

23* *Fey, P.:* Informationstheorie. Akademie-Verlag, Berlin 1963

24* *Dokter, F.* und *J. Steinhauer:* Digitale Elektronik in der Meßtechnik und Datenverarbeitung, Bd. I: Theor. Grundlagen und Schaltungstechnik. Philips Fachbücher, Deutsche Philips GmbH, Hamburg 1969

25*   *Semrad, H.:* Numerisch gesteuerte Maschinen. VEB-Verlag Technik, Berlin 1968

26   *Seliger, W.* und *A. Vogel:* Das Multiprismat – eine neuartige Positioniereinrichtung für Werkzeugmaschinen. Maschinenmarkt *71* (1965), H. 19

27   *Rosen, L.:* Characteristics of digital codes. Control engineering *6* (1959), H. 12, S. 115/119

28   *Küpfmüller, K.:* Nachricht und Energie. Regelungstechnik *6* (1958), H. 2, S. 52

29   *Simon, W.:* Steuerungsprinzipien an Werkzeugmaschinen. Werkst. u. Betr. *90* (1957), H. 11, S. 791/798

30*   *Sewig, R.:* Neuartige Fertigungsverfahren in der Feinwerktechnik. C. Hanser-Verlag, München 1969

31   N.N.: Selector guide NC-systems. Amer. Mach. 1967, Nov. 20, S. 122/136

32   Firmendruckschrift: Plessey Company Ltd.: Component technology *3* (1968), H. 1 (Mai 1968)

33   Normblatt DIN 44300: Informationsverarbeitung – Begriffe. Beuth-Vertrieb, Berlin

34   *Kuhrt, F.* und *H.J. Lippmann:* Hallgeneratoren – Eigenschaften und Anwendungen. Springer-Verlag, Berlin 1968

35*   *Spenke, E.:* Elektronische Halbleiter. Springer-Verlag, Berlin, 2. Aufl. 1965

36*   *Eisler, P.:* Gedruckte Schaltungen – Technologie der Folienätztechnik. C. Hanser-Verlag, München 1961

37*   *Schwartz, S.:* (Hrsg.) Integrated Circuit Technology. McGraw-Hill Book Company, New York 1967

38*   *Davis, W.L.* und *H.R. Weed:* Grundlagen der Elektronik. Berliner Union, Stuttgart, 2. Aufl. 1964

39*   *Appels, J.* und *B. Geels:* Handbuch der Relaisschaltungstechnik. Philips Technische Bibliothek, Eindhoven 1967

40   *Hofmann, R.:* Elektronische Steuerung von Werkzeugmaschinen nach dem Ferranti-System. Werkst. u. Betr. *92* (1959), H. 8, S. 494/500

41*   *Borucki/Dittmann:* Digitale Meßtechnik. Springer-Verlag, Berlin 1966

42   *Politsch, H.-W.:* Über die Anwendung einer numerischen Zählersteuerung für einen neuzeitlichen Revolverdrehautomaten. Diss. Darmstadt 1961

43*   *Thiedeken, R.:* Glasfaseroptik. Akademische Verlagsgesellschaft Geest & Portig KG, Leipzig 1967

44   *Deeg, Emil:* Glasfaseroptik erschließt völlig neue Möglichkeiten. „Umschau", *62* (1962), H. 16, S. 510/512

45   *Novotny, G.:* Fibre-Optic Techniques and their applications. Engineers Digest *23* (1962), H. 7, S. 89/105

46   Firmendruckschrift: Ferranti Co-Ordinate Inspection Machine Installation and Operation. Ferranti Ltd. Edinburgh 1960

47   Philips-Druckschrift: Linear-Meßsystem PE 2271. Philips Industrie Elektronik Deutschland, Hamburg 1968

48*   *Apel, K.:* Elektronische Zählschaltungen. Franck'sche Verlagshandlung, Stuttgart 1961

49   *Wang, A.* und *Way Dong Woo:* Static Magnetic Storage and Delay Time. Journal of Applied Physics, *21* (1950), H. 1, S. 49/54

50*   *Dosse, J.:* Der Transistor. Oldenbourg-Verlag, München, 4. Aufl. 1962

51*   *Madelung, O.:* Einführung in die Halbleiterphysik. Springer-Verlag, Berlin 1970

52   *Wagner, K.:* Ringzähler für Vorwärts- und Rückwärtszählung mit Transistoren. Elektron. Rundschau *14* (1960), H. 4, S. 121/125

53   *Gehring, H.:* Ein lineares Wegmeßsystem für Werkzeugmaschinen. Techn. Rundschau, Bern 1967, H. 41

54*   *Bühler, H.:* Einführung in die Anwendung kontaktloser Schaltelemente. Birkhäuser-Verlag, Basel/Stuttgart 1966

55   *Cooney, J. D.* und *B. K. Ledgerwood:* 31 Numerically controlled point-to-point positioning systems Control Engineering *5* (1958), Nr. 1 bis 3

56   *Lott, H.G.:* Lagemessung bei der Lageregelung mit digitalem Sollwert. AEG-Mitt. *51* (1961), H. 1/2, S. 45/49

57*   *Speiser, A.P.:* Digitale Rechenanlagen. Springer-Verlag, Berlin 1961

58*   *Volk, P.:* Antriebstechnik in der Metallverarbeitung – Einführung in die Automatisierung. Springer-Verlag, Berlin 1966

59   Firmendruckschrift: Inductosyn – Grundbegriffe und Anwendungen. Sonderdruck der Inductosyn Corporation, Carson City, Nev./USA, März 1960

60   *Brewer, R.C.:* The Numerical Control of Machine-Tools. The Engineers Digest Survey No. 5, 1959

61   *Brewer, R. C.:* Recent Developments in the Numerical Control of Machine-Tools. The Engineers Digest Survey No. 10, 1961

62   Firmendruckschrift Accupin-Elektromagnetisches Präzisionswegmeßsystem für lineare oder rotatorische Wegmessung. Druckschrift GEK – 4748 B (G) der General Electric Technical Services Company Inc., Zweigniederlassung Frankfurt a. M.

63   Firmendruckschrift: Accupin – Position Measuring System – Mark Century Numerical Control – Optional Equipment Data Sheet GEZ 3687. General Electric, Speciality Control Department, Waynesboro, Va/USA

64   *Barber, B.T.:* Servo Modulators. Control Eng. *4* (1957) H. 8, 10, 11, 12

65   *Cordes, H.:* Verfahren zur Einzelpunktsteuerung bei Werkzeugmaschinen. BBC-Nachr. *43* (1961), H. 10, S. 616/623

66   *Kirkham, E. E.:* Digital positioning control for precision jig-borers. Electr. Manufacturing 1957, H. 3 (März), S. 118/125

67   *Walker, D.F.:* Numerical control of machine-tools. Journal of the Inst. of El. Eng. *9* (1963), Febr.

68*   *Korn, A.* und *Th. M. Korn:* Elektronische Analogierechenmaschinen. Berliner Union, Stuttgart 1960

69*   *Haley, H.* und *W. E. Scott:* Analogue and Digital Computers. George Newnes Ltd., London 1960

70*   *Wagner, B.:* Elektronische Verstärker, 3. Aufl. VEB-Verlag Technik, Berlin 1961

71*   *Schwerd, F.:* Spanende Werkzeugmaschinen – Grundlagen und Konstruktionen. Springer-Verlag, Berlin 1956

72*   *Goldsche, J.:* Zerspanende Formgebung mit elektrischen Geräten. C. Hanser-Verlag, München 1967

73*   *Stephan, E.:* Optimale Stufenrädergetriebe für Werkzeugmaschinen. Springer-Verlag, Berlin 1958

74   *Bock, L.:* Elektrische Vorschubantriebe für numerisch gesteuerte Werkzeugmaschinen. Antriebstechnik *5* (1966), Nr. 12

75*   AEG-Handbücher, Bd. 2: Gleichstrommaschinen. 2. Aufl. Allgemeine Electricitäts-Gesellschaft Berlin/Ffm. 1964

76   *Volk, P.:* Zuverlässigkeit von elektrischen Steuerungen. VDI-Berichte *89* (1965)

77   *Waller, S.:* Betriebszuverlässigkeit elektronischer Steuerungen bei Werkzeugmaschinen. TZ f. prakt. Metallbearb. *59* (1965), H. 6

78*   Kielgas, H.: Transduktoren. A. Hüthig-Verlag, Heidelberg 1960

79*   *Baehr, H.:* Regeln und Steuern durch magnetische Verstärker. Fr. Vieweg & Sohn, Braunschweig 1960

80*   *Bühler, H.:* Einführung in die Theorie geregelter Gleichstromantriebe. Birkhäuser-Verlag, Basel/ Stuttgart 1962

81*   VDE-Buchreihe, Bd. 11: Energieelektronik und geregelte elektrische Antriebe. VDE-Verlag, Berlin 1966

82*   *Dürr, A.* und *O. Wachter:* Hydraulik in Werkzeugmaschinen – Hydraulische und elektrohydraulische Antriebe. C. Hanser-Verlag, München, 6. Aufl. 1968

83*   *Krug, H.:* Das Flüssigkeitsgetriebe bei Werkzeugmaschinen. Springer-Verlag, Berlin, 2. Aufl. 1959

84*   *Guillon, M.:* Hydraulische Regelkreise und Servosteuerungen – Grundlagen, Berechnungen und Anwendungen. C. Hanser-Verlag, München 1968

85   *Schaller, W.:* Hydraulische Steuerung automatischer Fertigungseinrichtungen. Beitrag BW 110 zum Lehrgangshandbuch „Automatisierung" des VDI-Bildungswerks, Düsseldorf 1960

86   Normblatt DIN 24300 Ölhydraulik und Pneumatik – Benennungen und Sinnbilder – Blatt 1–6. Beuth-Vertrieb, Berlin–Köln–Ffm 1966

87   Firmendruckschrift: AEG-Elektrohydraulische Bausteine – Tauchspulenregler, Druck-Nr. 2415 101 EMG/0564. Allgemeine Electricitäts-Gesellschaft, Frankfurt/Berlin

88   *Lück, J.:* Einflußgrößen auf das Zeitverhalten elektrohydraulischer Vorschubantriebe Diss. Aachen 1968

89   *Hofmann, W.:* Untersuchungen an Vorschubantrieben für numerisch gesteuerte Werkzeugmaschinen. Diss. Aachen 1969

90   *Uhrmeister, H.:* Die dynamischen Eigenschaften hydraulischer Vorschubmotoren für Werkzeugmaschinen. Diss. Aachen 1961

91   *Simon, W.:* Gegenüberstellung von elektrischen und hydraulischen Antriebssystemen bei numerisch gesteuerten Werkzeugmaschinen. Beitrag BW 371 zum Lehrgangshandbuch „Numerische Steuerung von Werkzeugmaschinen". VDI-Bildungswerk, Düsseldorf 1964

92   *Bell, R.* und *F. Koenigsberger:* Numerisch gesteuerte Werkzeugmaschinen auf der Olympia-Ausstellung 1968. Werkst. u. Betr. *101* (1968), H. 10, S. 603/609

93   Autorenkollektiv: Spanende und abtragende Werkzeugmaschinen auf der 11. EWA in Paris. Bericht des Laboratoriums für Werkzeugmaschinen und Betriebslehre der TH Aachen, Ind. Anz. *91* (1969), H. 72, S. 1700/1770 und H. 79, S. 1930/1935

94   Firmendruckschrift: Ro-tran Power Control Units. Inland Motor Corporation of Virginia – Subsidiary of Kollmorgen, Radford, Virg./USA 1969

95   *Thomas, A.G.* und *F.J. Fleishauer:* The power stepping motor – a new digital actuator. Contr. Eng. *4* (1957), H. 1, S. 74/81

96   *Izobotenko, B.A.:* Die Schrittmotoren (Bericht aus dem Moskauer Institut für Energetik). Elektricestvo No. 8, 1960, S. 54/61

97   *Bailey, S.J.:* Incremental Servos, Part IV – Todays Hardware. Contr. Eng. *8* (1961), H. 3, S. 133/135

98   Firmendruckschrift: Bulletin B – 51016 E „Electro-hydraulic pulse motor". Fuji Communication Apparatus Mfg. Co. Ltd., Kawasaki/Japan

99   *Punga, F.:* Vorlesungen über Elektromaschinenbau. Darmstadt 1931

100   *Meyringer, V.:* Elektrische Schrittmotoren. Diss. Aachen 1968

101   *Bock, L.:* Elektrische und elektrohydraulische Schrittmotoren. Steuerungstechnik *1* (1968), H. 1, S. 13/18

102   *Stüben, H.* und *W. Thron:* Der Minertia-Motor als reaktionsschneller Antrieb für Werkzeugmaschinen. „Antriebstechnik" *7* (1968), H. 6, S. 208/212

103   Firmendruckschrift „Gleichstrom-Scheibenläufer-Motoren AXEM-SERVALCO". Fa. Cie Electro-Mécanique (CEM-SEA, Frankreich), Ausgabe Nov. 1969

104   Bendix-US-Patente Nr. 2.939.287, 3006.550, 3.011.110, 3.002.115, 3.278.817, 3.128.374, 3.122.691, 3.267.344, 2.820.187, 3.069.608

105   Firmendruckschrift: Bosch-Bendix N/PNC 1 TB 0011 – Beschreibung der Bendix Serie Hundred

106   Firmendruckschrift: Bosch-Bendix N/PNC 1 TB 0047 – Beschreibung der Bendix Dynapath 20

107   Firmendruckschrift: Bosch-Bendix N/PNC 1 TB 0045 – Beschreibung der Bendix Directapath 20

108   Firmendruckschrift: Fortuna – Numerisch gesteuerte Rundschleifmaschinen. Fortuna-Werke Maschinenfabrik Stuttgart/Bad Cannstadt 1968

109* *Opitz, H.* (Hrsg.): Auslegung von Vorschubantrieben für NC-Maschinen. Bericht über die VDW-Konstrukteur-Arbeitstagung am 7. und 8.2.1969. Laboratorium für Werkzeugmaschinen und Betriebslehre der TH Aachen

110   *Geyer, W.:* Die Datenverarbeitung bei numerisch gesteuerten Werkzeugmaschinen. „Messen und Prüfen" *3* (1967), H. 9, S. 419/422

111   Patent-Auslegeschrift 1291515 des Deutschen Patentamtes, Kl. 42b–24 „Meßgerät für die Meßsteuerung von Werkzeugmaschinen, insbesondere der Schleifmaschinen nach Absolutmeßwerten". Fortuna-Werke Maschinenfabrik AG, Stuttgart-Bad Cannstadt

112* *Haidekker, A.:* Meßsteuerung-Kontrolle des Fertigungsablaufs durch Meßwerte. R. v. Decker's Verlag, G. Schenck, Hamburg 1958

113   *Stal, H.P.:* Entwurf einer numerischen Werkzeugmaschinensteuerung aus Fluidic-Elementen. Ind. Anz. *4* (1967), S. 21/26

114   Firmendruckschrift Moog Hydra-Point: Mehrzweck-Fräsmaschine mit numerischer Steuerung 1968. Moog Hydra-Point Ltd., Cheltenham, Gloucestershire/GB

115   Firmendruckschriften: Moog – Elektrohydraulischer Stellantrieb Moog – Elektrohydraulische Durchfluß-Steuerventile – Baureihen 72 und 73 für die Industrie. Moog GmbH, Böblingen 1969

116* VDI-Veröffentlichungen (Hrsg.): Neue pneumatische Logikelemente, Strömungsmechanische Logikelemente und -schaltungen (Fluidik). VDI-Verlag, Düsseldorf 1968

117* *Multrus, V.:* Fluidik-Pneumatische Logikelemente und Steuerungssysteme. Krausskopf-Verlag, Mainz 1970

118   *Pawlowitz, K.R.:* Eigenschaften pneumatischer logischer Schaltungen, Aufsatzfolge. TZ f. prakt. Metallbearb. *62* (1968), H. 4 bis 11

119   Firmendruckschrift „Plessey-Automation-Fluidic 220 Numerical Control System". Publication No. 3025/1 1969

120   *Müller, K.:* Über den Einfluß der Oberflächengestaltung und Oberflächenbehandlung auf das Verhalten von Gußeisen in Gleitführungen. Diss. Darmstadt 1958

121 *Spiess, D.:* Das Steifigkeits- und Reibungsverhalten unterschiedlich gestalteter Kugelschraubantriebe mit vorgespannten und nicht vorgespannten Muttersystemen. Diss. Berlin 1970

122 *Stromberger, C.* und *H. Schulz:* Die Kugelumlaufspindel als Antriebs- und Meßelement. Werkst. u. Betr. *99* (1966), H. 5, S. 311/315

123 *Bögelsack, G.:* Zur Konstruktion von Kugelschraubtrieben für Meßzwecke. Feingerätetechnik *15* (1966), S. 65/71

124* *Mitthof, Fritz:* Numerisch gesteuerte Fertigung – Einsatz von numerisch gesteuerten Werkzeugmaschinen und direkte Fertigungssteuerung über DVA – das neue Konzept der Fertigung im Mittelbetrieb. Krausskopf-Bücherei der Produktivitätssteigerung Band 4, Krausskopf-Verlag, Mainz 1969

125* *Wilson, F. W.* (Hrsg.): Numerical Control in Manufacturing. McGraw-Hill-Verlag, New York 1963

126 *Koenigsberger, F.:* Forschungsbericht über Probleme hydraulisch geschmierter Führungen für genaue Positionierung. Maschinenmarkt/Werkzeugmaschinenpraxis, Sammelband 1960

127 *Stromberger, C.:* Die numerische Steuerung von Werkzeugmaschinen. VDW-Informationsbericht März 1961

128 *Müller, K.:* Wälzkörpergelagerte Längsführungen. Werkst. u. Betr. *91* (1958), H. 2, S. 73/76

129 *Maßberg, W.:* Der Einfluß der Vielgestaltigkeit eines Werkstücks auf den wirtschaftlichen Einsatz numerisch gesteuerter Werkzeugmaschinen und maschinelle Programmierverfahren. Diss. Aachen 1965

130* *Atscherkan, N. F.:* Werkzeugmaschinen, Berechnung und Konstruktion. Bd. 1. VEB-Verlag Technik, Berlin 1963

131 *Cornely, H.:* Hilfsmittel zur Werkzeugvoreinstellung und Organisation des planmäßigen Werkzeugschnellwechsels. Werkst.-Techn. *50* (1960), H. 3, S. 154/158

132 Firmendruckschrift: Hartman Planetary Rol-Vane-Motors-PRV-Series 1968, Hartmann Hydraulics-Division of Koehring Company, Racine/Wisc. USA

133 *Jaeger, H.:* Drehautomaten. C. Hanser-Verlag, München 1967

134 *Schaumjan, G. A.:* Automaten (Übersetzung aus dem Russischen), VEB-Verlag Technik, Berlin, 2. Aufl. 1961

135 Firmendruckschriften PIMAT (1968) und PINUMAT (1969) Pittler Maschinenfabrik AG, Langen bei Frankfurt a. M.

136 Firmendruckschrift: Vertikale Revolverdrehmaschine PFVR 100 III. Gebr. Heller Maschinenfabrik GmbH., Nürtingen

137 Firmendruckschriften Heynumat Formendrehmaschine (1967) heyco-Mitteilungen 3/69 und 4/69 der Fa. Heyligenstaedt & Comp. GmbH., Gießen

138 Firmendruckschrift: Numerisch gesteuerter Einspindel-Drehautomat mit extrem hoher Drehgenauigkeit NA 300 (1968), A. Monforts Maschinenfabrik, Mönchengladbach

139 Firmendruckschrift WMW, VEB Großdrehmaschinenbau „8. Mai", Karl Marxstadt/DDR, Technische Informationen, Sonderheft 4, Drehmaschinen mit numerischer Steuerung 1966

140 Firmendruckschrift HÜBOMAT, Karl Hüller GmbH, Ludwigsburg/Württ.

141 Firmendruckschriften AUCTOR CNZ. Ing. C. Olivetti & C., Divisione Controllo Numerico, San Bernardo d'Ivrea/Italien

142 Firmendruckschriften P 500 NC und P 800 NC, VDF 193/1 d, Vereinigte Drehbank-Fabriken, Göppingen/Hamburg/Hannover

143 Firmendruckschrift MIN-E-CENTRE-MEC 3. Marwin Machine Tools Ltd., Anstey, Leicester/UK

144 Firmendruckschrift: Wadkin HDM 1000, Horizontale Bohr- und Fräsmaschine mit numerischer Steuerung (1969). Wadkin Ltd. Green Lane Works, Leicester/UK

145 Firmendruckschriften: Bearbeitungszentren M 2, 3, 4, 5 der Fa. Burkhard & Weber KG, Reutlingen/Württ.

146 Firmendruckschrift: Bohle Bearbeitungszentrum Type BOMOMATIC *65* (1969), Reinhard Bohle KG, Jöllenbeck bei Bielefeld

147 Firmendruckschrift: Transfer-Center TC mit numerischer Bahn- oder Streckensteuerung. Ludwigsburger Maschinenbau GmbH (Burr), Ludwigsburg/Württ., 1970

148 Firmendruckschriften: Bearbeitungszentrum NCMC-100 (1969) und Programmierung desselben. Karl Hüller GmbH, Ludwigsburg/Württ.

149 Firmendruckschriften: Kalimat-Bohrstangen-Einstellgeräte, Transport- und Lagerstättengeräte für Werkzeuge. Kelch & Co., Schorndorf/Württ.

150   Firmendruckschrift: Numerisch gesteuerter Plandreh- und Ausbohrkopf, Scharmann & Co., Rheydt/Rhld.

151   Firmendruckschriften: Molins System *24* (1969). Molins Machine Comp. Ltd., Machine Tool Division London/UK

152   *Williamson, D.N.T.:* Ein neues Fertigungsverfahren TZ f. prakt. Metallbearb. *61* (1967), H. 9 und 10

153   *Grüning, K.* und *E. Mayrhofer:* Automatisches Bohren von Lochfeldern in Platten. Werkstatt-techn. *59* (1969), H. 3, S. 102/107

154   *Krimphove, H.:* Beitrag zur Automatisierung der Fertigungsplanung von Wärmeaustauscher-blechen. Diss. Berlin 1969

155   Firmendruckschriften: Bohrmaschinen KBNP, KBNE50, MPN8 (1969) Herm. Kolb, Köln-Ehrenfeld

156   Firmendruckschriften: Blechbearbeitungszentrum Trumatic 20 mit Arbeitsbeispielen: Trumpf & Co., Stuttgart-Weilimdorf

157   Firmendruckschrift: Suprarex SM. Kjellberg-Eberle GmbH, Frankfurt a.M.

158   *Hirschberg, H.:* Die größte Brennschneidmaschine der Welt. „betriebstechnik" 1967, H. 9

159   Firmendruckschriften: Automation des Brennschneidens – Vergleich der Informationsträger (Schablonen, Zeichnung, Magnetband) für Brennschneidmaschinen. Messer-Griesheim GmbH Schweißtechnik, Frankfurt a.M.

160   Firmendruckschrift: NC-Information Nr. 3 – Die zukünftige Entwicklung auf dem Gebiet des numerisch gesteuerten Drehens. Vereinigte Drehbank-Fabriken (VDF), Göppingen/Hamburg/Hannover

161   Firmendruckschrift: Fraiseuse à commande numérique – 6 axes – UN800BRC: Société Ratier-Forest, Capdenac/Frankreich

162   Firmendruckschrift: Scharmann-Arbeitssystem (1968). Scharmann & Co., Rheydt/Rhld.

163   *Vogt, H.* und *G. Engel:* Geregelte Vorschubantriebe für numerisch gesteuerte Werkzeugmaschi-nen – eine konstruktive Aufgabe. „antriebstechnik" *8* (1969), H. 6, S. 201/207

164   *Vogt, H.* und *K. Jüstel:* Der Einsatz bahngesteuerter Horizontal Bohr- und Fräswerke. TZ f. prakt. Metallbearb. *63* (1969), H. 10, S. 573/579

165   Firmendruckschrift: Automatische, numerisch gesteuerte Drehmaschine DN 300. Gebr. Heine-mann, St. Georgen/Schwarzw.

166* RKW-Bericht A30: Numerische Steuerung von Werkzeugmaschinen. Bericht einer RKW-Stu-diengruppe über eine USA-Reise. Beuth-Vertrieb Berlin/Köln 1964

167   *Clark, C.D.:* Selection and performance criteria for electro hydraulic servodrives. Moog-Tech-nical Bulletin 122/1969, Moog. Inc., East Aurora, N.Y./USA

168   Normblatt DIN 66025 (z.Z. noch Entwurf): Programmaufbau für numerisch gesteuerte Arbeits-maschinen (1969).
Blatt 1 Allgemeines; Blatt 2 Wegbedingungen und Zusatzfunktionen; Blatt 3 Vorschübe und Spindeldrehzahlen; Blatt 4 Beschreibung der numerisch gesteuerten Arbeitsmaschine

169   VDI-Richtlinie 3255: „Programmieren numerisch gesteuerter Werkzeugmaschinen – Fest-legung der Koordinatenachsen und Zuordnung der Bewegungsrichtungen" VDI-Düsseldorf 1968

170   *Ackerknecht, B.:* Grundlagen des Programmierens numerisch gesteuerter Werkzeugmaschinen. Lehrgangsbeitrag BW342 des VDI-Bildungswerkes, Düsseldorf 1969

171   *Jüstel, K.:* Numerisch gesteuerter Plandreh- und Ausbohrkopf. Werkst. u. Betr. *101* (1968), H. 7, S. 413/414

172   *Spur, G.* und *H. Michels:* Über die Auslegung numerisch gesteuerter Drehmaschinen. ZWF *61* (1966), H. 10, S. 507/514

173   Firmendruckschrift: Bearbeitungszentrum MC5RV 1969. Fritz Werner GmbH., Berlin 48

174   Firmendruckschrift: AGIECUT-Drahtcodiermaschine mit numerischer Bahnsteuerung. AGIE Losone/Schweiz

175   *Spur, G.:* Betrachtungen zur Optimierung des Fertigungssystems Werkzeugmaschine. Werkst. Techn. *57* (1967), H. 9, S. 411/417

176   *Kopperschläger, F.D.:* Über die Auslegung mechanischer Übertragungselemente an numerisch gesteuerten Werkzeugmaschinen. Diss. Aachen 1969

177   *Schöbel, M.:* Fragen des Werkzeugeinsatzes bei numerisch gesteuerten Drehmaschinen. TZ f. prakt. Metallbearb. *63* (1969), H. 6

178   *Hölz, H.* und *M. Roßkopf:* Maschinen zur automatischen Bearbeitung von kleinen und mittel-
      großen Serien. Werkst. u. Betr. *96* (1963), H. 9, S. 589/594

179   *Michaelis, H.:* Interpolationsverfahren mit natürlichen Polynomen im Hinblick auf die Steuerung
      von Werkzeugmaschinen. Z-VDI *104* (1962), H. 35, S. 1806/1813

180*   *Stahl, K.:* Industrielle Steuerungstechnik in schaltalgebraischer Behandlung. Oldenbourg-Verlag,
      München 1965

181   *Blüsch, H.:* Analogie-Interpolator als Rechenwerk für programm-gesteuerte Werkzeugmaschinen.
      Maschinenbautechnik *8* (1959). H. 11

182   *Farmer, P.J.:* Analogue Control. Aircr. Prod. *18* (1956), H. 7

183   *Farmer, P.J.:* Continuous Analogue Control. Aircr. Prod. *19* (1957), H. 3

184   *Goetz, E.:* Digital arbeitende Interpolatoren für numerische Bahnsteuerungen. AEG-Mitt. *51*
      (1961), H. 1/2, S. 34/44

185*   *Haas, G.:* Grundlagen und Bauelemente elektrischer Ziffernrechenanlagen. Philips Technische
      Bibliothek, Eindhoven 1961

186*   *Rechenberg, P.:* Grundzüge digitaler Rechenautomaten. Oldenbourg Verlag, München, 2. Aufl.
      1968

187   *Pease, V.N.:* Numerical Control Aircr. Prod. *14* (1952), H. 10, S. 356/358

188   *Herger, H.:* Der digitale Interpolator – eine Steuereinrichtung hoher Genauigkeit für Werkzeug-
      maschinen. elektr. ausr. *4* (1963), H. 1, S. 11/16

189   *Lott, H.-G.:* Ein Digital-Analog-Umsetzer zur numerischen Steuerung von Synchros. Regelungs-
      techn. *8* (1960), H. 5, S. 157/162

190   *Herger, H.:* Kopieren ohne Modell-Wegregelung bei Werkzeugmaschinen. BBC-Nachr. 1961,
      H. 10

191   *Fischer, H., R. Klinge* und *C. Schendel:* Interpolationsverfahren für Bahnsteuerungen bei Werk-
      zeugmaschinen. Siemens-Z 1966, H. 2

192   *Morely, A.:* Interpolationsverfahren bei numerischen Bahnsteuerungen mit Inneninterpolatoren.
      Klepzigs Fachberichte 1968, Sept.

193   *Boese, P.:* Numerische Bahnsteuerungen mit Inneninterpolatoren. VDI-Bildungswerk-Lehrgangs-
      handbuch, Beitr. BW 358

194   *Krägeloh, W.:* Die Werkzeugdurchmesser-Kompensation bei der numerischen Bahnsteuerung.
      Z-VDI. *107* (1965), Nr. 25

195   *Maier, K.:* Adaptive Control bei Werkzeugmaschinen. Steuerungstechn. *2* (1969), H. 4, S. 128/
      130

196   *Maier, K.:* Anpaß- und Optimierregeleinrichtungen an Werkzeugmaschinen. Steuerungstechn.
      2 (1969), H. 6, S. 220/224

197   *Acs, M.:* Adaptive Regelung. Werkst. u. Betr. *101* (1968), H. 11, S. 683/687

198   *Meyer, J.* und *G. Sautter:* SINUMERIK mit Adaption für Werkzeugmaschinen. Siemens-Z. *43*
      (1969), H. 12, S. 949/955

199   *N.N.:* Adaptive Control „package" debuts. Amer. Mach. *112* (1968), Nr. 13, S. 104/105

200   *Stocker, W.M.:* The production man's guide to numerical control. Amer. Mach. *101* (1957),
      No. 14, S. 133/154 b

201   *Brose, G.:* Manuelles Programmieren einer Dreharbeit. TZ f. prakt. Metallbearb. (Numerik) *63*
      (1969), H. 12, S. 689/697

202   *Mitthof, F.:* Fallbeispiel dvb/f 04. Bohr- und Feinbohrarbeit manuell programmiert für numerisch
      gesteuerte Bohrmaschine mit automatischem Werkzeugwechsel am Spindelstockmagazin. TZ
      f. prakt. Metallbearb. *60* (1966), H. 6, S. 407/412

203   Firmendruckschrift: Scharmann Dekamat-System-Programmier-Lehrgang. Scharmann & Co.,
      Abt. Beratungsdienst, Rheydt 1969

204*   *Stute, G.* (Hrsg.): Lageregelung von Werkzeugmaschinenschlitten. C. Hanser-Verlag, München
      1970

205   *Mathias, R.A.:* An effective system for adaptive control of the milling process. Technical paper
      MS68-202 of the ASTME-Engineering Conferences. American Society of Tool and Manufac-
      turing Engineers, Dearborn/Mich.

206   *Kuplent, F.:* Weiterentwicklung und zukünftiger Einsatz von NC-Maschinen. Werkst. u. Betr.
      103 (1970), H. 1, S. 13/22

207   *Fetzer, H.:* Maschinelles Programmieren von NC-Maschinen mit Kleinrechnern. TZ f. prakt.
      Metallbearb. *64* (1970), H. 2, S. 89/93

208   *Brümmer, J.:* Teilautomatisches Programmieren von numerisch gesteuerten Werkzeugmaschinen. Werkst. Techn. *60* (1970), H. 3, S. 151/155

209   *Saljé, E.:* Über den Einfluß der numerischen Steuerung auf die Entwicklung der Werkzeugmaschinen. Klepzigs Fachberichte *76* (1968), H. 9, S. 533/535

210   Firmendruckschrift: Adaption an Werkzeugmaschinen mit SINUMERIK. TS 31/SG 503.650-6/69

211   Firmendruckschrift: High Spindle Speed Thread-Cutting Option for Mark Century Contouring Controls. GEI-98102B der Fa. General Electric Company, Speciality Control Department, Waynesboro, Va/USA

212*   Dubbels Taschenbuch für den Maschinenbau, S. 736 ff. 12. Aufl., Springer-Verlag, Berlin 1966

213   Firmendruckschrift: SINUMERIK 520/42, Siemens AG, Erlangen–Berlin 1969

214   Firmendruckschrift: AEG-Numeric 311 – Betriebsanweisung. Allgemeine Elektricitäts-Gesellschaft, Seligenstadt – Berlin 1969

215*   Hütte-Taschenbuch für Betriebsingenieure (Betriebshütte), Bd. I – Fertigung. Verlag Wilhelm Ernst & Sohn, Berlin, 4. Aufl. 1954

216   *Simon, W.:* Werkzeugmaschinen und integrierte Datenverarbeitung. Werkst. Techn. *55* (1965), H. 10, S. 477/482

217   Firmendruckschrift: A new concept in multi-axis control for point-to-point and straight-cut machine-tool applications: UMAC V. Sperry Gyroscope Company of Canada, Montreal 1964

218   Firmendruckschrift: Köping SMT/Kongsberg CNC. 3D-Bearbeitungszentrum. AS Kongsberg Vapenfabrikk, Kongsberg/Norwegen

219*   *Mayer, R.:* Die Problematik des Einsatzes numerisch gesteuerter Werkzeugmaschinen – eine betriebliche Untersuchung. Arbeitsstudium – Industrial Engineering – Beispiele für die praktische Anwendung, Bd. 12. Beuth-Vertrieb, Berlin/Köln 1970

220   *Stöferle, Th.:* Entwicklungstendenzen beim Bau von numerisch gesteuerten Bohr- und Fräsmaschinen. Werkst. u. Betr. *101* (1968), H. 10

221   *de Beauclair, W.:* Datenträger für automatische Werkzeugmaschinensteuerungen. Werkst. Techn. *50* (1960), H. 3, S. 131/136

222   *Kirschdorfer, H.* und *G. Schaad:* Manuelle Dateneingabe. Firmendruckschrift der Fa. Ghielmetti/Solothurn, Schweiz

223   *Wehner, M.* und *R. Georgi:* Fräsmaschinen mit Programmeinrichtung. Firmendruckschrift VEB Fritz-Heckert-Werk, Karl-Marx-Stadt, Techn. Informationen – Fräsmaschinen, H. 1/1967

224   Firmendruckschriften: „PIMAT" und „PINUMAT" mit „PITTLER-NUMERIK". Pittler Maschinenfabrik AG, Langen/Hessen 1966

225   Firmendruckschrift: Beschreibung der numerischen Steuerung und der Programmierung des Horizontal-Lehrenbohrwerkes DIXI 75, DIXI S.A., Le Locle/Schweiz

226*   *Kohring, G.:* Grundlagen und Praxis numerisch gesteuerter Werkzeugmaschinen. C. Hanser-Verlag, München 1966

227*   Autorenkollektiv: Rationalisierung der Fertigung in der metallverarbeitenden Industrie. Vortragssammlung zum 2. wissenschaftlichen Kongreß Leipzig 1966 – Kammer der Technik, Berlin 1966

228*   Halbleiter-Lexikon – Fachausdrücke. AEG-Telefunken Fachbuch, Franzis-Verlag, München 1965

229*   *Puckle, D. S.* und *J. R. Arrowsmith:* An Introduction to Numerical Control of Machine-Tools. Chapman and Hall Ltd., London 1964

230*   *Korn, A.* und *Th. Korn:* Electronic Analog and Hybrid Computers. McGraw Hill Book Company, New York 1964

231*   *Berdecke/Ptassek/Rothenbach/Vaske:* Elektrische Antriebe und Steuerungen. B. G. Teubner, Stuttgart 1969

232   *Mitthof, F.* (Hrsg.): Programmieren von NC-Werkzeugmaschinen. Sonderveröffentlichung 1970 der Zeitschrift „Steuerungstechnik", Krausskopf-Verlag, Mainz

233   *Fischer, H.:* Beitrag zur Untersuchung des thermischen Verhaltens von Bohr- und Fräsmaschinen. Diss. Berlin 1970

234   Firmendruckschriften über Photocell Punched Tape Reader/Spooler. Remex Electronics/Division of Ex-Cell-O-Corp. Hawthorne, Calif. USA

235   *Tafel, H. J.* (Hrsg.): Strömungsmechanische Logikelemente und -schaltungen (Fluidik), VDI-Verlag Düsseldorf 1968

236   *Wiesner, H.* (Hrsg.): Neue pneumatische Logikelemente. VDI-Verlag Düsseldorf 1968

237   *Williams, L.J.:* Fluid contamination effects on servovalve performance. Moog Technical Bulletin 115/1967 Moog Inc., East Aurora, N.Y./USA

238   Firmendruckschrift: Logische Steuerungen in Maschinenbau und Verfahrenstechnik – Programmsteuerungs-Einheiten Festo-Fluidic, Festo-Pneumatik, Berkheim-Esslingen 1969

239   *Farmer, P.J.:* Digital control. Aircr. Prod. *18* (1956), S. 256/267

240   *Opitz/Uhrmeister/Jüstel:* Aufbau und Wirkungsweise einer Magnetbandsteuerung. Forschungsberichte des Wirtschafts- und Verkehrsministers des Landes NRW. Westdeutscher Verlag Köln/Opladen 1958

241   Firmendruckschrift: Ferranti-MK4 continuous path control system. Ferranti Ltd., Thorneybank, Dalkeith, Scotland 1968

242   Firmendruckschrift: Ferranti-MULTIAX continuous path control system. Ferranti Ltd., Thorneybank, Dalkeith, Scotland 1968

243   *Ankeney, H.E.* und *D.H. Bingham:* Production proved numerical control; three steps from print to part. Amer. Mach. *101* (1957), H. 23, S. 145/156

244   *Walker, D.F.:* Das Magnetband und seine Zukunft bei der Steuerung von Werkzeugmaschinen – Ein kritischer Vergleich der Steuerung von Werkzeugmaschinen durch Magnetband und Lochstreifen. technica 1967, H. 11, S. 993/998

245   *Walker, D.F.:* Warum das Magnetband als Informationsträger für numerisch gesteuerte Werkzeugmaschinen? technica 1967, H. 26, S. 2647/2650

246   *Kamp, A.-W.:* Ausblick auf weitere Entwicklungen auf dem Gebiet numerischer Steuerungen in der Fertigungstechnik. VDI-Bildungswerk, Lehrgangsbeitrag BW 1481, Düsseldorf 1969

247   *Spur, G., H. Debler* und *J. Kurth:* Zukunftsperspektiven für die Automatisierung in der betrieblichen Fertigung. analysen und prognosen, Berlin, Januar 1970

248   *Carlson, R.D.* und *G.E. Houston:* Taking a plunge in DNC. Amer. Mach. *113* (1969), July 28, S. 84/90

249   *Kirkham, E.E.* und *J.P. Bunce:* DNC with dual computers. Amer. Mach. *113* (1969), August 11, S. 51/54

250   Sundstrand Corporation: Direkte Steuerung von Werkzeugmaschinen durch Rechner. TZ f. prakt. Metallbearb. *62* (1968), H. 7, S. 358/360

251   Sundstrand Corporation: Eine neue Daten-Steuereinheit zur Anpassung der numerischen Steuerung an den Rechner für das on-line-Verfahren. TZ f. prakt. Metallbearb. *63* (1969) H. 3, S. 136/137

252   NC-Society: From Tape to time-sharing. Annual conference paper, USA 1969

253   *Fidrich, P.:* Die Fertigungsregelung in der metallverarbeitenden Industrie mit Hilfe von Datenverarbeitungsanlagen als spezieller Fall allgemeiner Prozeßregelungen. Diss. Berlin 1969

254   Firmendruckschriften: Slo-syn-Photoelectric Tape-Readers. The Superior Electric Company, Bristol/Conn., USA 1968

255   *Dyke, R.M.:* Die numerische Steuerung in den USA – heute. technica 1970, Nr. 11, S. 907/920

256* *Schüring, H.:* Datenverarbeitende Maschinen und Systeme für kleine und mittlere Betriebe – Auswahl, Arbeitsweise und Verwendungsmöglichkeiten. Bertelsmann-Fachverlag, Gütersloh 1969

257   *Müller-Traut, H.:* Grundzüge des maschinellen Programmierens numerisch gesteuerter Werkzeugmaschinen mit Symbolsprachen. Lehrgangsbeitrag BW 968 des VDI-Bildungswerkes Düsseldorf 1968

258   DIN 66003 (Entwurf 1965) Informationsverarbeitung: 7-Bit-Code. Beuth-Vertrieb, Berlin

259   DIN 66024 (Entwurf 1967) Numerische Steuerung von Arbeitsmaschinen – Code für 8-Spur-Lochstreifen. Beuth-Vertrieb, Berlin

260* *Leslie, W.H.P.* (Hrsg.): Numerical Control Users' Handbook McGraw-Hill, London–New York 1970

261   *Mitthof, F.* (Hrsg.): Programmieren von NC-Maschinen – Einführung und 13 Beispiele für das manuelle Programmieren, das Programmieren über Kleinrechner, das Programmieren über Großrechner. Sonderveröffentlichung der Zeitschrift „Steuerungstechnik" 1968/69, Krausskopf-Verlag, Mainz 1970

262   Firmendruckschrift: NC-Drehmaschinenhersteller vereinheitlichen Programmierunterlagen 1968. VDF, Dörries, Pittler, Froriep, Max Müller, Gildemeister, Heyligenstaedt, Gebr. Heinemann, Schaerer, Index, Schieß

263* VDI-Lehrgangshandbuch: Numerisch gesteuerte Werkzeugmaschinen – Aufbau und Anwendung. VDI-Bildungswerk, Düsseldorf 1969

264   Firmendruckschrift: 2376 Flexowriter – Codierplatz für Werkzeugmaschinensteuerung mit EIA/ISO-Doppelcodierung. Singer/Friden Division/USA 1969

265   *Kips. P.:* Praxis mit NC-Werkzeugmaschinen. Werkst. u. Betr. *103* (1970), H. 1

266   *Stöckmann, P.* und *W. Küster:* NC-Drehmaschinen in der Praxis – Eine Auswertung von Kundenerfahrungen. Werkst. u. Betr. *103* (1970), H. 1, S. 7/10

267   *Stöckmann, P.* und *G. Richter:* Maschinelles Erstellen von Lochstreifen für NC-Drehmaschinen. Werkst. u. Betr. *101* (1968), H. 10, S. 577/584

268   Autorenkollektiv: Manuelles Programmieren von numerisch gesteuerten Drehmaschinen – Eine programmierte Unterweisung. Maschinenbau-Verlag GmbH, Frankfurt a. M. 1970

269   *Kasischke, K.:* Ein neuer Programmierplatz. TZ f. prakt. Metallbearb. *61* (1967), H. 9

270   Firmendruckschrift: VDF-Autoprogramer. Vereinigte Drehbank Fabriken/Göppingen, Hannover, Hamburg 1969

271   Firmendruckschrift: NC-Information Nr. 2. Numerisch gesteuerte Werkzeugmaschinen in Europa. Vereinigte Drehbank Fabriken/Göppingen, Hannover, Hamburg 1969

272* VDI-Lehrgangshandbuch: Manuelles Programmieren numerisch gesteuerter Werkzeugmaschinen. VDI-Bildungswerk, Düsseldorf 1969

273   *Fetzer, H.:* Untersuchungen zur Erweiterung des Einsatzbereiches der Programmiersprache EXAPT 1. Diss. Berlin 1968

274   Firmendruckschrift: Olivetti Te 318 – Endeinrichtung für Nachrichtenverbindungen – Allgemeines Handbuch. Deutsche Olivetti GmbH, Frankfurt a. M. 1968

275   TGL 28 – 210 Blatt 2 (1964). Programmcode 8 C – Aufbau – Eigenschaften. DDR-VVB Werkzeugmaschinen, Karl-Marx-Stadt

276   Firmendruckschrift: Série SP 5000 Equipments de Dactylocodage SPERAC. Systèmes et Périphérique associés aux calculateur. Paris 1969

277   Firmendruckschrift: ADDO – Programmierplatz – Lochstreifengewinnung für Numerische Werkzeugmaschinensteuerung. Addo Deutschland, Frankfurt a. M. 1965

278   CETOP-Comité Européen des Transmission Oleohydrauliques et Pneumatiques. Vorläufige Empfehlung RP2H – Gliederung von Hydrogeräten und -anlagen 1964. Sekretariat Frankfurt a. M., VDMA

279* *Güntsch, F. R.:* Einführung in die Programmierung digitaler Rechenautomaten. W. de Gruyter, Berlin 1960

280* *Chapin, N.:* Einführung in die elektronische Datenverarbeitung. R. Oldenbourg Verlag, München, 3. Aufl. 1967

281* *Dotzauer, E.:* Einführung in die Grundlagen der Datenverarbeitung. C. Hanser Verlag, München 1968

282* *Kaufmann, H.:* Nachrichtenverarbeitung und Automatisierung. R. Oldenbourg Verlag, München 1961

283* *Germain, C. B.:* Das Programmier-Handbuch der IBM/360. C. Hanser Verlag, München 1970

284* *McCracken, D. D.:* FORTRAN in der technischen Anwendung – Ein Lehrbuch in 29 Fallstudien aus der Praxis. C. Hanser Verlag, München 1970

285* *Bates, F.* und *M. L. Douglas:* PL/1 – Einführung in die Programmiersprache für den Selbstunterricht. C. Hanser Verlag, München 1967

286* *Saxon, J. H.:* Einführung in COBOL – Ein Leitfaden zum Selbststudium. C. Hanser Verlag, München 1969

287* PDP 10 – Reference Handbook. Programming Department (1969) of the Digital Equipment Corporation, Maynard, Mass./USA

288* Brunel University: Papers of the International Symposium Computer Graphics 1970. Brunel University, Department of Computer Science, Uxbrigde, Middlesex, England 1970

289* *McCracken, D. D.:* A Guide to FORTRAN-Programming. J. Wiley & Sons, Inc., New York/London 1962

290* *Hopgood, F. R. A.:* Compiling Techniques. MacDonald, London 1969

291* *McCracken, D. D.:* A Guide to ALGOL-Programming. J. Wiley & Sons, Inc., New York/London 1962

292* *Spieß, W. E.* und *F. G. Rheingans:* Einführung in das Programmieren mit FORTRAN. Walter de Gruyter & Co., Berlin 1970

293*   *Colman, H. L., C. Smallwood* and *G. W. Brown:* FORTRAN – Problemorientierte Programmiersprache – Ein Selbstunterrichtslehrgang. Verlag Kunst und Wissen (Erich Bieber), Stuttgart 1966

294*   *Müller, K. H.* und *I. Strecker:* FORTRAN IV – Programmieranleitung. Bibliographisches Institut, Mannheim 1967 Hochschultaschenbücher-Verlag

295*   *Herschel, R.:* Anleitung zum praktischen Gebrauch von ALGOL 60. Telefunken-Fachbuch, 3. Aufl. 1968

296*   *Bolle, K.:* COBOL-Fibel. R. v. Deckers Verlag, G. Schenck, Hamburg 1968

297*   *Ascher Opler:* Das IBM-System/360 und seine Programmiertechniken. R. Oldenbourg-Verlag, München 1968

298*   *Martin, J.:* Programming Real-Time Computer Systems. Prentice-Hall International, Inc., London 1965

299*   *Schöne, A.:* Prozeß-Rechensysteme der Verfahrensindustrie. C. Hanser Verlag, München 1970

300*   *Niederberger, A. R. V.:* Leistungsanalyse elektronischer Rechenanlagen. R. v. Deckers Verlag, G. Schenck, Hamburg 1963

301*   *Villiger, R. M.:* Möglichkeiten, Probleme und Auswirkungen der Datenfernverarbeitung. R. V. Decker's Verlag, G. Schenck, Hamburg 1969

302   Firmendruckschriften: Hartman Planetary Rol-Vane-Motors PRV-Series, HMC-Rol-Vane-Motors. Hartman Hydraulics, Division of Koehring Company, Racine, Wisc./USA 1969

303   *Lavick, J. J.:* Computer Aided Design at McDonnell-Douglas Section 2 – Paper of the International Symposium Computer Graphics 1970. Brunel University, Department of Computer Science, Uxbrigde, Middlesex, England

304*   *Chorafas, D. N.:* Programmiersysteme für elektronische Rechenanlagen. R. Oldenbourg Verlag, München 1967

305*   *Schulz, A.:* Strukturanalyse der maschinellen betrieblichen Informationsbearbeitung. Walter de Gruyter & Co., Berlin 1970

306   *Berndt, H.* und *H. Spreen:* Ein PL/1 – Dialekt zur Beschreibung funktioneller Eigenschaften von Datenverarbeitungssystemen. elektron. datenverarb. 1970, H. 5, S. 227/235

307*   Autorenkollektiv: NC-Maschinen – Datenverarbeitungsanlagen – Maschinelle Programmierung. Herausgegeben von EXAPT-Verein. Techn. Verlag Günter Großmann, Stuttgart 1968

308*   VDI-Bildungswerk: Lehrgangshandbuch: Maschinelles Programmieren numerisch gesteuerter Werkzeugmaschinen. (15 Beiträge). VDI-Düsseldorf 1969

309*   *Stute, G.* (Hrsg.): EXAPT – Möglichkeiten und Anwendung der automatischen Programmierung für NC-Maschinen. C. Hanser Verlag, München 1969

310*   Autorenkollektiv: Das Programmiersystem EXAPT – Grundlagen von EXAPT 3 – Anwendung von EXAPT 1, 2, 3 – Sonderdrucke aus „Steuerungstechnik 1968". Krausskopf-Verlag, Mainz 1968

311   *Krimphove, H.:* Einführung in die Programmiersprache APT. VDI-Bildungswerk (in [308] enthalten); Lehrgangsbeitrag BW 951. Düsseldorf 1969

312   *Krimphove, H.:* APT – Mehrdimensionale Bahnsteuerungen VDI-Bildungswerk (in [308] enthalten). Lehrgangsbeitrag BW 758. Düsseldorf 1968

313   *Brown, S. A., C. E. Drayton* und *B. Mittman:* A Description of the APT-Language. Communications of the ACM, Vol. 6 (1963), No. 11, S. 649/658

314*   IITRI: APT-Part Programming Manual. IIT Research Institute, Chicago 1964

315   *Engelskirchen, W. H.:* Anpassung rechnergestützter Programmierverfahren der Fertigungstechnik an numerisch gesteuerten Werkzeugmaschinen. Diss. Aachen 1968

316   IITRI: Introduction to Part Programming APT. IIT Research Institute, Chicago 1969

317   *Chingari, G.:* The numerical control information utility: Concepts and considerations. Proceedings ACM National Meeting 1967, S. 541/551

318   *Fucyman, B.:* IFAPT – Eine Programmiersprache für die Anfertigung von Numeric-Lochstreifen. Die Maschine 22 (1968), Nr. 6, S. 31/32

319   *Kelley, R. A.:* The productions man's guide to APT – ADAPT. Amer. Mach. 1969, June 22, S. 97/112

320   *Lange, D.:* Maschinelles Programmieren mit der Symbolsprache ADAPT. Beitrag BW 355 im VDI-Bildungswerk-Lehrgangshandbuch „Numerisch gesteuerte Werkzeugmaschinen", VDI Düsseldorf 1967

321   *Grupe, U.:* Sprachen und Übersetzer für die maschinelle Programmierung der Numerikmaschinen. VDI-Bildungswerk Lehrgangsbeitrag BW 853, Düsseldorf 1969

322   *Krimphove, H.:* Die automatische Programmierung numerisch gesteuerter Werkzeugmaschinen bei Einsatz des SYMPAC-Compilers. „Die Lochkarte" (1963), H. 192, Zeitschrift der Remington Rand Univac.

323*   *Mc. Waters, J.F.:* NC user's guide to the international computer programes – 2CL. McGraw – Hill Publishing Co., Ltd., High Wycombe, London

324   *Müller-Traut, H.:* Maschinelles Programmieren mit der Symbolsprache AUTOSPOT. VDI-Bildungswerk Lehrgangsbeitrag 354, Düsseldorf 1967

325   *Müller-Traut, H.:* Grundzüge des maschinellen Programmierens numerisch gesteuerter Werkzeugmaschinen mit Symbolsprachen. VDI-Bildungswerk Lehrgangsbeitrag BW 968, Düsseldorf 1969

326   *Stöckmann, P.* und *G. Richter:* Maschinelles Erstellen von Lochstreifen für NC-Drehmaschinen Werkst. u. Betr. *101* (1968), H. 10, S. 577/584

327*   EXAPT 1: Sprachbeschreibung, Mai 1969. Verein zur Förderung des EXAPT-Programmiersystems e.V., 51 Aachen, Postfach 587

328*   EXAPT 2: Sprachbeschreibung, Mai 1969. Verein zur Förderung des EXAPT-Programmiersystems e.V., 51 Aachen, Postfach 587

329   *Ackerknecht, B.:* Organisation des Betriebes bei Verwendung von problemorientierten Programmsprachen. VDI-Bildungswerk Lehrgangsbeitrag BW 770, Düsseldorf 1968

330*   *Horn, G.:* NC-Software – MINIAPT, 5 Bände. G. Horn – NC-Software GmbH, Frankfurt a. M. 1969

331*   *Parslow, R.D., R.W.Prowse* und *R. Elliot Green* (Hrsg.): Computer Graphics, Techniques and Applications. Plenum Press, London, New York 1969

332   NN: Interactive Graphics in Data Processing. IBM Systems Journal Vol. 7 (1968), Nr. 3 und 4

333*   Autorenkollektiv: Report on Computer Graphics. Ministry of Technology, Computer-aided Design Committee. London, Jan. 1969

334   *Reckziegel, D.* und *H. Zöller:* Stand und Entwicklung von EXAPT. VDI-Nachr. Nr. 6/Febr.1970

335   *Smith, D.N.* und *J.D. McCarroll:* Gegenwärtige Trends bei der numerischen Steuerung. technica 1969, Nr. 12, S. 1067/1077

336   *Soubiès-Camy, H.:* Les Langages de Programmation en Commande Numérique. Automatisme, Tome XIII (1968), No. 4, S. 173/179

337   *Tully, H.* und *J. Herrmann:* Praktischer Einsatz numerisch gesteuerter Werkzeugmaschinen aus der Sicht der Arbeitsvorbereitung. Industrie-Anzeiger *90* (1968), H. 5, S. 19/23 und H. 14, S. 23/26

338   *Mangold, W.E.:* Status of N/C-Language Standardization in ISO. IFIP-IFAC-PROLAMAT-Congress, Rom 1969. North-Holland-Publishing Company, Amsterdam–London 1969

339   *Budde, W., B. Hirsch, F. Tannenberg* und *R. Wutzo:* Programmiersprache für Drehbearbeitung EXAPT 2 mit Vokabelliste (in [308] enthalten). VDI-Bildungswerk, Lehrgangsbeitrag BW 1012, Düsseldorf 1969

340   *Horn, G.:* MINIAPT – Information Nr. 1. G. Horn NC-Software GmbH, Frankfurt a. M. 1970

341   *Horn, G.:* MINIAPT – Postprocessor-Anschluß. G. Horn NC-Software GmbH, Frankfurt a. M. 1970

342   *Miller, R.M.:* Look time-sharing in preparing N/C-tapes. Automation, Aug. 1969, S. 102/105

343   *Jantzen, K.:* Die symbolische Programmierung von numerisch gesteuerten Werkzeugmaschinen mit APT III. Werkst. Technik *57* (1967), H. 3, S. 107/113

344   *Herzog, M.* und *H. Glasauer:* Methoden und Probleme bei der Erstellung von Postprocessoren. TZ f. prakt. Metallbearb. *62* (1968), H. 9, S. 486/491

345   *Mittman, B.:* Development of numerical control programming languages in Europe. Proceedings A.C.M. National Meeting 1967, S. 479/482

346   *Koloc, J., J. Preisler* und *J. Vymer:* Das AUTOPROG-System – Ein Beitrag zur Integration des Fertigungsprozesses. TZ f. prakt. Metallbearb. *62* (1968), H. 2, S. 83/87

347   *Koloc, J., J. Preisler* und *J. Vymer:* A Contribution to the Manufacturing-System Concept in Production Engineering Research. Annals of the C.I.R.P., Vol. XIV, S. 65/82, 1966

348   *Opitz, H., W. Eversheim* und *J. Schleppegrell:* Merkmalanalyse für datenverarbeitende Fertigungsplanung 1968. Forschungsbericht des Landes Nordrhein-Westfalen, Köln-Opladen 1968

349   *Mitrofanow, S.B.:* Wissenschaftliche Grundlagen der Gruppentechnologie. VEB-Verlag Technik Berlin 1960

350   Firmendruckschrift: QUICKPOINT-8; PDP-8 Part Program Compiler – User's Guide. Digital Equipment Corporation, Maynard, Mass./USA. DEC-08-AUAA-D 1969

351 Firmendruckschrift: EASYPROG – Das integrierte System für maschinelles Programmieren. Max Müller – Brinker Maschinenfabrik – Hannover 1969

352 Firmendruckschrift: NC-System Nixdorf 820 – Konturberechnungsprogramm. Nixdorf Computer AG, Paderborn/Berlin 1970

353 Firmendruckschrift: QUICKPOINT-8 – numerical control tape preparation system that incorporates a low cost computer. Digital Equipment Corporation, Maynard, Mass./USA 1969

354 *Martin, M.* et *Ph. Gabrini:* Le système de fabrication programmée MECA – Réalisation d'un système pour commande numérique sur petites calculateurs. Séminaires de programmation de l'I.M.A.G., Grenoble

355 *Moore-Welti:* Produktionssteuerung, Produktionskontrolle. verlag moderne industrie, München 1967

356 *Warsewa, H.R.:* Datengesteuerte Produktion. Verlagsgesellschaft Rudolf Müller, Köln 1967

357 *Sandford, J.E.:* Metalworking Guide to N/C-Software. IRON AGE, April *18*, 1968, S. 95/118

358 *Williamson, D.N.T.:* Computer Control of Machine-Tools. Control 1958, H. 7/8

359 *Ogden, H.:* Electronic Control of Machine-Tools. Research (Butterworth) *15* (1962), November, S. 472/476

360 Informations SNECMA Nr. 96, 1961: Mechanische Herstellung, ausgehend von mathematischen Gleichungen. Société Nationale d'Etudes et de Constructions de Moteurs Aéronautiques, Paris 1961

361 *Tannenberg, F.:* Automatische Ermittlung des Arbeitsablaufes bei der maschinellen Programmierung numerisch gesteuerter Drehmaschinen. Diss. Berlin 1970

362 *Fischer, H.:* Beitrag zur Untersuchung des thermischen Verhaltens von Bohr- und Fräsmaschinen. Diss. Berlin 1970

363 *Balogh, L.:* Ein Beschreibungssystem für rotationssymmetrische Werkstücke unter besonderer Berücksichtigung des Rechnereinsatzes in Konstruktion und Fertigungsplanung. Diss. Berlin 1969

364 *Mai, E.:* Formales Beschreibungssystem für ebene Werkstückgeometrien und der Codierung des Informationsgehaltes technischer Einzelteilzeichnungen. Diss. Berlin 1970

365 *Wilkinson, D.G.:* Cam manufacture using 2C,L. IFIP-IFAC-PROLAMAT-Congress, Rom 1969. North-Holland Publishing Company, Amsterdam–London 1969

366 *McWaters, J.F.* und *W.T.K. Henderson:* The NEL 2C,L processor. IFIP-IFAC-PROLAMAT-Congress, Rom 1969. North-Holland-Publishing Company, Amsterdam–London 1969

367 *Weill, R.:* IFAPT, a unified system of modular design of NC-Languages. IFIP-IFAC-PROLAMAT-Congress, Rom 1969. North-Holland-Publishing Company, Amsterdam–London 1969

368 *Gabrini, Ph.:* Etude d'un système de programmation en commande numérique. Automatisme, XV (1970), H. 4, S. 176/180

369 *Hirsch, B.:* Ein System zur Ermittlung von Zerspanungs-Vorgabewerken, insbesondere bei rechnergestützter Programmierung von numerisch gesteuerten Drehmaschinen. Diss. Aachen 1969

370 *Depiereux, W.-R.:* Die Ermittlung optimaler Schnittbedingungen, insbesondere im Hinblick auf die wirtschaftliche Nutzung numerisch gesteuerter Werkzeugmaschinen. Diss. Aachen 1970

371* *Villiger, R.M.:* Möglichkeiten, Probleme und Auswirkungen der Datenfernverarbeitung. R.v.Decker's Verlag, G. Schenck, Hamburg–Berlin 1969

372 *Simon, W.:* On the Change in Premise Conditioning Investment Decisions Due to the Influence of Integral Date-Processing in the Metal Working Industry. Summary Report Third Training Seminar for Senior Staff of National Productivity Centres, RKW Frankfurt a.M. 1967

373 *Ulrich, E.:* Stufung und Messung der Mechanisierung und Automatisierung. Mitt. d. Institutes für Arbeitsmarkt- und Berufsforschung (IAB) Erlangen 1968, Nr. 2 und 3

374 *Fidrich, P.:* Vorschläge zur Anwendung eines normierten Maschinenstundensatzes für Kostenvergleiche bei Werkzeugmaschinen. Rationalisierung *17* (1966), H. 12, S. 293/298

375 *Hall, E.:* Die Automatisierung der Konstruktion mit Rechner und elektronischer Zeichenmaschine. TZ f. prakt. Metallbearb. *62* (1968), H. 4, S. 202/205

376 *Smith, D.N.* und *J.D. McCaroll:* Gegenwärtige Trends bei der numerischen Steuerung. Technica 1969, Nr.12

377 *Brödner, P.* und *F. Hamke:* Automatisierung und Arbeitsplatzstrukturen – Bericht über Methoden und Ergebnisse von Untersuchungen in der Einzel- und Kleinserienfertigung. Mitt. des IAB 1969, Nr. 8 Erlangen

378   *Brödner, P.* und *F. Hamke:* Automatisierung und Arbeitsplatzstrukturen – Bericht über eine Prognose der mutmaßlichen Entwicklung in der Einzel- und Kleinserienfertigung. Mitt. des IAB Erlangen (1970), H. 2, S. 137/172

379   *Stehle, P.:* Eine Methode zur Wirtschaftlichkeitsrechnung unter besonderer Berücksichtigung des Einsatzes numerisch gesteuerter Werkzeugmaschinen. Diss. Aachen 1966

380   *Krause, F.-L., R. Langebartels* und *V. Vassilacopoulos:* Geräte zum rechnergestützten Konstruieren – technischer Stand 1969. Konstruktion *22* (1970), H. 4, S. 121/132

381*   *Leslie, W. H. P.* (Hrsg.): Numerical Control Programming Languages. Proceedings of the PROLA-MAT-Conference, Rom 1969. North-Holland Publishing Company – Amsterdam, London 1970

382   *Bright, J. R.:* How to Evaluate Automation. Harvard Business Review, July/Aug. 1955

383   *Simon, W.:* Die numerische Steuerung von Werkzeugmaschinen – Technischer Stand 1966. ETZ-A *87* (1966), S. 904/910

384   *Simon, W.:* Der Datenfluß-Verflechtungsgrad – ein neues Ordnungsprinzip automatischer Werkzeugmaschinen. msr *10* (1967), H. 5, S. 163/166

385   *Schulz, H.:* Strukturanalyse der maschinellen betrieblichen Informationsverarbeitung. Schriftenreihe kommerzielle Datenverarbeitung. W. de Gruyter Verlag, Berlin 1969

386   *Hahn, R., W. Kunèrth* und *K.-H. Roschmann:* Elektronische Datenverarbeitung in der Fertigung – Stand und Entwicklungstendenzen. ADL-Nachrichten 1969, Heft 59, S. 751/764

387   *Dolezalek, C. M., R. Hahn* und *K.-H. Roschmann:* Elektronische Datenverarbeitung – ein Werkzeug für die Unternehmensführung. WT *57* (1967), H. 3, S. 102/105

388   *Hahn, R.* und *K. Roschmann:* Kurzfristige Terminierungsverfahren der Fertigungssteuerung bei Werkstattfertigung. WT *57* (1967), H. 4, S. 179/186

389   *Baginski, P.:* Fertigungsregelung mit datenverarbeitenden Real-Time-Systemen. VDI-Bericht Nr. *101* (1966), S. 105/113

390   *Hahn, R.* und *K. Roschmann:* Mathematische Methoden der Fertigungssteuerung. WT *56* (1966), H. 2, S. 67/69, H. 3, S. 123/128

391*   *Kamp, A.-W.:* NC-Maschinen, Fachwörter und Definitionen, Deutsch, Englisch, Französisch, Italienisch. VDI-Verlag, Düsseldorf 1970

Index

Absolute measuring method 47, 49, 105, 111
Access time 475
Accupin 133, 136
Accuracy 5
Actuator 170, 190, et seq, 195
Adaption 322, 327
Address 266, 389 et seq
Address method 266, 360, 398
ALGOL 429
All-purpose computer 474
Alphabet 19
Amplidyne 181
Amplifier 181, 194
Amplifier machine 181
Analogue 44, 113 et seq
Analogue computer 162
Analogue/digital converter 111
Analogue measurement 44, 168
Angular measurement 113 et seq, 119
APT 432, 437, 475
Automation, degree of 479 et seq
AUTOPROG 459, 464, 476
Auxiliary function 393, 396
Axial piston motor 129, 201, 221
Axial piston pump 187, 188
Axis direction definitions 399 et seq

Basic systems 477
BCD Code 23, 26
Belt speed 354
Belt conveyors 352
Bendix system 153
Binary-coded decimal system 26
Binary systems 21 et seq
Bit 21 et seq
Block 348, 390
Block diagram 371 et seq
Block scanning 377
Boring and milling machine 270

Cam disc 230
Characters required 380

Code 28, 376
Code conversion 26, 386
Code number 392 et seq
Code selection 19, 29
Coding 394
Coding theory 15
Coincidence testing 105 et seq
Column 347
Commutator 81
Comparator 77, 105, 106 et seq, 113 144
Compiler 427 et seq
Computer circuits 331 et seq
Computing unit 293 et seq
Continuous-path control 33 et seq, 43, 280
Contour control system 33
Control circuit 8, 93, 163
Control consoles 339
Controllable pumps 186
Coordinate scanning 287
Coordinatograph 286, 287
Copying control system 241, 278, 480
Cost determination 491
Cross-coding (of punched tape) 348
Cutter diameter correction 293
Cutting conditions 392

D.C. motor 175
Data carrier, information carrier 346 et seq
Data processing 376, 424
Data processing, external 30, 43
Data processing, internal 30, 35, 43
Decimal code, decimal system 99
De-coder 357
De-coding 357 et seq
Differences summator 299 et seq
Differential rotary indicator 148
Digital 45
Digital-absolute method 95, 164
Digital/analogue converter 138 et seq
Digital computer 162

Digital differential analyser (DDA)
  299 et seq
Digital incremental process 54, 60
Dipping coil regulator 194
Direct numerical control (DNC, On-line
  control) 360, 400
Directional discriminator 57
Displacement element 6
Displacement information 32, 36, 390
  et seq
Displacement measurement 49, 114
Displacement measurement principles
  44, 50
Displacement measurement system 50,
  61, 113, 197
Double-gripping system 251 et seq
Drilling jig 231
Drive element 200
Drive systems 170 et seq, 201, 211
Drum turret 238
Dual 18
Dual-code, dual system 17, 20, 22

EIA-Code 384 et seq, 421
Efficiency 12, 478 et seq, 499
Electro-magnetic clutch 171
Emitter 81
Energy control position 190
Energy converter 194
EXAPT 438 et seq, 477
Exclusive OR 60, 358
External interpolator 42, 471

Feedback 7
Feed drives 196, 202, 220, 235
Feed spindle 47, 226
Flame-cutting machine 282, 285
Flip-flop 82, 83
Flow controller 189
Follow-on control circuit 154
Form data 13
Form store 230
FORTRAN 429, 439
Four-layer triode 184
Frequency range 52

Gear pump 187
Glass scale 67, 70
Grating spacing 74 et seq
Gray-Code 21, 100

Hall generators 65
Hydraulic cylinder 195
Hydraulic fluid 185, 202
Hydraulic motor 198
Hydrostatically lubricated slideway
  228

ISO Code 381 et seq, 421
Increment 529
Incremental process 47 et seq, 67,
  77, 94
Inductosyn 136
Information distributor 360
Information flow 1 et seq, 499
Information store 230 et seq, 242,
  345, 366
Input unit 340, 341
Inspection machine 11, 150, 526
Internal interpolator 42, 336
Interpolation 297 et seq, 336 et seq

Jig Borer 511

Lathe 239, 323 et seq
Linear Accupin 134, 135
Linear Inductosyn 126, 129
Logic elements, pneumatic 361, 365
Longitudinal coding (of punched tape)
  348

Machining centre 274
Magnetic core stores 331, 426
Magnetic tape 366 et seq
Magnetic tape control systems 471
Manuscript 407
Measuring impulse marking 38
Measuring location 7, 49
Measuring machine 159, 516
Measuring systems 149
Milling machine, automatic milling
  machine 244, 329
MINIAPT 438, 458
MIT 431
Model structures 40, 483, 496
Moiré pattern process 70
Multi-axis machines 268 et seq
Multiprismat 73
Multivibrator 81
Multi-way valves 190

Number 398
Numerical 1, 156, 159, 221
Numerical indication 90

Off-line system 480, 482
Oil circuit 128
Oil compressibility 196
Oil pressure 187
Oil pumps 186
On-line system 156, 371, 483
Output signal 109, 144

Parallel scanning 348

Personnel questions 492 et seq
Photodiode 62
Plant organisation 288, 291
Playback process 364, 511
Polygon mirror process 74
Positional measuring systems 103, 105
Positioning 77, 210 et seq
Positioning control 34, 36, 44
Power drives 195
Preselector counter 77 et seq
Pressure-adjusting elements 189
Process information 15
Profile tools 230
Programme codes 387 et seq
Programme sentence 397
Programme symbolic language 398
Programmers 406 et seq, 428 et seq
Programming 379, 403 et seq, 421, 423
Programming effort 450 et seq
Programming languages, machine-orientated 397 et seq, 428
Programming languages, problem-orientated 423, 428
Programming languages, production-engineering 430
Punched tape 345, 376
Punched tape coding 347 et seq, 382 et seq
Punched tape dimensions 347
Punched tape reader 349, 353, 361 et seq, 376

Radial piston pump 187
Rapid traverse 393
Recirculating ball nut 226
Redundancy 19, 154
Reflected light method 70
Relative measuring method 49
Reliability 3
Remote-controlled clutch 171
Repeating control system 343
Resolution capacity 147
Resolver 115, 120
Response sensitivity 115
Reversing operation 179, 199
Rigidity 25 et seq
Ring counter 84
Roll-vane motors 198
Rotary indicator rotating field systems 118, 122, 147, 154

Scale graduations 68
Scanning 62, 351
Scanning, directional 67
Scanning, electro-magnetic 63, 102
Scanning, electro-mechanical 62, 102

Scanning, inductive 63
Scanning, photoelectric 62, 63, 102
Scanning zone 96
Seals 196
Semi-conductor components 220
Sentence 393
Series scanning 352
Silicon cells (SCR) 184 et seq, 220
Simulator 418
Simultaneous machining 520
Simultaneous method of examination 491
Single point control 37, 43
Slideways 227 et seq
Slideway displacement measurement control 8 et seq
Small computer 331, 335, 408, 466
Spurious pulses 56
Star turret 243
Step switching method 360, 361
Stepping motor 207 et seq, 221
Stepping switches 360
Stibitz Code 24
Stick-slip 226 et seq
Straight-line control systems 34 et seq, 43
Switch-off circuit 8, 163
Switching information 32, 36
Switching information, direct 32, 391
Switching information, indirect 33
Switching valve 190
Sub-programmes 436
SYMAP 459 et seq, 476
Synchronous 117
Synchronous operation 147
Systems engineering 40, 492 et seq

Tabulator method 266, 398 et seq
Tapping transformer 144
Technological information 13 et seq
Telex machine 383
Template 230
Tetrade 24
Thread cutting 317 et seq
Three-Excess Code 24, 25
Three-from-five Code 25
Three-point controller 164
Three-position valve 191
Thyristor 183, 221
Time of passage 487 et seq
Timing track 347
Tool alignment 258
Tool changing 248 et seq, 253
Tool correction 294
Tool magazine, tool store 288
Tool presetting 258 et seq

Tool wear 258, 294
Transductor 183
Translator 80, 83, 91
Triple Inductosyn 131
Turret drilling machine 234
Turret head 232
Turret lathe 235
Turret punching press 281
Two-part measuring systems 149 et seq
Typewriter, punched-tape controlled
  408

Utilisation factor 490

V-scanning 98, 99

Vertical lathe 515

Ward-Leonard set, Ward-Leonard control
  system 180
Word 398, 399
Work information 13 et seq, 36, 340,
  341
Work schedule 383
Workpiece measurement control system
  10

Zero-point fixing 60, 92, 111, 148
Zero-point shift 60, 148, 166
Zone scanning 96, 97.